Interdisciplinary Applied Mathematics

T0155925

Interdisciplinary Applied Mathematics

Volume 24

Editors
S.S. Antman J.E. Marsden
L. Sirovich S. Wiggins

Geophysics and Planetary Sciences

Mathematical Biology
L. Glass, J.D. Murray

Mechanics and Materials
R.V. Kohn

Systems and Control
S.S. Sastry, P.S. Krishnaprasad

Problems in engineering, computational science, and the physical and biological sciences are using increasingly sophisticated mathematical techniques. Thus, the bridge between the mathematical sciences and other disciplines is heavily traveled. The correspondingly increased dialog between the disciplines has led to the establishment of the series: *Interdisciplinary Applied Mathematics*.

The purpose of this series is to meet the current and future needs for the interaction between various science and technology areas on the one hand and mathematics on the other. This is done, firstly, by encouraging the ways that mathematics may be applied in traditional areas, as well as point towards new and innovative areas of applications; and, secondly, by encouraging other scientific disciplines to engage in a dialog with mathematicians outlining their problems to both access new methods and suggest innovative developments within mathematics itself.

The series will consist of monographs and high-level texts from researchers working on the interplay between mathematics and other fields of science and technology.

Nonholonomic Mechanics and Control

A.M. Bloch

With the Collaboration of J. Baillieul, P. Crouch, and J. Marsden

With Scientific Input from P.S. Krishnaprasad, R.M. Murray, and D. Zenkov

 Springer

A.M. Bloch
Department of Mathematics
University of Michigan
Ann Arbor, MI 48109-1109
USA
abloch@umich.edu

Editors
S.S. Antman
Department of Mathematics
and
Institute for Physical Science
  and Technology
University of Maryland
College Park, MD 20742
USA
ssa@math.umd.edu

J.E. Marsden
Control and Dynamical Systems
Mail Code 107-81
California Institute of Technology
Pasadena, CA 91125
USA
marsden@cds.caltech.edu

L. Sirovich
Division of Applied Mathematics
Brown University
Providence, RI 02912
USA
chico@camelot.mssm.edu

S. Wiggins
School of Mathematics
University of Bristol
Bristol BS8 1TW
UK
s.wiggins@bris.ac.uk

Mathematics Subject Classification (2000): 70F25, 58FXX, 93B27, 49K15, 34A34, 93D15

Library of Congress Cataloging-in-Publication Data
Bloch, Anthony, 1955–
        Nonholonomic mechanics and control / Anthony M. Bloch
            p. cm. — (Interdisciplinary applied mathematics; 24)
        Includes bibliographical references and index.

        1. Nonholonomic dynamical systems.   2. Control theory.   I. Title.
        II. Interdisciplinary applied mathematics; v. 24.
        QA614.833. B56 2003
        514′.74—dc21   2002042738

ISBN: 978-1-4419-3043-9   e-ISBN: 978-0-387-21644-7

Printed in the United States of America.

9 8 7 6 5 4 3 2

springer.com

Preface

Our goal in this book is to explore some of the connections between control theory and geometric mechanics; that is, we link control theory with a geometric view of classical mechanics in both its Lagrangian and Hamiltonian formulations and in particular with the theory of mechanical systems subject to motion constraints. This synthesis of topics is appropriate, since there is a particularly rich connection between mechanics and nonlinear control theory. While an introduction to many important aspects of the mechanics of nonholonomically constrained systems may be found in such sources as the monograph of Neimark and Fufaev [1972], the geometric view as well as the control theory of such systems remains largely scattered through various research journals. Our aim is to provide a unified treatment of nonlinear control theory and constrained mechanical systems that will incorporate material that has not yet made its way into texts and monographs.

Mechanics has traditionally described the behavior of free and interacting particles and bodies, the interaction being described by potential forces. It encompasses the Lagrangian and Hamiltonian pictures and in its modern form relies heavily on the tools of differential geometry (see, for example, Abraham and Marsden [1978] and Arnold [1989]). From our own point of view, our papers Bloch, Krishnaprasad, Marsden, and Murray [1996], Bloch and Crouch [1995], and Baillieul [1998] have been particularly influential in the formulations presented in this book.

Control Theory and Nonholonomic Systems. Control theory is the theory of prescribing motion for dynamical systems rather than describing their observed behavior. These systems may or may not be mechanical in nature, and in fact traditionally, the underlying system is not assumed to be mechanical. Modern control theory began largely as linear theory, having its roots in electrical engineering and using linear algebra, complex variable theory, and functional analysis as its principal tools. The nonlinear theory of control, on the other hand, relies to a large extent again on differential geometry.

Nonholonomic mechanics describes the motion of systems constrained by nonintegrable constraints, i.e., constraints on the system velocities that do not arise from constraints on the configurations alone. Classic examples are rolling and skating motion. Nonholonomic mechanics fits uneasily into the classical mechanics, since it is not variational in nature; i.e., it is neither Lagrangian nor Hamiltonian in the strict sense of the word. It has a close cousin (variational axiomatic mechanics—a term coined by Arnold, Kozlov, and Neishtadt [1988]); which is variational for systems subject to nonintegrable constraints but does not describe the motion of mechanical systems. It is important, however, for the theory of optimal control, as will be developed in the main text.

There is a close link between nonholonomic constraints and controllability of nonlinear systems. Nonholonomic constraints are given by nonintegrable distributions; i.e., taking the bracket of two vector fields in such a distribution may give rise to a vector field not contained in this distribution. It is precisely this property that one wants in a nonlinear control system so that we can drive the system to as large a part of the state space as possible.

A key concept for studying the control and geometry of nonholonomic systems, as well as many other mechanical systems, is the notion of a fiber bundle and an associated connection. The bundle point of view not only gives us a way of organizing variables in a physically meaningful way, but gives us basic ideas on how the system behaves physically, and on how to prescribe controls. A bundle connection relates base and fiber variables in the system, and in this sense one can take a gauge theoretic point of view of nonholonomic control systems.

Optimal Control. There is a beautiful link between optimal control of nonholonomic systems and so-called sub-Riemannian geometry. For a large class of physically interesting systems, the optimal control problem reduces to finding geodesics with respect to a singular (sub-Riemannian) metric. The geometry of such geodesic flows is exceptionally rich and provides guidance for designing control laws. See Montgomery [2002] for additional information.

Physical Examples. One of the aims of this book is to illustrate the elegant mathematics behind many simple, interesting, and useful mechanical examples. Among these are the rigid body and rolling rigid body, the rolling ball on a rotating turntable, the rattleback top, the rolling penny, and the satellite with momentum wheels. There are clearly a number of points in common between these systems, among them the fact that rotational motion and the existence of constraints, either externally imposed or dynamically generated (conserved momenta), play a key role. In one sense these notions—rotation and constraints—form the heart of the book and are vital to studying both the dynamics and control of these systems. Further, one of the delights of this subject is that although these systems may have many features in common, their behavior is quite different and often quite unexpected. Why does a rattleback top rotate in only one direction? What is the behavior of a ball on a rotating turntable? Why does a tennis racket not want to spin about its middle axis? How do I roll a penny to a particular point on a table, parallel to an edge, and with Lincoln's head in the upright position?

While we have attempted to cover a substantial amount of material, this book is very much written from the authors' perspective, and there is much fascinating work in this area that we have had to omit.

A Path Through the Book. This book can be read on many different levels. On the one hand, there are numerous physical examples that are analyzed in elementary terms, as in Chapter 1. On the other hand, there are theoretical sections that use some sophisticated analysis and geometry. There are also sections on the background mathematics used, and our hope is that this book mixes these ingredients in an instructive and useful way. Depending on one's background and preferences, this book can be read in a linear or nonlinear fashion (in the sense of progression through the pages).

Many of the examples are returned to later in the book as illustrations of the general theory. These later returns to the examples vary in difficulty from again quite elementary to more sophisticated demonstrations of the theory. We urge the reader to use them to understand the theory.

The theory itself varies greatly in difficulty—some of it is again quite elementary and easy to read—usually at the beginning of each section or chapter, but some of it quite technical and based on various pieces of the research work of the authors and sometimes their collaborators. Many technical sections may be omitted on first reading or without loss of continuity; we also refer the reader to the Internet supplement for additional material. This is available on the book's web site, where errata, reprint data, and other information may be found:

http://www.cds.caltech.edu/mechanics_and_control

We have gone to some trouble to fill in the necessary background for the general theory and to put in elementary illustrations of it. For example, we discuss the theory of connections and the geodesic flow on the line.

Scope of the Book. We should also emphasize that while this book cuts quite a large swath through an area of mechanics and nonlinear control, it is very much mechanics and control as seen by the authors. There is a huge and exciting literature on mechanics, nonholonomic mechanics, nonlinear control, and optimal control that we have not discussed at all and indeed have not even been able to reference in many cases. We urge the reader to follow up on related areas both through the references that are here and through references in those papers.

Prerequisites.   This book is intended for graduate students who wish to learn this subject as well as for researchers in the area who wish to enhance their techniques. Chapter 1 is written in a way that assumes rather few prerequisites and is intended to motivate people to read further and to acquire the needed background for that task. Chapters 2, 3, and 4 contain some of the needed background in geometry, mechanics, and control. They are necessarily brief in nature and are meant, in part, to summarize topics that are treated in other courses. There is, however, new expository material on various topics that we describe in more detail below. In addition, we have collected material that is hard to find in any one source. From Chapter 5 on, we assume that the reader is knowledgeable about these topics.

A knowledge of basic mechanics, as in the well-known book of Goldstein [1980], is helpful, although we do in fact develop both Lagrangian and Hamiltonian mechanics as well as the theory of control from first principles. So for a reader who knows nothing of these fields but has the usual dose of "mathematical sophistication" it is quite possible to read and benefit from this book. Similarly, it it not necessary to know anything about nonholonomic mechanics.

Some parts of Chapters 2 and 3 are based on *Mechanics and Symmetry* (Marsden and Ratiu [1999]) and can be skipped by the knowledgeable reader or consulted as the need arises. Another piece of useful background is the recently published Beijing lecture notes of Roger Brockett (see Brockett [2000]), whose spirit certainly pervades much of the nonlinear control theory in this book. Similarly, the collection *Mathematical Control Theory*, written in honor of Roger Brockett's 60th birthday, is a useful adjunct (see Baillieul and Willems [1999]).

This book can be viewed as somewhere between a research monograph and a textbook. It has been successfully used as a textbook for courses at Caltech and Michigan as well as for lecture series at the Technical University of Vienna and the IIIe Cycle Romand de Mathématique, Les Diablerets, Switzerland. In this regard there are a number of exercises, particularly in the earlier chapters, meant to help readers gauge their understanding, but this is not a main focus of the book.

A Brief Rundown of the Chapters in This Book.   We will now give a brief synopsis of the various chapters in the book. Specific citations to the works of authors that are mentioned here are given in the main text.

Chapter 1 consists of a little preliminary mechanics but mainly of examples that are used later in the book. Key examples include the vertical and falling rolling disks and various versions of a skate on ice as well as the rolling ball. More complicated examples include the roller racer and rattleback top. There are also mechanical examples that are holonomic but that are used later to illustrate basic Lagrangian and Hamiltonian mechanics and control. These include the Toda lattice, the free and controlled rigid body, and the pendulum on a cart.

Chapter 2 is devoted to various mathematical preliminaries and can be used as desired according to the mathematical expertise of the reader. It goes all the way from the basic theory of manifolds and ordinary differential equations to the theory of Ehresmann connections. The most original part and that which is least likely to be familiar to readers is that on connections. We have gone to a great deal of trouble here to analyze how Ehresmann connections specialize to principal connections and to affine connections and Riemannian connections. These ideas are also illustrated with very simple examples such as the geodesic flow on the line! Connections play a vital role in mechanics and in particular nonholonomic mechanics, where they arise from constraints.

Chapter 3 gives general background in geometric mechanics, and parts of it can again be skipped by the knowledgeable reader. There are, however, new things here: a new exposition of the theory of forces; a description of the Murray-Ostrowski view of mechanical systems and the mechanical connection, together with the example of the spacecraft with rotors; a detailed description of coupled planar rigid body motion as developed by Oh, Sreenath, Krishnaprasad, and Marsden; and a description of phases and holonomy as developed by Marsden and Ostrowski.

Chapter 4 gives general background in nonlinear control theory including basic definitions of controllability and accessibility, some theory on averaging and motion planning (including work of Leonard and Krishnaprasad), a proof of Brockett's necessary condition for stabilization, and some of the theory of Hamiltonian and Lagrangian control systems following work of Brockett, van der Schaft, Willems, and others.

Chapter 5 is the basic chapter on nonholonomic mechanics and owes much in exposition to the paper of Bloch, Krishnaprasad, Marsden, and Murray as well as to work of Bloch and Crouch. We discuss the basic geometric approach in these papers. The basic interaction of symmetries and constraints in nonholonomic systems and how they lead to the nonholonomic momentum equation is discussed. Explicit examples of the momentum map are given. In addition the role of "almost" Hamiltonian structure is discussed, building on work of Bates and Sniatycki, van der Schaft and Maschke, as well as that of Marsden and Koon.

Chapter 6 discusses various aspects of control and stabilization of nonholonomic systems both for kinematic and dynamic systems. Open loop controls are discussed for the Brockett canonical form following Murray and Sastry, and its discontinuous stabilization is discussed based on the work of Bloch, Drakunov, and Kinyon. The Coron approach to smooth time-varying stabilization is also briefly discussed. Following the work of Bloch, Reyhanoglu, and McClamroch, control and stabilization of dynamic nonholonomic control systems is described. Control of nonholonomic systems on Riemannian manifolds is discussed following work of Bloch and Crouch.

Chapter 7 is devoted to optimal control. It begins by discussing the relationship of variational nonholonomic control systems and the classical Lagrange problem to optimal control. A brief introduction to the maximum principle is given. We then discuss sub-Riemannian (kinematic) optimal control problems based on the work of Bloch, Crouch, and Ratiu and building on the work of Brockett and Baillieul. We give a brief discussion of abnormal extremals following work of Montgomery and Sussmann. Dynamic optimal control is discussed following work of Bloch and Crouch and Silva, Leite, and Crouch. Related work on integrable systems is discussed in the internet supplement.

Chapter 8 discusses an energy–momentum-based approach to the stability of nonholonomic systems. This is based on the thesis work of Zenkov and related work with Bloch and Marsden. Also described are notions of asymptotic stability in Euler–Poincaré–Suslov systems following work of Kozlov and its connection to the Toda lattice following work of Bloch.

Chapter 9 discusses some recent and still developing research on energy-based techniques for mechanical and nonholonomic systems. A brief description of the controlled Lagrangian or matching technique of Bloch, Leonard, and Marsden is given with some recent applications to certain nonholonomic systems based on work with D. Zenkov. Finally, work of Baillieul is described on second-order averaging methods and their connections with classical geometry.

Acknowledgments. We are very grateful to many colleagues for their collaboration and for their input, directly or indirectly. We are especially grateful to P. S. Krishnaprasad, Richard Murray, and Dmitry Zenkov, as well as to Roger Brockett, Chris Byrnes, Sergey Drakunov, Sameer Jalnapurkar, Michael Kinyon, Wang-Sang Koon, Naomi Leonard, Harris McClamroch, Jim Ostrowski, Tudor Ratiu, Mahmut Reyhanoglu and Hans Troger, among others. Dmitry Zenkov's work on checking the manuscript is greatly appreciated. Support from the NSF, AFOSR, the Institute for Advanced Study and the University of Michigan during the writing of this book is also most appreciated.

A largely complete copy of this book was submitted to the publisher in August 2001 and was circulated around this time. Earlier versions were circulated to students and colleagues in 2000. Thanks to all our students and colleagues who have used notes associated with the book and provided advice and corrections. We would like to thank Wendy McKay, Michael Jeffries, and Matt Haigh for the invaluable help with the typesetting and graphics for the book. Thanks also to the anonymous reviewers of the book and to the staff at Springer-Verlag especially Achi Dosanjh, Elizabeth Young, David Kramer, and Mary Ann Brickner.

About the Authors

Anthony Bloch is the Alexander Ziwet Collegiate Professor of Mathematics at the University of Michigan, where he is currently Chair of the Department. He received a B.Sc.Hons from the University of the Witwatersrand, Johannesburg in 1978, an M.S. from the California Institute of Technology in 1979, an M.Phil from Cambridge University in 1981, and a Ph.D. from Harvard University in 1985. His research interests are in mechanics and nonlinear control, Hamiltonian systems and integrable systems, and related areas of nonlinear dynamics. He has held visiting positions at the Mathematical Sciences Institute at Cornell University, the Mathematical Sciences Research Institute at Berkeley, and the Fields Institute in Canada, and has been a member of the Institute for Advanced Study in Princeton. He was a T. H. Hildebrandt Assistant Professor at Michigan from 1985 to 1988 and was a faculty member at The Ohio State University before returning to Michigan in 1994. He has received a Presidential Young Investigator Award and a Guggenheim Fellowship and is a Fellow of the IEEE. He has also received an Excellence in Research and a Faculty Recognition award from the University of Michigan. He was an Associate Editor at Large for the *IEEE Transactions on Automatic Control* and is an Associate Editor of *Mathematics of Controls, Signals and Systems, Nonlinear Science, Systems and Control Letters, Dynamical Systems*, the *Electronic Journal of Differential Equations* and the *Electronic Journal of Mathematical and Physical Sciences*.

John Baillieul holds a joint appointment as Professor of Aerospace/Mechanical Engineering and Professor of Manufacturing Engineering at Boston University, and he is currently Chairman of Aerospace/Mechanical Engineering. He has also served as Associate Dean for Academic Programs in the B.U. College of Engineering. After receiving the Ph.D. from Harvard University in 1975, he joined the Mathematics Department of Georgetown University. During the academic year 1983–84 he was the Vinton Hayes Visiting Scientist in Robotics at Harvard University, and in 1991 he was visiting scientist in the Department of Electrical Engineering at MIT. He was Editor-in-Chief of the *IEEE Transactions on Automatic Control*, 1992–2002. Currently, he is on the editorial boards of the *Proceedings of the IEEE*, the *IEEE Transactions on Automatic Control, Communications in Information and Systems*, and *Robotics and Computer Integrated Manufacturing*. He is a Fellow of the IEEE for his contributions to nonlinear control theory, robotics, and the control of complex mechanical systems. He is a recent recipient of the IEEE Third Millennium Medal for various professional contributions. At the level of the corporate IEEE, Professor Baillieul served as *TAB Transactions* Chair (1998–2001), member at large of the Publications Services and Products Board (PSPB) (1999–), and Chair of the PSPB Strategic Planning Committee (2001–). His research deals with robotics, the control of mechanical systems, and mathematical system theory.

Peter Crouch is the Dean of the University of Hawaii College of Engineering and was the Dean of the College of Engineering and Applied Sciences at Arizona State University from 1995 to 2006. From 1992 to 1995 he was Chair of the Electrical Engineering Department at Arizona State and from 1989 to 1995 Director of the Center for Systems Engineering. He was on the faculty at Arizona State since 1984 and prior to that taught at the University of Warick, England. He received his Ph.D. from Harvard University in Applied Sciences in 1977, an M.Sc. in Control Theory from Warwick University in 1974 and a B.Sc. in Engineering Science in 1973 also from Warwick. His research interests lie in control theory, nonlinear systems theory and dynamical systems, their applications to problems in electrical, mechanical engineering, aerospace engineering and semiconductor manufacturing, and related problems on numerical simulation. He is a Fellow of the *IEEE Transactions on Automatic Control* and Associate Editor of the *Journal of Dynamical and Control Systems, Systems and Control Letters*, and the *Journal of Mathematical Control and Information*.

Jerrold Marsden is the Carl F. Braun Professor of Engineering and Control and Dynamical Systems at the California Institute of Technology. He received his B.Sc. at Toronto in 1965 and his Ph.D. in 1968, both in Applied Mathematics. He has done extensive research in the area of geometric mechanics, with applications to rigid body systems, fluid mechanics, elasticity theory, plasma physics as well as to general field theory. His primary current interests are in applied dynamics and control theory, especially how these subjects relate to mechanical systems, systems with symmetry and multiscale systems. He was the recipient of SIAM's John von Neumann award in 2005, of the Norbert Wiener prize of the AMS and SIAM in 1990, was elected a fellow of the American Academy of Arts and Sciences in 1997, received a Max Planck research award in 2000 and was elected a Fellow of the Royal Society of London in 2006. He has been a Carnegie Fellow at Heriot–Watt University (1977), a Killam Fellow at the University of Calgary (1979), recipient of the Jeffrey–Williams prize of the Canadian Mathematical Society in 1981, a Miller Professor at the University of California, Berkeley (1981–1982), a recipient of the Humboldt Prize in Germany (1991 and 1999), and a Fairchild Fellow at Caltech (1992). He has served in several administrative capacities, such as the Advisory Panel for Mathematics at NSF, the Advisory committee of the Mathematical Sciences Institute at Cornell, and as director of the Fields Institute, 1990–1994. He served as the Chair of the Board of Trustees of IPAM for 2002–2004, the Institute for Pure and Applied Mathematics, and is currently the director of CIMMS, the Center for Integrative Multiscale Modeling and Simulation at Caltech. He has been an editor for Springer-Verlag's *Applied Mathematical Sciences* series since 1982 and serves on the editorial boards of several journals in mechanics, dynamics, and control.

Contents

A Diagram for the Book

In Chapter 1 we will introduce many of the basic ideas of mechanics, nonholonomic mechanics, control, and optimal control, together with a number of illustrative physical examples. These basic ideas will be explained, expanded on, and made rigorous in subsequent chapters. In addition, the examples discussed will occur in different contexts as the exposition develops. A detailed path through this book is given in the preface.

In the following two pages we present a diagram which attempts to tie together the various key threads in our exposition. These topics and the links between them will be clarified in the upcoming chapters of the book.

NEWTONIAN MECHANICS (Ch 3)

$$F = ma$$

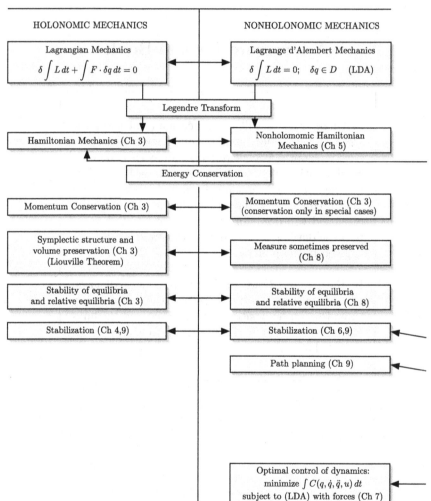

HOLONOMIC MECHANICS

NONHOLONOMIC MECHANICS

Lagrangian Mechanics

$$\delta \int L \, dt + \int F \cdot \delta q \, dt = 0$$

Lagrange d'Alembert Mechanics

$$\delta \int L \, dt = 0; \quad \delta q \in D \quad \text{(LDA)}$$

Legendre Transform

Hamiltonian Mechanics (Ch 3)

Nonholomomic Hamiltonian
Mechanics (Ch 5)

Energy Conservation

Momentum Conservation (Ch 3)

Momentum Conservation (Ch 3)
(conservation only in special cases)

Symplectic structure and
volume preservation (Ch 3)
(Liouville Theorem)

Measure sometimes preserved
(Ch 8)

Stability of equilibria
and relative equilibria (Ch 3)

Stability of equilibria
and relative equilibria (Ch 8)

Stabilization (Ch 4,9)

Stabilization (Ch 6,9)

Path planning (Ch 9)

Optimal control of dynamics:
minimize $\int C(q, \dot{q}, \ddot{q}, u) \, dt$
subject to (LDA) with forces (Ch 7)

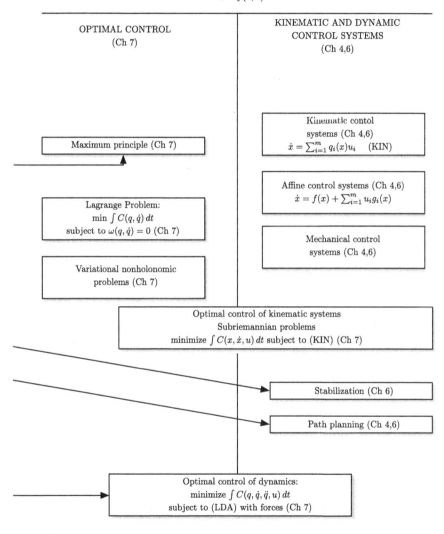

CONTROL SYSTEMS (Ch 6,7)
$$\dot{x} = f(x, u)$$

OPTIMAL CONTROL
(Ch 7)

KINEMATIC AND DYNAMIC
CONTROL SYSTEMS
(Ch 4,6)

Kinematic contol
systems (Ch 4,6)
$$\dot{x} = \sum_{i=1}^{m} q_i(x) u_i \quad \text{(KIN)}$$

Maximum principle (Ch 7)

Affine control systems (Ch 4,6)
$$\dot{x} = f(x) + \sum_{i=1}^{m} u_i g_i(x)$$

Lagrange Problem:
$\min \int C(q, \dot{q})\, dt$
subject to $\omega(q, \dot{q}) = 0$ (Ch 7)

Mechanical control
systems (Ch 4,6)

Variational nonholonomic
problems (Ch 7)

Optimal control of kinematic systems
Subriemannian problems
minimize $\int C(x, \dot{x}, u)\, dt$ subject to (KIN) (Ch 7)

Stabilization (Ch 6)

Path planning (Ch 4,6)

Optimal control of dynamics:
minimize $\int C(q, \dot{q}, \ddot{q}, u)\, dt$
subject to (LDA) with forces (Ch 7)

1

Introduction

The purpose of this chapter is to quickly introduce enough theory so that we can present some examples that will then be used throughout the course of the book to illustrate the theory and how to use it. These examples are simple to write down in general and to understand at an elementary level, but they are also useful for the understanding of deeper parts of the theory.

Two main classes of systems considered in the book are *holonomic systems* and *nonholonomic systems*. This terminology may be found in Hertz [1894]. Holonomic systems are mechanical systems that are subject to constraints that limit their possible configurations. As Hertz explains, the word holonomic (or holonomous) is comprised of the Greek words meaning "integral" (or "whole") and "law", and refers to the fact that such constraints, given as constraints on the velocity, may be integrated and reexpressed as constraints on the configuration variables. We make this idea precise as we move through the book. Examples of holonomic constraints are length constraints for simple pendula and rigidity constraints for rigid body motion.

The rolling disk and ball are archetypal nonholonomic systems: systems with *nonintegrable* constraints on their velocities. These examples have a long history going back for example to Vierkandt [1892] and Chaplygin [1897a]. In this chapter and the book in general we discuss both the rolling disk and ball, as well as many other nonholonomic systems such as the Chaplygin sleigh, the roller racer, and the rattleback. As pointed out in Sommerfeld [1952] a general analysis of the distinction between holonomic and nonholonomic constraints may be found as early as Voss [1885], while specific examples of nonholonomic systems were of course analyzed even earlier. For more on the history of nonholonomic systems see Chapter 5.

We remark that Hertz defines a holonomic system as a system "between whose possible positions all conceivable continuous motions are also possible motions" The point is that nonholonomic constraints restrict types of motion but not position. The meaning of Hertz's statment should become clearer as the reader continues through the book.

Other examples discussed here include the free rigid body and the somewhat more complex satellite with momentum wheels. These are (holonomic) examples of free and and coupled rigid body motion respectively—the motion of bodies with nontrivial spatial extent, as opposed to the motion of point particles. The latter is illustrated by the Toda lattice, which models a set of interacting particles on the line; we shall also be interested in some associated optimal control systems.

We also describe here the Heisenberg system, which was first studied by Brockett [1981] (see also Baillieul [1975], who studied some related systems). This does not model any particular physical system, but is a prototypical example for nonlinear kinematic control problems (both optimal and nonoptimal) and can be viewed as an approximation to a number of interesting physical systems; in particular, this example is basic for understanding more sophisticated optimal reorientation and locomotion problems, such as the falling cat theorem that we shall treat later. A key point about this system (and many others in this book) is that the corresponding linear theory gives little information.

1.1   Generalized Coordinates and Newton–Euler Balance

In this and subsequent sections in this chapter we discuss some ideas from mechanics in an informal fashion. This is intended to give context to the physical examples discussed in later sections. More formal derivations of many of the ideas discussed here are given in later chapters.

Coordinates and Kinematics.   The most basic goal of analytical mechanics is to provide a formalism for describing motion. This is often done in terms of a set of *generalized coordinates*, which may be interpreted as coordinates for the system's *configuration space*, often denoted by Q. This is a set of variables whose values uniquely specify the location in 3-space of each physical point of the mechanism. A set of generalized coordinates is minimal in the sense that no set of fewer variables suffices to determine the locations of all points on the mechanism. The number of variables in a set of generalized coordinates for a mechanical system is called the number of *degrees of freedom* of the system.

1.1.1 Example (A Simple Kinematic Chain).   Simple ideas along this line, which will be generalized to provide the foundation of most of the models studied in this book, may be illustrated using the simple kinematic chain shown in Figure 1.1.1.

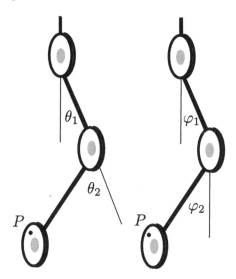

FIGURE 1.1.1. Kinematic chains.

Here there are drawn two copies of the same mechanism. This mechanism consists of planar rigid bodies connected by massless rods, and the joints are free to rotate in a fixed plane. In the first, the motion of a typical point P is described in terms of coordinate variables (θ_1, θ_2), where θ_2 is the relative angle between the two links in the chain. In Figure 1.1.1 (b), the motion of the typical point P is described in terms of coordinate variables (φ_1, φ_2), which are the (absolute) angles of the links with respect to the vertical direction.

Other choices of coordinate variables are, of course, possible. In any case, the coordinate variables serve the purpose of describing the location of typical points of the mechanism with respect to a privileged coordinate frame, which we may refer to as an ***inertial frame***. A thorough axiomatic discussion of inertial frames is beyond the scope of this book, but roughly speaking, these are frames that are "nonaccelerating relative to the distant stars." For the purposes of our discussions here it suffices to consider them as "fixed" coordinate systems.

Specifically, in this case, the inertial frame is chosen so that its origin is at the hinge point of the upper link. The y-axis is directed parallel and opposite to the gravitational field, and the x-axis is chosen so as to give the coordinate frame the standard orientation. Suppose the point P is located on the second link, as depicted. If this has coordinates (x_ℓ, y_ℓ) with respect to a local frame fixed in the second link, then the coordinates with respect to the inertial frame are given by

$$\begin{bmatrix} x \\ y \end{bmatrix} = \begin{bmatrix} r_1 \sin\theta_1 + x_\ell \sin(\theta_1 + \theta_2) + y_\ell \cos(\theta_1 + \theta_2) \\ -r_1 \cos\theta_1 - x_\ell \cos(\theta_1 + \theta_2) + y_\ell \sin(\theta_1 + \theta_2) \end{bmatrix}, \qquad (1.1.1)$$

where r_1 is the length of the first link, or equivalently by

$$\begin{bmatrix} x \\ y \end{bmatrix} = \begin{bmatrix} r_1 \sin\varphi_1 + x_\ell \sin\varphi_2 + y_\ell \cos\varphi_2 \\ -r_1 \cos\varphi_1 - x_\ell \cos\varphi_2 + y_\ell \sin\varphi_2 \end{bmatrix}. \qquad (1.1.2)$$

The mappings $(\theta_1, \theta_2) \mapsto (x, y)$ are examples of functions that associate values of the generalized coordinate variables (θ_1, θ_2) (respectively (φ_1, φ_2)) to inertial coordinates of the point P. In this example, the configuration manifold is given by $Q = S^1 \times S^1$ and is parameterized by the two angles θ_1, θ_2, which serve as generalized coordinates. One can also make the alternative choice of φ_1, φ_2 as generalized coordinates that provide a different set of coordinates on Q.   ♦

Newton's Laws.   The most fundamental contribution to mechanics were Newton's three laws of motion for a particle (see Newton [1650], Book I, Section 3, Propositions XI, XII, XIII) and, for example, Chorlton [1983]).
    They are as follows:

(1) Every particle continues in its state of rest or of uniform velocity in a straight line unless compelled to do otherwise by a force acting on it.

(2) The rate of change of linear momentum is proportional to the impressed force and takes place in the direction of action of the force.

(3) To every action there is an equal and opposite reaction.

For a particle of constant mass m, Newton's second law can be written as:

$$m\ddot{\mathbf{x}}(t) = \mathbf{F}(t), \qquad (1.1.3)$$

where $\mathbf{x} \in \mathbb{R}^3$ is the position vector of the particle and $\mathbf{F}(t)$ is the impressed force, both measured with respect to an inertial frame.

Remarks on Rigid Body Mechanics.   As a preliminary to describing general rigid body mechanics, one procedure is to consider the special case of a finite number of point masses constrained so that the distance between points is constant, with each point mass experiencing internal forces of interaction (equal in magnitude and acting in opposite directions along straight lines joining the points) together with external forces. In this case, Newton's laws lead to the equations of rigid body dynamics for this special type of rigid body.
    For a rigid body that is a continuum or for a system of point particles or rigid bodies mechanically linked, one may derive the equations of motion

by an application of Newton's law for the motion of the center of mass and Euler's law for motion about the center of mass, i.e.,

$$I\dot{\omega}(t) = T(t), \tag{1.1.4}$$

where I is the moment of inertia of the rigid body about its center of mass, ω is the angular velocity about the center of mass, and $T(t)$ is the applied torque about the center of mass, all measured with respect to an inertial frame.

It turns out, however, that the equations of motion for the special case of a finite number of constrained point masses described above may be derived solely from Newton's laws.

The rigid body also provides a nice example of a system whose configuration space is a manifold. In fact, it is the set $Q = \mathrm{SE}(3)$ of Euclidean motions, that is, transformations of $\mathbb{R}^3$ consisting of rotations and translations. Each element of Q gives a placement of all the particles in the rigid body relative to a reference position, all in an inertial frame. We will return to the rigid body from a more advanced point of view later.

Newton–Euler Balance Laws. More generally, for a system of interconnected rigid bodies, such as the kinematic chain described earlier, one can derive the equations of motion from Newton's laws together with Euler's law giving the rate of change of angular momentum about a pivot point in terms of applied torques, as in equation (1.1.4). It is interesting to note that these equations cannot (without further assumptions) be derived from Newton's laws alone; for an illuminating discussion of these relationships see Antman [1998].

So far, the examples mentioned are ones with holonomic constraints (the length of the pendula in the kinematic chain are assumed constrained to be constant, and the rigid body is constrained by rigidity). However, one of the purposes of this book is to study nonholonomic systems, wherein one has constraints on the velocities. Examples are systems such as rolling wheels. Even in this case, one can use Newton–Euler balance ideas to obtain the equations correctly.

For the bulk of this book, however, we will not take the point of view of Newton–Euler balance laws. One reason for this is that there is a more useful alternative given by Hamilton's principle (and the associated Euler–Lagrange equations) for holonomic systems and by the Lagrange–d'Alembert principle in the nonholonomic case. We shall briefly study these principles in the next sections and return to them in more detail later. In addition, the Hamilton principle and Lagrange–d'Alembert formalism are covariant, in the sense that they use only the intrinsic configuration manifold Q, and one may use any set of coordinates on it; in addition, there is a simple and elegant way to write the equations valid in any set of generalized coordinates. The covariant nature of the Euler–Lagrange formalism was one of the greatest discoveries of Lagrange and is the basis of the geometric approach to mechanics.

One should ask whether the Newton–Euler balance approach is equivalent to the Euler–Lagrange and Lagrange–d'Alembert approaches under general sets of hypotheses. This is a subtle question in general, which is, unfortunately, not systematically addressed in most books, including this one. However, these approaches can be shown to be equivalent in many concrete situations, such as interconnected rigid bodies and rolling rigid bodies, which we will come to later. See Jalnapurkar [1994] for one such exposition of this equivalence. We will confine ourselves to proving the equivalence in one concrete nonholonomic situation later, namely, a system called the Chaplygin sleigh; see Section 1.7.

1.2   Hamilton's Principle

In this section we give a brief introduction to the Euler–Lagrange equations of motion for holonomic systems from the point of view of variational principles. We return to this later in Chapter 3 from a more abstract point of view. The reader for whom this is familiar may, of course, skip ahead.

Let Q be the configuration space[1] of a system with (generalized) coordinates q^i, $i = 1, \ldots, n$. We are given a real-valued function $L(q^i, \dot{q}^i)$, called a **Lagrangian**. Often we choose L to be $L = K - V$, where K is the **kinetic energy** of the system and $V(q)$ is the **potential energy**.

1.2.1 Definition.   *Hamilton's principle singles out particular curves $q(t)$ by the condition*

$$\delta \int_a^b L(q(t), \dot{q}(t))\, dt = 0, \tag{1.2.1}$$

where the variation is over smooth curves in Q with fixed endpoints.

To make this precise, let the **variation** of a trajectory $q(\cdot)$ with fixed endpoints satisfying $q(a) = q_a$ and $q(b) = q_b$ be defined to be a smooth mapping

$$(t, \epsilon) \mapsto q(t, \epsilon), \qquad a \leq t \leq b, \qquad \epsilon \in (-\delta, \delta) \subset \mathbb{R},$$

satisfying

(i) $q(t, 0) = q(t)$, $t \in [a, b]$,

---

[1]The configuration space of a system is best thought of as a differentiable manifold, and generalized coordinates as a coordinate chart on this manifold. To enable us to introduce some examples early on, we shall treat this rather informally at first and return to a more intrinsic approach later.

(ii) $q(a, \epsilon) = q_a$, $q(b, \epsilon) = q_b$.

Letting $\delta q(t) = (\partial/\partial\epsilon)q(t, \epsilon)|_{\epsilon=0}$ be the **virtual displacement** correspond-ing to the variation of q, we have

$$\delta q(a) = \delta q(b) = 0. \tag{1.2.2}$$

The precise meaning of Hamilton's principle is then the statement

$$\frac{d}{d\epsilon} \int_a^b L(q(t, \epsilon), \dot{q}(t, \epsilon))\, dt \bigg|_{\epsilon=0} = 0 \tag{1.2.3}$$

for all variations.

One can view Hamilton's principle in the following way: The quantity $\int_a^b L(q(t), \dot{q}(t))\, dt$ is being extremized among all curves with fixed end-points; that is, the particular curve $q(t)$ that is sought is a *critical point* of the quantity $\int_a^b L(q(t), \dot{q}(t))\, dt$ thought of as a function on the space of curves with fixed endpoints. Examples show that the quantity $\int_a^b L\, dt$ being extremized in (1.2.1) need not be minimized at a solution of the Euler–Lagrange equations, just as in calculus: Critical points of functions need not be minima.[2]

A basic result of the calculus of variations is:

1.2.2 Proposition. *Hamilton's principle for a curve $q(t)$ is equivalent to the condition that $q(t)$ satisfy the **Euler–Lagrange equations***

$$\frac{d}{dt} \frac{\partial L}{\partial \dot{q}^i} - \frac{\partial L}{\partial q^i} = 0. \tag{1.2.4}$$

The idea of the proof is as follows: Let δq be a virtual displacement of the curve $q(t)$ corresponding to the variation $q(t, \epsilon)$. We may compute the variation of the integral in Definition 1.2.1 corresponding to this variation of the trajectory q by differentiating with respect to ϵ and using the chain rule. We obtain

$$\int_a^b \left(\frac{\partial L}{\partial q^i} \delta q^i + \frac{\partial L}{\partial \dot{q}^i} \delta \dot{q}^i \right) dt = 0, \tag{1.2.5}$$

where $\delta \dot{q}^i = \frac{d}{dt} \delta q^i$. Integrating by parts and using the boundary conditions $\delta q^i = 0$ at $t = a$ and $t = b$ yields the identity

$$\int_a^b \left(-\frac{d}{dt} \frac{\partial L}{\partial \dot{q}^i} + \frac{\partial L}{\partial q^i} \right) \delta q^i\, dt = 0. \tag{1.2.6}$$

---

[2]Perhaps the simplest example of this comes up in the study of geodesics on a sphere where geodesics that "go the long way around the sphere" are critical points, but not minima. In this example, L is just the kinetic energy of a point particle on the sphere. See Gelfand and Fomin [1963] for further information.

Assuming a rich enough class of variations yields the result.[3]

A critical aspect of the Euler–Lagrange equations is that they may be regarded as a way to write Newton's second law in a way that makes sense in arbitrary curvilinear and even moving coordinate systems. That is, the Euler–Lagrange formalism is *covariant*. This is of enormous benefit, not only theoretically, but for practical problems as well.

Mechanical Systems with External Forces.   In the presence of external forces F_i, the equations are

$$\frac{d}{dt}\frac{\partial L}{\partial \dot{q}^i} - \frac{\partial L}{\partial q^i} = F_i \tag{1.2.7}$$

for $i = 1, \ldots, n$. Here we regard the quantities F_i as given by external agencies.[4] Note that if these forces are derivable from a potential U in the sense that $F_i = -\partial U/\partial q^i$, then these forces can be incorporated into the Lagrangian by adding $-U$ to the Lagrangian. Thus, this way of adding forces is consistent with the Euler–Lagrange equations themselves.

These equations can be derived from a variational-like principle, the **Lagrange–d'Alembert principle** for systems with external forces, as follows:

$$\delta \int_a^b L(q^i, \dot{q}^i)\, dt + \int_a^b F \cdot \delta q\, dt = 0, \tag{1.2.8}$$

where $F \cdot \delta q = \sum_{i=1}^n F_i \delta q^i$ is the **virtual work** done by the force field F with a virtual displacement δq as defined above.

A rigorous analysis of virtual work and integral laws of motion for continuum mechanics in Euclidean space may be found in Antman and Osborn [1979].

Remarks on the History of Variational Principles.   The history of variational principles and the so-called principle of least action is quite complicated, and we leave most of the details to other references. Some of this history can be gleaned, for example, from Whittaker [1988] and Marsden and Ratiu [1999]. An interesting historical note is that the currently accepted notion of the "principle of least action" is regarded by some as being synonymous with "Hamilton's principle." Indeed Feynman [1989] advocates this point of view. However, both historically and factually, **Hamilton's principle** and the **principle of least action** (which

---

[3]Again, further geometric insight into the notion of the variation operation is something we will return to later; for example, the equality $\delta \dot{q}^i = \frac{d}{dt}\delta q^i$ is self-evident from our definition of the virtual displacement and equality of mixed partials.

[4]In elementary books on mechanics external forces are often regarded as a given vector field, but in fact, they should be regarded as a given one-form field. Such distinctions are not important just now, but this is a crucial distinction in the geometric formulation of mechanics that will be important for us later on.

should really be called the **principle of critical action**) are slightly different. Hamilton's principle involves varying the integral of the Lagrangian over all curves with fixed endpoint and fixed time. The principle of least action, on the other hand, involves variation of the quantity

$$\int_a^b \sum_i \dot{q}^i \frac{\partial L}{\partial \dot{q}^i}\, dt$$

over all curves with fixed energy.

The principle of critical action originated in Maupertuis's work (Maupertuis [1740]), which attempted to obtain for the corpuscular theory of light a theorem analogous to Fermat's **principle of least time.** Briefly put, the latter involves taking the variations of

$$\int n\, dl, \tag{1.2.9}$$

where n is the refractive index over the path of the light. This gives rise to Snel's law.[5] Maupertuis's principle was established by Euler [1744] for the case of a single particle and in more generality by Lagrange [1760].

It is curious that Lagrange dealt with the more difficult principle of critical action already in 1760, yet Hamilton's principle, which is simpler, came only much later in Hamilton [1834, 1835].

Another bit of interesting history is that Lagrange [1788] did not derive the Lagrange equations of motion by variational methods, but he did so by requiring that simple force balance be *covariant*, that is, expressible in arbitrary generalized coordinates. For further information on the history of variational principles and the precise formulation of the principle of least action, see Marsden and Ratiu [1999].

Energy and Hamilton's Equations.   If the matrix $\partial^2 L/\partial \dot{q}^i \partial \dot{q}^j$ is nonsingular, we call L a **nondegenerate** or **regular** Lagrangian, and in this case we can make (at least locally) the change of variables from $(q^i, \dot{q}^i)$ to the variables (q^i, p_i), where the momentum is defined by

$$p_i = \frac{\partial L}{\partial \dot{q}^i}.$$

This change of variables is commonly referred to as the **Legendre transformation.** We shall see how to write it in a coordinate-free way in Chapter 3. Introducing the Hamiltonian

$$H(q^i, p_i) = \sum_{i=1}^n p_i \dot{q}^i - L(q^i, \dot{q}^i),$$

---

[5]A simple derivation of Snel's law from the variational point of view can be found, for example, in Feynman [1989]. This law was discovered by the Dutch mathematician and geodesist Willebord Snel van Royen. (Because his name in Latin is "Snellius" the law is often called Snell's law.)

one checks, by a *careful* use of the chain rule, that the Euler–Lagrange equations become **Hamilton's equations**

$$\dot{q}^i = \frac{\partial H}{\partial p_i}, \quad \dot{p}_i = -\frac{\partial H}{\partial q^i},$$

where $i = 1, \ldots, n$. If we think of the Hamiltonian as a function of $(q^i, \dot{q}^i)$, then we write it as $E(q^i, \dot{q}^i)$ and still refer to it as the **energy**. If the Lagrangian is of the form kinetic minus potential, then the energy and Hamiltonian are kinetic plus potential.

If one introduces the **Poisson bracket** of two functions K, L of (q^i, p_i) by the definition

$$\{K, L\} = \sum_{i=1}^{n} \frac{\partial K}{\partial q^i} \frac{\partial L}{\partial p_i} - \frac{\partial L}{\partial q^i} \frac{\partial K}{\partial p_i},$$

then one checks, again using the chain rule, that Hamilton's equations may be written concisely as

$$\dot{F} = \{F, H\}$$

for all functions F. In particular, since the Poisson bracket is clearly skew symmetric in K, L, we see that $\{H, H\} = 0$, and so H has zero time derivative (conservation of energy). The corresponding statement for the energy E can be verified directly to be a consequence of the Euler–Lagrange equations (and this holds even if L is degenerate).

Exercises

$\diamond$ **1.2-1.**   Consider the Lagrangian

$$L(x, y, z, \dot{x}, \dot{y}, \dot{z}) = \frac{1}{2} m \left(\dot{x}^2 + \dot{y}^2 + \dot{z}^2 \right) - mgz.$$

Compute the equations of motion in both Lagrangian and Hamiltonian form. Verify that the Hamiltonian (energy) is conserved along the flow. Are there other conserved quantities?

$\diamond$ **1.2-2.**   Consider a Lagrangian of the form $L = \frac{1}{2} \sum_{k,l=1}^{n} g_{kl}(q) \dot{q}^k \dot{q}^l$, where g_{kl} is a symmetric matrix. Show that the Lagrange equation of motion are

$$\sum_{s} g_{rs} \ddot{q}^s + \sum_{l,m} \Gamma_{rlm} \dot{q}^l \dot{q}^m = 0$$

for suitable symbols Γ. Verify conservation of energy directly for this system.

1.3   The Lagrange–d'Alembert Principle

Holonomic and Nonholonomic Constraints. Suppose the system constraints are given by the following m equations, linear in the velocity field, where $m < n$:

$$\sum_{k=1}^{n} a_k^j(q^i)\dot{q}^k = 0, \tag{1.3.1}$$

where $j = 1, \ldots, m$.

If one can find m constraints on the positions alone, that is, constraints of the form $b^j(q^i) = 0$, such that their time derivatives, namely

$$\sum_{k=1}^{n} \frac{\partial b^j}{\partial q^k}\dot{q}^k = 0,$$

determine the same constraint distribution as the constraints (1.3.1), then one says that the constraints are **holonomic**. Otherwise, they are called **nonholonomic**. For example, the length constraint on a pendulum is a holonomic constraint, whereas a constraint of rolling without slipping (which we shall discuss in the next section) is nonholonomic.

It is also sometimes useful to distinguish between constraints that are dependent or independent of time. Those that are independent of time are called **scleronomic**, and those that depend on time are called **rheonomic**. This terminology can also be applied to the mechanical system itself; see, e.g., Greenwood [1977]. For example, a bead on a hoop is a rheonomic system. For more details on such "moving" systems see Marsden and Ratiu [1999].

The Frobenius theorem and differential forms, which we shall review in Chapter 2, give necessary and sufficient conditions under which a given set of constraints is integrable. We shall return to these ideas in a more geometric form in Chapter 5.

Dynamic Nonholonomic Equations of Motion. We will now sketch the derivation of the equations of motion of a nonholonomic mechanical system using Newton's laws and Lagrange's equations.[6] We omit external forces for the moment. Later on in the text we shall derive the equations of motion from other points of view.

We regard the system as being acted on by just those forces F_i, $i = 1, \ldots, n$, that have to be exerted by the constraints in order that the system satisfy the nonholonomic constraints (1.3.1). Let $F_1 \delta q^1 + F_2 \delta q^2 + \cdots + F_n \delta q^n$ be the work done by these forces when the system undergoes an arbitrary virtual displacement $(\delta q^1, \ldots, \delta q^n)$. One *assumes* that with these

---

[6]See also, for example, Whittaker [1988] and references therein, Ferrers [1871], Neumann [1888], and Vierkandt [1892].

forces, the system is described by a holonomic system subject to the forces of constraint; therefore, the equations of motion are given by (1.2.7). To determine these forces of constraint, we make the following fundamental assumption:

> **Assumption.** In any virtual displacement consistent with the constraints, the constraint forces F_i do no work, i.e., we assume that the identity
> $$F_1 \delta q^1 + F_2 \delta q^2 + \cdots + F_n \delta q^n = 0$$
> holds for all virtual displacements δq^i satisfying the constraints (1.3.1).

Assuming that the m vectors $(a_1^1, \ldots, a_n^1)$, $(a_1^2, \ldots, a_n^2)$, $\ldots$, $(a_1^m, \ldots, a_n^m)$ are linearly independent, it follows from the same linear algebra used to prove the Lagrange multiplier theorem that the forces of constraint have the form $F_i = \lambda_1 a_i^1 + \cdots + \lambda_m a_i^m$ for $i = 1, \ldots, n$.

In summary, the **dynamic nonholonomic equations of motion** are

$$\frac{d}{dt} \frac{\partial L}{\partial \dot{q}^i} - \frac{\partial L}{\partial q^i} = \sum_{j=1}^{m} \lambda_j a_i^j, \qquad (1.3.2)$$

where $i = 1, \ldots, n$, together with the constraint equations (1.3.1). One determines the Lagrange multipliers λ_i by imposing the constraints in much the same way as one solves constrained maximum and minimum problems in calculus.

The dynamic nonholonomic equations of motion (1.3.2) are also known as the **Lagrange–d'Alembert equations**. These equations are the correct equations for mechanical dynamical systems and in many cases (such as rolling bodies in contact) can be shown to be equivalent to Newton's law $F = ma$ with reaction forces.[7] We shall see this explicitly in the context of some simple and concrete examples shortly.

Lagrange–d'Alembert Principle. The generalization of Hamilton's principle to the nonholonomic context is as follows:

1.3.1 Definition. *The principle*

$$\delta \int_a^b L(q(t), \dot{q}(t)) \, dt = 0, \qquad (1.3.3)$$

where the virtual displacements δq are assumed to satisfy the constraints 1.3.1, that is,

$$\sum_{k=1}^{n} a_k^j \delta q^k = 0, \qquad (1.3.4)$$

---

[7]See, for example, Vershik and Gershkovich [1988], Bloch and Crouch [1998a], and Jalnapurkar [1994].

where $j = 1, \ldots, m$, is called the **Lagrange–d'Alembert principle.**

As with Hamilton's principle, one can check that the following propositions are true:

1.3.2 Proposition. *The Lagrange–d'Alembert principle given in Definition 1.3.1, together with the constraints (1.3.1), is equivalent to the Lagrange–d'Alembert equations of motion (1.3.2).*

This is a fundamental principle, and we shall return to it later in more detail.

Energy. We introduce the energy in the same way as with holonomic systems, namely

$$E(q^i, \dot{q}^i) = \frac{\partial L}{\partial \dot{q}^i} \dot{q}^i - L(q^i, \dot{q}^i). \qquad (1.3.5)$$

1.3.3 Proposition. *Energy is conserved for nonholonomic systems; that is, for solutions of (1.3.2) subject to the constraints (1.3.1), we have*

$$\frac{dE}{dt} = 0.$$

Proof. We begin by taking the time derivative of the energy expression (1.3.5) and using the equations of motion (1.3.2):

$$\frac{d}{dt} E(q^i, \dot{q}^i) = \frac{d}{dt} \left(\frac{\partial L}{\partial \dot{q}^i} \dot{q}^i - L(q^i, \dot{q}^i) \right)$$

$$= \frac{d}{dt} \left(\frac{\partial L}{\partial \dot{q}^i} \right) \dot{q}^i + \frac{\partial L}{\partial \dot{q}^i} \ddot{q}^i - \frac{\partial L}{\partial q^i} \dot{q}^i - \frac{\partial L}{\partial \dot{q}^i} \ddot{q}^i$$

$$= \sum_{j=1}^{m} \lambda_j a_i^j \dot{q}^i.$$

But this vanishes by virtue of the constraints (1.3.1).   ∎

This proposition is consistent with the fact that the forces of constraint do no work. Of course, this result is under the assumptions that the Lagrangian is not explicitly time-dependent and that the constraints are time-independent.

Nonholonomic Mechanical Systems with External Forces. If external forces F^e, such as control forces, are added to the system, then one adds these forces to the right-hand side of the equations, just as we did earlier for the Lagrange equations of motion. Namely, the equations are

$$\frac{d}{dt} \frac{\partial L}{\partial \dot{q}^i} - \frac{\partial L}{\partial q^i} = \sum_{j=1}^{m} \lambda_j a_i^j + F_i^e, \qquad (1.3.6)$$

where $i = 1, \ldots, n$, together with the constraint equations (1.3.1). One determines the Lagrange multipliers λ_i by imposing the constraints as before.

The corresponding Lagrange–d'Alembert principle is

$$\delta \int_a^b L(q(t), \dot{q}(t)) \, dt + \int_a^b F^e \cdot \delta q \, dt = 0, \qquad (1.3.7)$$

where the virtual displacements δq now are assumed to satisfy the constraints (1.3.4).

Variational Nonholonomic Equations. It is interesting to compare the dynamic nonholonomic equations, that is, the Lagrange–d'Alembert equations with the corresponding variational nonholonomic equations. The distinction between these two different systems of equations has a long and distinguished history going back to the review article of Korteweg [1899] and is discussed in a more modern context in Arnold, Kozlov, and Neishtadt [1988]. (For Kozlov's work on vakonomic systems see, e.g., Kozlov [1983] and Kozlov [1992]).[8] The upshot of the distinction is that the Lagrange–d'Alembert equations are the correct mechanical dynamical equations, while the corresponding variational problem is asking a different question, namely one of optimal control.

Perhaps it is surprising, at least at first, that *these two procedures give different equations.* What, exactly, is the difference in the two procedures? The distinction is one of whether the constraints are imposed before or after taking variations. These two operations do not, in general, commute. We shall see this explicitly with the vertical rolling disk in the next section. *With the dynamic Lagrange–d'Alembert equations, we impose constraints only on the variations, whereas in the variational problem we impose the constraints on the velocity vectors of the class of allowable curves.*

The variational equations are obtained by *using Lagrange multipliers with the Lagrangian* rather than Lagrange multipliers with the equations, as we did earlier. Namely, we consider the modified Lagrangian

$$L(q, \dot{q}) + \sum_{k=1}^{n} \sum_{j=1}^{m} \mu_j a_k^j \dot{q}^k. \qquad (1.3.8)$$

Notice that there are as many Lagrange multipliers μ_j as there are constraints, just as in the Lagrange–d'Alembert equations. Then one forms the Euler–Lagrange equations from this modified Lagrangian and determines

---

[8] As Korteweg points out, there were many confusions and mistakes in the literature because people were using the incorrect equations, namely the variational equations, when they should have been using the Lagrange–d'Alembert equations; some of these misunderstandings persist, remarkably, to the present day. What Arnold et al. call the *vakonomic* equations, we will call the *variational nonholonomic* equations. This terminology will be useful in distinguishing the system from the *dynamic nonholonomic* equations we introduced above.

the Lagrange multipliers, to the extent possible, from the constraints and initial conditions. We shall see explicitly how this works in the context of examples in the next section and return to the general theory later on.

Exercises

◇ **1.3-1.** Consider the Lagrangian

$$L(x, y, z, \dot{x}, \dot{y}, \dot{z}) = \frac{1}{2}m\left(\dot{x}^2 + \dot{y}^2 + \dot{z}^2\right) - mgz$$

with the constraints

$$y\dot{x} - x\dot{y} = 0.$$

(a) Are these constraints holonomic or nonholonomic?
(b) Write down the dynamic nonholonomic equations.
(c) Write down the variational nonholonomic equations.
(d) Are these two sets of equations the same?

◇ **1.3-2.** [Rosenberg [1977]] Consider the Lagrangian

$$L(x, y, z, \dot{x}, \dot{y}, \dot{z}) = \frac{1}{2}\left(\dot{x}^2 + \dot{y}^2 + \dot{z}^2\right)$$

with the constraints

$$\dot{z} - y\dot{x} = 0.$$

(a) Write down the dynamic nonholonomic equations.
(b) Write down the variational nonholonomic equations.
(c) Are these two sets of equations the same?

◇ **1.3-3.** Derive a formula for dE/dt for nonholonomic systems with forces.

1.4 The Vertical Rolling Disk

Geometry and Kinematics. The vertical rolling disk is a basic and simple example of a system subject to nonholonomic constraints: a homogeneous disk rolling without slipping on a horizontal plane. In the first instance we consider the "vertical" disk, a disk that, unphysically of course, may not tilt away from the vertical; it is not difficult to generalize the situation to the "falling" disk. It is helpful to think of a coin such as a penny, since we are concerned with orientation and the roll angle (the position of Lincoln's head, for example) of the disk.[9]

---

[9]Other references that treat this example (including the falling disk) are, for example, Vierkandt [1892], Bloch, Reyhanoglu, and McClamroch [1992], Bloch and Crouch [1995], Bloch, Krishnaprasad, Marsden, and Murray [1996], O'Reilly [1996], Cushman, Hermans, and Kemppainen [1996], and Zenkov, Bloch, and Marsden [1998].

Let S^1 denote the circle of radius 1 in the plane. It is parameterized by an angular variable (that is, a variable that is 2π-periodic. The configuration space for the vertical rolling disk is $Q = \mathbb{R}^2 \times S^1 \times S^1$ and is parameterized by the (generalized) coordinates $q = (x, y, \theta, \varphi)$, denoting the position of the contact point in the xy-plane, the rotation angle of the disk, and the orientation of the disk, respectively, as in Figure 1.4.1.

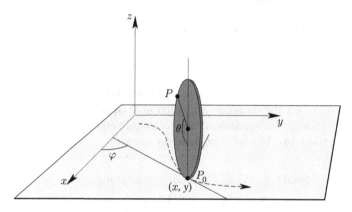

FIGURE 1.4.1. The geometry of the rolling disk.

The variables (x, y, φ) may also be regarded as giving a translational position of the disk together with a rotational position; that is, we may regard (x, y, φ) as an element of the **Euclidean group** in the plane. This group, denoted by SE(2), is the group of translations and rotations in the plane, that is, the group of rigid motions in the plane. Thus, SE(2) $= \mathbb{R}^2 \times S^1$ (as a set). This group and its three-dimensional counterpart in space, SE(3), play an important role throughout this book. They will be treated via their coordinate descriptions for the moment, but later on we will return to them in a more geometric and intrinsic way.

In summary, the configuration space of the vertical rolling disk is given by $Q = \text{SE}(2) \times S^1$, and this space has coordinates (generalized coordinates) given by $((x, y, \varphi), \theta)$.

The Lagrangian for the vertical rolling disk is taken to be the total kinetic energy of the system, namely

$$L(x, y, \varphi, \theta, \dot{x}, \dot{y}, \dot{\varphi}, \dot{\theta}) = \frac{1}{2}m(\dot{x}^2 + \dot{y}^2) + \frac{1}{2}I\dot{\theta}^2 + \frac{1}{2}J\dot{\varphi}^2, \qquad (1.4.1)$$

where m is the mass of the disk, I is the moment of inertia of the disk about the axis perpendicular to the plane of the disk, and J is the moment of inertia about an axis in the plane of the disk (both axes passing through the disk's center).

For the derivation of kinetic energy formulas of this sort we refer to any basic mechanics book, such as Synge and Griffiths [1950]. We shall derive such formulas from a slightly more advanced point of view in Section 3.13.

If R is the radius of the disk, the nonholonomic constraints of rolling without slipping are

$$\dot{x} = R(\cos\varphi)\dot{\theta}\,,$$
$$\dot{y} = R(\sin\varphi)\dot{\theta}\,,$$

(1.4.2)

which state that the point P_0 fixed on the rim of the disk has zero velocity at the point of contact with the horizontal plane. Notice that these constraints have the form (1.3.1) if we write them as

$$\dot{x} - R(\cos\varphi)\dot{\theta} = 0\,,$$
$$\dot{y} - R(\sin\varphi)\dot{\theta} = 0\,.$$

We can write these equations in the form of the equations (1.3.1), namely as the two constraint equations

$$a^1 \cdot (\dot{x}, \dot{y}, \dot{\varphi}, \dot{\theta})^T = 0\,,$$
$$a^2 \cdot (\dot{x}, \dot{y}, \dot{\varphi}, \dot{\theta})^T = 0\,,$$

where T denotes the transpose and where

$$a^1 = (1, 0, 0, -R\cos\varphi)\,, \quad a^2 = (0, 1, 0, -R\sin\varphi)\,.$$

In the notation used in (1.3.1),

$$a_1^1 = 1\,, \quad a_2^1 = 0\,, \quad a_3^1 = 0\,, \quad a_4^1 = -R\cos\varphi\,,$$

and similarly for a^2:

$$a_1^2 = 0\,, \quad a_2^2 = 1\,, \quad a_3^2 = 0\,, \quad a_4^2 = -R\sin\varphi\,.$$

We will compute the dynamical equations for this system with controls in the next section. In particular, when there are no controls, we will get the dynamical equations for the uncontrolled disk. As we shall see, these free equations can be explicitly integrated.

Dynamics of the Controlled Disk.   Consider the case where we have two controls, one that can steer the disk and another that determines the roll torque. Now we shall use the general equations (1.3.6) to write down the equations for the controlled vertical rolling disk. According to these equations, we add the forces to the right-hand side of the Euler–Lagrange equations for the given Lagrangian along with Lagrange multipliers to enforce the constraints and to represent the reaction forces. In our case, L is cyclic in the configuration variables $q = (x, y, \varphi, \theta)$, and so the required dynamical equations become

$$\frac{d}{dt}\left(\frac{\partial L}{\partial \dot{q}}\right) = u_\varphi f^\varphi + u_\theta f^\theta + \lambda_1 a^1 + \lambda_2 a^2,$$

(1.4.3)

where, from (1.4.1), we have

$$\frac{\partial L}{\partial \dot{q}} = (m\dot{x}, m\dot{y}, J\dot{\varphi}, I\dot{\theta}),$$

and where

$$f^\varphi = (0,0,1,0), \quad f^\theta = (0,0,0,1),$$

corresponding to assumed controls in the directions of the two angles φ and θ, respectively. Here u_φ and u_θ are control functions, so the external control forces are $F = u_\varphi f^\varphi + u_\theta f^\theta$, and the λ_i are Lagrange multipliers, chosen to ensure satisfaction of the constraints (1.4.2).

We eliminate the multipliers as follows. Consider the first two components of (1.4.3) and substitute the constraints (1.4.2) to eliminate $\dot{x}$ and $\dot{y}$ to give

$$\lambda_1 = m\frac{d}{dt}\left(R\cos\varphi\,\dot{\theta}\right),$$
$$\lambda_2 = m\frac{d}{dt}\left(R\sin\varphi\,\dot{\theta}\right).$$

Substitution of these expressions for λ_1 and λ_2 into the last two components of (1.4.3) and noticing the simple identities

$$\lambda_1 a_3^1 + \lambda_2 a_3^2 = 0,$$
$$\lambda_1 a_4^1 + \lambda_2 a_4^2 = -mR^2\ddot{\theta},$$

gives the dynamic equations

$$J\ddot{\varphi} = u_\varphi,$$
$$(I + mR^2)\ddot{\theta} = u_\theta, \tag{1.4.4}$$

which, together with the constraints

$$\dot{x} = R(\cos\varphi)\dot{\theta},$$
$$\dot{y} = R(\sin\varphi)\dot{\theta}, \tag{1.4.5}$$

(and some specification of the control forces), determine the dynamics of the system.

The *free equations*, in which we set $u_\varphi = u_\theta = 0$, are easily integrated. In fact, in this case, the dynamic equations (1.4.4) show that $\dot{\varphi}$ and $\dot{\theta}$ are constants; calling these constants ω and Ω, respectively, we have

$$\varphi = \omega t + \varphi_0,$$
$$\theta = \Omega t + \theta_0.$$

Using these expressions in the constraint equations (1.4.5) and integrating again gives

$$x = \frac{\Omega}{\omega} R \sin(\omega t + \varphi_0) + x_0,$$

$$y = -\frac{\Omega}{\omega} R \cos(\omega t + \varphi_0) + y_0.$$

Consider next the controlled case, with nonzero controls u_1, u_2. Call the variables θ and ϕ "base" or "controlled" variables and the variables x and y "fiber" variables. The distinction is that while θ and φ are controlled directly, the variables x and y are controlled indirectly via the constraints.[10]

It is clear that the base variables are controllable in any sense we can imagine. One may ask whether the full system is controllable. Indeed it is, in a precise sense as we shall show later, by virtue of the nonholonomic nature of the constraints.

The Kinematic Controlled Disk.   It is also useful to define and study a related system, the so-called kinematic controlled rolling disk. In this case we imagine we have direct control over velocities rather than forces, and accordingly, we consider the most general first-order system satisfying the constraints or lying in the "constraint distribution." In the present case of the vertically rolling disk, this system is

$$\dot{q} = u_1 X_1 + u_2 X_2, \tag{1.4.6}$$

where $X_1 = (R \cos\varphi, R \sin\varphi, 0, 1)^T$ and $X_2 = (0, 0, 1, 0)^T$ and where $\dot{q} = (\dot{x}, \dot{y}, \dot{\varphi}, \dot{\theta})^T$.

In fact, X_1 and X_2 constitute a maximal set of independent vector fields on Q satisfying the constraints, in the sense that the components of X_1 and X_2 satisfy the equations (1.4.5), as is easily checked. As we shall see, it is instructive to analyze both the control and optimal control of such systems.

The Variational Controlled System.   As we indicated in the last section, the variational system is obtained by using Lagrange multipliers with the Lagrangian rather than Lagrange multipliers with the equations, as we did earlier. Namely, we consider the Lagrangian

$$L = \frac{1}{2} m(\dot{x}^2 + \dot{y}^2) + \frac{1}{2} I \dot{\theta}^2 + \frac{1}{2} J \dot{\varphi}^2 + \mu_1 (\dot{x} - R \dot{\theta} \cos\varphi) + \mu_2 (\dot{y} - R \dot{\theta} \sin\varphi),$$

where, because of the Lagrange multipliers, we relax the constraints and take variations over all curves. In other words, we write down the Euler–Lagrange equations for this Lagrangian and determine the multipliers from the constraints and initial conditions to the extent possible.

---

[10]The notation "base" and "fiber" comes from the fact that the configuration space Q splits naturally into the base and fiber of a trivial fiber bundle, as we shall see later.

The Euler–Lagrange equations for this Lagrangian, including external forces in the φ and θ equations, are

$$m\ddot{x} + \dot{\mu}_1 = 0, \tag{1.4.7}$$

$$m\ddot{y} + \dot{\mu}_2 = 0, \tag{1.4.8}$$

$$J\ddot{\varphi} - R\mu_1 \dot{\theta} \sin\varphi + R\mu_2 \dot{\theta} \cos\varphi = u_\varphi, \tag{1.4.9}$$

$$I\ddot{\theta} - R\frac{d}{dt}(\mu_1 \cos\varphi + \mu_2 \sin\varphi) = u_\theta. \tag{1.4.10}$$

From the constraint equations (1.4.5) and integrating equations (1.4.7) and (1.4.8) once, we have

$$\mu_1 = -mR\dot{\theta}\cos\varphi + A,$$

$$\mu_2 = -mR\dot{\theta}\sin\varphi + B,$$

where A and B are integration constants. Substituting these into equations (1.4.9) and (1.4.10) and simplifying, we obtain

$$J\ddot{\varphi} = R\dot{\theta}(A\sin\varphi - B\cos\varphi) + u_\theta,$$

$$(I + mR^2)\ddot{\theta} = R\dot{\varphi}(-A\sin\varphi + B\cos\varphi) + u_\varphi.$$

These equations, together with the constraints, define the dynamics. Notice that for nonzero A and B, *they are different from the dynamic nonholonomic (Lagrange–d'Alembert) equations.* As we have indicated, the motion determined by these equations is not that associated with physical dynamics in general, but is a model of the type of problem that is relevant to optimal control problems, as we shall see later.

Note also that the constants of motion A and B are *not determined* by the constraints or initial data. Thus in this instance there are many variational nonholonomic trajectories with a given set of initial conditions; the choice of $A = B = 0$ yields the nonholonomic (i.e., the Lagrange–d'Alembert) case. Interestingly, it is not always true that the nonholonomic trajectories are special cases of the variational nonholonomic trajectories, but it is possible to quantify when this occurs; see, e.g., Cardin and Favretti [1996].

Exercises

⋄ **1.4-1.** Write down an expression for the energy of the (dynamic nonholonomic) vertical rolling disk and compute its time rate of change under the action of the controls u_φ and u_θ.

⋄ **1.4-2.** Compute the dynamic nonholonomic and variational nonholonomic equations of motion of the upright rolling penny in the presence of a linear potential of the form $V(x, y, \varphi, \theta) = \alpha x$ for a real number α. Solve the equations if possible.

1.5   The Falling Rolling Disk

A more realistic disk is of course one that is allowed to fall over (i.e., it is permitted to deviate from the vertical). This turns out to be a very instructive example to analyze. See Figure 1.5.1. As the figure indicates, we denote the coordinates of contact of the disk in the xy-plane by (x, y) and let θ, φ, and ψ denote the angle between the plane of the disk and the vertical axis, the "heading angle" of the disk, and "self-rotation" angle of the disk, respectively.[11] Note that the notation ψ for the falling rolling disk corresponds to the notation θ in the special case of the vertical rolling disk.

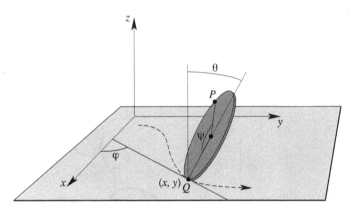

FIGURE 1.5.1. The geometry for the rolling disk.

For the moment, we just give the Lagrangian and constraints, and return to this example in Chapter 8, where we work things out in detail. While the equations of motion are straightforward to develop, as in the vertical case, they are somewhat messy, so we will defer these calculations until the later discussion. We will also show in Chapter 8 that this is a system that exhibits stability but not asymptotic stability.

Denote the mass and radius of the disk by m and R, respectively; let I be, as in the case of the vertical rolling disk, the moment of inertia about the axis through the disk's "axle" and J the moment of inertia about any

---

[11]A classical reference for the rolling disk is Vierkandt [1892], who showed something very interesting: On an appropriate symmetry-reduced space, namely, the constrained velocity phase space modulo the action of the group of Euclidean motions of the plane, all orbits of the system are periodic. Modern references that treat this example are Hermans [1995], O'Reilly [1996], Cushman, Hermans, and Kemppainen [1996], and Zenkov, Bloch, and Marsden [1998].

diameter. The Lagrangian is given by the kinetic minus potential energies:

$$L = \frac{m}{2} \left[(\xi - R(\dot{\varphi}\sin\theta + \dot{\psi}))^2 + \eta^2 \sin^2\theta + (\eta\cos\theta + R\dot{\theta})^2 \right]$$
$$+ \frac{1}{2} \left[J(\dot{\theta}^2 + \dot{\varphi}^2 \cos^2\theta) + I(\dot{\varphi}\sin\theta + \dot{\psi})^2 \right] - mgR\cos\theta,$$

where $\xi = \dot{x}\cos\varphi + \dot{y}\sin\varphi + R\dot{\psi}$ and $\eta = -\dot{x}\sin\varphi + \dot{y}\cos\varphi$, while the constraints are given by

$$\dot{x} = -\dot{\psi}R\cos\varphi,$$
$$\dot{y} = -\dot{\psi}R\sin\varphi.$$

Note that the constraints may also be written as $\xi = 0$, $\eta = 0$.

Unicycle with Rotor.   An interesting generalization of the falling disk is the "unicycle with rotor," analyzed in Zenkov, Bloch, and Marsden [2002b], (see Figure 1.5.2).

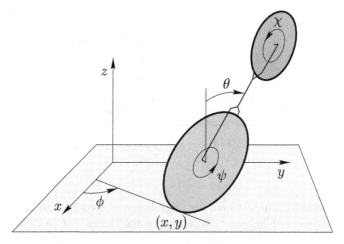

FIGURE 1.5.2. The configuration variables for the unicycle with rotor.

This is a homogeneous disk on a horizontal plane with a rotor. The rotor is free to rotate in the plane orthogonal to the disk. The rod connecting the centers of the disk and rotor keeps the direction of the radius of the disk through the contact point with the plane. We may view this system as a simple model of unicycle with rider whose arms are represented by the rotor. Stabilization is discussed in Chapter 9. A unicycle with pendulum is discussed in Zenkov, Bloch, and Marsden [2002b] and the web supplement.

The configuration space for this system is $Q = S^1 \times S^1 \times S^1 \times \mathrm{SE}(2)$, which we parameterize with coordinates $(\theta, \chi, \psi, \phi, x, y)$. As in Figure 1.5.2, θ is the tilt of the unicycle itself, and ψ and χ are the angular positions of the wheel of the unicycle and the rotor, respectively. The variables (ϕ, x, y), regarded as a point in SE(2), represent the angular orientation of the overall system and position of the point of contact of the wheel with the ground.

Further details are given in Chapter 9.

1.6   The Knife Edge

A simple and basic example of the behavior of a system with nonholonomic constraints is a knife edge or skate on an inclined plane.[12]

To set up the problem, consider a plane slanted at an angle α from the horizontal and let (x, y) represent the position of the point of contact of the knife edge with respect to a fixed Cartesian coordinate system on the plane (see Figure 1.6.1). The angle φ represents the orientation of the knife edge with respect to the xy-axis. The knife edge is moving under the influence of gravity with the acceleration due to gravity denoted by g. It also has mass m, and the moment of inertia of the knife edge about a vertical axis through its contact point is denoted by J.

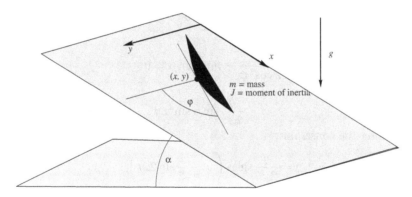

FIGURE 1.6.1. Motion of a knife edge on an inclined plane.

With this notation, the **knife edge Lagrangian** is taken to be

$$L = \frac{1}{2}m\left(\dot{x}^2 + \dot{y}^2\right) + \frac{1}{2}J\dot{\varphi}^2 + mgx\sin\alpha \tag{1.6.1}$$

with the constraint

$$\dot{x}\sin\varphi = \dot{y}\cos\varphi. \tag{1.6.2}$$

As for the rolling penny, we will compare the mechanical nonholonomic equations and the variational equations. In contrast to the penny we cannot solve the equations explicitly in general, but this is possible for certain

---

[12]This example is analyzed in, for example, Neimark and Fufaev [1972] and Arnold, Kozlov, and Neishtadt [1988].

initial data of interest. In particular, we shall be concerned with the initial data corresponding to the knife edge spinning about a point on the plane with zero initial velocity along the plane. The question is, what is the motion of the point of contact? We shall not consider the addition of controls for the moment.

The Nonholonomic Case. The equations of motion given in general by (1.3.2) become, in this case,

$$m\ddot{x} = \lambda \sin\varphi + g\sin\alpha \,,$$
$$m\ddot{y} = -\lambda \cos\varphi \,,$$
$$J\ddot{\varphi} = 0 \,.$$

We assume the initial data $x(0) = \dot{x}(0) = y(0) = \dot{y}(0) = \varphi(0) = 0$ and $\dot{\varphi}(0) = \omega$. The energy is given, according to the general formula (1.3.5), by

$$E = \frac{1}{2}m\left(\dot{x}^2 + \dot{y}^2\right) + \frac{1}{2}J\dot{\varphi}^2 - mgx\sin\alpha$$

and is preserved along the flow. Since it is preserved, it equals its initial value

$$E(0) = \frac{1}{2}J\omega^2 \,.$$

Hence, we have

$$\frac{1}{2}\frac{\dot{x}^2}{\cos^2\varphi} - mgx\sin\alpha = 0 \,.$$

Solving, we obtain

$$x = \frac{g}{2\omega^2}\sin\alpha\sin^2\omega t$$

and, using the constraint,

$$y = \frac{g}{2\omega^2}\sin\alpha\left(\omega t - \frac{1}{2}\sin 2\omega t\right) \,.$$

Hence the point of contact of the knife edge undergoes a cycloid motion along the plane, but does not slide down the plane.

The Variational Nonholonomic Case. Now we consider, in contrast, the variational nonholonomic equations of motion. We consider the constrained Lagrangian

$$L_C = \frac{1}{2}m\left(\dot{x}^2 + \dot{y}^2\right) + \frac{1}{2}J\dot{\varphi}^2 + mgx\sin\alpha - \lambda\left(\dot{x}\sin\varphi - \dot{y}\cos\varphi\right) \,.$$

As in the general theory, define the momenta by $p_i = \partial L/\partial\dot{q}^i$, which becomes, in this case,

$$p_x = \frac{\partial L_C}{\partial\dot{x}} = m\dot{x} - \lambda\sin\varphi \,,$$
$$p_y = \frac{\partial L_C}{\partial\dot{y}} = m\dot{y} + \lambda\cos\varphi \,.$$

Now assume initial data satisfying $p_x(0) = p_y(0) = \varphi(0) = \dot\varphi(0) = 0$. Then from Lagrange's equations, we get

$$\dot p_x = mg \sin \alpha \,,$$
$$\dot p_y = 0 \,,$$

and hence from the initial data

$$p_x = (mg \sin \alpha)t \,,$$
$$m\dot y + \lambda \cos \varphi = 0 \,.$$

Now the equation for $\dot\varphi$ is

$$J\ddot\varphi = -\lambda \dot x \cos \varphi - \lambda \dot y \sin \varphi = (\lambda g \sin \alpha \cos \varphi)t \,,$$

using the above expressions for p_x and p_y to solve for $\dot x$ and $\dot y$.
Again using the expressions for p_x and p_y we have

$$\lambda = p_y \cos \varphi - p_x \sin \varphi = -(mg \sin \alpha \sin \varphi)t \,.$$

Using this expression for λ gives

$$\dot x = (g \sin \alpha)t + \lambda/m \sin \varphi = (g \sin \alpha)t - (g \sin \alpha \sin^2 \varphi)t$$
$$= (mg \sin \alpha \cos^2 \varphi)t \,,$$
$$\dot y = -\lambda/m \cos \varphi = (g \sin \alpha \sin \varphi \cos \varphi)t \,,$$
$$\ddot\varphi = \left(\frac{m}{J}g^2 \sin^2 \alpha \sin \varphi \cos \varphi\right) t^2 \,.$$

Hence, in the variational formulation the point of contact of the knife edge slides monotonically down the plane, in contrast to the nonholonomic mechanical setting (see, e.g., Kozlov [1983]).

1.7   The Chaplygin Sleigh

One of the simplest mechanical systems that illustrates the possible "dissipative nature" of nonholonomic systems, even though they are energy-preserving, is the Chaplygin sleigh.[13]
  We now derive the equations of motion both using balance of forces as in Ruina [1998] and by the Lagrange multiplier approach, following the general theory. This system consists of a rigid body in the plane that is supported at three points, two of which slide freely without friction while

---

[13]The system is discussed in the original work of Chaplygin (see the references) as well as in Neimark and Fufaev [1972] and Ruina [1998].

the third is a knife edge, a constraint that allows no motion perpendicular to its edge.

To analyze the system, use a coordinate system Oxy fixed in the plane and a coordinate system $A\xi\eta$ fixed in the body with its origin at the point of support of the knife edge and the axis $A\xi$ through the center of mass C of the rigid body. The configuration of the body is described by the coordinates x, y and the angle θ between the moving and fixed sets of axes. Let m be the mass and I the moment of inertia about the center of mass. Let a be the distance from A to C. See Figure 1.7.1.

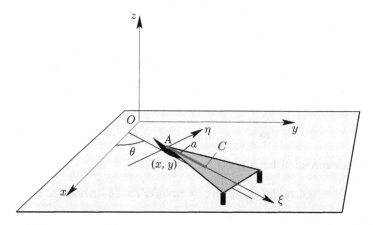

FIGURE 1.7.1. The Chaplygin sleigh is a rigid body moving on two sliding posts and one knife edge.

Denote the unit vectors along the axes $A\xi$ and $A\eta$ in the body by $\mathbf{e}_1$ and $\mathbf{e}_2$. The knife edge constraint can then be expressed as follows: The velocity at A is given by $\mathbf{v}_A = v_1\mathbf{e}_1$, where v_1 is the velocity in the direction $\mathbf{e}_1$.

The force at A is written as $R\mathbf{e}_2$; that is, the force is normal to the direction of motion at A. The position of point C is $\mathbf{r}_C = \mathbf{r}_A + a\mathbf{e}_1$, where the vectors $\mathbf{r}$ are in the fixed frame.

Since $\dot{\mathbf{e}}_1 = \dot{\theta}\mathbf{e}_2$ and $\dot{\mathbf{e}}_2 = -\dot{\theta}\mathbf{e}_1$, the velocity and acceleration of the point C are given by

$$\mathbf{v}_C = v\mathbf{e}_1 + \dot{\theta}a\mathbf{e}_2 ,$$
$$\mathbf{a}_C = \dot{v}\mathbf{e}_1 + v\dot{\theta}\mathbf{e}_2 + \ddot{\theta}a\mathbf{e}_2 - \dot{\theta}^2 a\mathbf{e}_1. \tag{1.7.1}$$

The balance of linear and angular momentum at the point A then gives

$$R\mathbf{e}_2 = m\mathbf{a}_C ,$$
$$0 = (\mathbf{r}_C - \mathbf{r}_A) \times (m\mathbf{a}_C) + I\ddot{\theta}\mathbf{e}_3, \tag{1.7.2}$$

where $\mathbf{e}_3$ is the normal vector to the plane. Setting $\dot{\theta} = \omega$ we find that equations (1.7.1), (1.7.2) yield the equations

$$\dot{v} = a\omega^2,$$
$$\dot{\omega} = -\frac{ma}{I + ma^2}v\omega. \tag{1.7.3}$$

The equations above are examples of **momentum equations** in nonholonomic mechanics, which we shall study in general in Chapter 5 and which will play an important role in the book. In the absence of nonholonomic constraints, this equation would yield conservation of angular momentum.

This set of equations has a family of equilibria (i.e., points at which the right-hand side vanishes) given by $\{(v, \omega) \mid v = \text{const}, \omega = 0\}$.

Linearizing about any of these equilibria one finds that one has one zero eigenvalue together with a negative eigenvalue if $v > 0$ and a positive eigenvalue if $v < 0$. In fact, the solution curves are ellipses in the $v\omega$ plane with the positive v-axis attracting all solutions; see below. (See Figure 1.7.2). We shall discuss this further in Section 8.6.

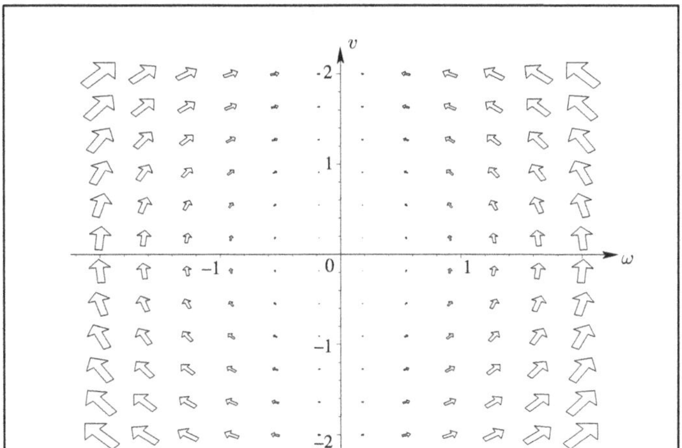

FIGURE 1.7.2. Chaplygin sleigh phase portrait.

We can also derive the equations from the Lagrangian (Lagrange–d'Alembert) point of view. The Lagrangian is given by

$$L(x_C, y_C, \theta) = \frac{1}{2}m\left(\dot{x}_C^2 + \dot{y}_C^2\right) + \frac{1}{2}I\dot{\theta}^2,$$

where x_C and y_C are the coordinates of the center of mass. We rewrite this in terms of the coordinates of the knife edge $x = x_C - a\cos\theta$ and

$y = y_C - a\sin\theta$. Hence we may rewrite the Lagrangian as

$$L(x, y, \theta) = \frac{1}{2}m\left(\frac{d}{dt}(x + a\cos\theta)^2 + \frac{d}{dt}(y + a\sin\theta)^2\right) + \frac{1}{2}I\dot{\theta}^2$$

$$= \frac{1}{2}m\left(\left(\dot{x} - a\sin\theta\dot{\theta}\right)^2 + \left(\dot{y} + a\cos\theta\dot{\theta}\right)^2\right) + \frac{1}{2}I\dot{\theta}^2$$

$$= \frac{1}{2}\left(m\dot{x}^2 + m\dot{y}^2 + \left(I + ma^2\right)\dot{\theta}^2 - 2ma\dot{x}\dot{\theta}\sin\theta + 2ma\dot{y}\dot{\theta}\cos\theta\right).$$

$$(1.7.4)$$

The knife edge constraint is

$$\dot{y}\cos\theta - \dot{x}\sin\theta = 0. \tag{1.7.5}$$

Hence the nonholonomic equations of motion are

$$m\frac{d}{dt}\left(\dot{x} - a\sin\theta\dot{\theta}\right) = -\lambda\sin\theta,$$

$$m\frac{d}{dt}\left(\dot{y} + a\cos\theta\dot{\theta}\right) = \lambda\cos\theta,$$

$$\frac{d}{dt}\left(I\dot{\theta} + ma^2\dot{\theta} - ma\dot{x}\sin\theta + ma\dot{y}\cos\theta\right)$$
$$-\left(-ma\dot{x}\dot{\theta}\cos\theta - ma\dot{y}\dot{\theta}\sin\theta\right) = 0;$$

that is,

$$\ddot{x} - a\cos\theta\dot{\theta}^2 - a\sin\theta\ddot{\theta} = -\frac{\lambda\sin\theta}{m},$$

$$\ddot{y} - a\sin\theta\dot{\theta}^2 + a\cos\theta\ddot{\theta} = \frac{\lambda\cos\theta}{m}, \tag{1.7.6}$$

$$(I + ma^2)\ddot{\theta} + ma\dot{\theta}\left(\dot{x}\cos\theta + \dot{y}\sin\theta\right) = 0,$$

where in the third of equations (1.7.6) we used the constraint (1.7.5). Now the velocity in the direction of motion is given by

$$v = \dot{x}\cos\theta + \dot{y}\sin\theta. \tag{1.7.7}$$

Hence the last of equations (1.7.6) becomes

$$\ddot{\theta} = \dot{\omega} = -\frac{ma}{I + ma^2}v\omega \tag{1.7.8}$$

and

$$\dot{v} = \ddot{x}\cos\theta + \ddot{y}\sin\theta - \dot{x}\dot{\theta}\sin\theta + \dot{y}\dot{\theta}\cos\theta$$
$$= a(\cos^2\theta + \sin^2\theta)\dot{\theta}^2 = a\dot{\theta}^2 = a\omega^2. \tag{1.7.9}$$

Thus we obtain our earlier sleigh equations.

Exercises

⋄ **1.7-1.**

(a) Compute the nonholonomic equations of motion for the Chaplygin sleigh on an incline.

(b) **Project.** Simulate the equations on the computer and discuss the nature of the dynamics; is the sleigh stable going down the incline "forwards" or "backwards"?

⋄ **1.7-2.**   Compute the variational equations of motion of the Chaplygin sleigh. Say what you can about the qualitative behavior of the system.

1.8   The Heisenberg System

The Heisenberg Algebra.   The Heisenberg algebra is the algebra one meets in quantum mechanics, wherein one has two operators q and p that have a nontrivial commutator, in this case a multiple of the identity. Thereby, one generates a three-dimensional Lie algebra. The system studied in this section has an associated Lie algebra with a similar structure, which is the reason the system is called the Heisenberg system. There is no intended relation to quantum mechanics per se other than this.

In Lie algebra theory this sort of a Lie algebra is of considerable interest. One refers to it as an example of a **central extension** because the element that one extends by (in this case a multiple of the identity) is in the center of the algebra; that is, it commutes with all elements of the algebra.

The Heisenberg system has played a significant role as an example in both nonlinear control and nonholonomic mechanics.

The Dynamic Heisenberg System.   As with the previous example, the dynamic Heisenberg system comes in two forms, one associated with the Lagrange–d'Alembert principle and one with an optimal control problem. As in the previous examples, the equations in each case are different.

In the dynamic setting, we consider the following standard kinetic energy Lagrangian on Euclidean three-space $\mathbb{R}^3$:

$$L = \frac{1}{2}(\dot{x}^2 + \dot{y}^2 + \dot{z}^2)$$

subject to the constraint

$$\dot{z} = y\dot{x} - x\dot{y}. \tag{1.8.1}$$

Controls u_1 and u_2 are given in the x and y directions. Letting $q = (x, y, z)^T$, the dynamic nonholonomic control system is[14]

$$\ddot{q} = u_1 X_1 + u_2 X_2 + \lambda W, \qquad (1.8.2)$$

where $X_1 = (1, 0, 0)^T$ and $X_2 = (0, 1, 0)^T$ and $W = (-y, x, 1)^T$. Eliminating λ we obtain the dynamic equations

$$(1 + x^2 + y^2)\ddot{x} = (1 + x^2)u_1 + xyu_2,$$
$$(1 + x^2 + y^2)\ddot{y} = (1 + y^2)u_2 + xyu_1,$$
$$(1 + x^2 + y^2)\ddot{z} = yu_1 - xu_2. \qquad (1.8.3)$$

Optimal Control for the Heisenberg System.   The control and optimal control of the corresponding kinematic problem have been quite important historically, and we shall return to them later on in the book in connection with, for example, the falling cat problem and optimal steering problems.[15]

The system may be written as

$$\dot{q} = u_1 g_1 + u_2 g_2, \qquad (1.8.4)$$

where $g_1 = (1, 0, y)^T$ and $g_2 = (0, 1, -x)^T$. As in the rolling disk example, g_1 and g_2 are a maximal set of independent vector fields satisfying the constraint

$$\dot{z} = y\dot{x} - x\dot{y}. \qquad (1.8.5)$$

Written out in full, these equations are

$$\dot{x} = u_1, \qquad (1.8.6)$$
$$\dot{y} = u_2, \qquad (1.8.7)$$
$$\dot{z} = yu_1 - xu_2. \qquad (1.8.8)$$

Aside on the Jacobi–Lie Bracket.   A notion that is important in mechanics and control theory is that of the ***Jacobi–Lie bracket*** $[f, g]$ of two vector fields f and g on $\mathbb{R}^n$ that are given in components by $f = (f^1, \ldots, f^n)$ and $g = (g^1, \ldots, g^n)$. It is defined to be the vector field with components

$$[f, g]^i = \sum_{j=1}^{n} \left(f^j \frac{\partial g^i}{\partial x^j} - g^j \frac{\partial f^i}{\partial x^j} \right),$$

---

[14]This example with controls was analyzed in Bloch and Crouch [1993]. A related nonholonomic system, but with slightly different constraints, may be found in Rosenberg [1977], Bates and Sniatycki [1993], and Bloch, Krishnaprasad, Marsden, and Murray [1996].

[15]As we mentioned earlier, this example was introduced in Brockett [1981].

or in vector calculus notation

$$[f, g] = (f \cdot \nabla) g - (g \cdot \nabla) f.$$

Later on, in Chapter 2, we will define the Jacobi–Lie bracket intrinsically on manifolds. An important geometric interpretation of this bracket is as follows.

Suppose we follow the vector field g (i.e., flow along the solution of the equation $\dot{x} = g(x)$) from point $x(0) = x_0$ for t units of time, then beginning with this as initial condition, we flow along the vector field f for time t; then along the vector field $-g$, and finally along $-f$ all for t units of time. Formally, we arrive at the point

$$(\exp -tf)(\exp -tg)(\exp tf)(\exp tg)(x_0), \tag{1.8.9}$$

where $(\exp tg)$ represents the flow of the vector field g for t units of time. Flows of vector fields will be described in more detail in Chapter 2.

Locally, expanding the exponential and in turn expanding each occurrence of g in the exponential in a Taylor series about x_0 we have along the flow of the equation $\dot{x} = g(x)$,

$$x(t) = x_0 + tg(x_0) + \frac{t^2}{2}g(x_0) \cdot \nabla g(x_0) + O(t^3). \tag{1.8.10}$$

(Here we compute the second derivative of $x(t)$ at $t = t_0$ by differentiating $\dot{x}(t) = g(x(t))$ with respect to t.) Hence, after a short computation using additional Taylor expansions, one finds that (1.8.9) becomes

$$x_0 - t^2 [f, g] (x_0)x_0 + O(t^3), \tag{1.8.11}$$

where $[f, g]$ is the Lie bracket as defined above. Thus, if $[f, g]$ is not in the span of vector fields g and f, then by concatenating the flows of f and g, we obtain motion in a new independent direction.

Return to the Heisenberg System. In the Heisenberg example, one verifies that the Jacobi–Lie bracket of the vector fields g_1 and g_2 is

$$[g_1, g_2] = 2g_3,$$

where $g_3 = (0, 0, 1)$. In fact, the three vector fields g_1, g_2, g_3 span all of $\mathbb{R}^3$ and, as a Lie algebra, is just the Heisenberg algebra described earlier.

By general controllability theorems that we shall discuss in Chapter 4 (Chow's theorem), one knows now that one can, with suitable controls, steer trajectories between any two points in $\mathbb{R}^3$. The above geometric interpretation makes this plausible. In particular, we are interested in the following optimal steering problem (see Figure 1.8.1).

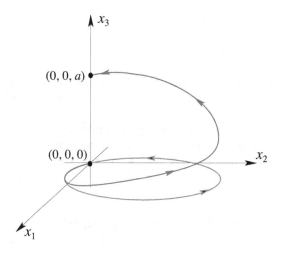

FIGURE 1.8.1. An optimal steering problem.

Optimal Steering Problem. Given a number $a > 0$, find time-dependent controls u_1, u_2 that steer the trajectory starting at $(0,0,0)$ at time $t = 0$ to the point $(0,0,a)$ after a given time $T > 0$ and that among all such controls minimizes

$$\frac{1}{2} \int_0^T (u_1^2 + u_2^2) \, dt \,.$$

An equivalent formulation is the following: Minimize the integral

$$\frac{1}{2} \int_0^T (\dot{x}^2 + \dot{y}^2) \, dt$$

among all curves $q(t)$ joining $q(0) = (0,0,0)$ to $q(T) = (0,0,a)$ that satisfy the constraint

$$\dot{z} = y\dot{x} - x\dot{y} \,.$$

As before, any solution must satisfy the Euler–Lagrange equations for the Lagrangian with a Lagrange multiplier inserted:

$$L\left(x, \dot{x}, y, \dot{y}, z, \dot{z}, \lambda, \dot{\lambda}\right) = \tfrac{1}{2}\left(\dot{x}^2 + \dot{y}^2\right) + \lambda\left(\dot{z} - y\dot{x} + x\dot{y}\right) \,.$$

The corresponding Euler–Lagrange equations are given by

$$\ddot{x} - 2\lambda\dot{y} = 0 \,, \tag{1.8.12}$$

$$\ddot{y} + 2\lambda\dot{x} = 0 \,, \tag{1.8.13}$$

$$\dot{\lambda} = 0 \,. \tag{1.8.14}$$

From the third equation λ is a constant, and the first two equations state that *the particle $(x(t), y(t))$ moves in the plane in a constant magnetic field (pointing in the z direction, with charge proportional to the constant λ).* For more on these ideas, see Chapter 7 on optimal control.

Some remarks are in order here:

1. The fact that this optimal steering problem gives rise to an interesting mechanical system is not an accident; we shall see this in much more generality in Chapter 7 and the Internet Supplement.

2. Since particles in constant magnetic fields move in circles with constant speed, they have a sinusoidal time dependence, and hence so do the controls. This has led to the "steering by sinusoids" approach in many nonholonomic steering problems (see, for example, Murray and Sastry [1993] and Section 6.1).

Equations (1.8.12) and (1.8.13) are linear first-order equations in the velocities and are readily solved:

$$\begin{bmatrix} \dot{x}(t) \\ \dot{y}(t) \end{bmatrix} = \begin{bmatrix} \cos(2\lambda t) & \sin(2\lambda t) \\ -\sin(2\lambda t) & \cos(2\lambda t) \end{bmatrix} \begin{bmatrix} \dot{x}(0) \\ \dot{y}(0) \end{bmatrix}. \tag{1.8.15}$$

Integrating once more and using the initial conditions $x(0) = 0$, $y(0) = 0$ gives

$$\begin{bmatrix} x(t) \\ y(t) \end{bmatrix} = \frac{1}{2\lambda} \begin{bmatrix} \cos(2\lambda t) - 1 & \sin(2\lambda t) \\ -\sin(2\lambda t) & \cos(2\lambda t) - 1 \end{bmatrix} \begin{bmatrix} -\dot{y}(0) \\ \dot{x}(0) \end{bmatrix}. \tag{1.8.16}$$

The other boundary condition $x(T) = 0$, $y(T) = 0$ gives

$$\lambda = \frac{n\pi}{T}$$

for an integer n. Using this information, we find z by integration: From $\dot{z} = y\dot{x} - x\dot{y}$ and the preceding expressions we get

$$\dot{z}(t) = \frac{T}{2n\pi} \left[\dot{x}(0)^2 + \dot{y}(0)^2 - \cos\left(\frac{2n\pi t}{T}\right)(\dot{x}(0)^2 + \dot{y}(0)^2) \right].$$

Integration from 0 to T and using $z(0) = 0$ gives

$$z(T) = \frac{T^2}{2n\pi} \left[\dot{x}(0)^2 + \dot{y}(0)^2 \right].$$

Thus, to achieve the boundary condition $z(T) = a$ one must choose

$$\dot{x}(0)^2 + \dot{y}(0)^2 = \frac{2\pi na}{T^2}.$$

One also finds that

$$\frac{1}{2} \int_0^T \left[\dot{x}(t)^2 + \dot{y}(t)^2 \right] dt = \frac{1}{2} \int_0^T \left[\dot{x}(0)^2 + \dot{y}(0)^2 \right] dt$$
$$= \frac{T}{2} \left[\dot{x}(0)^2 + \dot{y}(0)^2 \right]$$
$$= \frac{\pi n a}{T} ,$$

so that the minimum is achieved when $n = 1$.

Summary: The solution of the optimal control problem is given by choosing initial conditions such that $\dot{x}(0)^2 + \dot{y}(0)^2 = 2\pi a / T^2$ and with the trajectory in the xy-plane given by the circle

$$\begin{bmatrix} x(t) \\ y(t) \end{bmatrix} = \frac{1}{2\lambda} \begin{bmatrix} \cos(2\pi t/T) - 1 & -\sin(2\pi t/T) \\ \sin(2\pi t/T) & \cos(2\pi t/T) - 1 \end{bmatrix} \begin{bmatrix} -\dot{y}(0) \\ \dot{x}(0) \end{bmatrix} \qquad (1.8.17)$$

and with z given by

$$z(t) = \frac{ta}{T} - \frac{a}{2\pi} \sin\left(\frac{2\pi t}{T} \right).$$

Notice that any such solution can be rotated about the z axis to obtain another one.

Exercises

$\diamond$ **1.8-1.** For the standard kinetic energy Lagrangian on $\mathbb{R}^3$ and constraint (1.8.1) above, write down the variational nonholonomic problem. How does this compare with the optimal steering problem?

1.9   The Rigid Body

The Free Rigid Body. A key system in mechanics is the free rigid body. There are many excellent treatments of this topic; see, for example, Whittaker [1988], Arnold [1989], and Marsden and Ratiu [1999]. We restrict ourselves here to some essentials, although we shall return to the topic in detail in the context of nonholonomic mechanics and optimal control.

The configuration space of a rigid body moving freely in space is $\mathbb{R}^3 \times$ SO(3), describing the position of a coordinate frame fixed in the body and the orientation of the frame, the orientation of the frame given by an element of SO(3), i.e., an orthogonal 3×3 matrix with determinant 1. Since the three components of translational momentum are conserved, the body behaves as if it were rotating freely about its center of mass.[16]

Hence the phase space for the body may be taken to be $T\,SO(3)$—the tangent bundle of $SO(3)$—with points representing the position and velocity of the body, or in the Hamiltonian context we may choose the phase space to be the cotangent bundle $T^*\,SO(3)$, with points representing the position and momentum of the body. (This example may be equally well formulated for the group $SO(n)$ or indeed any compact Lie group.)

If I is the moment of inertia tensor computed with respect to a body fixed frame, which, in a *principal* body frame, we may represent by the diagonal matrix $\mathrm{diag}(I_1, I_2, I_3)$, the Lagrangian of the body is given by the kinetic energy, namely

$$L = \frac{1}{2}\Omega \cdot I\Omega, \qquad (1.9.1)$$

where Ω is the vector of angular velocities computed with respect to the axes fixed in the body.

The Euler–Lagrange equations of motion may be written as the system

$$\dot{A} = A\Omega, \qquad (1.9.2)$$

$$I\dot{\Omega} = I\Omega \times \Omega, \qquad (1.9.3)$$

where $A \in SO(3)$. Writing

$$\hat{\Omega} \equiv \begin{pmatrix} 0 & -\Omega_3 & \Omega_2 \\ \Omega_3 & 0 & -\Omega_1 \\ -\Omega_2 & \Omega_1 & 0 \end{pmatrix},$$

the dynamics may be rewritten

$$I\dot{\hat{\Omega}} = [I\hat{\Omega}, \hat{\Omega}], \qquad (1.9.4)$$

or, in terms of the angular momentum matrix $\hat{M} = I\hat{\Omega}$,

$$\dot{\hat{M}} = [\hat{M}, \hat{\Omega}]. \qquad (1.9.5)$$

The Rolling Ball. This paragraph considers the controlled rolling inhomogeneous ball on the plane, the kinematics of which were discussed in Brockett and Dai [1992], establishing the completely nonholonomic nature of the constraint distribution H. (A distribution is completely nonholonomic if the span of the iterated brackets of the vector fields lying in it has dimension equal to the dimension of the underlying manifold; see Chapter

---

[16]This is not the case with other systems, such as a rigid body moving in a fluid; even though the system is translation-invariant, its "center of mass" need not move on a straight line, so the configuration space must be taken to be the full Euclidean group.

4 for a full explanation.) The dynamics of the uncontrolled system are described, for example, in McMillan [1936] (see also Bloch, Krishnaprasad, Marsden, and Murray [1996], Jurdjevic [1993], Koon and Marsden [1997b], and Krishnaprasad, Yang and Dayawansa [1991]). We will use the coordinates x, y for the linear horizontal displacement and $P \in \mathrm{SO}(3)$ for the angular displacement of the ball. Thus P gives the orientation of the ball with respect to inertial axes $\mathbf{e}_1$, $\mathbf{e}_2$, $\mathbf{e}_3$ fixed in the plane, where the $\mathbf{e}_i$ are the standard basis vectors aligned with the x-, y-, and z-axes, respectively. See Figure 1.9.1.

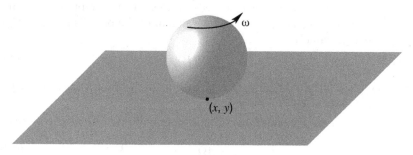

FIGURE 1.9.1. The rolling ball.

Let the ball have radius a and mass m and let $\boldsymbol{\omega} \in \mathbb{R}^3$ denote the angular velocity of the ball with respect to the inertial axes. In particular, the ball may spin freely about the z-axis, and the z-component of angular momentum is conserved. If J denotes the inertia tensor of the ball with respect to the body axes (i.e., fixed in the body), then $\mathbb{J} = P^T J P$ denotes the inertia tensor of the ball with respect to the inertial axes (i.e., fixed in space) and $\mathbb{J}\boldsymbol{\omega}$ is the angular momentum of the ball with respect to the inertial axes. The conservation law alluded to above is expressed as

$$\mathbf{e}_3^T \mathbb{J}\boldsymbol{\omega} = c \,. \tag{1.9.6}$$

The nonholonomic constraints of rolling without slipping may be expressed as

$$a\mathbf{e}_2^T \boldsymbol{\omega} + \dot{x} = 0 \,,$$
$$a\mathbf{e}_1^T \boldsymbol{\omega} - \dot{y} = 0 \,. \tag{1.9.7}$$

We may express the kinematics for the rotating ball as $\dot{P} = \hat{\Omega} P$, where $\Omega = P\boldsymbol{\omega}$ is the angular velocity in the body frame.

Appending the constraints via Lagrange multipliers we obtain the equations of motion

$$\hat{\Omega}P - (\widehat{J^{-1}\hat{\Omega}J\Omega})P = a\lambda_1(\widehat{J^{-1}P\mathbf{e}_1})P + a\lambda_2(\widehat{J^{-1}P\mathbf{e}_2})P \,,$$
$$m\ddot{x} = \lambda_2 + u_1 \,,$$
$$m\ddot{y} = -\lambda_1 + u_2 \,. \tag{1.9.8}$$

Using inertial coordinates $\boldsymbol{\omega} = P^T \Omega$, the system becomes

$$\dot{\boldsymbol{\omega}} = \mathbb{J}^{-1}\hat{\boldsymbol{\omega}}\mathbb{J}\boldsymbol{\omega} + a\lambda_1\mathbb{J}^{-1}\mathbf{e}_1 + a\lambda_2\mathbb{J}^{-1}\mathbf{e}_2 \,,$$
$$m\ddot{x} = \lambda_2 + u_1 \,,$$
$$m\ddot{y} = -\lambda_1 + u_2 \,,$$
$$\dot{P} = P\hat{\boldsymbol{\omega}} \,. \tag{1.9.9}$$

Also, from the constraints and the constants of motion we obtain the following expression for $\boldsymbol{\omega}$:

$$\boldsymbol{\omega} = \dot{x}(\alpha_2 \mathbf{e}_3 - \mathbf{e}_2) + \dot{y}(\mathbf{e}_1 - \alpha_1 \mathbf{e}_3) + \alpha_3 \mathbf{e}_3 \,,$$

where

$$\alpha_1 = \frac{\mathbf{e}_3^T \mathbb{J} \mathbf{e}_1}{a\mathbf{e}_3^T \mathbb{J} \mathbf{e}_3} \,, \qquad \alpha_2 = \frac{\mathbf{e}_3^T \mathbb{J} \mathbf{e}_2}{a\mathbf{e}_3^T \mathbb{J} \mathbf{e}_3} \,, \qquad \alpha_3 = \frac{c}{\mathbf{e}_3^T \mathbb{J} \mathbf{e}_3} \,. \tag{1.9.10}$$

Then the equations become

$$m\ddot{x} = \lambda_2 + u_1 \,,$$
$$m\ddot{y} = -\lambda_1 + u_2 \,, \tag{1.9.11}$$
$$\dot{P} = P(\dot{x}(\alpha_2 \mathbf{e}_3 - \mathbf{e}_2) + \widehat{\dot{y}(\mathbf{e}_1 - \alpha_1 \mathbf{e}_3) + \alpha_3 \mathbf{e}_3}) \,.$$

One can now eliminate the multipliers using the first three equations of (1.9.9) and the constraints. The resulting expressions are a little complicated in the general case (although they can be found in straightforward fashion), but become pleasingly simple in the case of a homogeneous ball, where say $J = mk^2$ (k is called the radius of gyration in the classical literature).

In the latter case, the equations of motion for ω_1 and ω_2 become simply

$$mk^2\dot{\omega}_1 = a\lambda_1 \,,$$
$$mk^2\dot{\omega}_2 = a\lambda_2 \,. \tag{1.9.12}$$

Rewriting these equations in terms of x and y using the multipliers and substituting the resulting expressions for the λ_i into the equations of motion for x and y yields the equations

$$m\ddot{x} = \frac{a^2}{a^2 + k^2}u_1 \,,$$
$$m\ddot{y} = \frac{a^2}{a^2 + k^2}u_2 \,. \tag{1.9.13}$$

A similar elimination argument works in the general nonhomogeneous case.

Note that the homogeneous ball moves under the action of external forces like a point mass located at its center but with force reduced by the ratio $a^2/(a^2 + k^2)$; see also the following subsection.

A Homogeneous Ball on a Rotating Plate.   A useful example is a model of a homogeneous ball on a rotating plate (see Neimark and Fufaev [1972] and Yang [1992] for the affine case and, for example, Bloch and Crouch [1992], Brockett and Dai [1992], and Jurdjevic [1993] for the linear case). As we mentioned earlier, Chaplygin [1897b, 1903] studied the motion of an *inhomogeneous* rolling ball on a fixed plane.

Let the plane rotate with constant angular velocity $\tilde{\Omega}$ about the z-axis. The configuration space of the sphere is $Q = \mathbb{R}^2 \times \mathrm{SO}(3)$, parameterized by $(x, y, R), R \in \mathrm{SO}(3)$, all measured with respect to the inertial frame. Let $\boldsymbol{\omega} = (\omega_1, \omega_2, \omega_3)$ be the angular velocity vector of the sphere measured also with respect to the inertial frame, let m be the mass of the sphere, mk^2 its inertia about any axis, and let a be its radius.

The Lagrangian of the system is

$$L = \frac{1}{2}m(\dot{x}^2 + \dot{y}^2) + \frac{1}{2}mk^2(\omega_1{}^2 + \omega_2{}^2 + \omega_3{}^2) \tag{1.9.14}$$

with the affine nonholonomic constraints

$$\begin{aligned} \dot{x} + a\omega_2 &= -\tilde{\Omega}y\,, \\ \dot{y} - a\omega_1 &= \tilde{\Omega}x. \end{aligned} \tag{1.9.15}$$

Note that the Lagrangian here is a metric on Q that is bi-invariant on $\mathrm{SO}(3)$, since the ball is homogeneous. Note also that $\mathbb{R}^2 \times \mathrm{SO}(3)$ is a principal bundle over $\mathbb{R}^2$ with respect to the right $\mathrm{SO}(3)$ action on Q given by

$$(x, y, R) \mapsto (x, y, RS) \tag{1.9.16}$$

for $S \in \mathrm{SO}(3)$. The action is on the *right*, since the symmetry is a material symmetry.

A brief calculation shows that the equations of motion become

$$\begin{aligned} \ddot{x} - \frac{k^2\tilde{\Omega}}{a^2 + k^2}\dot{y} &= 0\,, \\ \ddot{y} + \frac{k^2\tilde{\Omega}}{a^2 + k^2}\dot{x} &= 0. \end{aligned} \tag{1.9.17}$$

These equations are easily integrated to show that the ball simply oscillates on the plate between two circles rather than flying off as one might expect.

Set

$$\alpha = \frac{k^2\tilde{\Omega}}{a^2 + k^2}\,.$$

Then one can see that the equations are equivalent to

$$\dddot{x} + \alpha^2\dot{x} = 0\,, \tag{1.9.18}$$

$$\dddot{y} + \alpha^2\dot{y} = 0\,. \tag{1.9.19}$$

Hence
$$x = A \cos \alpha t + B \sin \alpha t + C$$
for constants A, B, C depending on the initial data, and similarly for y.

The Inverted Pendulum on a Cart. A useful classical system for testing control-theoretic ideas is the inverted pendulum on a cart, the goal being to stabilize the pendulum about the vertical using a force acting on the cart. In this book we will use this system to illustrate stabilization using the energy methods as discussed in Bloch, Marsden, and Alvarez [1997] and Bloch, Leonard, and Marsden [1997]. (see Chapter 9 for further references) Here we just write down the equations of motion.

First, we compute the Lagrangian for the cart–pendulum system. Let s denote the position of the cart on the s-axis and let ϕ denote the angle of the pendulum with the upright vertical, as in Figure 1.9.2.

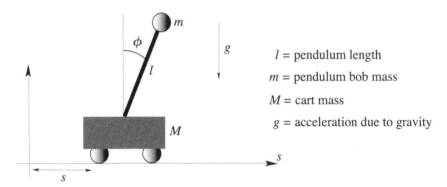

FIGURE 1.9.2. The pendulum on a cart.

Here, the configuration space is $Q = G \times S = \mathbb{R} \times S^1$ with the first factor being the cart position s, and the second factor being the pendulum angle ϕ. The velocity phase space TQ has coordinates $(s, \phi, \dot{s}, \dot{\phi})$.

The velocity of the cart relative to the lab frame is $\dot{s}$, while the velocity of the pendulum relative to the lab frame is the vector

$$v_{\text{pend}} = (\dot{s} + l \cos \phi \, \dot{\phi}, -l \sin \phi \, \dot{\phi}). \tag{1.9.20}$$

The kinetic energy of the coupled cart-pendulum system is given by:

$$K\left(s, \phi, \dot{s}, \dot{\phi}\right) = \frac{1}{2}(\dot{s}, \dot{\phi}) \begin{pmatrix} M + m & ml \cos \phi \\ ml \cos \phi & ml^2 \end{pmatrix} \begin{pmatrix} \dot{s} \\ \dot{\phi} \end{pmatrix}. \tag{1.9.21}$$

The Lagrangian is the kinetic minus potential energy, so we get

$$L(s, \phi, \dot{s}, \dot{\phi}) = K(s, \phi, \dot{s}, \dot{\phi}) - V(\phi), \tag{1.9.22}$$

where the potential energy is $V = mgl \cos \phi$. Note that there is a symmetry group G of the pendulum–cart system, that of translation in the s variable, so $G = \mathbb{R}$. We do not destroy this symmetry when doing stabilization in ϕ.

For convenience we rewrite the Lagrangian as

$$L(s, \phi, \dot{s}, \dot{\phi}) = \frac{1}{2}(\alpha \dot{\phi}^2 + 2\beta \cos \phi \dot{s} \dot{\phi} + \gamma \dot{s}^2) + D \cos \phi, \qquad (1.9.23)$$

where $\alpha = ml^2, \beta = ml, \gamma = M + m$, and $D = -mgl$ are constants. Note that $\alpha \gamma - \beta^2 > 0$. The momentum conjugate to s is $p_s = \gamma \dot{s} + \beta \cos \phi \dot{\phi}$, and the momentum conjugate to ϕ is $p_\phi = \alpha \dot{\phi} + \beta \cos \phi \dot{s}$. The relative equilibrium defined by $\phi = 0$, $\dot{\phi} = 0$, and $\dot{s} = 0$ is unstable, since $D < 0$.

The equations of motion of the cart–pendulum system with a control force u acting on the cart (and no direct forces acting on the pendulum) are, since s is a cyclic variable (i.e., L is independent of s),

$$\frac{d}{dt} \frac{\partial L}{\partial \dot{s}} = u,$$

$$\frac{d}{dt} \frac{\partial L}{\partial \dot{\phi}} - \frac{\partial L}{\partial \phi} = 0,$$

that is,

$$\frac{d}{dt} p_s = \frac{d}{dt}(\gamma \dot{s} + \beta \cos \phi \dot{\theta}) = u,$$

$$\frac{d}{dt} p_\phi + \beta \sin \phi \dot{s} \dot{\phi} + D \sin \phi = \frac{d}{dt}(\alpha \dot{\phi} + \beta \cos \phi \dot{s})$$

$$+ \beta \sin \phi \dot{s} \dot{\phi} + D \sin \phi = 0.$$

Rigid Body with a Rotor.   Following the work of Krishnaprasad [1985], Bloch, Krishnaprasad, Marsden, and Alvarez [1992], and Bloch, Leonard, and Marsden[1997, 2000], we consider a rigid body with a rotor aligned along the third principal axis of the body as in Figure 1.9.3. This is a model for a satellite. The rotor spins under the influence of a torque u acting on the rotor. The configuration space is $Q = \mathrm{SO}(3) \times S^1$, with the first factor being the rigid body attitude and the second factor being the rotor angle. The Lagrangian is total kinetic energy of the system (rigid carrier plus rotor), with no potential energy.

Again, this system will be used in Section 9.2 to illustrate the energy method in analyzing stabilization and stability.

The Lagrangian for this system (see Bloch, Krishnaprasad, Marsden, and Alvarez [1992] and Bloch, Leonard, and Marsden [2001]) is

$$L = \frac{1}{2}(\lambda_1 \Omega_1^2 + \lambda_2 \Omega_2^2 + I_3 \Omega_3^2 + J_3(\Omega_3 + \dot{\alpha})^2), \qquad (1.9.24)$$

where $I_1 > I_2 > I_3$ are the rigid body moments of inertia, $J_1 = J_2$ and J_3 are the rotor moments of inertia, $\lambda_i = I_i + J_i$, $\Omega = (\Omega_1, \Omega_2, \Omega_3)$ is the

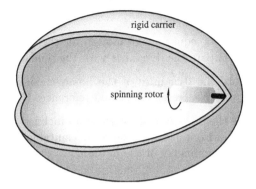

FIGURE 1.9.3. The rigid body with rotor.

body angular velocity vector of the carrier, and α is the relative angle of the rotor.

The body angular momenta are determined by the Legendre transform to be

$$\Pi_1 = \lambda_1 \Omega_1 \,,$$
$$\Pi_2 = \lambda_2 \Omega_2 \,,$$
$$\Pi_3 = \lambda_3 \Omega_3 + J_3 \dot{\alpha} \,,$$
$$l_3 = J_3 (\Omega_3 + \dot{\alpha}) \,.$$

The momentum conjugate to α is l_3.

The equations of motion with a control torque u acting on the rotor are

$$\lambda_1 \dot{\Omega}_1 = \lambda_2 \Omega_2 \Omega_3 - (\lambda_3 \Omega_3 + J_3 \dot{\alpha}) \Omega_2 \,,$$
$$\lambda_2 \dot{\Omega}_2 = -\lambda_1 \Omega_1 \Omega_3 + (\lambda_3 \Omega_3 + J_3 \dot{\alpha}) \Omega_1 \,,$$
$$\lambda_3 \dot{\Omega}_3 + J_3 \ddot{\alpha} = (\lambda_1 - \lambda_2) \Omega_1 \Omega_2 \,, \qquad (1.9.25)$$
$$\dot{l}_3 = u \,.$$

The equations may also be written in Hamiltonian form:

$$\dot{\Pi}_1 = \left(\frac{1}{I_3} - \frac{1}{\lambda_2} \right) \Pi_2 \Pi_3 - \frac{l_3 \Pi_2}{I_3} \,,$$
$$\dot{\Pi}_2 = \left(\frac{1}{\lambda_1} - \frac{1}{I_3} \right) \Pi_1 \Pi_3 + \frac{l_3 \Pi_1}{I_3} \,,$$
$$\dot{\Pi}_3 = \left(\frac{1}{\lambda_2} - \frac{1}{\lambda_1} \right) \Pi_1 \Pi_2 = a_3 \Pi_1 \Pi_2 \,,$$
$$\dot{l}_3 = u \,.$$

Here $\lambda_i = I_i + J_i$.

Exercises

⋄ **1.9-1.** Compute the equations of motion for the variational nonholonomic ball and compare the dynamics with the nonholonomic case.

⋄ **1.9-2.** Compute the dynamics of the homogeneous ball on a *freely* rotating table. (See Weckesser [1997] and references therein.)

⋄ **1.9-3.** Analyze the motion of the cart on an inclined plane making an angle of α to the horizontal. Show that with a suitable change of variable one can still find a symmetry of the motion.

1.10   The Roller Racer

We now consider a tricycle-like mechanical system called the ***roller racer***, or the ***Tennessee racer***, that is capable of locomotion by oscillating the front handlebars. This toy was studied using the methods of Bloch, Krishnaprasad, Marsden, and Murray [1996] in Tsakiris [1995] and Krishnaprasad and Tsakiris [2001] and by energy–momentum methods in Zenkov, Bloch, and Marsden [1998]. Analysis of this system may be a useful guide for modeling and studying the stability of other systems, such as aircraft landing gears and train wheels.

The roller racer is modeled as a system of two planar coupled rigid bodies (the main body and the second body) with a pair of wheels attached on each of the bodies at their centers of mass: a nonholonomic generalization of the coupled planar bodies discussed earlier. We assume that the mass and the linear momentum of the second body are negligible, but that the moment of inertia about the vertical axis is not. See Figure 1.10.1.

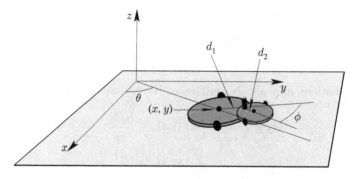

FIGURE 1.10.1. The geometry for the roller racer.

Let (x, y) be the location of the center of mass of the first body and denote the angle between the inertial reference frame and the line passing through the center of mass of the first body by θ, the angle between the bodies by ϕ, and the distances from the centers of mass to the joint by d_1 and d_2. The mass of body 1 is denoted by m, and the inertias of the two bodies are written as I_1 and I_2.

The Lagrangian and the constraints are

$$L = \frac{1}{2}m(\dot{x}^2 + \dot{y}^2) + \frac{1}{2}I_1\dot{\theta}^2 + \frac{1}{2}I_2(\dot{\theta} + \dot{\phi})^2$$

and

$$\dot{x} = \cos\theta\left(\frac{d_1\cos\phi + d_2}{\sin\phi}\dot{\theta} + \frac{d_2}{\sin\phi}\dot{\phi}\right);$$

$$\dot{y} = \sin\theta\left(\frac{d_1\cos\phi + d_2}{\sin\phi}\dot{\theta} + \frac{d_2}{\sin\phi}\dot{\phi}\right).$$

The configuration space is $\text{SE}(2) \times \text{SO}(2)$. The Lagrangian and the constraints are invariant under the left action of $\text{SE}(2)$ on the first factor of the configuration space.

We shall see later that the roller racer has a two-dimensional manifold of equilibria and that under a suitable stability condition some of these equilibria are stable modulo $\text{SE}(2)$ and in addition asymptotically stable with respect to $\dot{\phi}$.

1.11 The Rattleback

A rattleback is a convex asymmetric rigid body rolling without sliding on a horizontal plane. It is known for its ability to spin in one direction and to resist spinning in the opposite direction for some parameter values, and for other values to exhibit multiple reversals. See Figure 1.11.1.

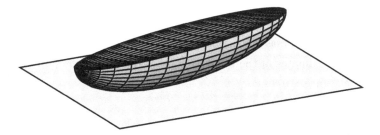

FIGURE 1.11.1. The rattleback.

Basic references on the rattleback are Walker [1896], Karapetyan [1980, 1981], Markeev [1983, 1992], Pascal [1983, 1986], and Bondi [1986]. We adopt the ideal model (with no energy dissipation and no sliding) of these references, and within that context no approximations are made. In particular, the shape need not be ellipsoidal. Walker did some initial stability and instability investigations by computing the spectrum, while Bondi extended this analysis and also used what we now recognize as the momentum equation. (See Chapter 5 for the general theory of the momentum equation and see Zenkov, Bloch, and Marsden [1998] and Section 8.5 for the explicit form of the momentum for the rattleback A discussion of the momentum equation for the rattleback may also be found in Burdick, Goodwine and Ostrowski [1994].) Karapetyan carried out a stability analysis of the relative equilibria, while Markeev's and Pascal's main contributions were to the study of spin reversals using small-parameter and averaging techniques. Energy methods were used to analyze the problem in Zenkov, Bloch, and Marsden [1998], and we return to this in Section 8.5

Introduce the Euler angles θ, ϕ, ψ using the principal axis body frame relative to an inertial reference frame. We use the same convention for the angles as in Arnold [1989] and Marsden and Ratiu [1999]. These angles together with two horizontal coordinates x, y of the center of mass are coordinates in the configuration space $SO(3) \times \mathbb{R}^2$ of the rattleback.

The Lagrangian of the rattleback is computed to be

$$
\begin{aligned}
L = \frac{1}{2} & \left[A \cos^2 \psi + B \sin^2 \psi + m(\gamma_1 \cos \theta - \zeta \sin \theta)^2 \right] \dot{\theta}^2 \\
& + \frac{1}{2} \left[(A \sin^2 \psi + B \cos^2 \psi) \sin^2 \theta + C \cos^2 \theta \right] \dot{\phi}^2 \\
& + \frac{1}{2} \left(C + m\gamma_2^2 \sin^2 \theta \right) \dot{\psi}^2 + \frac{1}{2} m \left(\dot{x}^2 + \dot{y}^2 \right) \\
& + m(\gamma_1 \cos \theta - \zeta \sin \theta)\gamma_2 \sin \theta \, \dot{\theta}\dot{\psi} + (A - B) \sin \theta \sin \psi \cos \psi \, \dot{\theta}\dot{\phi} \\
& + C \cos \theta \, \dot{\phi}\dot{\psi} + mg(\gamma_1 \sin \theta + \zeta \cos \theta),
\end{aligned}
$$

where

$A, B, C =$ the principal moments of inertia of the body,

$m =$ the total mass of the body,

$(\xi, \eta, \zeta) =$ coordinates of the point of contact relative to the body frame,

$\gamma_1 = \xi \sin \psi + \eta \cos \psi$,

$\gamma_2 = \xi \cos \psi - \eta \sin \psi$.

The shape of the body is encoded by the functions ξ, η, and ζ. The constraints are

$$
\dot{x} = \alpha_1 \dot{\theta} + \alpha_2 \dot{\psi} + \alpha_3 \dot{\phi}, \quad \dot{y} = \beta_1 \dot{\theta} + \beta_2 \dot{\psi} + \beta_3 \dot{\phi},
$$

where

$$\alpha_1 = -(\gamma_1 \sin \theta + \zeta \cos \theta) \sin \phi,$$
$$\alpha_2 = \gamma_2 \cos \theta \sin \phi + \gamma_1 \cos \phi,$$
$$\alpha_3 = \gamma_2 \sin \phi + (\gamma_1 \cos \theta - \zeta \sin \theta) \cos \phi,$$
$$\beta_k = -\frac{\partial \alpha_k}{\partial \phi}, \quad k = 1, 2, 3.$$

The Lagrangian and the constraints are SE(2)-invariant, where the action of an element $(a, b, \alpha) \in$ SE(2) is given by

$$(x, y, \phi) \mapsto (x \cos \alpha - y \sin \alpha + a, x \sin \alpha + y \cos \alpha + b, \phi + \alpha).$$

Corresponding to this invariance, ξ, η, and ζ are functions of the variables θ and ψ only.

1.12 The Toda Lattice

An important and beautiful mechanical system that describes the interaction of particles on the line (i.e., in one dimension) is the Toda lattice. We shall describe the nonperiodic finite Toda lattice following the treatment of Moser [1975].

This is a key example in integrable systems theory. Later on, in Chapter 8, we shall compare the behavior of this system to certain nonholonomic systems. In the Internet Supplement we also consider the Toda lattice from the point of view of optimal control theory.

The model consists of n particles moving freely on the x-axis and interacting under an exponential potential. Denoting the position of the kth particle by x_k, the Hamiltonian is given by

$$H(x, y) = \frac{1}{2} \sum_{k=1}^{n} y_k^2 + \sum_{k=1}^{n-1} e^{(x_k - x_{k+1})}.$$

The associated Hamiltonian equations are

$$\dot{x}_k = \frac{\partial H}{\partial y_k} = y_k,$$
$$\dot{y}_k = -\frac{\partial H}{\partial x_k} = e^{x_{k-1} - x_k} - e^{x_k - x_{k+1}}, \quad (1.12.1)$$

where we use the convention $e^{x_0 - x_1} = e^{x_n - x_{n+1}} = 0$, which corresponds to formally setting $x_0 = -\infty$ and $x_{n+1} = +\infty$.

This system of equations has an extraordinarily rich structure. Part of this is revealed by Flaschka's (Flaschka [1974]) change of variables given by

$$a_k = \frac{1}{2} e^{(x_k - x_{k+1})/2} \quad \text{and} \quad b_k = -\frac{1}{2} y_k. \quad (1.12.2)$$

In these new variables, the equations of motion then become

$$\dot{a}_k = a_k(b_{k+1} - b_k), \quad k = 1,\ldots,n-1,$$
$$\dot{b}_k = 2(a_k^2 - a_{k-1}^2), \quad k = 1,\ldots,n,$$

with the boundary conditions $a_0 = a_n = 0$. This system may be written in the following matrix form (called the **Lax pair representation**):

$$\frac{d}{dt}L = [B, L] = BL - LB, \tag{1.12.3}$$

where

$$L = \begin{pmatrix} b_1 & a_1 & 0 & \cdots & 0 \\ a_1 & b_2 & a_2 & \cdots & 0 \\ & & \ddots & & \\ & & b_{n-1} & a_{n-1} \\ 0 & & a_{n-1} & b_n \end{pmatrix}, \quad B = \begin{pmatrix} 0 & a_1 & 0 & \cdots & 0 \\ -a_1 & 0 & a_2 & \cdots & 0 \\ & & \ddots & & \\ & & 0 & a_{n-1} \\ 0 & & -a_{n-1} & 0 \end{pmatrix}.$$

If $O(t)$ is the orthogonal matrix solving the equation

$$\frac{d}{dt}O = BO, \quad O(0) = \text{Identity},$$

then from (1.12.3) we have

$$\frac{d}{dt}(O^{-1}LO) = 0.$$

Thus, $O^{-1}LO = L(0)$; i.e., $L(t)$ is related to $L(0)$ by a similarity transformation, and thus the eigenvalues of L, which are real and distinct, are preserved along the flow. This is enough to show that in fact this system is explicitly solvable or integrable.

Discussion. There is, however, much more structure in this example. For instance, if N is the matrix $\text{diag}[1, 2, \ldots, n]$, the Toda flow (1.12.3) may be written in the following **double bracket** form:

$$\dot{L} = [L, [L, N]]. \tag{1.12.4}$$

This was shown in Bloch [1990] and analyzed further in Bloch, Brockett, and Ratiu [1990], Bloch, Brockett, and Ratiu [1992], and Bloch, Flaschka, and Ratiu [1990]. This double bracket equation restricted to a level set of the integrals described above is in fact the gradient flow of the function $\text{Tr}LN$ with respect to the so-called normal metric; see Bloch, Brockett, and Ratiu [1990]. Double bracket flows are derived in Brockett [1994].

From this observation it is easy to show that the flow tends asymptotically to a diagonal matrix with the eigenvalues of $L(0)$ on the diagonal and ordered according to magnitude, recovering the observation of Moser, Symes [1982], and Deift, Nanda, and Tomei [1983].

A very important feature of the tridiagonal aperiodic Toda lattice flow is that it can be solved explicitly as follows: Let the initial data be given by $L(0) = L_0$. Given a matrix A, use the Gram–Schmidt process on the columns of A to factorize A as $A = k(A)u(A)$, where $k(A)$ is orthogonal and $u(A)$ is upper triangular. Then the explicit solution of the Toda flow is given by

$$L(t) = k(\exp(tL_0))L_0 k^T (\exp(tL_0)).$$

(1.12.5)

The reader can check this explicitly or refer for example to Symes [1980, 1982].

Four-Dimensional Toda.   Here we simulate the Toda lattice in four dimensions (see Bloch [2000]). The Hamiltonian is

$$H(a,b) = a_1^2 + a_2^2 + b_1^2 + b_2^2 + b_1 b_2.$$

(1.12.6)

and one has the equations of motion

$$\begin{aligned}
\dot{a}_1 &= -a_1(b_1 - b_2) & \dot{b}_1 &= 2a_1^2, \\
\dot{a}_2 &= -a_2(b_1 + 2b_2) & \dot{b}_2 &= -2(a_1^2 - a_2^2).
\end{aligned}$$

(1.12.7)

(setting $b_1 + b_2 + b_3 = 0$, for convenience, which we may do since the trace is preserved along the flow). In particular, Trace LN is, in this case, equal to b_2 and can be checked to decrease along the flow.

Figure 1.12.1 exhibits the asymptotic behavior of the Toda flow. We will return to this property in Chapter 8.

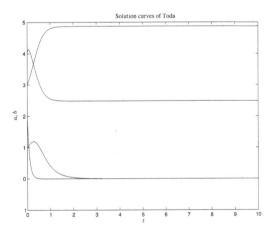

FIGURE 1.12.1. Asymptotic behavior of the solutions of the four-dimensional Toda lattice.

Exercises

◇ **1.12-1.** Show that Trace L^k for all k is conserved along the flow of the Toda lattice

◇ **1.12-2.** Characterize all the equilibria for the Toda flow (allowing a_i to take the value 0). Hint: Use the double bracket form of the equations.

2
Mathematical Preliminaries

This chapter covers a fairly wide array of topics from mathematics that we shall need in later chapters. We do not pretend to give all the needed background for the reader to learn these things in a comprehensive way from scratch. However, we hope that this summary will be helpful to set the notation, fill in some gaps the reader may have, and to provide a guide to the literature for needed background and proofs.

2.1 Vector Fields, Flows, and Differential Equations

This section introduces vector fields on Euclidean space and the flows they determine. This topic puts together and globalizes two basic ideas learned in undergraduate mathematics: vector fields and differential equations.

2.1.1 Example (A Basic Example). An example that illustrates many of the concepts of dynamical systems is the ball in a rotating hoop. Refer to Figure 2.1.1.

This system consists of a rigid hoop that hangs from the ceiling with a small ball resting in the bottom of the hoop. The hoop rotates with frequency ω about a vertical axis through its center.

Consider varying ω, keeping the other parameters (radius of the hoop, mass of the ball, ect.) fixed. For small values of ω, the ball stays at the bottom of the hoop, and correspondingly, that position is stable (Figure 2.1.1 (left)). Accept this in an intuitive sense for the moment; eventually,

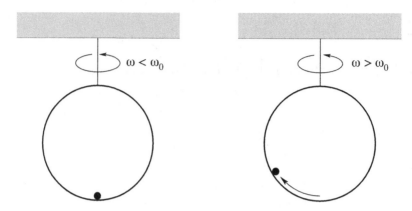

FIGURE 2.1.1. The ball in the hoop system; the equilibrium is stable for $\omega < \omega_c$ and unstable for $\omega > \omega_c$.

one has to define this concept carefully. However, when ω reaches a certain critical value ω_0, this point becomes unstable and the ball rolls up the side of the hoop to a new position $x(\omega)$, which is stable. The ball may roll to the left or to the right, depending perhaps upon the side of the vertical axis to which it was initially leaning. (See Figure 2.1.1 (right).) The position at the bottom of the hoop is still a fixed point, but it has become *unstable*. The solutions to the initial value problem governing the ball's motion are unique for all values of ω. Despite this uniqueness, because of uncertainties in the initial condition, for $\omega > \omega_0$ we cannot predict which way the ball will roll.

Using the basic principles of mechanics given in Chapter 1, we start with the Lagrangian function for this problem (the kinetic energy in an inertial frame minus the gravitational potential energy). Then the associated Euler–Lagrange equations *with forces* are given by

$$mR^2\ddot{\theta} = mR^2\omega^2 \sin\theta\cos\theta - mgR\sin\theta - \nu R\dot{\theta}, \qquad (2.1.1)$$

where R is the radius of the hoop, θ is the angle from the bottom vertical, m is the mass of the ball, g is the acceleration due to gravity, and ν is a coefficient of friction.[1]

---

[1]This does not represent a realistic friction law, but is an ad hoc one for illustration only; even for this simple problem friction laws are controversial, depending on the exact nature of the mechanical system. If one were to suppose Coulomb friction, one would make the tangential force proportional to the normal force.

To analyze the system (2.1.1), we use a ***phase plane analysis***; that is, we write the equation as a system:

$$\dot{x} = y,$$

$$\dot{y} = \frac{g}{R}(\alpha \cos x - 1)\sin x - \beta y,$$

where $\alpha = R\omega^2/g$ and $\beta = \nu/mR$. This system of equations produces for each initial point in the xy-plane a unique trajectory. That is, given a point (x_0, y_0) there is a unique solution $(x(t), y(t))$ of the equation that equals (x_0, y_0) at $t = 0$. This statement is proved by using general existence and uniqueness theory, which we give in Theorem 2.1.4. When we draw these curves in the plane, we get figures like those shown in Figure 2.1.2.   ◆

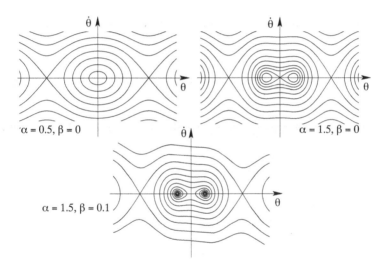

$\alpha = 0.5, \beta = 0$

$\alpha = 1.5, \beta = 0$

$\alpha = 1.5, \beta = 0.1$

FIGURE 2.1.2. The phase portrait for the ball in the hoop before and after the onset of instability for the case $g/R = 1$.

The above example has ***system parameters*** such as g, R, and ω. In many problems one takes the point of view, as we have done in the preceding discussion, of looking at how the phase portrait of the system changes as the parameters change. Sometimes these parameters are called ***control parameters***, since one can readily imagine changing them. However, this is still a *passive point of view*, since we imagine sitting back and watching the dynamics unfold for each value of the parameters.

In control theory, on the other hand, we take a more *active point of view*, and try to *intervene directly* with the dynamics to achieve a desired end. For example, we might imagine manipulating ω as a *function of time* to make the ball move in a desired way.

Dynamical Systems. More generally than in the above example, in Euclidean space $\mathbb{R}^n$, whose points are denoted by $x = (x^1, \ldots, x^n)$, we are concerned with a system of the form

$$\dot{x} = F(x), \qquad\qquad\qquad (2.1.2)$$

which, in components, reads

$$\dot{x}^i = F^i(x^1, \ldots, x^n), \qquad i = 1, \ldots, n,$$

where F is an n-component function of the n variables $x^1, \ldots, x^n$. Sometimes the function, or *vector field*, F depends on time or on other parameters (such as the mass or angular velocity in the example), and keeping track of this dependence is important. For general dynamical systems one needs some theory to develop properties of solutions; roughly, we draw curves in $\mathbb{R}^n$ emanating from initial conditions, just as we did in the preceding example.

Equilibrium Points, Stability, and Bifurcation. Equilibrium points are points where the right-hand side of the system (2.1.2) vanishes. In the ball in the hoop example, as ω increases, we see that the original stable fixed point becomes *unstable* and two *stable* fixed points split off at a critical value that we denoted above by ω_0, as indicated in Figure 2.1.2. One can use some basic stability theory that we shall develop to show that $\omega_0 = \sqrt{g/R}$. This is one of the simplest situations in which *symmetric problems can have asymmetric solutions* and in which *there can be multiple stable equilibria*, so there is nonuniqueness of equilibria (even though the solution of the initial value problem is unique).

This example shows that in some systems the phase portrait can change as certain parameters are changed. Changes in the qualitative nature of phase portraits as parameters are varied are called **bifurcations**. Consequently, the corresponding parameters are often called **bifurcation parameters**. These changes can be simple, such as the formation of new fixed points, called **static bifurcations**, or more complex **dynamic bifurcations** such as the formation of **periodic orbits**, that is, an orbit $x(t)$ with the property that $x(t+T) = x(t)$ for some T and all t, or even more complex dynamical structures. Thus, the ball in the hoop example exhibits a static bifurcation called a **pitchfork bifurcation** as the parameter ω crosses the critical value $\omega_0 = \sqrt{g/R}$.

Another important bifurcation, called the **Hopf bifurcation**, or more properly, the **Poincaré–Andronov–Hopf** bifurcation, occurs in a number of examples. This is a dynamic bifurcation in which, roughly speaking, a periodic orbit rather than another fixed point is formed when an equilibrium loses stability. In this case, too, there will be a bifurcation parameter, say μ, that crosses a critical value μ_0, as indicated in Figure 2.1.3, while the original critical point loses stability.

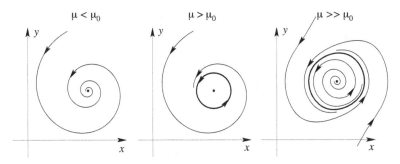

FIGURE 2.1.3. A periodic orbit appears for μ close to μ_0.

Depending on the nonlinear terms, in this bifurcation the periodic orbits can appear above (supercritical) or below (subcritical) the critical value. Unless a special degeneracy occurs, the subcritical case gives rise to unstable periodic orbits, and the supercritical case gives rise to stable orbits. See Figure 2.1.4.

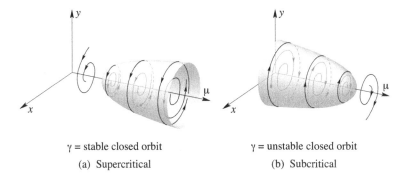

γ = stable closed orbit

(a) Supercritical

γ = unstable closed orbit

(b) Subcritical

FIGURE 2.1.4. The periodic orbit appears for μ close to μ_0 can be super- or subcritical.

An everyday example of a Hopf bifurcation is **flutter**. For example, when venetian blinds flutter in the wind or a television antenna "sings" as the wind velocity increases, there is probably a Hopf bifurcation occurring. A related example that is physically easy to understand is flow through a hose: Consider a straight vertical rubber tube conveying fluid. The lower end is a nozzle from which the fluid escapes. This is called a **follower-load** problem, since the water exerts a force on the free end of the tube that follows the movement of the tube. Those with any experience in a garden will not be surprised by the fact that the hose will begin to oscillate if the water velocity is high enough.

Vector Fields. With the above example as motivation, we can begin the more formal treatment of vector fields and their associated differential equations. Of course, we will eventually add the concept of controls to these vector fields, but we need to understand the notion of vector field itself first.

2.1.2 Definition. *Let $r \geq 0$ be an integer. A C^r **vector field** on an open set $U \subset \mathbb{R}^n$ is a mapping $X : U \to \mathbb{R}^n$ of class C^r from $U \subset \mathbb{R}^n$ to $\mathbb{R}^n$. The set of all C^r vector fields on U is denoted by $\mathfrak{X}^r(U)$, and the C^∞ vector fields by $\mathfrak{X}^\infty(U)$ or $\mathfrak{X}(U)$.*

We think of a vector field as assigning to each point $x \in U$ a vector $X(x)$ based (i.e., bound) at that same point.

Newton's Law of Gravitation.   Here the set U is $\mathbb{R}^3$ minus the origin, and the vector field is defined by

$$\mathbf{F}(x, y, z) = -\frac{mMG}{r^3}\mathbf{r},$$

where m is the mass of a test body, M is the mass of the central body, G is the constant of gravitation, $\mathbf{r}$ is the vector from the origin to (x, y, z), and $r = (x^2 + y^2 + z^2)^{1/2}$; see Figure 2.1.5.

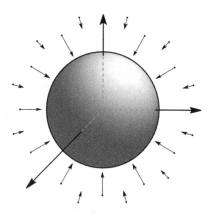

FIGURE 2.1.5. The gravitational force field.

Evolution Operators.   Consider a general physical system that is capable of assuming various "states" described by points in a set Z. For example, Z might be $\mathbb{R}^3 \times \mathbb{R}^3$, and a state might be the position and velocity $(q, \dot{q})$ of a particle. As time passes, the state evolves. If the state is $z_0 \in Z$ at time t_0 and this changes to z at a later time t, we set

$$F_{t,t_0}(z_0) = z$$

and call F_{t,t_0} the **evolution operator**; it maps a state at time t_0 to what the state would be at time t. "Determinism" is expressed by the law

$$F_{t_2,t_1} \circ F_{t_1,t_0} = F_{t_2,t_0}, \quad F_{t,t} = \text{identity},$$

sometimes called the **Chapman–Kolmogorov law.**

The evolution laws are called **time-independent** when F_{t,t_0} depends only on the elapsed time interval $t - t_0$; i.e.,

$$F_{t,t_0} = F_{s,s_0} \quad \text{if} \quad t - t_0 = s - s_0.$$

Setting $F_t = F_{t,0}$, the preceding law becomes the **group property**:

$$F_\tau \circ F_t = F_{\tau+t}, \quad F_0 = \text{identity}.$$

We call such an F_t a **flow** and F_{t,t_0} a **time-dependent flow**, or an evolution operator. If the system is defined only for $t \geq 0$, we speak of a **semiflow**.

It is usually not F_{t,t_0} that is given, but rather the **laws of motion**. In other words, differential equations are given that we must solve to find the flow. In general, Z is a manifold (a generalization of a smooth surface), but we confine ourselves for this section to the case that $Z = U$ is an open set in some Euclidean space $\mathbb{R}^n$. These equations of motion have the form

$$\frac{dx}{dt} = X(x), \quad x(0) = x_0,$$

where X is a (possibly time-dependent) vector field on U.

Newton's Second Law. The motion of a particle of mass m moving in $\mathbb{R}^3$ under the influence of the gravitational force field is determined by Newton's second law:

$$m\frac{d^2\mathbf{r}}{dt^2} = \mathbf{F},$$

i.e., by the ordinary differential equations

$$m\frac{d^2x}{dt^2} = -\frac{mMGx}{r^3},$$
$$m\frac{d^2y}{dt^2} = -\frac{mMGy}{r^3},$$
$$m\frac{d^2z}{dt^2} = -\frac{mMGz}{r^3}.$$

Letting $\mathbf{q} = (x, y, z)$ denote the position and $\mathbf{p} = m(d\mathbf{r}/dt)$ denote the linear momentum, these equations become

$$\frac{d\mathbf{q}}{dt} = \frac{\mathbf{p}}{m}, \quad \frac{d\mathbf{p}}{dt} = \mathbf{F}(\mathbf{q}).$$

The phase space here is the open set $U = (\mathbb{R}^3 \backslash \{\mathbf{0}\}) \times \mathbb{R}^3$. The right-hand side of the preceding equations define a vector field by

$$X(\mathbf{q}, \mathbf{p}) = ((\mathbf{q}, \mathbf{p}), (\mathbf{p}/m, \mathbf{F}(\mathbf{q}))).$$

In many courses on mechanics or differential equations, it is shown how to integrate these equations explicitly, producing trajectories, which are planar conic sections. These trajectories comprise the flow of the vector field.

Of course, these equations are special cases of the Euler–Lagrange equations, and so we see how dynamical systems are relevant to the study of mechanics, and this relevance is both for holonomic and nonholonomic systems of the sorts we saw in Chapter 1.

Integral Curves of Vector Fields.   Relative to a chosen set of Euclidean coordinates, we can identify a vector field X defined on an open set in $\mathbb{R}^n$ with an n-component vector function $(X^1(x), \dots, X^n(x))$, the components of X.

2.1.3 Definition.   *Let $U \subset \mathbb{R}^n$ be an open set and $X \in \mathfrak{X}^r(U)$ a vector field on U. An **integral curve** of X with initial condition x_0 is a differentiable curve c defined on some open interval $I \subset \mathbb{R}$ containing 0 such that $c(0) = x_0$ and $c'(t) = X(c(t))$ for each $t \in I$.*

Clearly, c is an integral curve of X when the following system of ordinary differential equations is satisfied:

$$
\begin{aligned}
\frac{dc^1}{dt}(t) &= X^1(c^1(t), \dots, c^n(t)), \\
&\;\;\vdots \\
\frac{dc^n}{dt}(t) &= X^n(c^1(t), \dots, c^n(t)).
\end{aligned}
$$

We shall often write $x(t) = c(t)$, an admitted abuse of notation. The preceding system of equations is called **autonomous** when X is time-independent. If X were time-dependent, time t would appear explicitly on the right-hand side. As we have already seen, the preceding system of equations includes equations of higher order (such as second-order Euler–Lagrange equations) by the usual reduction to first-order systems.

Existence and Uniqueness Theorems.   One of the basic theorems concerning the existence and uniqueness of solutions of ordinary differential equations of the above sort is the following.

2.1.4 Theorem (Local Existence, Uniqueness, and Smoothness).   *Let $U \subset \mathbb{R}^n$ be open and X be a C^r vector field on U for some $r \geq 1$. For each $x_0 \in U$, there is a curve $c : I \to U$ with $c(0) = x_0$ such that $c'(t) = X(c(t))$ for all $t \in I$. Any two such curves are equal on the intersection of their domains. Furthermore, there are a neighborhood U_0 of the point $x_0 \in U$, a real number $a > 0$, and a C^r mapping $F : U_0 \times I \to U$, where I is the open interval $]-a, a[$, such that the curve $c_u : I \to U$ defined by $c_u(t) = F(u, t)$ is a curve satisfying $c_u(0) = u$ and the differential equations $c'_u(t) = X(c_u(t))$ for all $t \in I$.*

This theorem has many variants. We refer to Coddington and Levinson [1955], Hartman [1982], and Abraham, Marsden, and Ratiu [1988] for these variants and for proofs.[2]

Here is an example of a variant: with just continuity of X one can get existence (the **Peano existence theorem**) but *without uniqueness*. The equation in one dimension given by $\dot{x} = \sqrt{x}, x(0) = 0$ has the two C^1 solutions $x_1(t) = 0$ and $x_2(t)$, which is defined to be 0 for $t \leq 0$ and $x_2(t) = t^2/4$ for $t > 0$. This example shows that one can indeed have existence without uniqueness for continuous vector fields.

The standard proof of Theorem 2.1.4 starts with a Lipschitz assumption on the vector field and proceeds to show existence and uniqueness by showing that there is a unique solution to the integral equation

$$x(t) = x_0 + \int_0^t X(x(s))\, ds.$$

One way to do this is by using the contraction mapping principle on a suitable space of curves[3] or by showing that the sequence of curves given by **Picard iteration** converges: Let $x_0(t) = x_0$ and define inductively

$$x_{n+1}(t) = x_0 + \int_{t_0}^t X(x_n(s))\, ds.$$

The existence and uniqueness theory also holds if X depends explicitly on t or on a parameter ρ, is jointly continuous in (t, ρ, x), and is Lipschitz or class C^r in x uniformly in t and ρ.

Dependence on Initial Conditions and Parameters.   The following inequality is of basic importance not only in existence and uniqueness theorems, but also in making estimates on solutions.

2.1.5 Theorem (Gronwall's Inequality).   *Let $f, g : [a, b[\to \mathbb{R}$ be continuous and nonnegative.[4] Suppose there is a constant $A \geq 0$ such that for all t satisfying $a \leq t \leq b$,*

$$f(t) \leq A + \int_a^t f(s)\, g(s)\, ds.$$

Then

$$f(t) \leq A \exp\left(\int_a^t g(s)\, ds\right) \quad \text{for all} \quad t \in [a, b[.$$

---

[2]This last reference also has a proof based directly on the implicit function theorem applied in suitable function spaces. This proof has a technical advantage: It works easily for other types of differentiability assumptions on X or on F_t, such as Hölder or Sobolev differentiability; this result is due to Ebin and Marsden [1970].

[3]The contraction mapping principle is a standard result in basic real analysis, with which we assume the reader is familiar; see, for example, Marsden and Hoffman [1993].

[4]We denote an interval that is open on the right and closed on the left by either $[a, b[$ or by $[a, b)$.

We refer to the preceding references for the proof. This result is one of the key ingredients in showing that the solutions depend in a Lipschitz or smooth way on initial conditions. Specifically, let $F_t(x_0)$ denote the solution (= integral curve) of $x'(t) = X(x(t)), x(0) = x_0$. Then for Lipschitz vector fields, $F_t(x)$ depends in a continuous, and indeed Lipschitz, manner on the initial condition x and is jointly continuous in (t, x). Again, the same result holds if X depends explicitly on t and on a parameter ρ, is jointly continuous in (t, ρ, x), and is Lipschitz in x uniformly in t and ρ. We let $F_{t,\lambda}^{\rho}(x)$ be the unique integral curve $x(t)$ satisfying $x'(t) = X(x(t), t, \rho)$ and $x(\lambda) = x$. Then $F_{t,t_0}^{\rho}(x)$ is jointly continuous in the variables (t_0, t, ρ, x), and is Lipschitz in x, uniformly in (t_0, t, ρ).

Additional work along these same lines shows that F_t is C^r if X is. Again, there is an analogous result for the evolution operator $F_{t,t_0}^{\rho}(x)$ for a time-dependent vector field $X(x, t, \rho)$, which depends on extra parameters ρ in some other Euclidean space, say $\mathbb{R}^m$. If X is C^r, then $F_{t,t_0}^{\rho}(x)$ is C^r in all variables and is C^{r+1} in t and t_0.

Suspension Trick.   The parameter ρ can be dealt with by suspending X to a new vector field obtained by appending the trivial differential equation $\rho' = 0$; this defines a vector field on $\mathbb{R}^n \times \mathbb{R}^m$, and the basic existence and uniqueness theorem may be applied to it. The flow on $\mathbb{R}^n \times \mathbb{R}^m$ is just $F_t(x, \rho) = (F_t^{\rho}(x), \rho)$.

Rectification.   An interesting result, called the *rectification theorem*, shows that near a point x_0 satisfying $X(x_0) \neq 0$, the flow can be transformed by a change of variables so that the integral curves become straight lines moving with unit speed.[5] This shows that in effect, nothing interesting happens with flows away from equilibrium points *as long as one looks at the flow only locally and for short time.*

The mapping F gives a locally unique integral curve c_u for each $u \in U_0$, and for each $t \in I, F_t = F|(U_0 \times \{t\})$ maps U_0 to some other set. It is convenient to think of each point u being allowed to "flow for time t" along the integral curve c_u (see Figure 2.1.6). This is a picture of a U_0 "flowing," and the system (U_0, a, F) is a local flow of X, or **flow box.**

Global Uniqueness.   The first global issue concerns uniqueness. Recall that *local* uniqueness was already addressed in Theorem 2.1.4; now we are concerned with *global* uniqueness. The following is readily proved by combining local uniqueness with a connectedness argument.

2.1.6 Proposition (Global Uniqueness).   *Suppose c_1 and c_2 are two integral curves of X in U and that for some time t_0, $c_1(t_0) = c_2(t_0)$. Then $c_1 = c_2$ on the intersection of their domains.*

---

[5]The proof can be found in Abraham and Marsden [1978], Arnold [1983], and Abraham, Marsden, and Ratiu [1988], but of course the result goes back to the classical literature.

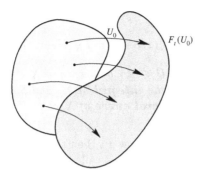

FIGURE 2.1.6. The flow of a vector field.

Completeness.   Other global issues center on considering the flow of a vector field as a whole, extended as far as possible in the t-variable.

2.1.7 Definition.   *Given an open set U and a vector field X on U, let $\mathcal{D}_X \subset U \times \mathbb{R}$ be the set of $(x, t) \in U \times \mathbb{R}$ such that there is an integral curve $c : I \to U$ of X with $c(0) = x$ with $t \in I$. The vector field X is **complete** if $\mathcal{D}_X = U \times \mathbb{R}$. A point $x \in U$ is called σ-**complete**, where $\sigma = +, -,$ or $\pm$, if $\mathcal{D}_X \cap (\{x\} \times \mathbb{R})$ contains all (x, t) for $t > 0$, $t < 0$, or $t \in \mathbb{R}$, respectively. Let $T^+(x)$ (respectively $T^-(x)$) denote the sup (respectively inf) of the times of existence of the integral curves through x; $T^+(x)$ respectively $T^-(x)$ is called the **positive (negative) lifetime of** x.*

Thus, X is complete iff each integral curve can be extended so that its domain becomes $]-\infty, \infty[$; i.e., $T^+(x) = \infty$ and $T^-(x) = -\infty$ for all $x \in U$.

2.1.8 Examples.

A.   For $U = \mathbb{R}^2$, let X be the constant vector field $X(x, y) = (0, 1)$. Then X is complete, since the integral curve of X through (x, y) is $t \mapsto (x, y + t)$.

B.   On $U = \mathbb{R}^2 \backslash \{0\}$, the same vector field is not complete, since the integral curve of X through $(0, -1)$ cannot be extended beyond $t = 1$; in fact, as $t \to 1$ this integral curve tends to the point $(0, 0)$. Thus $T^+(0, -1) = 1$, while $T^-(0, -1) = -\infty$.

C.   On $\mathbb{R}$ consider the vector field $X(x) = 1 + x^2$. This is not complete, since the integral curve c with $c(0) = 0$ is $c(\theta) = \tan \theta$, and thus it cannot be continuously extended beyond $-\pi/2$ and $\pi/2$; i.e., $T^\pm(0) = \pm \pi/2$.   ♦

2.1.9 Proposition.   *Let $U \subset \mathbb{R}^n$ be open and $X \in \mathfrak{X}^r(U)$, $r \geq 1$. Then:*

(i) $\mathcal{D}_X \supset U \times \{0\}$.

(ii) $\mathcal{D}_X$ *is open in* $U \times \mathbb{R}$.

(iii) *There is a unique C^r mapping $F_X : \mathcal{D}_X \to U$ such that the mapping $t \mapsto F_X(x, t)$ is an integral curve at x for all $x \in U$.*

(iv) *For* $(x,t) \in \mathcal{D}_X$, $(F_X(x,t),s) \in \mathcal{D}_X$ *iff* $(m, t + s) \in \mathcal{D}_X$; *in this case*

$$F_X(x, t + s) = F_X(F_X(x,t), s).$$

2.1.10 Definition. *Let* $U \subset \mathbb{R}^n$ *be open and* $X \in \mathfrak{X}^r(U)$, $r \geq 1$. *Then the mapping* F_X *is called the **integral** of* X, *and the curve* $t \mapsto F_X(x,t)$ *is called the **maximal integral curve** of* X *at* x. *If* X *is complete,* F_X *is called the **flow** of* X.

Thus, if X is complete with flow F, then the set $\{F_t \mid t \in \mathbb{R}\}$ is a group of diffeomorphisms on U, sometimes called a ***one-parameter group of diffeomorphisms***. Since $F_n = (F_1)^n$ (the nth power), the notation F^t is sometimes convenient and is used where we use F_t. For incomplete flows, (iv) says that $F_t \circ F_s = F_{t+s}$ wherever it is defined. Note that $F_t(x)$ is defined for $t \in]T^-(x), T^+(x)[$. The reader should write out similar definitions for the time-dependent case and note that the lifetimes depend on the starting time t_0.

Criteria for Completeness. A useful criterion for global existence or completeness is the following:

2.1.11 Proposition. *Let* X *be a* C^r *vector field on an open subset* U *of* $\mathbb{R}^n$, *where* $r \geq 1$. *Let* $c(t)$ *be a maximal integral curve of* X *such that for every finite open interval* $]a,b[$ *in the domain* $]T^-(c(0)), T^+(c(0))[$ *of* c, $c(]a,b[)$ *lies in a compact subset of* U. *Then* c *is defined for all* $t \in \mathbb{R}$. *If* $U = \mathbb{R}^n$, *this holds, provided that* $c(t)$ *lies in a bounded set.*

For example, this is used to prove the following:

2.1.12 Corollary. *A* C^r *vector field on an open set* U *with compact support contained in* U *is complete.*

Completeness corresponds to well-defined dynamics persisting eternally. In some circumstances (shock waves in fluids and solids, singularities in general relativity, etc.) one has to live with incompleteness, realize that one may be dealing with an overly idealized model, or overcome it in some other way.

2.1.13 Examples.
A. Let X be a C^r vector field, $r \geq 1$, on the open set $U \subset \mathbb{R}^n$ admitting a ***first integral***, i.e., a C^r function $f : U \to \mathbb{R}$ such that

$$\sum_{i=1}^{n} X^i(x^1, \ldots, x^n) \frac{\partial f}{\partial x^i}(x^1, \ldots, x^n) = 0.$$

If all level sets $f^{-1}(r)$, $r \in \mathbb{R}$, are compact, then X is complete. Indeed, by the chain rule, it follows that f is constant along integral curves of X, and so each integral curve lies on a level set of f. Thus, the result follows by the preceding proposition. Of course, in mechanics we often turn to quantities like energy and linear and angular momentum to find first integrals.

B. Suppose $X(x) = A \cdot x + B(x)$ where A is a linear operator of $\mathbb{R}^n$ to itself and B is *sublinear*; i.e., $B : \mathbb{R}^n \to \mathbb{R}^n$ is C^r with $r \geq 1$ and satisfies $\|B(x)\| \leq K\|x\| + L$ for constants K and L. We shall show that X is complete. Let $x(t)$ be an integral curve of X on the bounded interval $[0, T]$. Then

$$x(t) = x(0) + \int_0^t (A \cdot x(s) + B(x(s)))\, ds.$$

Hence

$$\|x(t)\| \leq \|x(0)\| + \int_0^t (\|A\| + K)\|x(s)\|\, ds + Lt.$$

By Gronwall's inequality,

$$\|x(t)\| \leq (Lt + \|x(0)\|)e^{(\|A\|+K)t}$$

for $0 \leq t \leq T$. Hence, $x(t)$ remains bounded on bounded t-intervals, so the result follows by Proposition 2.1.11. ◆

A further example on the global existence of solutions for a particle in a potential field is given in the web supplement.

The following is proved by a study of the local existence theory; we state it for completeness only.

2.1.14 Proposition. *Let X be a C^r vector field on U, $r \geq 1$, $x_0 \in U$, and $T^+(x_0)(T^-(x_0))$ the positive (negative) lifetime of x_0. Then for each $\varepsilon > 0$, there exists a neighborhood V of x_0 such that for all $x \in V$, $T^+(x) > T^+(x_0) - \varepsilon$ (respectively, $T^-(x) < T^-(x_0) + \varepsilon$). (One says that $T^+(x_0)$ is a lower semicontinuous function of x.)*

2.1.15 Corollary. *Let X_t be a C^r time-dependent vector field on U, $r \geq 1$, and let x_0 be an **equilibrium** of X_t; i.e., $X_t(x_0) = 0$, for all t. Then for any T there exists a neighborhood V of x_0 such that any $x \in V$ has integral curve existing for time $t \in [-T, T]$.*

Linear Equations. Flows of linear equations $\dot{x} = Ax$, where A is an $n \times n$ matrix, are given by $F_t(x) = e^{tA}$, where the exponential is defined, for example, by a power series

$$e^{tA} = I + tA + \frac{1}{2}t^2 A^2 + \frac{1}{3!}t^3 A^3 + \cdots.$$

Of course, one has to show that this series converges and is differentiable in t and that the derivative is given by Ae^{tA}, but this is learned in courses on real analysis. One also learns how to carry out exponentiation in courses in linear algebra by bringing A into a canonical form.

Exercises

◇ **2.1-1.** Derive equation (2.1.1) using Lagrangian mechanics.

◇ **2.1-2.** Is the flow of the vector field

$$X(x,y) = \left(x + y, \frac{1}{1 + x^2 + y^2}\right)$$

on $\mathbb{R}^2$ complete?

◇ **2.1-3.** Consider the matrix

$$A = \begin{bmatrix} -2 & -1 & 0 & 0 \\ 1 & 0 & 0 & 0 \\ 0 & 0 & 3 & -1 \\ 0 & 0 & 1 & 1 \end{bmatrix}.$$

Solve the system $\dot{x} = Ax$ for $x \in \mathbb{R}^4$ with initial condition $x = (1, -1, 2, 3)$.

2.2   Differentiable Manifolds

Modern analytical mechanics and nonlinear control theory are most naturally discussed in the mathematical language of differential geometry. The present chapter is meant to serve as an introduction to the elements of geometry that we shall use in the remainder of the book. Since this is not primarily a text on geometry, there is a great deal that must be left out.[6]

Studying the motion of physical systems leads immediately to the study of rates of change of position and velocity, i.e., to calculus. Differentiable manifolds provide the most natural setting in which to study calculus. Roughly speaking, differentiable manifolds are topological spaces that locally look like Euclidean space, but that may be globally quite different from Euclidean space. Since taking a derivative involves only a local computation—carried out in a neighborhood of the point of interest—it would appear that derivatives should be computable on any topological space that is infinitesimally indistinguishable from Euclidean space. This is indeed the case. What makes differentiable manifolds most important in the study of analytical mechanics, however, is the global features and their implications for the large-scale behavior of trajectories of the corresponding equations of motion.

With these remarks in mind, we begin with a definition of manifold that relates these objects to Euclidean space in small neighborhoods of each point. Questions about important global features of differentiable manifolds will be discussed in subsequent sections.

---

[6]Some references are Abraham, Marsden, and Ratiu [1988], Auslander and MacKenzie [1977], Boothby [1986], Dubrovin, Fomenko and Novikov [1984], and Warner [1983].

2.2.1 Definition.   *An n-dimensional **differentiable manifold** M is a set of points together with a finite or countably infinite set of subsets $U_\alpha \subset M$ and 1-to-1 mappings $\varphi_\alpha : U_\alpha \to \mathbb{R}^n$ such that:*

1. *$\bigcup_\alpha U_\alpha = M$.*

2. *For each nonempty intersection $U_\alpha \cap U_\beta$, the set $\varphi_\alpha(U_\alpha \cap U_\beta)$ is an open subset of $\mathbb{R}^n$, and the 1-to-1 and onto mapping $\varphi_\alpha \circ \varphi_\beta^{-1} : \varphi_\beta(U_\alpha \cap U_\beta) \to \varphi_\alpha(U_\alpha \cap U_\beta)$ is a smooth function.*

3. *The family $\{U_\alpha, \varphi_\alpha\}$ is maximal with respect to conditions 1 and 2.*

The situation is illustrated in Figure 2.2.1.

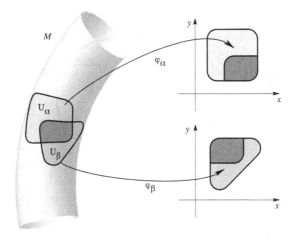

FIGURE 2.2.1. Coordinate charts on a manifold.

The sets U_α in the definition are called **coordinate charts**. The mappings φ_α are called **coordinate functions** or **local coordinates**. A collection of charts satisfying 1 and 2 is called an **atlas**. The notion of a C^k-differentiable (respectively analytic) manifold is defined similarly, wherein the **coordinate transformations** $\varphi_\alpha \circ \varphi_\beta^{-1}$ are required only to have continuous partial derivatives of all orders up to k (respectively be analytic). We remark that condition 3 is included merely to make the definition of manifold independent of a choice of atlas. A set of charts satisfying 1 and 2 can always be extended to a maximal set, and in practice, 1 and 2 define the manifold.

A **coordinate neighborhood** V of a point x in a manifold M is a subset of the domain U of a coordinate chart $\varphi : U \subset M \to \mathbb{R}^n$ such that $\varphi(V)$ is open in $\mathbb{R}^n$. Unions of coordinate neighborhoods define the open sets in M, and one checks that these open sets in M define a topology. *Usually we assume without explicit mention that the topology is Hausdorff:* Two different points x, x' in M have nonintersecting neighborhoods.

A useful viewpoint is to think of M as a set covered by a collection of coordinate charts with local coordinates $(x^1, \ldots, x^n)$ with the property that all mutual changes of coordinates are smooth maps.

We can also extend the definition of manifold to **manifold with boundary**, in which case we take the maps φ to be either into $\mathbb{R}^n$ or $\mathbb{R}^n_+ = \{(x_1, \ldots, x_n) \mid x_n \geq 0\}$ (see, e.g., Spivak [1979] or Abraham, Marsden, and Ratiu [1988] for details). In doing this, one must define the notion of a smooth map from a half-open set in $\mathbb{R}^n$ to another, and this is done by requiring the map to be the restriction of a smooth map on a containing open set. See Figure 2.2.2.

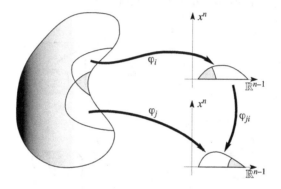

FIGURE 2.2.2. For a manifold with boundary, charts map into either open sets or half-open sets, and the overlap maps are still required to be smooth.

Level Sets as Differentiable Manifolds in $\mathbb{R}^n$.   A typical and important way that manifolds arise is as follows. Let $p_1, p_2, \ldots, p_m : \mathbb{R}^n \to \mathbb{R}$. The zero (or level) set

$$M = \{x \mid p_i(x) = 0, \, i = 1, \ldots, m\}$$

is called a **differentiable variety** in $\mathbb{R}^n$. If the $n \times m$ matrix

$$\left(\frac{\partial p_1}{\partial x^i} : \ldots : \frac{\partial p_m}{\partial x^i} \right)$$

has a (constant) rank ρ at each point $x \in M$, then M admits the structure of a differentiable manifold of dimension $n - \rho$. We call this the **rank condition**. In particular, if the rank is m (so that the matrix is onto $\mathbb{R}^m$) then we say that the map $p = (p_1, \ldots, p_m)$ is a **submersion**. This is a common case that is often encountered.

The idea of the proof of the rank criterion (or its special case, the submersion criterion) is that under our rank assumption an argument using the implicit function theorem shows that an $(n - \rho)$-dimensional coordinate chart may be defined in a neighborhood of each point on M. In this situation, we say that the level set is a **submanifold** of $\mathbb{R}^n$.

Matrix Groups.   We briefly discuss matrix groups as examples of differentiable manifolds. More details are given in Section 2.8.

Let $\mathbb{R}^{n \times n}$ be the set of $n \times n$ matrices with entries in $\mathbb{R}$, and let $\mathrm{GL}(n, \mathbb{R})$ denote the set of all $n \times n$ invertible matrices with entries in $\mathbb{R}$. Clearly, $\mathrm{GL}(n, \mathbb{R})$ is a group, called the **general linear group**. Let $A \in \mathrm{GL}(n, \mathbb{R})$ be symmetric (and invertible). Consider the subset

$$\mathcal{S} = \{X \in \mathrm{GL}(n, \mathbb{R}) \mid XAX^T = A\}.$$

It is easy to see that if $X \in \mathcal{S}$, then $X^{-1} \in \mathcal{S}$, and if $X, Y \in \mathcal{S}$, then the product XY is also in $\mathcal{S}$. Hence, $\mathcal{S}$ *is a subgroup of* $\mathrm{GL}(n, \mathbb{R})$.

We can also show that $\mathcal{S}$ *is a submanifold of* $\mathbb{R}^{n \times n}$. Indeed, $\mathcal{S}$ is the zero locus of the mapping $X \mapsto XAX^T - A$. Let $X \in \mathcal{S}$, and let δX be an arbitrary element of $\mathbb{R}^{n \times n}$. Then

$$(X + \delta X)A(X + \delta X)^T - A =$$
$$XAX^T - A + \delta XAX^T + XA\delta X^T + O(\delta X)^2.$$

We can conclude that $\mathcal{S}$ is a submanifold of $\mathbb{R}^{n \times n}$ if we can show that the linearization of the locus map, namely the linear mapping L defined by $\delta X \mapsto \delta XAX^T + XA\delta X^T$ of $\mathbb{R}^{n \times n}$ to itself, has constant rank for all $X \in \mathcal{S}$. We see that both the original map and the image of L lie in the subspace of $n \times n$ symmetric matrices. We claim that the map L is onto this space and hence the original map is a submersion. Indeed, given X and any symmetric matrix S we can find δX such that $(\delta X)AX^T + XA(\delta X)^T = S$, namely $\delta X = SA^{-1}X/2$. Thus, the original map to the space of symmetric matrices is a submersion. For a submersion, the dimension of the level set is the dimension of the domain minus the dimension of the range space. In this case, this dimension is $n^2 - n(n + 1)/2 = n(n - 1)/2$. In summary, we have established the following fact.

2.2.2 Proposition.   *Let $A \in \mathrm{GL}(n, \mathbb{R})$ be symmetric. Then the subgroup $\mathcal{S}$ of $\mathrm{GL}(n, \mathbb{R})$ defined by*

$$\mathcal{S} = \{X \in \mathrm{GL}(n, \mathbb{R}) \mid XAX^T = A\}$$

is a submanifold of $\mathbb{R}^{n \times n}$ of dimension $n(n - 1)/2$.

The Orthogonal Group.   Of special interest in mechanics is the case $A = I$. Here $\mathcal{S}$ specializes to $\mathrm{O}(n)$, the group of $n \times n$ orthogonal matrices. It is both a subgroup of $\mathrm{GL}(n, \mathbb{R})$ and a submanifold of the vector space $\mathbb{R}^{n \times n}$. $\mathrm{GL}(n, \mathbb{R})$ is an open, dense subset of $\mathbb{R}^{n \times n}$ that inherits the topology and manifold structure from $\mathbb{R}^{n \times n}$. Thus, $\mathrm{O}(n)$ (or any $\mathcal{S}$ defined as above) is both a subgroup and a submanifold of $\mathrm{GL}(n, \mathbb{R})$. Subgroups of $\mathrm{GL}(n, \mathbb{R})$ that are also submanifolds are called **matrix Lie groups**. We shall discuss Lie groups more abstractly later on.

Tangent Vectors to Manifolds. Two curves $t \mapsto c_1(t)$ and $t \mapsto c_2(t)$ in an n-manifold M are called **equivalent** at $x \in M$ if

$$c_1(0) = c_2(0) = x \quad \text{and} \quad (\varphi \circ c_1)'(0) = (\varphi \circ c_2)'(0)$$

in some chart φ, where the prime denotes the derivative with respect to the curve parameter. It is easy to check that this definition is chart independent. A **tangent vector** v to a manifold M at a point $x \in M$ is an equivalence class of curves at x. One proves that the set of tangent vectors to M at x forms a vector space. It is denoted by $T_x M$ and is called the **tangent space** to M at $x \in M$. Given a curve $c(t)$, we denote by $c'(s)$ the tangent vector at $c(s)$ defined by the equivalence class of $t \mapsto c(s + t)$ at $t = 0$.

Let U be a chart of an atlas for the manifold M with coordinates $(x^1, \ldots, x^n)$. The **components** of the tangent vector v to the curve $t \mapsto (\varphi \circ c)(t)$ are the numbers $v^1, \ldots, v^n$ defined by

$$v^i = \left. \frac{d}{dt} (\varphi \circ c)^i \right|_{t=0},$$

$i = 1, \ldots, n$. The **tangent bundle** of M, denoted by TM, is the differentiable manifold whose underlying set is the disjoint union of the tangent spaces to M at the points $x \in M$; that is,

$$TM = \bigcup_{x \in M} T_x M.$$

Thus, a point of TM is a vector v that is tangent to M at some point $x \in M$. To define the differentiable structure on TM, we need to specify how to construct local coordinates on TM. To do this, let $x^1, \ldots, x^n$ be local coordinates on M and let $v^1, \ldots, v^n$ be components of a tangent vector in this coordinate system. Then the $2n$ numbers $x^1, \ldots, x^n, v^1, \ldots, v^n$ give a local coordinate system on TM. Notice that $\dim TM = 2 \dim M$.

The Tangent Bundle Projection. The **tangent bundle**, or **natural projection**, is the map $\tau_M : TM \to M$ that takes a tangent vector v to the point $x \in M$ at which the vector v is attached (that is, $v \in T_x M$). The inverse image $\tau_M^{-1}(x)$ of a point $x \in M$ under the natural projection τ_M is the tangent space $T_x M$. This space is called the **fiber** of the tangent bundle over the point $x \in M$.

Manifolds with Boundary. A manifold M with boundary is the union of two other manifolds, the interior and the boundary, denoted by ∂M. The boundary has its own tangent space, which is a subspace of the tangent space to the entire manifold at that point. See Figure 2.2.3.

Tangent Spaces to Level Sets. Let $M = \{x \mid p_i(x) = 0, i = 1, \ldots, m\}$ be a differentiable variety in $\mathbb{R}^n$. For each $x \in M$,

$$T_x M = \left\{ v \in \mathbb{R}^n \; \middle| \; \frac{\partial p_i}{\partial x}(x) \cdot v = 0 \right\}$$

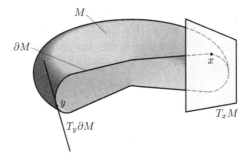

FIGURE 2.2.3. An example of a manifold M showing its boundary ∂M. In this example, M is two-dimensional, while the bounding curve is one-dimensional.

is called the ***tangent space to M at x***. Clearly, $T_x M$ is a vector space. If the rank condition holds, so that M is a differentiable manifold, then this definition may be shown to be equivalent to the one given earlier. For example, the reader may show that the tangent spaces to spheres are what they should intuitively be, as in Figure 2.2.4.

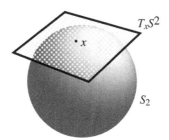

FIGURE 2.2.4. The tangent space to a sphere.

Tangent Spaces to Matrix Groups.   Let $A \in \mathrm{GL}(n, \mathbb{R})$ be a symmetric matrix. We wish to explicitly describe the tangent space at a typical point of the group $\mathcal{S} = \{X \in \mathrm{GL}(n, \mathbb{R}) \mid X^T A X = A\}$. Given our definition, it is clear that the tangent space $T_X \mathcal{S}$ is a subspace of the linear space of all $n \times n$ matrices, $\mathbb{R}^{n \times n}$. Let $V \in \mathbb{R}^{n \times n}$. Then V is in $T_X \mathcal{S}$ precisely when it is tangent to a curve in the group:

$$\frac{d}{d\epsilon}\Big|_{\epsilon=0} \left[(X + \epsilon V)^T A (X + \epsilon V) - A \right] = 0.$$

This condition is equivalent to $V^T A X + X^T A V = 0$.

This shows that if $X \in \mathcal{S}$, then

$$T_X \mathcal{S} = \{V \in \mathbb{R}^{n \times n} \mid V^T A X + X^T A V = 0\}.$$

Differentiable Maps.   Let E and F be vector spaces (for example, $\mathbb{R}^n$ and $\mathbb{R}^m$, respectively), and let $f : U \subset E \to V \subset F$, where U and V are open sets, be of class C^{r+1}. We define the **tangent map** [7] of f to be the map $Tf : TU = U \times E \to TV = V \times F$ defined by

$$Tf(u, e) = (f(u), Df(u) \cdot e), \qquad (2.2.1)$$

where $u \in U$ and $e \in E$. This notion from calculus may be generalized to the context of manifolds as follows. Let $f : M \to N$ be a map of a manifold M to a manifold N. We call f **differentiable** (or C^k) if in local coordinates on M and N it is expressed, or represented, by differentiable (or C^k) functions. The **derivative** of a differentiable map $f : M \to N$ at a point $x \in M$ is defined to be the linear map

$$T_x f : T_x M \to T_{f(x)} N$$

constructed in the following way. For $v \in T_x M$, choose a curve $c : \,]-\epsilon, \epsilon[\,\to M$ with $c(0) = x$, and velocity vector $dc/dt\,|_{t=0} = v$. Then $T_x f \cdot v$ is the velocity vector at $t = 0$ of the curve $f \circ c : \mathbb{R} \to N$; that is,

$$T_x f \cdot v = \frac{d}{dt} f(c(t)) \bigg|_{t=0}.$$

The vector $T_x f \cdot v$ does not depend on the curve c but only on the vector v. If M and N are manifolds and $f : M \to N$ is of class C^{r+1}, then $Tf : TM \to TN$ is a mapping of class C^r. Note that

$$\frac{dc}{dt}\bigg|_{t=0} = T_0 c \cdot 1.$$

Vector Fields and Flows.   Let us now interpret what we did with vector fields and flows in $\mathbb{R}^n$ in the context of manifolds. A **vector field** X on a manifold M is a map $X : M \to TM$ that assigns a vector $X(x)$ at the point $x \in M$; that is, $\tau_M \circ X = $ identity. An **integral curve** of X with initial condition x_0 at $t = 0$ is a (differentiable) map $c : \,]a, b[\,\to M$ such that $]a, b[$ is an open interval containing 0, $c(0) = x_0$, and

$$c'(t) = X(c(t))$$

for all $t \in \,]a, b[$. In formal presentations we usually suppress the domain of definition, even though this is technically important. The **flow** of X is the collection of maps

$$\varphi_t : M \to M$$

such that $t \mapsto \varphi_t(x)$ is the integral curve of X with initial condition x. Existence and uniqueness theorems from ordinary differential equations, as

---

[7]The tangent map is sometimes denoted by f_*.

reviewed in the last section, guarantee that φ is smooth in x and t (where defined) if X is. From uniqueness, we get the **flow property**

$$\varphi_{t+s} = \varphi_t \circ \varphi_s$$

along with the initial condition $\varphi_0 =$ identity. The flow property generalizes the situation where $M = V$ is a *linear* space, $X(x) = Ax$ for a (bounded) *linear* operator A, and

$$\varphi_t(x) = e^{tA}x$$

to the *nonlinear* case.

Differentials. If $f : M \to \mathbb{R}$ is a smooth function, we can differentiate it at any point $x \in M$ to obtain a map $T_x f : T_x M \to T_{f(x)}\mathbb{R}$. Identifying the tangent space of $\mathbb{R}$ at any point with itself (a process we usually do in any vector space), we get a linear map $\mathbf{d}f(x) : T_x M \to \mathbb{R}$. That is, $\mathbf{d}f(x) \in T_x^* M$, the dual of the vector space $T_x M$.

In coordinates, the **directional derivative** $\mathbf{d}f(x) \cdot v$, where $v \in T_x M$, is given by

$$\mathbf{d}f(x) \cdot v = \sum_{i=1}^{n} \frac{\partial f}{\partial x^i} v^i.$$

We will employ the **summation convention** and drop the summation sign when there are repeated indices. We also call $\mathbf{d}f$ the **differential** of f.

One can show that specifying the directional derivatives completely determines a vector, and so we can identify a basis of $T_x M$ using the operators $\partial/\partial x^i$. We write

$$(e_1, \ldots, e_n) = \left(\frac{\partial}{\partial x^1}, \ldots, \frac{\partial}{\partial x^n} \right)$$

for this basis, so that $v = v^i \partial/\partial x^i$.

If we replace each vector space $T_x M$ with its dual $T_x^* M$, we obtain a new $2n$-manifold called the **cotangent bundle** and denoted by $T^* M$. The dual basis to $\partial/\partial x^i$ is denoted by dx^i. Thus, relative to a choice of local coordinates we get the basic formula

$$\mathbf{d}f(x) = \frac{\partial f}{\partial x^i} dx^i$$

for any smooth function $f : M \to \mathbb{R}$.

Degree of a Map. As we shall see in Chapter 4, an important notion for understanding stabilization is the notion of the **degree** of a map (see, e.g., Milnor [1965]). Let M and N be oriented n-dimensional manifolds without boundary, M compact and N connected. Let $f : M \to N$ be a smooth map. Let $x \in M$ be a regular point of the map and let $T_x f : T_x M \to T_{f(x)} N$ denote the corresponding tangent map, which is thus a linear isomorphism.

Define the **sign** of $T_x f$ to be $+1$ or -1 according to whether or not it reverses orientation. Then for any regular value $y \in N$ we define

$$\deg(f; y) = \sum_{x \in f^{-1}(y)} \operatorname{sign} T_x f \qquad (2.2.2)$$

if $f^{-1}(y) \neq \varnothing$, 0 if $f^{-1}(y) = \varnothing$.

Now consider a smooth vector field X defined on an open set U of $\mathbb{R}^n$ with an isolated zero at $x \in U$. Consider the function

$$\frac{X(x)}{\|X(x)\|}, \qquad (2.2.3)$$

which maps a small sphere centered at x into the unit sphere regarded as the oriented boundary of the corresponding ball. Note that this is just the unit direction vector of the vector field. The degree of this mapping is called the **index** of the vector field.

It is not hard to see, for example, that if x is a nondegenerate zero of X, then the index of the vector field X at x is either $+1$ or -1: If X is orientation-preserving, we can locally smoothly deform X to the identity without introducing any new zeros, and if it is orientation-reversing, to a reflection. (Details of this smooth isotopy may be found in Milnor [1965].)

In the plane the index of a zero of a vector field simply measures how many times the vector field rotates in the anticlockwise direction as one traverses a small loop around the zero in the anticlockwise direction. One can easily check that a source, sink, or center has index $+1$, while the index of a saddle is -1. Similarly, the index of a zero of the linear differential equation on $\mathbb{R}^n$, $\dot{x} = Ax$, A nonsingular, is Index $=$ sign $(\det A)$. For example, for the stable system $\dot{x} = -x$ on $\mathbb{R}^n$, the index of zero is $(-1)^n$.

Exercises

⋄ **2.2-1.** Using the submersion criterion, show that the level set $x_1^2 + \cdots + x_n^2 - 1 = 0$ is a differentiable manifold of dimension $n - 1$.

⋄ **2.2-2.** Show that the set $\{(x, y) \mid x^2(x + 1) - y^2 = 0\}$ in $\mathbb{R}^2$ is *not* a differentiable manifold.

⋄ **2.2-3.** Let $\mathcal{S} = \{X \in \mathrm{GL}(n, \mathbb{R}) \mid X^T A X = A\}$, as in the text. Note that the $n \times n$ identity matrix I is in $\mathcal{S}$, and show that for any pair of matrices $V_1, V_2 \in T_I \mathcal{S}$ we have $V_1 V_2 - V_2 V_1 \in T_I \mathcal{S}$.

⋄ **2.2-4.** If $\varphi_t : S^2 \to S^2$ rotates points on S^2 about a fixed axis through an angle t, show that φ_t is the flow of a certain vector field on S^2.

⋄ **2.2-5.** Let $f : S^2 \to \mathbb{R}$ be defined by $f(x, y, z) = z$. Compute $\mathbf{d}f$ relative to spherical coordinates (θ, φ).

◇ **2.2-6.** One can show that the sum of the indices of the singular points of a vector field on a compact manifold without boundary is independent of the vector field and depends only on the manifold. This sum is called the ***Euler characteristic***. Use this fact to show that the Euler characteristic of the n-torus is zero and that of the n-sphere is $1+(-1)^n$. (Construct a vector field with no zeros on the torus and a vector field with the given index on the sphere.)

2.3   Stability

In this section we summarize some of the key notions of stability.

2.3.1 Definition. *Let x_0 be an equilibrium of the system of differential equations $\dot{x} = f(x)$. The point x_0 is said to be **nonlinearly** or **Lyapunov stable** if for any neighborhood U of x_0 there exists a neighborhood $V \subset U$ of x_0 such that any trajectory $x(t)$ of the system with initial point in V remains in U for all time. If in addition $x(t) \to x_0$ as $t \to \infty$, x_0 is said to be **asymptotically stable**.*

The basic notions of Lyapunov stability and asymptotic stability are illustrated in Figure 2.3.1. For the harmonic oscillator, the origin is Lyapunov stable, but not asymptotically stable.

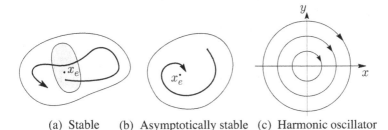

(a) Stable   (b) Asymptotically stable   (c) Harmonic oscillator

FIGURE 2.3.1. In Lyapunov stability points near the equilibrium point stay near, while for asymptotic stability, they also converge to the equilibrium point as $t \to +\infty$.

The notion of stability can be attached to invariant sets other than equilibrium points via a similar definition. In particular, the notion of a stable periodic orbit is illustrated in Figure 2.3.2.

Spectral Stability. There are some specific criteria for stability. The most basic one is the classical spectral test of Lyapunov.

2.3.2 Definition. *Let x_0 be an equilibrium of the system of differential equations $\dot{x} = f(x)$. The point x_0 is said to be **spectrally stable** if all the*

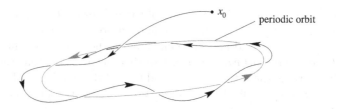

FIGURE 2.3.2. A periodic orbit is asymptotically stable when nearby orbits wind towards it.

eigenvalues of the linearization of f at x_0, i.e., of the matrix

$$A_{ij} = \frac{\partial f^i}{\partial x^j}(x_0),$$

lie in the left half-plane.

A basic result on stability is the following:

2.3.3 Theorem (Lyapunov). *Spectral stability implies asymptotic stability.*

Invariant Manifolds. At a general equilibrium x_0, if one computes that spectrum of the linearization A and finds a number of eigenvalues in the left half-plane, then there is an invariant manifold (i.e., a manifold that is invariant under the flow and that is simply the graph of a mapping in this case) that is tangent to the corresponding (generalized) eigenspace; it is called the local **stable manifold**. All trajectories on this stable manifold are asymptotic to the point x_0 as $t \to \infty$.

Similarly, associated with the eigenvalues in the right half-plane is an **unstable manifold**. The basic notion of invariant manifolds is illustrated schematically in Figure 2.3.3.

If none of the eigenvalues associated with an equilibrium are on the imaginary axis, then the equilibrium is called **hyperbolic**. In this case, the tangent spaces to the stable and unstable manifolds span the whole of $\mathbb{R}^n$. This is the situation shown in Figure 2.3.3. When there are eigenvalues on the imaginary axis, one introduces the notion of the **center manifold** as well; this is discussed in the next section.

One can also have invariant manifolds attached to other invariant sets, and in particular to periodic orbits. This is illustrated in Figure 2.3.4.

Much more on how to analyze and achieve stability through control will be discussed in later parts of this book.

The LaSalle Invariance Principle. A key ingredient in proving asymptotic stability of controlled or uncontrolled systems is the LaSalle invariance principle.

This main theorem may be stated as follows:

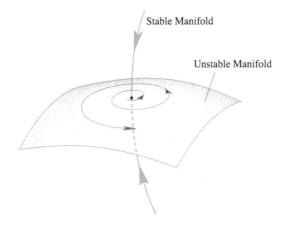

FIGURE 2.3.3. Invariant manifolds for an equilibrium point.

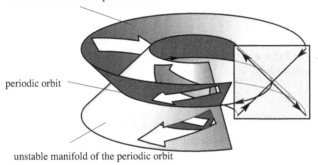

FIGURE 2.3.4. The stable and unstable manifolds of a periodic orbit.

2.3.4 Theorem. *Consider the smooth dynamical system on $\mathbb{R}^n$ given by $\dot{x} = f(x)$ and let Ω be a compact set in $\mathbb{R}^n$ that is (positively) invariant under the flow of f. Let $V : \Omega \to \mathbb{R}$, $V \geq 0$, be a C^1 function such that*

$$\dot{V}(x) = \frac{\partial V}{\partial x} \cdot f \leq 0$$

in Ω. Let M be the largest invariant set in Ω where $\dot{V}(x) = 0$. Then every solution with initial point in Ω tends asymptotically to M as $t \to \infty$. In particular, if M is an isolated equilibrium, it is asymptotically stable.

The invariance principle is due to Barbashin and Krasovskii [1952], Lasalle and Lefschetz [1961], and Krasovskii [1963]. The book of Khalil [1992] has a nice treatment.

Note that in the statement of the theorem, $V(x)$ need not be positive definite, but rather only semidefinite, and that if in particular M is an equilibrium, the theorem proves that the equilibrium is asymptotically stable. The set Ω in the LaSalle theorem also gives us an estimate of the region of attraction of an equilibrium. This is one of the reasons that this is a more attractive methodology than that of spectral stability tests, which could in principle give a very small region of attraction.

Exercises

$\diamond$ **2.3-1.**   Derive the critical stability value $\omega_0 = \sqrt{g/R}$ for the particle in the rotating hoop.

$\diamond$ **2.3-2.**   Consider the following vector field in $\mathbb{R}^3$:

$$\dot{x} = -x + y + f\,,$$
$$\dot{y} = -y + g\,,$$
$$\dot{z} = z\,,$$

where $f(x, y, z) = -\left(x + \frac{1}{2}y\right)^3$ and $g(x, y, z) = -\left(y + \frac{1}{2}x\right)^3$.

(a) Compute the linearized system at the origin and write it in the form $\dot{\mathbf{x}} = A\mathbf{x}$ for a suitable 3×3 matrix A and where $\mathbf{x}$ is the vector with components (x, y, z).

(b) Sketch the phase portrait of this linear system.

(c) To what extent is the phase portrait of the nonlinear system similar to that of the linear system in a neighborhood of the origin?

(d) Consider the function

$$V(x, y, z) = \frac{1}{2}\left[x^2 + y^2 + xy\right]\,.$$

Compute its time derivative along the flow of the given vector field.

(e) Show that the plane $z = 0$ is invariant.

(f) Is the origin globally attracting *within* the plane $z = 0$?

(g) Describe the invariant manifolds of the origin for this system.

(h) Can this vector field have any periodic orbits?

2.4   Center Manifolds

Here we discuss some results in center manifold theory and show how they relate to the Lyapunov–Malkin theorem, which plays an important role in the stability analysis of nonholonomic systems. The center manifold theorem provides useful insight into the existence of invariant manifolds. These invariant manifolds will play a crucial role in our analysis. Lyapunov's original proof of the Lyapunov–Malkin theorem used a different approach to proving the existence of local integrals, as we shall discuss below. Malkin extended the result to the nonautonomous case.

Center Manifold Theory in Stability Analysis.   We consider firstly center manifold theory and its applications to the stability analysis of non-hyperbolic equilibria.

Consider a system of differential equations

$$\dot{x} = Ax + X(x, y), \tag{2.4.1}$$
$$\dot{y} = By + Y(x, y), \tag{2.4.2}$$

where $x \in \mathbb{R}^m$, $y \in \mathbb{R}^n$, and A and B are constant matrices. It is supposed that all eigenvalues of A have nonzero real parts, and all eigenvalues of B have zero real parts. The functions X, Y are smooth, and satisfy the conditions $X(0, 0) = 0$, $dX(0, 0) = 0$, $Y(0, 0) = 0$, $dY(0, 0) = 0$. We now recall the following definition:

2.4.1 Definition.   *A smooth invariant manifold of the form $x = h(y)$ where h satisfies $h(0) = 0$ and $dh(0) = 0$ is called a **center manifold**.*

We are going to use the following version of the center manifold theorem following the exposition of Carr [1981] (see also Chow and Hale [1982]).

2.4.2 Theorem (The center manifold theorem).   *Suppose that the functions $X(x, y)$, $Y(x, y)$ are C^k, $k \geq 2$. Then there exist a (local) center manifold for (2.4.1), (2.4.2), $x = h(y)$, $|y| < \delta$, where h is C^k. The flow on the center manifold is governed by the system*

$$\dot{y} = By + Y(h(y), y). \tag{2.4.3}$$

The basic idea of realizing the center manifold as a graph over the linear center subspace is shown in Figure 2.4.1.

The next theorem explains that the reduced equation (2.4.3) contains information about stability of the zero solution of (2.4.1), (2.4.2).

2.4.3 Theorem.   *Suppose that the zero solution of (2.4.3) is stable (resp. asymptotically stable) and that the eigenvalues of A are in the left half-plane. Then the zero solution of (2.4.1), (2.4.2) is stable (resp. asymptotically stable). If either the zero solution of (2.4.3) is unstable, or if any eigenvalues of A are in the right half plane, then the zero solution of (2.4.1), (2.4.2) is also unstable.*

Let us now look at the special case of (2.4.2) in which the matrix B vanishes. Equations (2.4.1), (2.4.2) become

$$\dot{x} = Ax + X(x, y), \tag{2.4.4}$$
$$\dot{y} = Y(x, y). \tag{2.4.5}$$

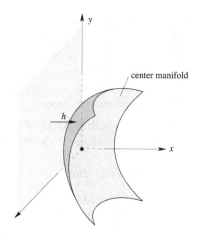

FIGURE 2.4.1. The center manifold realized as the graph of the function h.

2.4.4 Theorem. *Consider the system of equations* (2.4.4), (2.4.5). *If* $X(0, y) = 0$, $Y(0, y) = 0$, *and all of the eigenvalues of the matrix A have negative real parts, then the system* (2.4.4), (2.4.5) *has n local integrals in a neighborhood of $x = 0$, $y = 0$.*

Proof. The center manifold in this case is given by $x = 0$. Each point of the center manifold is an equilibrium of the system (2.4.4), (2.4.5). For each equilibrium point $(0, y_0)$ of our system, consider the associated m-dimensional stable manifold $S^s(y_0)$. The center manifold and these manifolds $S^s(y_0)$ can be used for a (local) substitution $(x, y) \mapsto (\bar{x}, \bar{y})$ such that in the new coordinates the system of differential equations becomes

$$\dot{\bar{x}} = \bar{A}\bar{x} + \bar{X}(\bar{x}, \bar{y}), \quad \dot{\bar{y}} = 0.$$

The last system of equations has n integrals $\bar{y} = $ const, so that the original equation has n smooth local integrals. Observe that the tangent spaces to the common level sets of these integrals at the equilibria are the planes $y = y_0$. Therefore, the integrals are of the form

$$y = f(x, k), \quad \text{where } \partial_x f(0, k) = 0. \qquad \blacksquare$$

The Lyapunov–Malkin Theorem. The following theorem gives stability conditions for equilibria of the system (2.4.4), (2.4.5).

2.4.5 Theorem (Lyapunov–Malkin). *Consider the system of differential equations* (2.4.4), (2.4.5), *where $x \in \mathbb{R}^m$, $y \in \mathbb{R}^n$, A is an $m \times m$ matrix, and $X(x, y)$, $Y(x, y)$ represent nonlinear terms. If all eigenvalues of the matrix A have negative real parts, and $X(x, y)$, $Y(x, y)$ vanish when $x = 0$, then the solution $x = 0$, $y = c$ of the system* (2.4.4), (2.4.5) *is stable with respect to x, y, and asymptotically stable with respect to x. If a solution*

$x(t)$, $y(t)$ of (2.4.4), (2.4.5) is close enough to the solution $x = 0$, $y = 0$, then

$$\lim_{t \to \infty} x(t) = 0, \quad \lim_{t \to \infty} y(t) = c.$$

Proof. (See Lyapunov [1907], Malkin [1938], and Zenkov, Bloch, and Marsden [1998].) From Theorem 2.4.4, the phase space of system (2.4.4), (2.4.5) is locally represented as a union of invariant leaves

$$Q_c = \{(x, y) \mid y = f(x, c)\}.$$

Use x as local coordinates on these leaves. On each leaf we have a reduced system that can be written as $\dot{x} = F(x)$, where $F(x) = Ax + X(x, f(x, c))$. Since $\det A \neq 0$, each reduced system has an isolated equilibrium $x = 0$ on a corresponding leaf. The equilibrium of the system reduced to a leaf passing through $x = 0, y = 0$ is asymptotically stable, since the matrix of the linearization of the reduced system is A. To finish the proof, we notice that the equilibria of systems on nearby leaves are asymptotically stable as well because the corresponding matrices A_c, by continuity, also have all eigenvalues in the left half-plane, since locally, A_c will be close to A. ■

Historical Note. The proof of the Lyapunov–Malkin theorem uses the fact that the system of differential equations has local integrals, as discussed above. To prove existence of these integrals, Lyapunov used a theorem of his own about the existence of solutions of PDEs. He did this assuming that the nonlinear terms on the right-hand sides are series in x and y with time-dependent bounded coefficients. Malkin generalized Lyapunov's result for systems for which the matrix A is time-dependent. We consider the nonanalytic case, and to prove existence of these local integrals, we use center manifold theory. This simplifies the arguments to some extent as well as showing how the results are related.

A Class Satisfying the Lyapunov–Malkin Theorem. The following lemma specifies a class of systems of differential equations that satisfy the conditions of the Lyapunov–Malkin theorem.

2.4.6 Lemma. *Consider a system of differential equations of the form*

$$\dot{w} = Aw + By + \mathcal{U}(w, y), \quad \dot{y} = \mathcal{Y}(w, y), \tag{2.4.6}$$

where $w \in \mathbb{R}^n$, $y \in \mathbb{R}^m$, $\det A \neq 0$, and where $\mathcal{U}$ and $\mathcal{Y}$ represent higher-order nonlinear terms. There is a change of variables of the form $w = x + \phi(y)$ such that:

(i) *In the new variables x, y, the system (2.4.6) has the form*

$$\dot{x} = Ax + X(x, y), \quad \dot{y} = Y(x, y).$$

(ii) *If $Y(0, y) = 0$, then $X(0, y) = 0$ as well.*

Proof.   Put $w = x + \phi(y)$, where $\phi(y)$ is defined by

$$A\phi(y) + By + \mathcal{U}(\phi(y), y) = 0.$$

Then the system (2.4.6) written in terms of the variables x, y becomes

$$\dot{x} = Ax + X(x, y), \quad \dot{y} = Y(x, y),$$

where

$$X(x, y) = A\phi(y) + By + \mathcal{U}(x + \phi(y), y) - \frac{\partial\phi}{\partial y}Y(x, y),$$

$$Y(x, y) = \mathcal{Y}(x + \phi(y), y).$$

Note that $Y(0, y) = 0$ implies $X(0, y) = 0$.   ∎

Exercise

⋄ **2.4-1.**   Sketch the phase portrait of the system

$$\dot{x} = -x + xy,$$
$$\dot{y} = xy.$$

Verify the conclusions of the Lyapunov–Malkin theorem. See Figure 2.4.2.

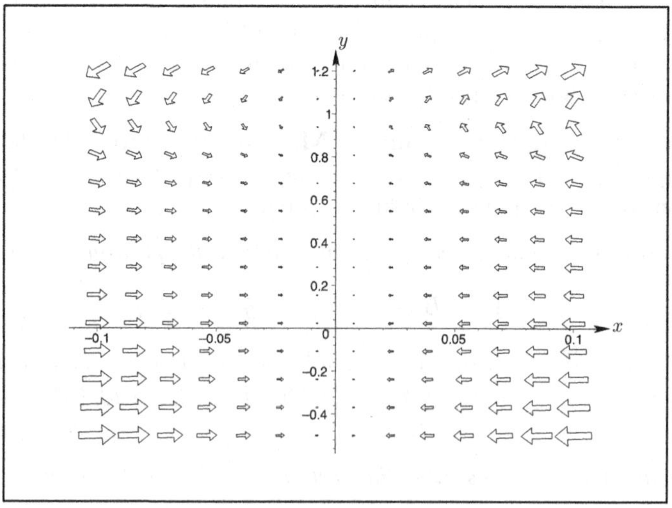

FIGURE 2.4.2. Phase portrait of the system $\dot{x} = -x + xy; \dot{y} = xy$.

2.5   Differential Forms

We next review some of the basic definitions, properties, and operations on differential forms, without proofs (see Abraham, Marsden, and Ratiu [1988] and references therein). *The main idea of differential forms is to provide a generalization of the basic operations of vector calculus,* div, grad, *and* curl, *of Green, Gauss, and Stokes to manifolds of arbitrary dimension.*

Basic Definitions.   Let V be a vector space. A map $\alpha : V \times \cdots \times V$ (where there are k factors) $\to \mathbb{R}$ is **multilinear** when it is linear in each of its factors, that is,

$$\alpha(v_1, \ldots, av_j + bv'_j, \ldots, v_k)$$
$$= a\alpha(v_1, \ldots, v_j, \ldots, v_k) + b\alpha(v_1, \ldots, v'_j, \ldots, v_k)$$

for all j with $1 \leq j \leq k$. A k-multilinear map $\alpha : V \times \cdots \times V \to \mathbb{R}$ is **skew** (or **alternating**) when it changes sign whenever two of its arguments are interchanged, that is, for all $v_1, \ldots, v_k \in V$,

$$\alpha(v_1, \ldots, v_i, \ldots, v_j, \ldots, v_k) = -\alpha(v_1, \ldots, v_j, \ldots, v_i, \ldots, v_k).$$

A **2-form** Ω on a manifold M is, for each point $x \in M$, a smooth skew-symmetric bilinear mapping $\Omega(x) : T_x M \times T_x M \to \mathbb{R}$. More generally, a **k-form** α (sometimes called a **differential form of degree** k) on a manifold M is a function $\alpha(x) : T_x M \times \cdots \times T_x M$ (there are k factors)$\to \mathbb{R}$ that assigns to each point $x \in M$ a smooth skew-symmetric k-multilinear map on the tangent space $T_x M$ to M at x.

Without the skew-symmetry assumption, α would be referred to as a $(0, k)$-**tensor**.

Let $x^1, \ldots, x^n$ denote coordinates on M, let

$$\{e_1, \ldots, e_n\} = \left\{ \frac{\partial}{\partial x^1}, \ldots, \frac{\partial}{\partial x^n} \right\}$$

be the corresponding basis for $T_x M$, and let $\{e^1, \ldots, e^n\} = \{dx^1, \ldots, dx^n\}$ be the dual basis for $T_x^* M$. Then at each $x \in M$, we can write a 2-form as

$$\Omega_x(v, w) = \Omega_{ij}(x)v^i w^j, \quad \text{where} \quad \Omega_{ij}(x) = \Omega_x \left(\frac{\partial}{\partial x^i}, \frac{\partial}{\partial x^j} \right).$$

Here the **summation convention** is used; that is,

$$\Omega_x(v, w) = \Omega_{ij}(x)v^i w^j = \sum_{i,j=1}^{n} \Omega_{ij}(x)v^i w^j.$$

More generally, a k-form can be written

$$\alpha_x(v_1, \ldots, v_k) = \alpha_{i_1 \ldots i_k}(x)v_1^{i_1} \ldots v_k^{i_k},$$

where there is a sum on $i_1, \ldots, i_k$, where

$$\alpha_{i_1 \ldots i_k}(x) = \alpha_x \left(\frac{\partial}{\partial x^{i_1}}, \ldots, \frac{\partial}{\partial x^{i_k}} \right),$$

and where $v_i = v_i^j \partial / \partial x^j$, with a sum on j.

Tensor and Wedge Products. If α is a $(0, k)$-tensor on a manifold M, and β is a $(0, l)$-tensor, their **tensor product** $\alpha \otimes \beta$ is the $(0, k + l)$-tensor on M defined by

$$(\alpha \otimes \beta)_x(v_1, \ldots, v_{k+l}) = \alpha_x(v_1, \ldots, v_k) \beta_x(v_{k+1}, \ldots, v_{k+l}) \qquad (2.5.1)$$

at each point $x \in M$.

If t is a $(0, p)$-tensor, define the **alternation operator** $\mathbf{A}$ acting on t by

$$\mathbf{A}(t)(v_1, \ldots, v_p) = \frac{1}{p!} \sum_{\pi \in S_p} \mathrm{sgn}(\pi) t(v_{\pi(1)}, \ldots, v_{\pi(p)}), \qquad (2.5.2)$$

where $\mathrm{sgn}(\pi)$ is the **sign** of the permutation π,

$$\mathrm{sgn}(\pi) = \begin{cases} +1 & \text{if } \pi \text{ is even,} \\ -1 & \text{if } \pi \text{ is odd,} \end{cases} \qquad (2.5.3)$$

and S_p is the group of all permutations of the numbers $1, 2, \ldots, p$. A permutation is called **odd** if it can be written as the product of an odd number of transpositions (that is, a permutation that interchanges just two objects) and otherwise is **even**. The operator $\mathbf{A}$ therefore skew-symmetrizes p-multilinear maps.

If α is a k-form and β is an l-form on M, their **wedge product** $\alpha \wedge \beta$ is the $(k + l)$-form on M defined by[8]

$$\alpha \wedge \beta = \frac{(k + l)!}{k! \, l!} \mathbf{A}(\alpha \otimes \beta). \qquad (2.5.4)$$

For example, if α and β are one-forms, then

$$(\alpha \wedge \beta)(v_1, v_2) = \alpha(v_1)\beta(v_2) - \alpha(v_2)\beta(v_1),$$

while if α is a 2-form and β is a 1-form,

$$(\alpha \wedge \beta)(v_1, v_2, v_3) = \alpha(v_1, v_2)\beta(v_3) + \alpha(v_3, v_1)\beta(v_2) + \alpha(v_2, v_3)\beta(v_1).$$

We state the following without proof:

---

[8]The numerical factor in (2.5.4) agrees with the convention of Abraham and Marsden [1978], Abraham, Marsden, and Ratiu [1988], and Spivak [1979], but *not* that of Arnold [1989], Guillemin and Pollack [1974], or Kobayashi and Nomizu [1963]; it is the Bourbaki [1971] convention.

2.5.1 Proposition. *The wedge product has the following properties:*

(i) $\alpha \wedge \beta$ *is* **associative:** $\alpha \wedge (\beta \wedge \gamma) = (\alpha \wedge \beta) \wedge \gamma$.

(ii) $\alpha \wedge \beta$ *is* **bilinear** *in* α, β :

$$(a\alpha_1 + b\alpha_2) \wedge \beta = a(\alpha_1 \wedge \beta) + b(\alpha_2 \wedge \beta),$$
$$\alpha \wedge (c\beta_1 + d\beta_2) = c(\alpha \wedge \beta_1) + d(\alpha \wedge \beta_2).$$

(iii) $\alpha \wedge \beta$ *is* **anticommutative:** $\alpha \wedge \beta = (-1)^{kl}\beta \wedge \alpha$, *where α is a k-form and β is an l-form.*

In terms of the dual basis dx^i, any k-form can be written locally as

$$\alpha = \alpha_{i_1 \ldots i_k} dx^{i_1} \wedge \cdots \wedge dx^{i_k} ,$$

where the sum is over all i_j satisfying $i_1 < \cdots < i_k$.

Pull Back and Push Forward. Let $\varphi : M \to N$ be a C^∞ map from the manifold M to the manifold N and let α be a k-form on N. Define the **pull back** $\varphi^*\alpha$ of α by φ to be the k-form on M given by

$$(\varphi^*\alpha)_x(v_1, \ldots, v_k) = \alpha_{\varphi(x)}(T_x\varphi \cdot v_1, \ldots, T_x\varphi \cdot v_k). \tag{2.5.5}$$

If φ is a diffeomorphism, the **push forward** φ_* is defined by $\varphi_* = (\varphi^{-1})^*$.

Here is another basic property.

2.5.2 Proposition. *The pull back of a wedge product is the wedge product of the pull backs:*

$$\varphi^*(\alpha \wedge \beta) = \varphi^*\alpha \wedge \varphi^*\beta. \tag{2.5.6}$$

Interior Products and Exterior Derivatives. Let α be a k-form on a manifold M, and X a vector field. The **interior product** $\mathbf{i}_X\alpha$ (sometimes called the contraction of X and α, and written $\mathbf{i}(X)\alpha$) is defined by

$$(\mathbf{i}_X\alpha)_x(v_2, \ldots, v_k) = \alpha_x(X(x), v_2, \ldots, v_k). \tag{2.5.7}$$

2.5.3 Proposition. *Let α be a k-form and β an l-form on a manifold M. Then*

$$\mathbf{i}_X(\alpha \wedge \beta) = (\mathbf{i}_X\alpha) \wedge \beta + (-1)^k \alpha \wedge (\mathbf{i}_X\beta). \tag{2.5.8}$$

The **exterior derivative** $\mathbf{d}\alpha$ of a k-form α on a manifold M is the $(k+1)$-form on M determined by the following proposition:

2.5.4 Proposition. *There is a unique mapping $\mathbf{d}$ from k-forms on M to $(k+1)$-forms on M such that:*

(i) *If α is a 0-form $(k = 0)$, that is, $\alpha = f \in C^\infty(M)$, then $\mathbf{d}f$ is the one-form that is the differential of f.*

(ii) $\mathbf{d}\alpha$ is **linear** in α; that is, for all real numbers c_1 and c_2,

$$\mathbf{d}(c_1\alpha_1 + c_2\alpha_2) = c_1\mathbf{d}\alpha_1 + c_2\mathbf{d}\alpha_2.$$

(iii) $\mathbf{d}\alpha$ satisfies the **product rule**; that is,

$$\mathbf{d}(\alpha \wedge \beta) = \mathbf{d}\alpha \wedge \beta + (-1)^k \alpha \wedge \mathbf{d}\beta,$$

where α is a k-form and β is an l-form.

(iv) $\mathbf{d}^2 = 0$; that is, $\mathbf{d}(\mathbf{d}\alpha) = 0$ for any k-form α.

(v) $\mathbf{d}$ is a **local operator**; that is, $\mathbf{d}\alpha(x)$ depends only on α restricted to any open neighborhood of x; in fact, if U is open in M, then

$$\mathbf{d}(\alpha|U) = (\mathbf{d}\alpha)|U.$$

If α is a k-form given in coordinates by

$$\alpha = \alpha_{i_1\ldots i_k} dx^{i_1} \wedge \cdots \wedge dx^{i_k} \quad (\text{sum on } i_1 < \cdots < i_k),$$

then the coordinate expression for the exterior derivative is

$$\mathbf{d}\alpha = \frac{\partial \alpha_{i_1\ldots i_k}}{\partial x^j} dx^j \wedge dx^{i_1} \wedge \cdots \wedge dx^{i_k} \quad (\text{sum on all } j \text{ and } i_1 < \cdots < i_k). \tag{2.5.9}$$

Formula (2.5.9) can be taken as the definition of the exterior derivative, provided one shows that (2.5.9) has the above-described properties and, correspondingly, is independent of the choice of coordinates.

Next is a useful proposition that, in essence, rests on the chain rule:

2.5.5 Proposition. *Exterior differentiation commutes with pull back:*

$$\mathbf{d}(\varphi^*\alpha) = \varphi^*(\mathbf{d}\alpha), \tag{2.5.10}$$

where α is a k-form on a manifold N and φ is a smooth map from a manifold M to N.

A k-form α is called **closed** if $\mathbf{d}\alpha = 0$ and **exact** if there is a $(k-1)$-form β such that $\alpha = \mathbf{d}\beta$. By Proposition 2.5.4 every exact form is closed. The exercises give an example of a closed nonexact one-form.

2.5.6 Proposition (Poincaré Lemma). *A closed form is locally exact; that is, if $\mathbf{d}\alpha = 0$, there is a neighborhood about each point on which $\alpha = \mathbf{d}\beta$.*

The proof is given in the exercises.

Vector Calculus. The table below entitled "Vector Calculus and Differential Forms" summarizes how forms are related to the usual operations of vector calculus. We now elaborate on a few items in this table. In item 4, note that

$$\mathbf{d}f = \frac{\partial f}{\partial x}dx + \frac{\partial f}{\partial y}dy + \frac{\partial f}{\partial z}dz = (\text{grad}f)^\flat = (\nabla f)^\flat,$$

which is equivalent to $\nabla f = (\mathbf{d}f)^\sharp$. (The $\flat$ and $\sharp$ terminology are explained in the table below.)

The Hodge star operator on $\mathbb{R}^3$ maps k-forms to $(3 - k)$-forms and is uniquely determined by linearity and the properties in item 2.[9]

In item 5, if we let $F = F_1\mathbf{e}_1 + F_2\mathbf{e}_2 + F_3\mathbf{e}_3$, so

$$F^\flat = F_1\,dx + F_2\,dy + F_3\,dz,$$

we obtain

$$\mathbf{d}(F^\flat) = \mathbf{d}F_1 \wedge dx + F_1\mathbf{d}(dx) + \mathbf{d}F_2 \wedge dy + F_2\mathbf{d}(dy) + \mathbf{d}F_3 \wedge dz + F_3\mathbf{d}(dz),$$

which equals

$$\left(\frac{\partial F_1}{\partial x}dx + \frac{\partial F_1}{\partial y}dy + \frac{\partial F_1}{\partial z}dz\right) \wedge dx$$
$$+ \left(\frac{\partial F_2}{\partial x}dx + \frac{\partial F_2}{\partial y}dy + \frac{\partial F_2}{\partial z}dz\right) \wedge dy$$
$$+ \left(\frac{\partial F_3}{\partial x}dx + \frac{\partial F_3}{\partial y}dy + \frac{\partial F_3}{\partial z}dz\right) \wedge dz.$$

This becomes

$$-\frac{\partial F_1}{\partial y}dx \wedge dy + \frac{\partial F_1}{\partial z}dz \wedge dx + \frac{\partial F_2}{\partial x}dx \wedge dy - \frac{\partial F_2}{\partial z}dy \wedge dz$$
$$-\frac{\partial F_3}{\partial x}dz \wedge dx + \frac{\partial F_3}{\partial y}dy \wedge dz$$
$$= \left(\frac{\partial F_2}{\partial x} - \frac{\partial F_1}{\partial y}\right)dx \wedge dy + \left(\frac{\partial F_1}{\partial z} - \frac{\partial F_3}{\partial x}\right)dz \wedge dx$$
$$+ \left(\frac{\partial F_3}{\partial y} - \frac{\partial F_2}{\partial z}\right)dy \wedge dz.$$

---

[9]This operator can be defined on general Riemannian manifolds; see Abraham, Marsden, and Ratiu [1988].

Hence, using item 2,

$$*(\mathbf{d}(F^\flat)) = \left(\frac{\partial F_2}{\partial x} - \frac{\partial F_1}{\partial y}\right) dz + \left(\frac{\partial F_1}{\partial z} - \frac{\partial F_3}{\partial x}\right) dy$$
$$+ \left(\frac{\partial F_3}{\partial y} - \frac{\partial F_2}{\partial z}\right) dx,$$

$$(*(\mathbf{d}(F^\flat)))^\sharp = \left(\frac{\partial F_3}{\partial y} - \frac{\partial F_2}{\partial z}\right) \mathbf{e}_1 + \left(\frac{\partial F_1}{\partial z} - \frac{\partial F_3}{\partial x}\right) \mathbf{e}_2$$
$$+ \left(\frac{\partial F_2}{\partial x} - \frac{\partial F_1}{\partial y}\right) \mathbf{e}_3$$
$$= \operatorname{curl} F = \nabla \times F.$$

With reference to item 6, let

$$F = F_1 \mathbf{e}_1 + F_2 \mathbf{e}_2 + F_3 \mathbf{e}_3,$$

so that

$$F^\flat = F_1 \, dx + F_2 \, dy + F_3 \, dz.$$

Thus,

$$*(F^\flat) = F_1 \, dy \wedge dz + F_2(-dx \wedge dz) + F_3 \, dx \wedge dy,$$

and so

$$\mathbf{d}(*(F^\flat)) = \mathbf{d}F_1 \wedge dy \wedge dz - \mathbf{d}F_2 \wedge dx \wedge dz + \mathbf{d}F_3 \wedge dx \wedge dy$$
$$= \left(\frac{\partial F_1}{\partial x} dx + \frac{\partial F_1}{\partial y} dy + \frac{\partial F_1}{\partial z} dz\right) \wedge dy \wedge dz$$
$$- \left(\frac{\partial F_2}{\partial x} dx + \frac{\partial F_2}{\partial y} dy + \frac{\partial F_2}{\partial z} dz\right) \wedge dx \wedge dz$$
$$+ \left(\frac{\partial F_3}{\partial x} dx + \frac{\partial F_3}{\partial y} dy + \frac{\partial F_3}{\partial z} dz\right) \wedge dx \wedge dy$$
$$= \frac{\partial F_1}{\partial x} dx \wedge dy \wedge dz + \frac{\partial F_2}{\partial y} dx \wedge dy \wedge dz + \frac{\partial F_3}{\partial z} dx \wedge dy \wedge dz$$
$$= \left(\frac{\partial F_1}{\partial x} + \frac{\partial F_2}{\partial y} + \frac{\partial F_3}{\partial z}\right) dx \wedge dy \wedge dz$$
$$= (\operatorname{div} F) \, dx \wedge dy \wedge dz.$$

Therefore, $*(\mathbf{d}(*(F^\flat))) = \operatorname{div} F = \nabla \cdot F$.

The definition and properties of vector-valued forms are direct extensions of these for usual forms on vector spaces and manifolds. One can think of a vector-valued form as an array of usual forms.

The following table should serve as a useful reference for future computations.

Vector Calculus and Differential Forms

1. **Sharp and Flat** (Using standard coordinates in $\mathbb{R}^3$)

    (a) $v^\flat = v^1\, dx + v^2\, dy + v^3\, dz = $ one-form corresponding to the vector $v = v^1\mathbf{e}_1 + v^2\mathbf{e}_2 + v^3\mathbf{e}_3$.

    (b) $\alpha^\sharp = \alpha_1\mathbf{e}_1 + \alpha_2\mathbf{e}_2 + \alpha_3\mathbf{e}_3 = $ vector corresponding to the one-form $\alpha = \alpha_1\, dx + \alpha_2\, dy + \alpha_3\, dz$.

2. **Hodge Star Operator**

    (a) $*1 = dx \wedge dy \wedge dz$.

    (b) $*dx = dy \wedge dz$, $*dy = -dx \wedge dz$, $*dz = dx \wedge dy$,
    $*(dy \wedge dz) = dx$, $*(dx \wedge dz) = -dy$, $*(dx \wedge dy) = dz$.

    (c) $*(dx \wedge dy \wedge dz) = 1$.

3. **Cross Product and Dot Product**

    (a) $v \times w = [*(v^\flat \wedge w^\flat)]^\sharp$.

    (b) $(v \cdot w)dx \wedge dy \wedge dz = v^\flat \wedge *(w^\flat)$.

4. **Gradient** $\nabla f = \operatorname{grad} f = (\mathbf{d}f)^\sharp$.

5. **Curl** $\nabla \times F = \operatorname{curl} F = [*(\mathbf{d}F^\flat)]^\sharp$.

6. **Divergence** $\nabla \cdot F = \operatorname{div} F = *\mathbf{d}(*F^\flat)$.

Exercises

$\diamond$ **2.5-1.** Let $\varphi : \mathbb{R}^3 \to \mathbb{R}^2$ be given by $\varphi(x, y, z) = (x + z, xy)$. For

$$\alpha = e^v\, du + u\, dv \in \Omega^1(\mathbb{R}^2) \quad \text{and} \quad \beta = u\, du \wedge dv$$

compute $\alpha \wedge \beta, \varphi^*\alpha, \varphi^*\beta$, and $\varphi^*\alpha \wedge \varphi^*\beta$.

$\diamond$ **2.5-2.** Given

$$\alpha = y^2\, dx \wedge dz + \sin(xy)\, dx \wedge dy + e^x\, dy \wedge dz \in \Omega^2(\mathbb{R}^3)$$

and

$$X = 3\frac{\partial}{\partial x} + \cos z\frac{\partial}{\partial y} - x^2\frac{\partial}{\partial z} \in \mathfrak{X}(\mathbb{R}^3),$$

compute $\mathbf{d}\alpha$ and $\mathbf{i}_X\alpha$.

⋄ **2.5-3.**   (a) Denote by $\Lambda^k(\mathbb{R}^n)$ the vector space of all skew-symmetric k-linear maps on $\mathbb{R}^n$. Prove that this space has dimension $n!/k!\,(n-k)!$ by showing that a basis is given by $\{e^{i_1} \wedge \cdots \wedge e^{i_k} \mid i_1 < \cdots < i_k\}$, where $\{e_1, \ldots, e_n\}$ is a basis of $\mathbb{R}^n$ and $\{e^1, \ldots, e^n\}$ is its dual basis; that is, $e^i(e_j) = \delta^i_j$.

   (b) If $\mu \in \Lambda^n(\mathbb{R}^n)$ is nonzero, prove that the map $v \in \mathbb{R}^n \mapsto \mathbf{i}_v\mu \in \Lambda^{n-1}(\mathbb{R}^n)$ is an isomorphism.

   (c) If M is a smooth n-manifold and $\mu \in \Omega^n(M)$ is nowhere vanishing (in which case it is called a volume form), show that the map $X \in \mathfrak{X}(M) \mapsto \mathbf{i}_X\mu \in \Omega^{n-1}(M)$ is a module isomorphism over $\mathcal{F}(M)$.

⋄ **2.5-4.**   Let $\alpha = \alpha_i\,dx^i$ be a closed one-form in a ball around the origin in $\mathbb{R}^n$. Show that $\alpha = \mathbf{d}f$ for

$$f(x^1, \ldots, x^n) = \int_0^1 \alpha_j(tx^1, \ldots, tx^n)x^j\,dt.$$

⋄ **2.5-5.**   (a) Let U be an open ball around the origin in $\mathbb{R}^n$ and $\alpha \in \Omega^k(U)$ a closed form. Verify that $\alpha = \mathbf{d}\beta$, where

$$\beta(x^1, \ldots, x^n)$$
$$= \left(\int_0^1 t^{k-1}\alpha_{ji_1\ldots i_{k-1}}(tx^1, \ldots, tx^n)x^j\,dt \right) dx^{i_1} \wedge \cdots \wedge dx^{i_{k-1}},$$

and where the sum is over $i_1 < \cdots < i_{k-1}$. Here,

$$\alpha = \alpha_{j_1\ldots j_k}\,dx^{j_1} \wedge \cdots \wedge dx^{j_k},$$

where $j_1 < \cdots < j_k$ and where α is extended to be skew-symmetric in its lower indices.

   (b) Deduce the Poincaré lemma from **(a)**.

2.6   Lie Derivatives

The *dynamic definition* of the Lie derivative is as follows. Let α be a k-form and let X be a vector field with flow φ_t. The **Lie derivative** of α along X is given by

$$\pounds_X\alpha = \lim_{t \to 0} \frac{1}{t}[(\varphi_t^*\alpha) - \alpha] = \frac{d}{dt}\varphi_t^*\alpha\bigg|_{t=0}. \qquad (2.6.1)$$

This definition together with properties of pull backs yields the following.

2.6.1 Theorem (Lie Derivative Theorem).   *Using the above notation, we have*

$$\frac{d}{dt}\varphi_t^* \alpha = \varphi_t^* \pounds_X \alpha. \qquad (2.6.2)$$

This fundamental formula holds also for *time-dependent* vector fields.

If f is a real-valued function on a manifold M and X is a vector field on M, the **Lie derivative of f along** X is the **directional derivative**

$$\pounds_X f = X[f] := \mathbf{d}f \cdot X. \qquad (2.6.3)$$

In coordinates on M,

$$\pounds_X f = X^i \frac{\partial f}{\partial x^i}. \qquad (2.6.4)$$

If Y is a vector field on a manifold N and $\varphi : M \to N$ is a diffeomorphism, the **pull back** $\varphi^* Y$ is a vector field on M defined by

$$(\varphi^* Y)(x) = T_x \varphi^{-1} \circ Y \circ \varphi(x). \qquad (2.6.5)$$

Two vector fields X on M and Y on N are said to be φ-**related** if

$$T\varphi \circ X = Y \circ \varphi. \qquad (2.6.6)$$

Clearly, if $\varphi : M \to N$ is a diffeomorphism and Y is a vector field on N, $\varphi^* Y$ and Y are φ-related. For a diffeomorphism φ, the **push forward** is defined, as for forms, by $\varphi_* = (\varphi^{-1})^*$.

Jacobi–Lie Brackets.   In Section 1.8 we discussed the Jacobi–Lie bracket for vector fields in $\mathbb{R}^n$ and saw its importance for the analysis of control systems.

We now extend this operation to vector fields on manifolds. If M is a finite-dimensional (smooth) manifold, then the set of vector fields on M coincides with the set of derivations on $\mathcal{F}(M)$.[10] This identification is as follows. Given a vector field $X \in \mathfrak{X}(M)$ define the map $\theta_X : \mathcal{F}(M) \to \mathcal{F}(M)$ by $f \to X[f]$, where $X[f](x) = \mathbf{d}f(x) \cdot X(x)$, which in coordinates is just the directional derivative

$$X[f] = X^i \frac{\partial f}{\partial x^i}$$

with, as usual, a sum understood on the index i. This map θ_X is a derivation in that it is linear and satisfies the Leibniz rule for products. Conversely, any derivation is given in this fashion.

Given two vector fields X and Y on M, one can check that the map $f \mapsto X[Y[f]] - Y[X[f]]$ is a derivation; thus, it determines a unique vector field denoted by $[X, Y]$ and called the **Jacobi–Lie bracket.** of X and Y.

---

[10]The same result is true for C^k manifolds and vector fields if $k \geq 2$. This property is false for infinite-dimensional manifolds; see Abraham, Marsden, and Ratiu [1988].

In coordinates,

$$[X, Y]^j = X^i \frac{\partial Y^j}{\partial x^i} - Y^i \frac{\partial X^j}{\partial x^i} = (X \cdot \nabla) Y^i - (Y \cdot \nabla) X^j, \qquad (2.6.7)$$

and in general, where we identify X, Y with their local representatives,

$$[X, Y] = \mathbf{D} Y \cdot X - \mathbf{D} X \cdot Y. \qquad (2.6.8)$$

There is an interesting link between the Jacobi–Lie bracket and the Lie derivative as follows. Defining $\pounds_X Y = [X, Y]$ gives the **Lie derivative** of Y along X. Then the Lie derivative theorem holds with α replaced by Y.

The Lie bracket of two vector fields has a geometric meaning in terms of successive applications of the flows of the two vector fields in the forward and reverse directions. We discussed this in Section 1.8. We invite the reader to generalize it to the context of manifolds.

If a set of vector fields X_i is such that there exist functions γ_{ijk} such that

$$[X_i, X_j] = \gamma_{ijk} X_k,$$

then the set is said to be **involutive**. As we shall see later, it is in this case that one generates no new directions by bracketing, and so this is an impediment to showing controllability. This may be a good time to reread the Heisenberg example in Section 1.8.

Algebraic Definition of the Lie Derivative. The *algebraic approach* to the Lie derivative on forms or tensors proceeds as follows. Extend the definition of the Lie derivative from functions and vector fields to differential forms, by requiring that the Lie derivative be a derivation; for example, for one-forms α, write

$$\pounds_X \langle \alpha, Y \rangle = \langle \pounds_X \alpha, Y \rangle + \langle \alpha, \pounds_X Y \rangle, \qquad (2.6.9)$$

where X, Y are vector fields and $\langle \alpha, Y \rangle = \alpha(Y)$. More generally,

$$\pounds_X(\alpha(Y_1, \ldots, Y_k)) = (\pounds_X \alpha)(Y_1, \ldots, Y_k) + \sum_{i=1}^{k} \alpha(Y_1, \ldots, \pounds_X Y_i, \ldots, Y_k), \qquad (2.6.10)$$

where $X, Y_1, \ldots, Y_k$ are vector fields and α is a k-form.

2.6.2 Proposition. *The dynamic and algebraic definitions of the Lie derivative of a differential k-form are equivalent.*

Cartan's Magic Formula. A very important formula for the Lie derivative is given by the following.

2.6.3 Theorem. *For X a vector field and α a k-form on a manifold M, we have*

$$\pounds_X \alpha = \mathbf{d} \mathbf{i}_X \alpha + \mathbf{i}_X \mathbf{d} \alpha. \qquad (2.6.11)$$

This is proved by a lengthy but straightforward calculation.

Another property of the Lie derivative is the following: If $\varphi : M \to N$ is a diffeomorphism, then

$$\varphi^* \pounds_Y \beta = \pounds_{\varphi^* Y} \varphi^* \beta$$

for $Y \in \mathfrak{X}(N)$, $\beta \in \Omega^k(N)$. More generally, if $X \in \mathfrak{X}(M)$ and $Y \in \mathfrak{X}(N)$ are ψ related, that is, $T\psi \circ X = Y \circ \psi$ for $\psi : M \to N$ a smooth map, then

$$\pounds_X \psi^* \beta = \psi^* \pounds_Y \beta \qquad \text{for all } \beta \in \Omega^k(N).$$

Volume Forms and Divergence. An n-manifold M is said to be *orientable* if there is a nowhere-vanishing n-form μ on it; μ is called a *volume form*, and it is a basis of $\Omega^n(M)$ over $\mathcal{F}(M)$. Two volume forms μ_1 and μ_2 on M are said to define the same *orientation* if there is an $f \in \mathcal{F}(M)$ with $f > 0$ and such that $\mu_2 = f\mu_1$. Connected orientable manifolds admit precisely two orientations. A basis $\{v_1, \ldots, v_n\}$ of $T_m M$ is said to be *positively oriented* relative to the volume form μ on M if $\mu(m)(v_1, \ldots, v_n) > 0$. Note that the volume forms defining the same orientation form a convex cone in $\Omega^n(M)$; that is, if $a > 0$ and μ is a volume form, then $a\mu$ is again a volume form, and if $t \in [0, 1]$ and μ_1, μ_2 are volume forms, then $t\mu_1 + (1 - t)\mu_2$ is again a volume form. The first property is obvious. To prove the second, let $m \in M$ and let $\{v_1, \ldots, v_n\}$ be a positively oriented basis of $T_m M$ relative to the orientation defined by μ_1, or equivalently (by hypothesis) by μ_2. Then $\mu_1(m)(v_1, \ldots, v_n) > 0$, $\mu_2(m)(v_1, \ldots, v_n) > 0$, so that their convex combination is again strictly positive.

If $\mu \in \Omega^n(M)$ is a volume form, since $\pounds_X \mu \in \Omega^n(M)$, there is a function, called the *divergence* of X relative to μ and denoted $\mathrm{div}_\mu(X)$ or simply $\mathrm{div}(X)$, such that

$$\pounds_X \mu = \mathrm{div}_\mu(X)\mu. \tag{2.6.12}$$

From the dynamic approach to Lie derivatives it follows that $\mathrm{div}_\mu(X) = 0$ iff $F_t^* \mu = \mu$, where F_t is the flow of X. This condition says that F_t is *volume-preserving*. If $\varphi : M \to M$, since $\varphi^* \mu \in \Omega^n(M)$, there is a function, called the *Jacobian* of φ and denoted by $J_\mu(\varphi)$ or simply $J(\varphi)$, such that

$$\varphi^* \mu = J_\mu(\varphi)\mu. \tag{2.6.13}$$

Thus, φ is volume-preserving iff $J_\mu(\varphi) = 1$. The inverse function theorem shows that φ is a local diffeomorphism iff $J_\mu(\varphi) \neq 0$ on M.

There are a number of valuable identities relating the Lie derivative, the exterior derivative, and the interior product. For example, if Θ is a one-form and X and Y are vector fields, identity 6 in the following table gives

$$\mathbf{d}\Theta(X, Y) = X[\Theta(Y)] - Y[\Theta(X)] - \Theta([X, Y]). \tag{2.6.14}$$

The following list of identities will be a useful reference for the remainder of the text.

Identities for Vector Fields and Forms

1. Vector fields on M with the bracket $[X, Y]$ form a **Lie algebra**; that is, $[X, Y]$ is real bilinear and skew-symmetric, and **Jacobi's identity** holds:
$$[[X, Y], Z] + [[Z, X], Y] + [[Y, Z], X] = 0.$$
Locally,
$$[X, Y] = \mathbf{D}Y \cdot X - \mathbf{D}X \cdot Y = (X \cdot \nabla)Y - (Y \cdot \nabla)X,$$
and on functions,
$$[X, Y][f] = X[Y[f]] - Y[X[f]].$$

2. For diffeomorphisms φ and ψ,
$$\varphi_*[X, Y] = [\varphi_* X, \varphi_* Y] \quad \text{and} \quad (\varphi \circ \psi)_* X = \varphi_* \psi_* X.$$

3. The forms on a manifold constitute a real associative algebra with $\wedge$ as multiplication. Furthermore, $\alpha \wedge \beta = (-1)^{kl} \beta \wedge \alpha$ for k- and l-forms α and β, respectively.

4. For maps φ and ψ,
$$\varphi^*(\alpha \wedge \beta) = \varphi^* \alpha \wedge \varphi^* \beta \quad \text{and} \quad (\varphi \circ \psi)^* \alpha = \psi^* \varphi^* \alpha.$$

5. $\mathbf{d}$ is a real linear map on forms, $\mathbf{dd}\alpha = 0$, and
$$\mathbf{d}(\alpha \wedge \beta) = \mathbf{d}\alpha \wedge \beta + (-1)^k \alpha \wedge \mathbf{d}\beta$$
for α a k-form.

6. For α a k-form and $X_0, \ldots, X_k$ vector fields,
$$(\mathbf{d}\alpha)(X_0, \ldots, X_k) = \sum_{i=0}^{k} (-1)^i X_i[\alpha(X_0, \ldots, \hat{X}_i, \ldots, X_k)]$$
$$+ \sum_{0 \leq i < j \leq k} (-1)^{i+j} \alpha([X_i, X_j], X_0, \ldots, \hat{X}_i, \ldots, \hat{X}_j, \ldots, X_k),$$
where $\hat{X}_i$ means that X_i is omitted. Locally,
$$\mathbf{d}\alpha(x)(v_0, \ldots, v_k) = \sum_{i=0}^{k} (-1)^i \mathbf{D}\alpha(x) \cdot v_i(v_0, \ldots, \hat{v}_i, \ldots, v_k).$$

7. For a map φ, $\varphi^* \mathbf{d}\alpha = \mathbf{d}\varphi^* \alpha$.

8. **Poincaré Lemma.** If $\mathbf{d}\alpha = 0$, then α is locally exact; that is, there is a neighborhood U about each point on which $\alpha = \mathbf{d}\beta$. The same result holds globally on a contractible manifold.

9. $\mathbf{i}_X \alpha$ is real bilinear in x, α, and for $h : M \to \mathbb{R}$,

$$\mathbf{i}_{hX}\alpha = h\mathbf{i}_X\alpha = \mathbf{i}_X h\alpha.$$

Also, $\mathbf{i}_X \mathbf{i}_X \alpha = 0$ and

$$\mathbf{i}_X(\alpha \wedge \beta) = \mathbf{i}_X\alpha \wedge \beta + (-1)^k \alpha \wedge \mathbf{i}_X\beta$$

for α a k-form.

10. For a diffeomorphism φ,

$$\varphi^*(\mathbf{i}_X\alpha) = \mathbf{i}_{\varphi^* X}(\varphi^*\alpha);$$

if $f : M \to N$ is a mapping and Y is f-related to X, i.e., $Tf \circ X = Y \circ f$, then

$$\mathbf{i}_Y f^*\alpha = f^*\mathbf{i}_X\alpha.$$

11. $\mathcal{L}_X \alpha$ is real bilinear in x, α, and

$$\mathcal{L}_X(\alpha \wedge \beta) = \mathcal{L}_X\alpha \wedge \beta + \alpha \wedge \mathcal{L}_X\beta.$$

12. **Cartan's Magic Formula:** $\mathcal{L}_X\alpha = \mathbf{d}\mathbf{i}_X\alpha + \mathbf{i}_X\mathbf{d}\alpha$.

13. For a diffeomorphism φ,

$$\varphi^*\mathcal{L}_X\alpha = \mathcal{L}_{\varphi^* X}\varphi^*\alpha;$$

if $f : M \to N$ is a mapping and Y is f-related to X, then

$$\mathcal{L}_Y f^*\alpha = f^*\mathcal{L}_X\alpha.$$

14. For vector fields $X, X_1, \ldots, X_k$ and a k-form α,

$$(\mathcal{L}_X\alpha)(X_1, \ldots, X_k) = X[\alpha(X_1, \ldots, X_k)]$$
$$- \sum_{i=1}^{k} \alpha(X_1, \ldots, [X, X_i], \ldots, X_k).$$

Locally,

$$(\mathcal{L}_X\alpha)(x) \cdot (v_1, \ldots, v_k) = (\mathbf{D}\alpha_x \cdot X(x))(v_1, \ldots, v_k)$$
$$+ \sum_{i=0}^{k} \alpha_x(v_1, \ldots, \mathbf{D}X_x \cdot v_i, \ldots, v_k).$$

15. The following identities hold:

   (a) $\pounds_{fX}\alpha = f\pounds_X\alpha$, $\pounds_X f\alpha = f\pounds_X\alpha + df \wedge \mathbf{i}_X\alpha$;

   (b) $\pounds_{[X,Y]}\alpha = \pounds_X\pounds_Y\alpha - \pounds_Y\pounds_X\alpha$;

   (c) $\mathbf{i}_{[X,Y]}\alpha = \pounds_X\mathbf{i}_Y\alpha - \mathbf{i}_Y\pounds_X\alpha$;

   (d) $\pounds_X d\alpha = d\pounds_X\alpha$;

   (e) $\pounds_X\mathbf{i}_X\alpha = \mathbf{i}_X\pounds_X\alpha$.

16. If M is a finite-dimensional manifold, $X = X^l \partial/\partial x^l$, and

$$\alpha = \alpha_{i_1\ldots i_k}\, dx^{i_1} \wedge \cdots \wedge dx^{i_k},$$

where $i_1 < \cdots < i_k$, then the following formulas hold:

$$d\alpha = \left(\frac{\partial \alpha_{i_1\ldots i_k}}{\partial x^l}\right) dx^l \wedge dx^{i_1} \wedge \cdots \wedge dx^{i_k},$$

$$\mathbf{i}_X\alpha = X^l \alpha_{li_2\ldots i_k} dx^{i_2} \wedge \cdots \wedge dx^{i_k},$$

$$\pounds_X\alpha = X^l \left(\frac{\partial \alpha_{i_1\ldots i_k}}{\partial x^l}\right) dx^{i_1} \wedge \cdots \wedge dx^{i_k}$$

$$+ \alpha_{li_2\ldots i_k}\left(\frac{\partial X^l}{\partial x^{i_1}}\right) dx^{i_1} \wedge dx^{i_2} \wedge \cdots \wedge dx^{i_k} + \cdots .$$

Exercises

◇ **2.6-1.** Consider the two-form β on $\mathbb{R}^3$ given by

$$\beta = x\, dy \wedge dz + y\, dx \wedge dz + z dx \wedge dy$$

and the vector fields X, Y on $\mathbb{R}^3$ defined by

$$X = x\frac{\partial}{\partial x} + y\frac{\partial}{\partial y} + z\frac{\partial}{\partial z}; \qquad Y = \frac{\partial}{\partial x} + \frac{\partial}{\partial y} + \frac{\partial}{\partial z}.$$

   (a) Is β closed? exact?

   (b) Compute $\mathbf{i}_X\beta$.

   (c) Find the flow F_t of X.

(d) Compute $\dfrac{d}{dt}\Big|_{t=0} F_t^*\beta$ and $\dfrac{d}{dt}\Big|_{t=0} F_t^*Y$.

◇ **2.6-2.**   Let M be an n-manifold, $\omega \in \Omega^n(M)$ a volume form, $X, Y \in \mathfrak{X}(M)$, and $f, g : M \to \mathbb{R}$ smooth functions such that $f(m) \neq 0$ for all m. Prove the following identities:

(a) $\operatorname{div}_{f\omega}(X) = \operatorname{div}_\omega(X) + X[f]/f$;

(b) $\operatorname{div}_\omega(gX) = g\operatorname{div}_\omega(X) + X[g]$;

(c) $\operatorname{div}_\omega([X, Y]) = X[\operatorname{div}_\omega(Y)] - Y[\operatorname{div}_\omega(X)]$.

◇ **2.6-3.**   Show that the partial differential equation

$$\frac{\partial f}{\partial t} = \sum_{i=1}^n X^i(x^1, \ldots, x^n)\frac{\partial f}{\partial x^i}$$

with initial condition $f(x, 0) = g(x)$ has the solution $f(x, t) = g(F_t(x))$, where F_t is the flow of the vector field $(X^1, \ldots, X^n)$ in $\mathbb{R}^n$ whose flow is assumed to exist for all time. Show that the solution is *unique*. Generalize this exercise to the equation

$$\frac{\partial f}{\partial t} = X[f]$$

for X a vector field on a manifold M.

◇ **2.6-4.**   Show that if M and N are orientable manifolds, so is $M \times N$.

2.7   Stokes's Theorem, Riemannian Manifolds, Distributions

The basic idea behind the definition of the integral of an n-form ω on an oriented n-manifold M is to pick a covering by coordinate charts and to sum up the ordinary integrals of $f(x^1, \ldots, x^n)\, dx^1 \cdots dx^n$ in these charts, where

$$\omega = f(x^1, \ldots, x^n)\, dx^1 \wedge \cdots \wedge dx^n$$

is the local representative of ω, being careful not to count overlaps twice. The change of variables formula guarantees that the result, denoted by $\int_M \omega$, is well-defined. Literally carrying this out as stated would involve some fairly serious combinatorial problems in keeping track of overlaps of coordinate charts. Thus, an alternative approach using a tool called partitions of unity (a bunch of positive functions that add up to one) is often used, since it makes the bookkeeping fairly easy. See Abraham, Marsden, and Ratiu [1988] for details.

If one has an oriented manifold with boundary, then the boundary ∂M inherits a compatible orientation. This proceeds in a way that generalizes the relation between the orientation of a surface and its boundary that one learns in the classical Stokes's theorem in $\mathbb{R}^3$.

2.7.1 Theorem (Stokes's Theorem). *Suppose that M is a compact, oriented k-dimensional manifold with boundary ∂M. Let α be a smooth $(k-1)$-form on M. Then*

$$\int_M \mathbf{d}\alpha = \int_{\partial M} \alpha. \tag{2.7.1}$$

Special cases of Stokes's theorem are as follows:

The Integral Theorems of Calculus.　Stokes's theorem generalizes and synthesizes the classical theorems:

(a) Fundamental Theorem of Calculus.

$$\int_b^a f'(x)\,dx = f(b) - f(a). \tag{2.7.2}$$

(b) Green's Theorem.　For a region $\Omega \subset \mathbb{R}^2$,

$$\iint_\Omega \left(\frac{\partial Q}{\partial x} - \frac{\partial P}{\partial y} \right) dx\,dy = \int_{\partial \Omega} P\,dx + Q\,dy. \tag{2.7.3}$$

(c) Divergence Theorem.　For a region $\Omega \subset \mathbb{R}^3$,

$$\iiint_\Omega \operatorname{div} \mathbf{F}\,dV = \iint_{\partial \Omega} \mathbf{F} \cdot \mathbf{n}\,dA. \tag{2.7.4}$$

(d) Classical Stokes's Theorem.　For a surface $S \subset \mathbb{R}^3$,

$$\iint_S \left\{ \left(\frac{\partial R}{\partial y} - \frac{\partial Q}{\partial z} \right) dy \wedge dz \right.$$
$$\left. + \left(\frac{\partial P}{\partial z} - \frac{\partial R}{\partial x} \right) dz \wedge dx + \left(\frac{\partial Q}{\partial x} - \frac{\partial P}{\partial y} \right) dx \wedge dy \right\}$$

$$= \iint_S \mathbf{n} \cdot \operatorname{curl} \mathbf{F}\,dA = \int_{\partial S} P\,dx + Q\,dy + R\,dz, \tag{2.7.5}$$

where $\mathbf{F} = (P, Q, R)$.

Notice that the Poincaré lemma generalizes the vector calculus theorems in $\mathbb{R}^3$ saying that if $\operatorname{curl} \mathbf{F} = 0$, then $\mathbf{F} = \nabla f$, and if $\operatorname{div} \mathbf{F} = 0$, then $\mathbf{F} = \nabla \times \mathbf{G}$. Recall that it states, *If α is closed, then locally α is exact; that is, if $\mathbf{d}\alpha = 0$, then locally $\alpha = \mathbf{d}\beta$ for some β.*

Change of Variables.   Another basic result in integration theory is the global change of variables formula.

2.7.2 Theorem (Change of Variables). *Suppose that M and N are oriented n-manifolds and $F : M \to N$ is an orientation-preserving diffeomorphism. If α is an n-form on N (with, say, compact support), then*

$$\int_M F^* \alpha = \int_N \alpha.$$

Riemannian Manifolds.   A differentiable manifold with a positive definite symmetric quadratic form $\langle \cdot, \cdot \rangle$ on every tangent space TM_x is called a **Riemannian manifold**. The quadratic form $\langle \cdot, \cdot \rangle$ itself, often denoted by $g(\,,)$ is called a **Riemannian metric**.

In local coordinates q^i on M and the associated tangent coordinates $\dot{q}^i$ the length of a vector $v = v^i e_i$ is then given by

$$g(v, v) = g_{ij}(q) v^i v^j, \qquad g_{ij} = g_{ji},$$

where as above, the summation convention is in force.

Let f be a smooth function on M. The **gradient vector field** associated with f, which is denoted by grad f or ∇f, is defined by

$$\mathbf{d}f(v) = \langle \operatorname{grad} f, v \rangle$$

for any $v \in TM$. The flow of the vector field grad f is called the **gradient flow** of f.

Frobenius's Theorem.   A basic result called **Frobenius's theorem** plays a critical role in control theory, and we shall have much to say about it later in the book. For now we just state it briefly, since it is normally regarded as part of the theory of differentiable manifolds. The theory of distributions plays a key role in both the theory of nonholonomic systems and nonlinear control theory. Two useful references (from the control-theoretic point of view) are Sussmann [1973] and Isidori [1995].

2.7.3 Definition.   *A smooth distribution on a manifold M is the assignment to each point $x \in M$ of the subspace spanned by the values at x of a set of smooth vector fields on M; i.e., it is a "smooth" assignment of a subspace to the tangent space at each point, also called a **vector subbundle**. We denote the distribution by Δ and the subspace at $x \in M$ by $\Delta_x \subset T_x M$.*

A distribution is said to be **involutive** if for any two vector fields X, Y on M with values in Δ, $[X, Y]$ is also a vector field with values in Δ. The subbundle Δ is said to be **integrable** if for each point $x \in M$ there is a local submanifold of M containing x such that its tangent bundle equals Δ restricted to this submanifold. See Figure 2.7.1.

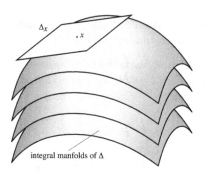

FIGURE 2.7.1. The integral manifolds of a distribution.

If Δ is integrable, the local integral manifolds can be extended to get, through each $x \in M$, a maximal integral manifold, which is an immersed submanifold of M. The collection of all maximal integral manifolds through all points of M forms a *foliation*.

2.7.4 Theorem (Frobenius's Theorem). *Involutivity of Δ is equivalent to the integrability of Δ, which in turn is equivalent to the existence of a foliation on M whose tangent bundle equals Δ.*

Given a set of smooth vector fields $X_1, \ldots, X_d$ on M we denote the distribution defined by their span by

$$\Delta = \mathrm{span}\{X_1, \ldots, X_d\}.$$

The distribution at any point is denoted by Δ_x. A distribution Δ on M is said to be *nonsingular* (or *regular*) on M if there exists an integer d such that $\dim(\Delta_x) = d$ for all $x \in M$. A point $x \in M$ is said to be a regular point is there exists a neighborhood U of x such that Δ is nonsingular on U. Otherwise, the point is said to be *singular*.

Note that the Frobenius theorem as stated above applies to nonsingular or regular distributions. For generalized distributions in the sense of Sussmann, see for example, Vaisman [1994].

Exercises

◇ **2.7-1.** Let Ω be a closed bounded region in $\mathbb{R}^2$. Use Green's theorem to show that the area of Ω equals the line integral

$$\frac{1}{2} \int_{\partial\Omega} (x\,dy - y\,dx).$$

◇ **2.7-2.** On $\mathbb{R}^2 \backslash \{(0,0)\}$ consider the one-form

$$\alpha = (x\,dy - y\,dx)/(x^2 + y^2).$$

(a) Show that this form is closed.

(b) Using the angle θ as a variable on S^1, compute $i^*\alpha$, where $i : S^1 \to \mathbb{R}^2$ is the standard embedding.

(c) Show that α is not exact.

⬦ **2.7-3.** Suppose that a set of linearly independent vector fields X_i has the property that there are functions γ_{ijk} such that

$$[X_i, X_j] = \gamma_{ijk} X_k.$$

Show that the span of these vector fields defines an integrable distribution.

⬦ **2.7-4. The magnetic monopole.** Let $\mathbf{B} = g\mathbf{r}/r^3$ be a vector field on Euclidean three-space minus the origin, where $r = \|\mathbf{r}\|$. Show that $\mathbf{B}$ cannot be written as the curl of something.

⬦ **2.7-5.** Let M be a manifold and ω a two-form on M.

(a) Consider the distribution D on M defined at $x \in M$ by

$$D_x = \{v_x \in T_x M \mid \mathbf{i}_{v_x}\omega = 0\}.$$

Develop a condition(s) that guarantees that this distribution is integrable.

(b) Let ω on $\mathbb{R}^4$, with coordinates (x, y, z, w), be given by

$$\omega = dx \wedge dy + dx \wedge dz + dx \wedge dw.$$

Compute the distribution D in this case. Does your condition hold?

(d) Find an explicit example of such a vector field X for the example in part (b).

2.8 Lie Groups

Lie groups arise in discussing conservation laws for mechanical and control systems and in the analysis of systems with some underlying symmetry. There is a huge literature on the subject. Useful references include Abraham and Marsden [1978], Marsden and Ratiu [1999], Sattinger and Weaver [1986], and Libermann and Marle [1987].

2.8.1 Definition. *A **Lie group** is a smooth manifold G that is a group and for which the group operations of multiplication, $(g, h) \mapsto gh$ for $g, h \in G$, and inversion, $g \mapsto g^{-1}$, are smooth.*

Before giving a brief description of some of the theory of Lie groups we mention an important example: the group of linear isomorphisms of $\mathbb{R}^n$ to itself. This is a Lie group of dimension n^2 called the general linear group and denoted by $GL(n, \mathbb{R})$. The conditions for a Lie group are easily checked: This is a manifold, since it is an open subset of the linear space of all linear maps of $\mathbb{R}^n$ to itself; the group operations are smooth, since they are algebraic operations on the matrix entries.

2.8.2 Definition. *A matrix Lie group is a set of invertible $n \times n$ matrices that is closed under matrix multiplication and that is a submanifold of $\mathbb{R}^{n \times n}$.*

A theorem in Lie group theory shows that (although this is by no means obvious) one could equivalently define a matrix Lie group to be a (topologically) closed subgroup of $GL(n, \mathbb{R})$. All of the Lie groups discussed in this book will be matrix Lie groups.

Lie Algebras.   Lie groups are frequently studied in conjunction with *Lie algebras*, which are associated with the tangent spaces of Lie groups as we now describe. To begin with, we state a generalization of the result established in Exercise 2.2-3.

2.8.3 Proposition.   *Let G be a matrix Lie group, and let $A, B \in T_I G$ (the tangent space to G at the identity element). Then $AB - BA \in T_I G$.*

Our proof makes use of the following lemma.

2.8.4 Lemma.   *Let R be an arbitrary element of a matrix Lie group G, and let $B \in T_I G$. Then $RBR^{-1} \in T_I G$.*

Proof.   Let $R_B(t)$ be a curve in G such that $R_B(0) = I$ and $R'_B(0) = B$. Then $S(t) = R R_B(t) R^{-1} \in G$ for all t, and $S(0) = I$. Hence $S'(0) \in T_I G$, proving the lemma.   ▼

Proof of Proposition.   Let $R_A(s)$ be a curve in G such that $R_A(0) = I$ and $R'_A(0) = A$. Thus, by the preceding lemma, $S(t) = R_A(t) B R_A(t)^{-1} \in T_I G$. Hence $S'(t) \in T_I G$, and in particular, $S'(0) = AB - BA \in T_I G$.   ■

2.8.5 Definition.   *For any pair of $n \times n$ matrices A, B we define the matrix Lie bracket $[A, B] = AB - BA$.*

2.8.6 Proposition.   *The matrix Lie bracket operation has the following two properties:*

(i) *For any $n \times n$ matrices A and B, $[B, A] = -[A, B]$ (this is the property of **skew-symmetry**).*

(ii) *For any $n \times n$ matrices A, B, and C,*

$$[[A, B], C] + [[B, C], A] + [[C, A], B] = 0.$$

*(This is known as the **Jacobi identity**.)*

The proof of this proposition involves a straightforward calculation and is left to the reader.

2.8.7 Definition. *A (matrix)* **Lie algebra** $\mathfrak{g}$ *is a set of $n \times n$ matrices that is a vector space with respect to the usual operations of matrix addition and multiplication by real numbers (scalars) and that is closed under the matrix Lie bracket operation $[\cdot\,,\cdot]$.*

2.8.8 Proposition. *For any matrix Lie group G, the tangent space at the identity $T_I G$ is a Lie algebra.*

Proof.   This is an immediate consequence of the fact that $T_I G$ is a vector space and the preceding proposition. ∎

One can also define a Lie algebra $\mathfrak{g}$ abstractly as a vector space over a field F on which a Lie bracket operation $[\cdot\,,\cdot]$ is defined such that $\mathfrak{g}$ is closed under this operation; $[A, \alpha B + \beta C] = \alpha[A, B] + \beta[A, C]$ for any $\alpha, \beta \in F$ and $A, B, C \in \mathfrak{g}$; and properties (i) and (ii) in theorem 2.8.6 hold.

For $A \in \mathfrak{g}$ we define the operator ad_X to be the operator that maps $B \in \mathfrak{g}$ to $[A, B]$. We write $\mathrm{ad}_A B = [A, B]$.

2.8.9 Definition. *A* **representation** *of a Lie algebra $\mathfrak{g}$ on a vector space V is a mapping ρ from $\mathfrak{g}$ to the linear transformations of V such that for $A, B \in \mathfrak{g}$*

(i) $\rho(\alpha A + \beta B) = \alpha \rho(A) + \beta \rho(B)$

(ii) $\rho([A, B]) = \rho(A)\rho(B) - \rho(B)\rho(A)$.

If the map ρ is 1-1 the representation is said to be **faithful***.*

For a Lie algebra $\mathfrak{g}$ the map $A \to \mathrm{ad}_A$ is a representation of the Lie algebra $\mathfrak{g}$, with $\mathfrak{g}$ itself the vector space of the representation. This is called the **adjoint representation**. The ad-action of the Lie algebra on itself is the infinitesimal action of the Adjoint action of the group—see later in this section and, for example, Arnold [1989].

The **Killing form** of a Lie algebra is the symmetric bilinear form defined by

$$\kappa(A, B) = \mathrm{Trace}(\mathrm{ad}_A \, \mathrm{ad}_B). \tag{2.8.1}$$

One can show that a Lie algebra is **semi-simple**, i.e. it contains no abelian ideals other than $\{0\}$, if and only if the Killing form is nondegenerate. Further, the group G corresponding to $\mathfrak{g}$ is compact if and only if the Killing form is negative definite. See, for example, Sattinger and Weaver [1986] for proofs.

A great deal of the structure of a Lie group may be inferred from studying the Lie algebra. Before discussing important general relationships between Lie groups and Lie algebras, we describe several examples that play an important role in mechanics and control.

The Special Orthogonal Group.   The set of all elements of $O(n)$ having determinant 1 is a subgroup called the *special orthogonal group*, denoted by $SO(n)$. Because any $X \in O(n)$ satisfies $XX^T = I$, it follows that $\det X = \pm 1$. We could also characterize $SO(n)$ as the connected component of the identity element in $O(n)$. Thus, $T_I\, SO(n) = T_I\, O(n)$. From this observation and the calculation carried out for $GL(n, \mathbb{R})$ in Section 2.2, $T_I\, SO(n)$ is just the set of $n \times n$ skew-symmetric matrices, which we denote by $\mathfrak{so}(n)$.

The Symplectic Group.   Suppose $n = 2l$ (that is, n is even) and consider the nonsingular skew-symmetric matrix

$$J = \begin{pmatrix} 0 & I \\ -I & 0 \end{pmatrix},$$

where I is the $l \times l$ identity matrix. It is an exercise left to the reader to verify that

$$Sp(l) = \{X \in GL(2l) \mid XJX^T = J\}$$

is a group. It is called the *symplectic group*. Again referring to the example of $GL(n, \mathbb{R})$ in Section 2.2, we find that this matrix Lie algebra $T_I\, Sp(l)$ is the set of $n \times n$ matrices satisfying $JY^T + YJ = 0$. We denote this Lie algebra by $sp(l)$.

The Heisenberg Group.   Consider the set of all 3×3 matrices of the form

$$\begin{pmatrix} 1 & x & y \\ 0 & 1 & z \\ 0 & 0 & 1 \end{pmatrix}$$

where x, y, and z are real numbers. It is straightforward to show that this is a group, and since it is a submanifold of the set of all 3×3 matrices, it is a Lie group. Call it H. The corresponding Lie algebra may be written down from the definition. Specifically,

$$X_1(t) = \begin{pmatrix} 1 & t & 0 \\ 0 & 1 & 0 \\ 0 & 0 & 1 \end{pmatrix}, \quad X_2(t) = \begin{pmatrix} 1 & 0 & 0 \\ 0 & 1 & t \\ 0 & 0 & 1 \end{pmatrix},$$

$$X_3(t) = \begin{pmatrix} 1 & 0 & t \\ 0 & 1 & 0 \\ 0 & 0 & 1 \end{pmatrix}$$

are three curves in H that pass through the identity when $t = 0$. The derivatives $X_i'(0)$ are elements of the Lie algebra:

$$A = \begin{pmatrix} 0 & 1 & 0 \\ 0 & 0 & 0 \\ 0 & 0 & 0 \end{pmatrix}, \quad B = \begin{pmatrix} 0 & 0 & 0 \\ 0 & 0 & 1 \\ 0 & 0 & 0 \end{pmatrix}, \quad C = \begin{pmatrix} 0 & 0 & 1 \\ 0 & 0 & 0 \\ 0 & 0 & 0 \end{pmatrix},$$

respectively. Since three independent parameters are used to specify H, H is three-dimensional. The matrices A, B, and C are linearly independent and span the Lie algebra. The commutation relations for the Lie brackets of these three basis elements are $[A, B] = C$, $[A, C] = 0$, and $[B, C] = 0$. This Lie algebra is called the **Heisenberg algebra**.

Recall that we encountered this algebra in Section 1.8 when we analyzed the Heisenberg system, and we shall encounter it several times again.

The Euclidean Group.   Consider the Lie group of all 4×4 matrices of the form

$$\begin{pmatrix} R & v \\ 0 & 1 \end{pmatrix},$$

where $R \in SO(3)$ and $v \in \mathbb{R}^3$. This group is usually denoted by $SE(3)$ and is called the **special Euclidean group**. Let the associated matrix be denoted by

$$E(R, v) = \begin{pmatrix} R & v \\ 0 & 1 \end{pmatrix}.$$

The corresponding Lie algebra, se(3), is six-dimensional and is spanned by

$$\begin{pmatrix} 0 & 0 & 0 & 0 \\ 0 & 0 & -1 & 0 \\ 0 & 1 & 0 & 0 \\ 0 & 0 & 0 & 0 \end{pmatrix}, \quad \begin{pmatrix} 0 & 0 & 1 & 0 \\ 0 & 0 & 0 & 0 \\ -1 & 0 & 0 & 0 \\ 0 & 0 & 0 & 0 \end{pmatrix}, \quad \begin{pmatrix} 0 & -1 & 0 & 0 \\ 1 & 0 & 0 & 0 \\ 0 & 0 & 0 & 0 \\ 0 & 0 & 0 & 0 \end{pmatrix},$$

$$\begin{pmatrix} 0 & 0 & 0 & 1 \\ 0 & 0 & 0 & 0 \\ 0 & 0 & 0 & 0 \\ 0 & 0 & 0 & 0 \end{pmatrix}, \quad \begin{pmatrix} 0 & 0 & 0 & 0 \\ 0 & 0 & 0 & 1 \\ 0 & 0 & 0 & 0 \\ 0 & 0 & 0 & 0 \end{pmatrix}, \quad \begin{pmatrix} 0 & 0 & 0 & 0 \\ 0 & 0 & 0 & 0 \\ 0 & 0 & 0 & 1 \\ 0 & 0 & 0 & 0 \end{pmatrix}.$$

The special Euclidean group is of central interest in mechanics since it describes the set of rigid motions and coordinate transformations of 3-space. More specifically, suppose there are two coordinate frames A and B located in space such that the origin of the B-frame has A-frame coordinates $v = (v_1, v_2, v_3)^T$ and such that the unit vectors in the principal B-frame coordinate directions are $(r_{11}, r_{21}, r_{31})^T$, $(r_{12}, r_{22}, r_{32})^T$, and $(r_{13}, r_{23}, r_{33})^T$ with respect to A-frame coordinates. The **rigid motion** that moves the A-frame into coincidence with the B-frame is specified by the rotation

$$\begin{pmatrix} r_{11} & r_{12} & r_{13} \\ r_{21} & r_{22} & r_{23} \\ r_{31} & r_{32} & r_{33} \end{pmatrix}$$

followed by the translation $v = (v_1, v_2, v_3)^T$. Thus a point with A-frame coordinates $x = (x_1, x_2, x_3)$ is moved under the rigid motion to a new location whose A-frame coordinates are $Rx + v$.

Remark. The above discussion is nice and concrete, but gives the impression that one needs coordinate frames to define the Euclidean group. More intrinsically, the Euclidean group SE(3) can also be defined simply as the set of all isometries of $\mathbb{R}^3$ to itself. (It is a famous theorem of Mazur and Ulam that such isometries are, in fact, affine maps.)

Resuming the previous discussion, we observe that the group SE(3) is also associated with the set of *rigid coordinate transformations* of $\mathbb{R}^3$ as follows. Suppose a point Q is located in space and has A-frame coordinates $(x_1^A, x_2^A, x_3^A)^T$ and B-frame coordinates $(x_1^B, x_2^B, x_3^B)^T$. The relationship between these coordinate descriptions is given by

$$x^A = Rx^B + v.$$

Let G be a matrix Lie group and let $\mathfrak{g} = T_I G$ be the corresponding Lie algebra. The dimensions of the differentiable manifold G and the vector space $\mathfrak{g}$ are of course the same, and there must be a one-to-one local correspondence between a neighborhood of 0 in $\mathfrak{g}$ and a neighborhood of the identity element I in G. One explicit local correspondence is provided by the exponential mapping $\exp : \mathfrak{g} \to G$, which we now describe.

Let $A \in \mathbb{R}^{n \times n}$ (the space of $n \times n$ matrices). We define $\exp(A)$ by the series

$$I + A + \frac{1}{2}A^2 + \frac{1}{3!}A^3 + \cdots. \qquad (2.8.2)$$

2.8.10 Proposition. *The series (2.8.2) is absolutely convergent.*

Proof. The ijth entry in the nth term of this matrix series is bounded in absolute value by $(n-1)\bar{a}^n/n!$, where $\bar{a} = \max_{ij}\{|a_{ij}|\}$. Hence, the ijth element in each term in the series is bounded in absolute value by the corresponding term in the absolutely convergent series $e^{\bar{a}n} = 1 + \bar{a}n + \frac{1}{2}\bar{a}^2 n^2 + \cdots$. Hence each entry in the series of matrices converges absolutely, proving the proposition. ∎

2.8.11 Proposition. *Let G be a matrix Lie group with corresponding Lie algebra $\mathfrak{g}$. If $A \in \mathfrak{g}$, then $\exp(At) \in G$ for all real numbers t.*

Group Actions. We now define the action of a Lie group G on a manifold M. Roughly speaking, a group action is a group of transformations of M indexed by elements of the group G and whose composition in M is compatible with group multiplication in G.

2.8.12 Definition. *Let M be a manifold and let G be a Lie group. A* **left action** *of a Lie group G on M is a smooth mapping $\Phi : G \times M \to M$ such that (i) $\Phi(e, x) = x$ for all $x \in M$, (ii) $\Phi(g, \Phi(h, x)) = \Phi(gh, x)$ for all $g, h \in G$ and $x \in M$, and (iii) $\Phi(g, \cdot)$ is a diffeomorphism for each $g \in G$.*

We often use the convenient notation gx for $\Phi(g, x)$ and think of the group element g *acting* on the point $x \in M$. The condition above then simply reads $(gh)x = g(hx)$.

Similarly, one can define a ***right action***, which is a map $\Psi : M \times G \rightarrow M$ satisfying $\Psi(x, e) = x$ and $\Psi(\Psi(x, g), h) = \Psi(x, gh)$.

Orbits.   Given a group action of G on M, for a given point $x \in M$, we let
$$\operatorname{Orb} x = \{gx \mid g \in G\},$$
called the ***group orbit*** through x. It can be shown that orbits are always smooth (possibly immersed) manifolds. This notion generalizes the notion of an orbit of a dynamical system, for the flow of a vector field on M can be thought of as an action of $\mathbb{R}$ on M, and in this case the general notion of orbit reduces to the familiar notion of orbit.

A simple example is the action of SO(3) on $\mathbb{R}^3$ given by matrix multiplication: The action of $A \in$ SO(3) on a point $x \in \mathbb{R}^3$ is simply the product Ax. In this case, the orbit of the origin is a single point (the origin itself), while the orbit of another point is the sphere through that point.

Infinitesimal Generator.   An important concept for mechanics is that of the infinitesimal generator of the group action:

2.8.13 Definition.   *Suppose $\Phi : G \times M \rightarrow M$ is an action. For $\xi \in \mathfrak{g}$, the map $\Phi^\xi : \mathbb{R} \times M \rightarrow M$ defined by $\Phi^\xi(t, x) = \Phi(\exp(t\xi), x)$ is an $\mathbb{R}$-action—that is, a flow—on M. The vector field on M that generates this flow, namely*
$$\xi_M(x) = \left.\frac{d}{dt}\right|_{t=0} \Phi^\xi(t, x). \tag{2.8.3}$$
*is called the **infinitesimal generator** of the action corresponding to ξ.*

A basic important identity relating the Jacobi–Lie bracket of generators to the Lie algebra bracket is as follows (see, for example, Marsden and Ratiu [1999] for the proof):
$$[\xi_Q, \eta_Q] = -[\xi, \eta]_Q. \tag{2.8.4}$$

Left- and Right-Invariant Vector Fields.   A Lie group acts on its tangent bundle by the tangent map. Given $\xi \in \mathfrak{g}$ we can consider the action of G on ξ either on the left or the right: $T_e L_g \xi$ or $T_e R_g \xi$, where L_g and R_g denote left and right translations, respectively; for example, $L_g : G \rightarrow G$ is the map given by $g' \mapsto gg'$. We can abbreviate these expressions and write $g\xi$ and ξg, respectively. For matrix Lie groups this action is just multiplication on the left or right.

Allowing $g \in G$ to vary over the group, the vectors $T_e L_g \xi$, $T_e R_g \xi$ define left- and right-invariant vector fields, that is, vector fields satisfying
$$(T_h L_g)\, X(h) = X(gh) \text{ or } (T_h R_g)\, X(h) = X(hg), \tag{2.8.5}$$
respectively. If we let $\xi_L(g) = T_e L_g \xi$, then the Jacobi–Lie bracket of two such left-invariant vector fields in fact gives the Lie algebra bracket:
$$[\xi_L, \eta_L](g) = [\xi, \eta]_L(g).$$

For the right-invariant case, one inserts a minus sign.

Spatial and Body Velocities.   There are two ways to pull back a tangent vector to a group to the identity. One can think of these as "body" or "spatial" velocities denoted by

$$\xi^b = \left(T_g L_{g^{-1}}\right) \dot{g} \quad \text{and} \quad \xi^s = \left(T_g R_{g^{-1}}\right) \dot{g}, \qquad (2.8.6)$$

respectively.

Adjoint and Coadjoint Actions.   We also define the **adjoint action** of G on its Lie algebra to be given by

$$\text{Ad}_g \xi = T_{g^{-1}} L_g \left(T_e R_{g^{-1}} \xi\right) \qquad (2.8.7)$$

for $\xi \in \mathfrak{g}$.

For matrix groups this is simply conjugation by the matrix g: $g\xi g^{-1}$. Thus $\xi^s = \text{Ad}_g \xi^b$.

The dual action $\text{Ad}^*_{g^{-1}}$ is called the **coadjoint action**.

Quotient Spaces and Equivariance.   If we have an action of a group G on M and the action is free (that is, if $gx = x$ for any x implies that g is the identity) and if the action is also proper (that is, the map $(g, x) \mapsto (g, gx)$ is a proper map: inverse images of compact sets are compact), then it can be shown (see, for example, Abraham and Marsden [1978]) that the space of orbits, denoted by M/G, is a smooth manifold and the natural projection $\pi : M \to M/G$ taking a point to its orbit is a smooth submersion.

If G acts on two manifolds M and N and if $f : M \to N$ is **equivariant**, that is, $f(gx) = gf(x)$, then f induces, in a natural way, a map of the quotients: $f_G : M/G \to N/G$.

There are similar statements for other equivariant objects. For example, let X be an **equivariant vector field** on M; that is, fixing g and denoting the map $x \mapsto gx$ by Φ_g, we have $\Phi^*_g X = X$. Then X induces, in a natural way, a vector field $X_{M/G}$ on M/G.

Exercises

◇ **2.8-1.**   A point P in $\mathbb{R}^3$ undergoes a rigid motion associated with $E(R_1, v_1)$ followed by a rigid motion associated with $E(R_2, v_2)$. What matrix element of SE(3) is associated with the composite of these motions in the given order?

◇ **2.8-2.**   A coordinate frame B is located with respect to a coordinate frame A as follows. B is initially coincident with A, but is displaced by the rigid motion associated with $E(R_1, v_1)$ and is then subsequently further displaced by $E(R_2, v_2)$. What matrix element of SE(3) is associated with the coordinate transformation from the A-frame to the B-frame? (That is, what matrix element of SE(3) is used to describe A-frame coordinates of a point in terms of the B-frame coordinates of the same point?)

⋄ **2.8-3.** Let $Y \in \mathfrak{sp}(l)$ be partitioned into $l \times l$ blocks,

$$Y = \left(\begin{array}{cc} A & B \\ C & D \end{array} \right).$$

Write down a complete set of equations involving A, B, C, and D that must be satisfied if $Y \in \mathfrak{sp}(l)$. Deduce that the dimension of $\mathfrak{sp}(l)$ as a real vector space is $2l^2 + l = n(n+1)/2$, and consequently, $\dim \mathrm{Sp}(l) = 2l^2 + l$.

⋄ **2.8-4.** Suppose the $n \times n$ matrices A and M satisfy $AM + MA^T = 0$. Show that $\exp(At)M \exp(A^T t) = M$ for all t. This direct calculation shows that for $A \in \mathfrak{so}(n)$ or $A \in \mathfrak{sp}(l)$, we have $\exp(At) \in \mathrm{SO}(n)$ or $\exp(At) \in \mathrm{Sp}(l)$, respectively.

2.9 Fiber Bundles and Connections

In this section we give a somewhat brief but, we hope, instructive treatment of fiber bundles and related concepts. We describe both theory and some illustrative examples. Our exposition is somewhat more explicit than is usual. References are given to more comprehensive treatments.

Fiber Bundles. Fiber bundles provide a basic geometric structure for the understanding of many mechanical and control problems, in particular for nonholonomic problems. References include Abraham, Marsden, and Ratiu [1988], Steenrod [1951], and Schutz [1980].
  A fiber bundle essentially consists of a given space (the base) together with another space (the fiber) attached at each point, plus some compatibility conditions. More formally, we have the following:

2.9.1 Definition. *A **fiber bundle** is a space Q for which the following are given: a space B called the **base space**, a **projection** $\pi : Q \to B$ with fibers $\pi^{-1}(b), b \in B$, homeomorphic to a space F, a **structure group** G of homeomorphisms of F into itself, and a **covering** of B by open sets U_j, satisfying*

  (i) *the bundle is locally trivial, i.e., $\pi^{-1}(U_j)$ is homeomorphic to the product space $U_j \times F$ and*

  (ii) *if h_j is the map giving the homeomorphism on the fibers above the set U_j, for any $x \in U_j \cap U_k$, $h_j(h_k^{-1})$ is an element of the structure group G.*

If the fibers of the bundle are homeomorphic to the structure group, we call the bundle a **principal bundle**.
  If the fibers of the bundle are homeomorphic to a vector space, we call the bundle a **vector bundle**.

2.9.2 Example. A basic example of a vector bundle is TS^1, the tangent bundle of the circle. The base is S^1, the fibers are homeomorphic to $\mathbb{R}$, and since the tangent space can be represented by any nonzero real number, the structure group is ratios of nonzero real numbers and may be identified with $\mathbb{R}\backslash\{0\}$.

The frame bundle of a manifold has the same structure group as TM, but the fibers are the set of all **bases** for the tangent space. Hence for TS^1 the fibers of the frame bundle are homeomorphic to its structure group $\mathbb{R}\backslash\{0\}$, and hence the frame bundle is a principal bundle. In fact, all frame bundles are principal.   ◆

Connections. An important additional structure on a bundle is a **connection** or **Ehresmann connection**; see, for example, Kobayashi and Nomizu [1963], Marsden, Montgomery, and Ratiu [1990], or Bloch, Krishnaprasad, Marsden, and Ratiu [1996]. We follow the treatment in the last of these here.

However, before we give the precise mathematical definitions we will give a somewhat intuitive discussion of the nature of and need for connections. A nice reference in this regard is the book by Burke [1985].

Suppose we have a bundle and consider (locally) a section of this bundle, i.e., a choice of a point in the fiber over each point in the base. We call such a choice a "field."

The idea is to single out fields that are "constant." For vector fields on the plane, for example, it is clear what we want such fields to be—they should be *literally* constant. For vector fields on a manifold or an arbitrary bundle, we have to specify this notion. Such fields are called "horizontal" and are also key to defining a notion of derivative, or rate of change of a vector field along a curve.[11] A connection is used to single out horizontal fields, and is chosen to have other desirable properties, such as linearity. For example, the sum of two constant fields should still be constant. As we shall see below, we can specify horizontality by taking a class of fields that are the kernel of a suitable form. Note that we do not in general have a metric; given one, there is a natural choice of connection and horizontality on the tangent bundle, as we shall see below.

More formally, we consider a bundle with projection map π and as usual let $T_q\pi$ denote its tangent map at any point. We call the kernel of $T_q\pi$ at any point the **vertical space** and denote it by V_q.

2.9.3 Definition. *An **Ehresmann connection** A is a vector-valued one-form on Q that satisfies:*

(i) *A is **vertical valued**: $A_q : T_qQ \to V_q$ is a linear map for each point $q \in Q$.*

---

[11] Recall that we already have a notion of derivative, namely the Lie derivative. However, Lie derivatives do not give one a way of differentiating vector fields along curves.

(ii) *A is a **projection**:* $A(v_q) = v_q$ *for all* $v_q \in V_q$.

The key property of the connection is the following: If we denote by H_q or hor$_q$ the kernel of A_q and call it the **horizontal space**, the tangent space to Q is the direct sum of the V_q and H_q; i.e., we can split the tangent space to Q into horizontal and vertical parts. For example, we can project a tangent vector onto its vertical part using the connection. Note that the vertical space at Q is tangent to the fiber over q.

Later on when we discuss nonholonomic systems we shall choose the connection so that the constraint distribution is the horizontal space of the connection.

Now define the bundle coordinates $q^i = (r^\alpha, s^a)$ for the base and fiber. The coordinate representation of the projection π is just projection onto the factor r, and the connection A can be represented locally by a vector-valued differential form ω^a:

$$A = \omega^a \frac{\partial}{\partial s^a}, \quad \text{where} \quad \omega^a(q) = ds^a + A^a_\alpha(r, s) dr^\alpha.$$

We can see this as follows: Let

$$v_q = \sum_\beta \dot{r}^\beta \frac{\partial}{\partial r^\beta} + \sum_b \dot{s}^b \frac{\partial}{\partial s^b}$$

be an element of $T_q Q$. Then $\omega^a(v_q) = \dot{s}^a + A^a_\alpha \dot{r}^\alpha$ and

$$A(v_q) = (\dot{s}^a + A^a_\alpha \dot{r}^\alpha) \frac{\partial}{\partial s^a}.$$

This clearly demonstrates that A is a projection, since when A acts again only ds^a results in a nonzero term, and this has coefficient unity.

2.9.4 Example.   It may be helpful to the reader to keep in mind here the physical example of the vertical rolling disk from Chapter 1. There it is natural to choose $r^1 = \theta$, $r^2 = \varphi$, $s^1 = x$, $s^2 = y$. Then the connection given by the constraints gives $\omega_1 = dx - \cos\varphi \, d\theta$ and $\omega_2 = dy - \sin\varphi \, d\theta$.

Note that we use a different notation, namely ω^a, for the local coordinate representation of the connection A for three reasons. First, it is common in the literature to use ω to stand for constraint one-forms. Second, in the preceding formula, it is standard to define the components of the connection A by A^a_α as shown, reflecting the fact that the connection is a projection; to distinguish this use of indices on A from the use of indices on the constraint one-forms, it is convenient to use a different letter. Third, we want to regard ω^a as (coordinate-dependent) differential forms, as opposed to A, which is a vertical-valued form; again, a different letter emphasizes this fact. Note in particular that the exterior derivative of A is not defined, but we can (locally) take the exterior derivative of ω^a. In fact, this will give an easy way to compute the curvature, as we shall see. ◆

Horizontal Lift.   Given an Ehresmann connection A, a point $q \in Q$, and a vector $v_r \in T_r B$ tangent to the base at a point $r = \pi(q) \in B$, we can define the **horizontal lift** of v_r to be the unique vector v_r^h in H_q that projects to v_r under $T_q \pi$. If we have a vector $X_q \in T_q Q$, we shall also write its **horizontal part** as

$$\text{hor } X_q = X_q - A(q) \cdot X_q.$$

In coordinates, the vertical projection is the map

$$(\dot{r}^\alpha, \dot{s}^a) \mapsto (0, \dot{s}^a + A_\alpha^a(r, s)\dot{r}^\alpha),$$

while the horizontal projection is the map

$$(\dot{r}^\alpha, \dot{s}^a) \mapsto (\dot{r}^\alpha, -A_\alpha^a(r, s)\dot{r}^\alpha).$$

Curvature.   Next, we give the basic notion of curvature.

2.9.5 Definition.   *The **curvature** of A is the vertical-vector-valued two-form B on Q defined by its action on two vector fields X and Y on Q by*

$$B(X, Y) = -A([\text{hor } X, \text{hor } Y]),$$

where the bracket on the right hand side is the Jacobi–Lie bracket of vector fields obtained by extending the stated vectors to vector fields.

One can show that curvature is independent of the extension of the vector fields.

Notice that this definition shows that the curvature exactly measures the failure of the horizontal distribution to be integrable.

Recall from equation (2.6.14) that we have the following useful identity for the exterior derivative $\mathbf{d}\alpha$ of a one-form α (which could be vector-space valued) on a manifold M acting on two vector fields X, Y:

$$(\mathbf{d}\alpha)(X, Y) = X[\alpha(Y)] - Y[\alpha(X)] - \alpha([X, Y]).$$

This identity shows that *in coordinates*, one can evaluate the curvature by writing the connection as a form ω^a in coordinates, computing its exterior derivative (component by component), and restricting the result to horizontal vectors, that is, to the constraint distribution. In other words,

$$B(X, Y) = d\omega^a (\text{hor } X, \text{hor } Y) \frac{\partial}{\partial s^a},$$

so that the local expression for curvature is given by

$$B(X, Y)^a = B_{\alpha\beta}^a X^\alpha Y^\beta, \tag{2.9.1}$$

where the coefficients $B_{\alpha\beta}^a$ are given by

$$B_{\alpha\beta}^b = \left(\frac{\partial A_\alpha^b}{\partial r^\beta} - \frac{\partial A_\beta^b}{\partial r^\alpha} + A_\alpha^a \frac{\partial A_\beta^b}{\partial s^a} - A_\beta^a \frac{\partial A_\alpha^b}{\partial s^a} \right). \tag{2.9.2}$$

2.9.6 Example (Connections on $T\mathbb{R}^1$). The idea of a connection can be illustrated by considering the simplest possible example: a connection on the bundle $TQ = T\mathbb{R}^1$ with coordinates $(x, \dot{x})$. We may define the horizontal space to be the kernel of the form

$$d\dot{x} + A_1^1(x, \dot{x})dx,$$

where A_1^1 is a smooth function of x and $\dot{x}$. More specifically, we can choose a connection that is linear in the velocities:

$$d\dot{x} + a(x)\dot{x}dx.$$

Here A is the $\mathbb{R}$-valued form

$$(d\dot{x} + a(x)\dot{x}dx)\frac{\partial}{\partial \dot{x}}.$$

Elements of T_qQ are of the form

$$v_q = \dot{x}\frac{\partial}{\partial x} + \ddot{x}\frac{\partial}{\partial \dot{x}},$$

and their projection onto the vertical space is

$$A(v_q) = (\ddot{x} + a(x)\dot{x}^2)\frac{\partial}{\partial \dot{x}}.$$

The kernel of A, i.e., the horizontal vectors, is the span of

$$\frac{\partial}{\partial x} - a(x)\dot{x}\frac{\partial}{\partial \dot{x}}.$$

Note that the standard choice is $a(x) = 0$; i.e., the standard horizontal space is the span of the vectors $\partial/\partial x$. ♦

Linear Connections, Affine Connections, and Geodesics. Here we consider how Ehresmann connections specialize to linear connections and affine connections defined in the tangent bundle, and we shall derive the geodesic equations. (As above, a good related reference for some of these ideas, but with a rather different approach, is Burke [1985].)

As above, we use bundle coordinates (r^α, s^a), and we specify the connection by the one-forms

$$\omega^a(q) = ds^a + A^a_\alpha(r, s)dr^\alpha,$$

and the action of A on a tangent vector $v_q = (\dot{r}^\alpha, \dot{s}^a)$ is given by

$$A(v_q) = (\dot{s}^a + A^a_\alpha \dot{r}^\alpha)\frac{\partial}{\partial s^a}. \tag{2.9.3}$$

For linear connections we require that the sum of two (local) horizontal sections be horizontal; i.e., if $(\dot{r}^\alpha, \dot{s}^a(r))$ and $(\dot{r}^\alpha, \dot{\hat{s}}^a(r))$ are horizontal, then so should be $\left(\dot{r}^\alpha, \dot{s}^a(r) + \dot{\hat{s}}^a(r)\right)$. Thus if we have

$$\dot{s}^a + A^a_\alpha(r,s)\dot{r}^\alpha = \dot{\hat{s}}^a + A^a_\alpha(r,\hat{s})\dot{r}^\alpha = 0,$$

then we require

$$\dot{s}^a + \dot{\hat{s}}^a + A^a_\alpha(r, s + \hat{s})\dot{r}^\alpha = 0.$$

Hence we take the connection coefficients be of the form

$$A^a_\alpha(r,s) = \Gamma^a_{\alpha b}(r)s^b. \tag{2.9.4}$$

If the bundle is the tangent bundle, these are called the **components** of the **affine connection** in the tangent bundle.

In the tangent bundle we have $s^a = \dot{r}^a$. We define **geodesic motion** along a curve $r(t)$ as being one for which the tangent vector is **parallel transported** along the curve; i.e., v_q along the curve is always horizontal, or $A(v_q)$ is zero. Making use of (2.9.3), this condition is

$$\ddot{r}^a + \Gamma^a_{bc}\dot{r}^b\dot{r}^b = 0. \tag{2.9.5}$$

This is the equation of geodesic motion. We can also determine this equation by another method, developed in what follows.

2.9.7 Example (Connections on $T\mathbb{R}^1$ continued). Returning to our system on $T\mathbb{R}^1$ suppose now that we have a curve $x(t)$ such that its tangent vector is parallel transported along the curve. In this case v_q along the curve being horizontal, or having $A(v_q)$ equal to zero, gives

$$\ddot{x} + a(x)\dot{x}^2 = 0.$$

For $a(x) = 0$ this reduces to $\ddot{x} = 0$, the equation of motion for a free particle on the line. Our example gives the generalization of this equation for arbitrary connections. ◆

Affine Connections and the Covariant Derivative. In the tangent bundle we can specify a linear connection by its action on vector fields, or by a map from vector fields (X, Y) to the vector field $\nabla_X Y$ that satisfies for smooth functions f and g and a vector fields X, Y, Z:

(i) $\nabla_{fX+gY} Z = f\nabla_X Z + g\nabla_Y Z$.

(ii) $\nabla_X(Y + Z) = \nabla_X Y + \nabla_X Z$.

(iii) $\nabla_X(fY) = f\nabla_X Y + (df \cdot X)Y$,

where $d f \cdot X$ is the directional derivative of f along X, or Lie derivative.

Given a basis of vector fields $\frac{\partial}{\partial r_j}$ we can represent ∇ by

$$\nabla_{\partial/\partial r_i} \frac{\partial}{\partial r_j} = \Gamma^k_{ij} \frac{\partial}{\partial r_k}. \tag{2.9.6}$$

For X, Y vector fields given locally by $X = X^i(\partial/\partial r_i)$, $Y = Y^i(\partial/\partial r_i)$, (i) and (iii) imply

$$\nabla_X Y = \left(X^j \frac{\partial Y^i}{\partial r^j} + X^k Y^j \Gamma^i_{kj} \right) \frac{\partial}{\partial r^i}. \tag{2.9.7}$$

The geodesic equations above then may be written

$$\nabla_{\dot r} \dot r = 0. \tag{2.9.8}$$

We can see this directly by a simple computation, again using (i) and (iii):

$$\nabla_{\dot r^i(\partial/\partial r_i)} \dot r^j \frac{\partial}{\partial r_j} = \dot r^i \nabla_{\partial/\partial r_i} \dot r^j \frac{\partial}{\partial r_j}$$

$$= \dot r^i \frac{\partial}{\partial r_i} \dot r^j \frac{\partial}{\partial r_j} + \dot r^i \dot r^j \Gamma^k_{ij} \frac{\partial}{\partial r_k}$$

$$= (\ddot r^j + \Gamma^j_{ik} \dot r^i \dot r^k) \frac{\partial}{\partial r_j} \quad \text{(by the chain rule)}.$$

Sometimes we will write

$$\nabla_{\dot r} \dot r = \frac{D^2 r}{dt^2}, \qquad \frac{DX}{dt} = \nabla_{\dot r(t)} X. \tag{2.9.9}$$

We define DX/dt to be the **covariant derivative**.

By (2.9.7), in local coordinates

$$\frac{DX}{dt} = \nabla_{\dot r} X = \left(\dot r^j \frac{\partial X^i}{\partial r^j} + \Gamma^i_{kj} X^k \dot r^j \right) \frac{\partial}{\partial r^i} = \left(\dot X^i + \Gamma^i_{kj} X^k \dot r^j \right) \frac{\partial}{\partial r^i}, \tag{2.9.10}$$

where $\dot r(t) = \dot r^i(\partial/\partial r^i)$. For $X = \dot r$ we of course recover the geodesic equations.

Curvature and Torsion. For an affine connection we define the curvature and torsion as follows. For X, Y, and Z arbitrary vector fields on M, the **curvature tensor** R and the **torsion tensor** are defined by

$$R(X,Y)(Z) = \nabla_X(\nabla_Y Z) - \nabla_Y(\nabla_X Z) - \nabla_{[X,Y]}(Z)$$

and

$$T(X,Y) = \nabla_X Y - \nabla_Y X - [X,Y].$$

Riemannian Connections. Now suppose M is endowed with a Riemannian metric g. This means that we can define orthonormal bases of $T_p(M)$ at each $p \in M$, and can define a subbundle P' of the frame bundle P whose fibers are orthonormal bases, and P' has structure group $O(n)$. This subbundle is said to be a **reduced bundle** of P.

There exists a unique affine connection on M, called the **Riemannian connection** or **Levi-Civita connection**, such that $\nabla g = 0$ and the torsion tensor T vanishes. An affine connection is called a metric connection if $\nabla g = 0$.

If the metric is given by $g = \sum g_{ij} dx^i dx^j$, the connection coefficients, which are called **Christoffel symbols**, are given by

$$\Gamma^i_{jk} = \frac{1}{2} g^{il} \left\{ \frac{\partial g_{jl}}{\partial x^k} + \frac{\partial g_{lk}}{\partial x^j} - \frac{\partial g_{jk}}{\partial x^l} \right\},$$

where, as usual, there is a sum over the index l understood.

Principal Connections. We now consider the special case of principal connections. We start with a free and proper group action of a Lie group G with Lie algebra $\mathfrak{g}$ on a manifold Q and construct the projection map $\pi : Q \to Q/G$; this setup is also referred to as a **principal bundle**. The kernel $\ker T_q \pi$ (the tangent space to the group orbit through q) is called the vertical space of the bundle at the point q and is denoted by ver_q.

2.9.8 Definition. *A **principal connection** on the principal bundle $\pi : Q \to Q/G$ is a map (referred to as the **connection form**) $\mathcal{A} : TQ \to \mathfrak{g}$ that is linear on each tangent space (i.e., $\mathcal{A}$ is a $\mathfrak{g}$-valued one-form) and is such that*

(i) *$\mathcal{A}(\xi_Q(q)) = \xi$ for all $\xi \in \mathfrak{g}$ and $q \in Q$, and*

(ii) *$\mathcal{A}$ is equivariant:*

$$\mathcal{A}(T_q \Phi_g(v_q)) = \text{Ad}_g \mathcal{A}(v_q)$$

for all $v_q \in T_q Q$ and $g \in G$, where Φ_g denotes the given action of G on Q and where Ad denotes the adjoint action of G on $\mathfrak{g}$.

The **horizontal space** of the connection at $q \in Q$ is the linear space

$$\text{hor}_q = \{ v_q \in T_q Q \mid \mathcal{A}(v_q) = 0 \}.$$

Thus, at any point, we have the decomposition

$$T_q Q = \text{hor}_q \oplus \text{ver}_q.$$

Often one finds connections defined by specifying the horizontal spaces (complementary to the vertical spaces) at each point and requiring that they transform correctly under the group action. In particular, notice that a connection is uniquely determined by the specification of its horizontal spaces, a fact that we will use later on. We will denote the projections onto the horizontal and vertical spaces relative to the above decomposition using the same notation; thus, for $v_q \in T_q Q$, we write

$$v_q = \mathrm{hor}_q\, v_q + \mathrm{ver}_q\, v_q.$$

The projection onto the vertical part is given by

$$\mathrm{ver}_q\, v_q = (\mathcal{A}(v_q))_Q(q),$$

and the projection to the horizontal part is thus

$$\mathrm{hor}_q\, v_q = v_q - (\mathcal{A}(v_q))_Q(q).$$

The projection map at each point defines an isomorphism from the horizontal space to the tangent space to the base; its inverse is called the **horizontal lift**. Using the uniqueness theory of ODEs one finds that a curve in the base passing through a point $\pi(q)$ can be lifted uniquely to a horizontal curve through q in Q (i.e., a curve whose tangent vector at any point is a horizontal vector).

Since we have a splitting, we can also regard a principal connection as a special type of Ehresmann connection. However, Ehresmann connections are regarded as vertical-valued forms, whereas principal connections are regarded as Lie-algebra-valued. Thus, the Ehresmann connection A and the connection one-form $\mathcal{A}$ are different, and we will distinguish them; they are related in this case by

$$A(v_q) = (\mathcal{A}(v_q))_Q(q).$$

The general notions of curvature and other properties that hold for general Ehresmann connections specialize to the case of principal connections. As in the general case, given any vector field X on the base space, using the horizontal lift, there is a unique vector field X^h that is horizontal and that is π-related to X; that is, at each point q, we have

$$T_q\pi \cdot X^h(q) = X(\pi(q)),$$

and the vertical part is zero:

$$(\mathcal{A}(X_q^h))_Q(q) = 0.$$

It is well known (see, for example, Abraham, Marsden, and Ratiu [1988]) that the relation of being π-related is bracket-preserving; in our case, this means that

$$\mathrm{hor}\,[X^h, Y^h] = [X, Y]^h,$$

where X and Y are vector fields on the base.

2.9.9 Definition.   *The **covariant exterior derivative** D of a Lie-algebra-valued one-form α is defined by applying the ordinary exterior derivative d to the horizontal parts of vectors:*

$$\mathbf{D}\alpha(X,Y) = \mathbf{d}\alpha(\text{hor } X, \text{hor } Y).$$

*The **curvature** of a connection $\mathcal{A}$ is its covariant exterior derivative, and it is denoted by $\mathcal{B}$.*

Thus, $\mathcal{B}$ is the Lie-algebra-valued two-form given by

$$\mathcal{B}(X,Y) = \mathbf{d}\mathcal{A}(\text{hor } X, \text{hor } Y).$$

Using the identity

$$(\mathbf{d}\alpha)(X,Y) = X[\alpha(Y)] - Y[\alpha(X)] - \alpha([X,Y])$$

together with the definition of horizontal shows that for two vector fields X and Y on Q, we have

$$\mathcal{B}(X,Y) = -\mathcal{A}([\text{hor } X, \text{hor } Y]),$$

where the bracket on the right-hand side is the Jacobi–Lie bracket of vector fields. The **Cartan structure equations** say that if X and Y are vector fields that are invariant under the group action, then

$$\mathcal{B}(X,Y) = \mathbf{d}\mathcal{A}(X,Y) - [\mathcal{A}(X), \mathcal{A}(Y)],$$

where the bracket on the right-hand side is the Lie algebra bracket. This follows readily from the definitions, the fact that $[\xi_Q, \eta_Q] = -[\xi, \eta]_Q$ (see equation (2.8.4)), the first property in the definition of a connection, and writing hor $X = X - \text{ver } X$, and similarly for Y, in the preceding formula for the curvature. The proof of the structure equations is given in the Internet supplement.

Remark.   Given a general distribution $\mathcal{D} \subset TQ$ on a manifold Q one can also define its curvature in an analogous way directly in terms of its lack of integrability. Define vertical vectors at $q \in Q$ to be the quotient space $T_q Q / \mathcal{D}_q$ and define the curvature acting on two horizontal vector fields u, v (that is, two vector fields that take their values in the distribution) to be the projection onto the quotient of their Jacobi–Lie bracket. One can check that this operation depends only on the point values of the vector fields, so indeed defines a two-form on horizontal vectors.

The Maurer–Cartan Equations.   A consequence of the structure equations relates curvature to the process of left and right trivialization.

2.9.10 Theorem (Maurer–Cartan Equations).   *Let G be a Lie group and let $\rho: TG \to \mathfrak{g}$ be the map that right translates vectors to the identity:*

$$\rho(v_g) = T_g R_{g^{-1}} \cdot v_g.$$

Then

$$\mathbf{d}\rho - [\rho, \rho] = 0.$$

Proof. Note that ρ is literally a connection on G for the left action. In considering this, keep in mind that for the action by left multiplication we have $\xi_Q(q) = T_e R_g \cdot \xi$. On the other hand, the curvature of this connection must be zero, since the shape space G/G is a point. Thus, the result follows from the structure equations. ∎

Of course, there is a similar result for the left trivialization λ, and we get the identity

$$\mathbf{d}\lambda + [\lambda, \lambda] = 0.$$

Bianchi Identities. The Bianchi identities are a famous set of identities for the Riemann curvature tensor of a given Riemannian metric. We defined the Riemann curvature tensor above for general affine connections. The relation between the Riemannian connection and the present formalism is to use the frame bundle as the bundle Q and think of it as a principal bundle over the underlying manifold M and the group $\mathrm{SO}(n)$ as the structure group. Then the curvature as defined here coincides with the Riemann curvature tensor. We will not go into this in detail here, since it is not needed for our present purposes, and instead we refer to Spivak [1979] or Kobayashi and Nomizu [1963] for an exposition of this. It is interesting that in the context of principal connections, the general proof is rather easy.

2.9.11 Theorem (Bianchi Identities). *We have the identity $D\mathcal{B} = 0$, that is, for any vector fields u, v, w on Q,*

$$\mathbf{d}\mathcal{B}(\mathrm{hor}(u), \mathrm{hor}(v), \mathrm{hor}(w)) = 0.$$

Proof. From the structure equations and the fact that $\mathbf{d}^2\mathcal{A} = 0$ we find that $\mathbf{d}\mathcal{B} = \mathbf{d}[\mathcal{A}, \mathcal{A}]$. Using the identity relating the exterior derivative and the Jacobi–Lie bracket of vector fields, we get

$$\begin{aligned}
(\mathbf{d}[\mathcal{A}, \mathcal{A}])(\mathrm{hor}(u), \mathrm{hor}(v), \mathrm{hor}(w)) = {}& \mathrm{hor}(u)[\,[\mathcal{A}, \mathcal{A}](\mathrm{hor}(v), \mathrm{hor}(w))] + \text{cyclic} \\
& - ([\mathcal{A}, \mathcal{A}])([\mathrm{hor}(u), \mathrm{hor}(v)], \mathrm{hor}(w)) \\
& - \text{cyclic}.
\end{aligned}$$

But all the terms in this expression are zero, since $\mathcal{A}$ vanishes on horizontal vectors. ∎

Local Formulas for the Connection. Pick a local trivialization of the bundle; that is, locally in the base, we write $Q = Q/G \times G$, where the action of G is given by left translation on the second factor. We choose coordinates r^α on the first factor and a basis e_a of the Lie algebra $\mathfrak{g}$ of G. We write coordinates of an element ξ relative to this basis as ξ^a. Let tangent vectors in this local trivialization at the point (r, g) be denoted by (u, w). We will write the action of $\mathcal{A}$ on this vector simply as $\mathcal{A}(u,w)$. Using this notation, we can write the connection form in this local trivialization as

$$\mathcal{A}(u,w) = \mathrm{Ad}_g(w_b + \mathcal{A}_{\mathrm{loc}}(r) \cdot u), \tag{2.9.11}$$

where w_b is the left translation of w to the identity (that is, the expression of w in "body coordinates"). The preceding equation defines the expression $\mathcal{A}_{\mathrm{loc}}(r)$. We define the connection components by writing

$$\mathcal{A}_{\mathrm{loc}}(r) \cdot u = A^a_\alpha u^\alpha e_a.$$

We can also phrase this local representation in the following way:

2.9.12 Proposition. *In local coordinates* $q = (g, r)$ *a principal connection one-form can be written as*

$$\mathcal{A} = \mathrm{Ad}_g\left(g^{-1}dg + A(r)dr\right), \tag{2.9.12}$$

so that

$$\mathcal{A} \cdot \dot{q} = \mathrm{Ad}_g\left(g^{-1}\dot{g} + A(r)\dot{r}\right), \tag{2.9.13}$$

where $g^{-1}\dot{g}$ *denotes the lifted action of* g^{-1} *on the tangent vector* $\dot{g}$.

Proof. The infinitesimal generator of the action of the group on itself by the left action is of the form $\xi_G(g) = \xi g$, where again we are using shorthand for the lifted action. (Note that this is a push forward of ξ by the right action!)

Condition (i) of definition 2.9.8 then implies that

$$\mathcal{A}(r,g) \cdot (0, \xi g) = \xi.$$

Writing

$$(0, \xi g) = \xi g \frac{\partial}{\partial g}$$

we see that this holds if

$$\mathcal{A} = \mathrm{Ad}_g\left(g^{-1}dg\right) + A(r,g)dr.$$

Thus $\mathcal{A} \cdot \dot{q}$ must be of the form

$$\mathcal{A}(q) \cdot (\dot{r}, \dot{g}) = \dot{g}g^{-1} + A(r,g)\dot{r}.$$

Now the equivariance condition (ii) of definition 2.9.8 implies

$$\mathrm{Ad}_h A(g,r) = A(hg, r).$$

Setting $h = g^{-1}$ this gives $\mathrm{Ad}_{g^{-1}} A(r,g) = A(r,e)$ or $A(r,g) = \mathrm{Ad}_g A(r,e) \equiv \mathrm{Ad}_g A(r)$, giving the result. ∎

Local Formulas for the Curvature. Similarly, the curvature can be written in a local representation as

$$\mathcal{B}((u_1, w_1), (u_2, w_2)) = \text{Ad}_g(\mathcal{B}_{\text{loc}}(r) \cdot (u_1, u_2)),$$

which again serves to define the expression $\mathcal{B}_{\text{loc}}(r)$. We can also define the coordinate form for the local expression of the curvature by writing

$$\mathcal{B}_{\text{loc}}(r) \cdot (u_1, u_2) = \mathcal{B}^a_{\alpha\beta} u_1^\alpha u_2^\beta e_a.$$

Then one has the formula

$$\mathcal{B}^b_{\alpha\beta} = \left(\frac{\partial \mathcal{A}^b_\beta}{\partial r^\alpha} - \frac{\partial \mathcal{A}^b_\alpha}{\partial r^\beta} - C^b_{ac} \mathcal{A}^a_\alpha \mathcal{A}^c_\beta \right),$$

where C^b_{ac} are the structure constants of the Lie algebra defined by

$$[e_a, e_c] = C^b_{ac} e_b.$$

Parallel Translation and Holonomy Groups. Let P be a principal bundle with a connection and C a piecewise differentiable curve in its base space M with beginning point p and endpoint q. Suppose x is a point on the fiber over p. Then there is a unique curve C^*_x in P starting at x such that $\pi(C^*_x) = C$ and each tangent vector to C^*_x is horizontal. The curve C^*_x is said to be a *lift* of C that starts at x, and the map that takes x to the lift of q, the endpoint of the lifted curve, is said to be **parallel translation.**

Now suppose C is a closed curve starting at p. Parallel translation then maps the point x to a point in the same fiber over p, xa say, $a \in G$. Thus each closed curve at p and fiber point x determines an element of G, and the set of all such elements forms a subgroup of G called the **holonomy group** of the connection with reference point x.

Holonomy for the Heisenberg Control System. A nice example of holonomy in action is for the Heisenberg control system:

$$\dot{x} = u_1,$$
$$\dot{y} = u_2, \qquad\qquad (2.9.14)$$
$$\dot{z} = xu_2 - yu_1.$$

Here we consider the bundle $\mathbb{R}^3$ with base the xy-plane, fiber z, and connection

$$A = (dz - xdy + ydx)\frac{\partial}{\partial z}. \qquad\qquad (2.9.15)$$

A horizontal curve has tangent vectors $(\dot{x}, \dot{y}, x\dot{y} - y\dot{x})$.

Now suppose we consider a loop in the base with z starting at the point z_0. Then the final position in the fiber, z_f, is given by

$$z_f - z_0 = \oint x\,dy - y\,dx. \qquad\qquad (2.9.16)$$

By Green's theorem the right-hand side is just $2A$, where A is the area of the loop! Hence the term "nonholonomic integrator." This is also sometimes referred to as the **area rule**. Recalling also our analysis of Lie brackets, note that that if the loop is a square with sides of length ϵ, then $A = \epsilon^2$.

We will see more analysis of this in Chapter 6, where we analyze the control of nonholonomic systems. For more on holonomy and phases, see Chapter 3.

Exercises

◇ **2.9-1.**   Consider the trivial bundle $\mathbb{R}^4$ with base $\mathbb{R}^2$ parametrized by coordinates (θ, φ) and fibers $\mathbb{R}^2$ parametrized by coordinates (x, y). Compute the curvature of the connection given by the vertical rolling disk constraints $\omega_1 = dx - \cos\varphi\, d\theta$ and $\omega_2 = dy - \sin\varphi\, d\theta$.

◇ **2.9-2.**   Consider the same space as above but with connection given by the integrable constraints $\omega_1 = dx - \cos\theta\, d\theta$ and $\omega_2 = dy - \sin\theta\, d\theta$. Show that the curvature of this connection is zero.

3
Basic Concepts in Geometric Mechanics

In this chapter we develop and summarize some basic concepts in the geometric mechanics of holonomic systems (see Chapter 1), which provide background for parts of the rest of this book. Readers well versed in this material may omit part or all of it. Parts of the rest of the book can be read without this material, so the reader can return to it as the need arises.

Geometric Mechanics. The geometric view of mechanics has a rich history, going back to the founders of mechanics, but especially Euler, Lagrange, Hamilton, Jacobi, and Poincaré. After a historical lull starting about 1910, there have been many modern developments that blossomed starting about 1950. One of the goals, recognizing the fundamental work of Lagrange and Hamilton, was to make the covariance properties of mechanics explicit by working in the context of manifolds from the beginning. The two branches, Hamiltonian and Lagrangian mechanics, have led to two possible starting points for mechanics.

The Lagrangian side of mechanics focuses on variational principles for its basic formulation, while the Hamiltonian side focuses on geometric structures called symplectic or Poisson structures. These two sides interact with each other, at least on the most basic level, through what is called the Legendre transformation.

We shall begin (for no particular reason) with the Hamiltonian side of the story. The reader is referred to one of the standard books, such as Abraham and Marsden [1978], Arnold [1989], Guillemin and Sternberg [1984], Libermann and Marle [1987], or Marsden and Ratiu [1999], for proofs that we have omitted.

3.1 Symplectic and Poisson Manifolds and Hamiltonian Flows

3.1.1 Definition. *Let P be a manifold and let $\mathcal{F}(P)$ denote the set of smooth real-valued functions on P. Consider a bracket operation denoted by*

$$\{\,,\} : \mathcal{F}(P) \times \mathcal{F}(P) \to \mathcal{F}(P).$$

*The pair $(P, \{\,,\})$ is called a **Poisson manifold** if $\{\,,\}$ satisfies:*

(PB1) **bilinearity** $\{f, g\}$ *is bilinear in f and g.*
(PB2) **anticommutativity** $\{f, g\} = -\{g, f\}$.
(PB3) **Jacobi's identity** $\{\{f, g\}, h\} + \{\{h, f\}, g\} + \{\{g, h\}, f\} = 0$.
(PB4) **Leibniz's rule** $\{fg, h\} = f\{g, h\} + g\{f, h\}$.

Notice that conditions (PB1)–(PB3) make $(\mathcal{F}(P), \{\,,\})$ into a Lie algebra.

If $(P, \{\,,\})$ is a Poisson manifold, then one can show that because of (PB1) and (PB4), there is a tensor B on P assigning to each $z \in P$ a linear map $B(z) : T_z^* P \to T_z P$ such that

$$\{f, g\}(z) = \langle B(z) \cdot \mathbf{d}f(z), \mathbf{d}g(z) \rangle. \tag{3.1.1}$$

Here $\langle\,,\rangle$ denotes the natural pairing between vectors and covectors. Because of (PB2), $B(z)$ is antisymmetric. Letting z^I, $I = 1, \ldots, M$, denote coordinates on P, (3.1.1) becomes

$$\{f, g\} = B^{IJ} \frac{\partial f}{\partial z^I} \frac{\partial g}{\partial z^J}. \tag{3.1.2}$$

(By our summation convention, there is a summation understood on repeated indices). Antisymmetry means that $B^{IJ} = -B^{JI}$, and Jacobi's identity reads

$$B^{LI} \frac{\partial B^{JK}}{\partial z^L} + B^{LJ} \frac{\partial B^{KI}}{\partial z^L} + B^{LK} \frac{\partial B^{IJ}}{\partial z^L} = 0. \tag{3.1.3}$$

3.1.2 Definition. *The pair $(P, \{\,,\})$ is called an **almost Poisson manifold** if all the conditions of Definition 3.1.1 hold except (PB3) (Jacobi's identity).*

As we shall see later, in Chapter 5, the notion of almost Poisson structures comes up in nonholonomic systems, and the failure of Jacobi's identity is related to the nonintegrability of the constraints that we have seen already in Chapter 1. See Cannas Da Silva and Weinstein [1999] for more mathematical information about almost Poisson manifolds.

3.1.3 Definition. *Let $(P_1, \{\,,\}_1)$ and $(P_2, \{\,,\}_2)$ be Poisson manifolds. A mapping $\varphi : P_1 \to P_2$ is called **Poisson** if for all $f, h \in \mathcal{F}(P_2)$, we have*

$$\{f, h\}_2 \circ \varphi = \{f \circ \varphi, h \circ \varphi\}_1. \tag{3.1.4}$$

In other words, Poisson maps are maps that preserve the Poisson structure in the obvious way. We define almost Poisson maps in the same way relative to almost Poisson structures. Often, Poisson structures come from symplectic structures, which are defined next.

3.1.4 Definition. *Let P be a manifold and Ω a 2-form on P. The pair (P, Ω) is called a **symplectic manifold** if Ω satisfies*

(S1) $\mathbf{d}\Omega = 0$ *(i.e., Ω is closed) and*

(S2) Ω *is nondegenerate.*

In this context we can define the abstract notion of Hamilton's equations, whose solution curves will be integral curves of Hamiltonian vector fields.

3.1.5 Definition. *Let (P, Ω) be a symplectic manifold and let $f \in \mathcal{F}(P)$. Let X_f be the unique vector field on P satisfying*

$$\Omega_z(X_f(z), v) = \mathbf{d}f(z) \cdot v \quad \text{for all} \quad v \in T_z P. \tag{3.1.5}$$

*We call X_f the **Hamiltonian vector field** of f. **Hamilton's equations** are the differential equations on P given by*

$$\dot{z} = X_f(z). \tag{3.1.6}$$

*If (P, Ω) is a symplectic manifold, define the **Poisson bracket operation** $\{\cdot, \cdot\} : \mathcal{F}(P) \times \mathcal{F}(P) \to \mathcal{F}(P)$ by*

$$\{f, g\} = \Omega(X_f, X_g). \tag{3.1.7}$$

The construction (3.1.7) makes $(P, \{\,,\,\})$ into a Poisson manifold. In other words, if a manifold is symplectic, then it is also Poisson.

3.1.6 Proposition. *Every symplectic manifold is Poisson.*

The converse is not true; for example, the zero bracket makes any manifold Poisson. A nontrivial example of Poisson brackets that are not symplectic is the Lie–Poisson structure associated with the rigid body, a notion that is defined in Section 3.5.

Hamiltonian vector fields are defined on Poisson manifolds as follows.

3.1.7 Definition. *Let $(P, \{\,,\,\})$ be a Poisson manifold and let $f \in \mathcal{F}(P)$. Define X_f to be the unique vector field on P satisfying*

$$X_f[k] := \langle \mathbf{d}k, X_f \rangle = \{k, f\} \quad \text{for all} \quad k \in \mathcal{F}(P).$$

*We call X_f the **Hamiltonian vector field** of f.*

A check of the definitions shows that in the symplectic case, Definitions 3.1.5 and 3.1.7 of Hamiltonian vector fields coincide.

If $(P, \{\,,\,\})$ is a Poisson manifold, there are therefore three equivalent ways to write Hamilton's equations for $H \in \mathcal{F}(P)$, as is readily verified:

(i) $\dot{z} = X_H(z)$.

(ii) $\dot{f} = \mathbf{d}f(z) \cdot X_H(z)$ for all $f \in \mathcal{F}(P)$.

(iii) $\dot{f} = \{f, H\}$ for all $f \in \mathcal{F}(P)$.

The Flow of a Hamiltonian Vector Field. Hamilton's equations described in the abstract setting above are very general. They include not only what one normally thinks of as Hamilton's canonical equations in classical mechanics, but Schrödinger's equation in quantum mechanics as well. Despite this generality, the theory has a rich structure.

Let $H \in \mathcal{F}(P)$, where P is a Poisson manifold. Let φ_t be the flow of Hamilton's equations; thus $\varphi_t(z)$ is the integral curve of $\dot{z} = X_H(z)$ starting at z. (If the flow is not complete, restrict attention to its domain of definition.) There are two basic facts about Hamiltonian flows given in the next proposition.

3.1.8 Proposition. *The following hold for Hamiltonian systems on Poisson manifolds:*

(i) *Each φ_t is a Poisson map.*

(ii) *$H \circ \varphi_t = H$ (conservation of energy).*

We refer to the basic references given earlier for the proof. One should note that the first part of this proposition is true even if H is a time-dependent Hamiltonian, while the second part is true only when H is independent of time.

In case (P, Ω) is symplectic, a diffeomorphism $\varphi : P \to P$ is Poisson iff it is symplectic; that is, $\varphi^* \Omega = \Omega$. In particular, we get

3.1.9 Proposition. *The flow of a Hamiltonian vector field on a symplectic manifold consists of symplectic diffeomorphisms.*

One direct way to prove this is to use the Lie derivative and Cartan's magic formula:

$$\frac{d}{dt}\varphi_t^* \Omega = \varphi_t^* \pounds_{X_H} \Omega$$
$$= \varphi_t^* \left(\mathbf{i}_{X_H} \mathbf{d}\Omega + \mathbf{d}\mathbf{i}_{X_H} \Omega \right)$$
$$= \varphi_t^* \left(\mathbf{d}\mathbf{d}H \right) = 0.$$

Since φ_t is the identity at $t = 0$, we get $\varphi_t^* \Omega = \Omega$ for all t.

If P is $2n$ dimensional, $\mu = \Omega^n$ is a volume element and so we get

3.1.10 Corollary (Liouville's Theorem). *The flow of a Hamiltonian vector field consists of volume preserving maps.*

It is also instructive to prove this by noting that in canonical coordinates,

$$X_H(q, p) = \left(\frac{\partial H}{\partial p_i}, -\frac{\partial H}{\partial q_i} \right)$$

which has zero divergence by equality of mixed partial derivatives.

Exercises

◇ **3.1-1.** Consider $\mathbb{R}^{2n}$ with coordinates $(q^1, \ldots, q^n, p_1, \ldots, p_n)$. Show that $\omega = \sum_{i=1}^{n} dq^i \wedge dp_i$ is a symplectic form on $\mathbb{R}^{2n}$.

◇ **3.1-2.** Consider the subset $T \subset \mathbb{R}^{2n}$ defined to be the set of all points $(q^1, \ldots, q^n, p_1, \ldots, p_n) \in \mathbb{R}^{2n}$ such that $q^1 > 0, \ldots, q^n > 0$. Show that

$$\omega = \sum_{i=1}^{n} \sum_{j=i}^{n} \frac{dq^j}{q^j} \wedge dp_i$$

is a symplectic form on T. Write out the Poisson brackets of the coordinate functions and a formula for the Poisson bracket of two functions f, g of $(q^1, \ldots, q^n, p_1, \ldots, p_n)$.

3.2 Cotangent Bundles

Let Q be a given manifold (usually the configuration space of a mechanical system) and T^*Q its cotangent bundle. Coordinates q^i on Q induce, in a natural way, coordinates (q^i, p_j) on the cotangent bundle T^*Q, called the **canonical cotangent coordinates** of T^*Q.

3.2.1 Proposition. *There is a unique 1-form Θ on T^*Q such that in any choice of canonical cotangent coordinates,*

$$\Theta = p_i dq^i; \tag{3.2.1}$$

Θ *is called the* **canonical 1-form.** *We define the* **canonical 2-form** Ω *by*

$$\Omega = -\mathbf{d}\Theta = dq^i \wedge dp_i \quad (a \; sum \; on \; i \; is \; understood). \tag{3.2.2}$$

To clarify its covariant nature, one should find an intrinsic definition of Θ, and there are many such. One of these is to require the identity $\beta^*\Theta = \beta$ for any one-form $\beta : Q \to T^*Q$. Another is the definition

$$\Theta(w_{\alpha_q}) = \langle \alpha_q, T\pi_Q \cdot w_{\alpha_q} \rangle,$$

where $\alpha_q \in T_q^*Q$, $w_{\alpha_q} \in T_{\alpha_q}(T^*Q)$, and where $\pi_Q : T^*Q \to Q$ is the cotangent bundle projection.

In summary, we have the following result:

3.2.2 Proposition. (T^*Q, Ω) *is a symplectic manifold.*

In canonical coordinates the Poisson brackets on T^*Q have the classical form

$$\{f, g\} = \frac{\partial f}{\partial q^i} \frac{\partial g}{\partial p_i} - \frac{\partial g}{\partial q^i} \frac{\partial f}{\partial p_i}, \tag{3.2.3}$$

where summation on repeated indices is understood.

3.2.3 Theorem. (**Darboux's Theorem**) *Every symplectic manifold locally looks like* T^*Q; *in other words, on every finite-dimensional symplectic manifold there are local coordinates in which* Ω *has the form (3.2.2).*[1]

Hamilton's equations in these canonical coordinates have the classical form

$$\dot{q}^i = \frac{\partial H}{\partial p_i}, \qquad \dot{p}_i = -\frac{\partial H}{\partial q^i}, \qquad (3.2.4)$$

as one can readily check.

The local structure of Poisson manifolds is more complex than what one obtains in the symplectic case. However, every Poisson manifold is the union of **symplectic leaves**; to compute the bracket of two functions in P, one may do it **leaf-wise**. In other words, to calculate the bracket of f and g at $z \in P$, select the symplectic leaf S_z through z, and evaluate the bracket of $f|S_z$ and $g|S_z$ at z. We shall see a specific case of this method in Section 3.5.

Exercise

⋄ **3.2-1.** Compute the Lagrangian and Hamiltonian equations of motion for a particle on the surface on the n-sphere. Describe the phase space. Hint: To compute the motion it is helpful to view the system as an example of constrained (holonomic) motion.

3.3 Lagrangian Mechanics and Variational Principles

Let Q be a manifold and TQ its tangent bundle. Coordinates q^i on Q induce coordinates $(q^i, \dot{q}^i)$ on TQ, called **tangent coordinates**. A mapping $L : TQ \to \mathbb{R}$ is called a **Lagrangian**. Often we choose L to be $L = K - V$, where $K(v) = \frac{1}{2}\langle v, v \rangle$ is the **kinetic energy** associated with a given Riemannian metric and where $V : Q \to \mathbb{R}$ is the **potential energy**.

3.3.1 Definition. *Hamilton's principle singles out particular curves* $q(t)$ *by the condition*

$$\delta \int_b^a L(q(t), \dot{q}(t))dt = 0, \qquad (3.3.1)$$

where the variation is over smooth curves in Q *with fixed endpoints.*

The precise meaning of the variations was discussed in Section 1.2.

---

[1] See Marsden [1981] and Olver [1988] for a discussion of the infinite-dimensional case.

It is interesting to note that (3.3.1) is unchanged if we replace the integrand by $L(q, \dot{q}) - \frac{d}{dt}S(q, t)$ for any function $S(q, t)$. This reflects the **gauge invariance** of classical mechanics and is closely related to Hamilton–Jacobi theory. It is also interesting to note that if one keeps track of the boundary conditions in Hamilton's principle, they essentially *define* the canonical one-form $p_i dq^i$. This turns out to be a useful remark in numerical algorithms as well as in more complex field theories (see Marsden, Patrick, and Shkoller [1998] and Marsden and West [2001]).

If one prefers, the action principle may be stated as follows: The map I defined by $I(q(\cdot)) = \int_a^b L(q(t), \dot{q}(t))dt$ from the space of curves with prescribed endpoints in Q to $\mathbb{R}$ has a critical point at the curve in question. In any case, a basic and elementary result of the calculus of variations, whose proof was already sketched in Chapter 1, is contained in the following proposition:

3.3.2 Proposition. *The principle of critical action for a curve $q(t)$ is equivalent (assuming sufficient regularity) to the condition that $q(t)$ satisfies the **Euler–Lagrange equations***

$$\frac{d}{dt}\frac{\partial L}{\partial \dot{q}^i} - \frac{\partial L}{\partial q^i} = 0. \tag{3.3.2}$$

3.3.3 Definition. *Let L be a Lagrangian on TQ and let $\mathbb{F}L : TQ \to T^*Q$ be defined (in coordinates) by*

$$(q^i, \dot{q}^j) \mapsto (q^i, p_j),$$

*where $p_j = \partial L/\partial \dot{q}^j$. We call $\mathbb{F}L$ the **fiber derivative**. (Intrinsically, $\mathbb{F}L$ differentiates L in the fiber direction.)*

A Lagrangian L is called **hyperregular** *if $\mathbb{F}L$ is a diffeomorphism. If L is a hyperregular Lagrangian, we define the corresponding **Hamiltonian** by*

$$H(q^i, p_j) = p_i \dot{q}^i - L.$$

*The change of data from L on TQ to H on T^*Q is called the **Legendre transform**.*

One checks that the Euler–Lagrange equations for L are equivalent to Hamilton's equations for H.

Second-Order Systems. Consider the projection

$$\pi_Q : TQ \to Q$$

and assume that L is hyperregular. Define the function H_L on TQ by

$$H_L(q, \dot{q}) = \mathbb{F}L(q, \dot{q})\dot{q} - L(q, \dot{q}), \tag{3.3.3}$$

so that

$$H = H_L \circ (\mathbb{F}L)^{-1}. \tag{3.3.4}$$

We say that a vector field $X \in \Gamma(TTQ)$ is of *second order* if

$$\pi_{Q*}X_{(q,\dot{q})} = \dot{q}, \qquad \dot{q} \in T_q Q.$$

Setting $\Omega_L = \mathbb{F}L^*\Omega$, we obtain a symplectic form on TQ, since we assumed that $\mathbb{F}L$ is a diffeomorphism. We may now define the Lagrangian vector field $X_L \in \Gamma(TTQ)$, corresponding to the Hamiltonian (3.3.3) as the second-order vector field satisfying

$$\Omega_L(X_L, Z) = \mathbf{d}H_L(Z), \tag{3.3.5}$$

where Z is an arbitrary vector field on TQ. It is easily verified that X_L is related to the Hamiltonian vector field X_H, with Hamiltonian H defined in (3.3.4), by the relation

$$\mathbb{F}L_*X_L = X_H.$$

The system of Lagrangian equations (3.3.2) may now be abstracted by the dynamical system on TQ given by

$$\dot{v} = X_L(v), \qquad v \in TQ. \tag{3.3.6}$$

A vector field Z on TQ is said to be *vertical* if $\pi_{Q*}Z = 0$. Using local coordinates one easily shows that (3.3.5) is satisfied identically for all vertical vector fields and any second-order vector field X_L.

Additional Holonomic Constraints. In problem (3.3.1), if we assume that there are additional holonomic constraints of the form $\phi_i(q) = 0$, $1 \le i \le m$, then we can extend the principle of critical action to the augmented variational problem with Lagrange multipliers

$$\delta \int_a^b \left(L(q, \dot{q}) + \sum_{i=1}^m \lambda_i \phi_i(q) \right) dt = 0, \tag{3.3.7}$$

where the variations are subject to the condition that q is a smooth curve in Q satisfying $\phi_i(q) = 0$, $1 \le i \le m$, and $q(a) = q_a$, as well as the endpoint conditions $q(b) = q_b$, $\phi_i(q_a) = \phi_i(q_b) = 0$, $1 \le i \le m$. Set $(q(t), \dot{q}(t)) = \bar{q}(t)$.

Necessary conditions may therefore be written as

$$\mathbf{i}_{\dot{q}}\Omega_L - \mathbf{d}H_L + \sum_{i=1}^m \lambda_i \pi_Q^* \mathbf{d}\phi_i = 0 \tag{3.3.8}$$

with $\phi_i(q) = 0$, $1 \le i \le m$.

Suppose that the set

$$Q_c = \{ q \in Q \mid \phi_i(q) = 0 \} \subset Q$$

is a regular smooth submanifold of Q. If $i : Q_c \to Q$ is the inclusion mapping, then $\bar{i} = i_* : TQ_c \to TQ$ and $\bar{i}^* \pi_Q^* \mathbf{d}\phi_i = \mathbf{d}(\phi_i \circ \pi_Q \circ \bar{i}) \equiv 0$. It follows that the necessary conditions (3.3.8) are satisfied by simply pulling back via $\bar{i}$ to Q_c to obtain a Lagrangian system on Q_c with Lagrangian $L \circ \bar{i}$. Thus, as long as the holonomic constraints are sufficiently regular, we may remove the constraints by restricting to the space defined by the constraints.

More general nonholonomic constraints will be considered later on, in Chapter 5.

Lagrangian Submanifolds. There is another mechanism by which one may characterize Hamiltonian systems. Suppose that M is a manifold that admits a nondegenerate two-form Ω. A distribution D on a submanifold $N \subset M$ is said to be *isotropic (coisotropic)* if

$$D_x^\perp = \{Y_x \in T_x M \mid \Omega(D_x, Y_x) = 0\}, \quad x \in N,$$

satisfies $D_x^\perp \subset D_x$ $(D_x^\perp \supset D_x)$ for all $x \in N \subset M$. A distribution D is said to be *Lagrangian* if D is both isotropic and coisotropic. In this case $\mathrm{Dim}\, D_x = \frac{1}{2}\,\mathrm{Dim}\, M$. A submanifold $N \subset M$, with inclusion map $i : N \to M$, is said to be *Lagrangian* when Ω is a symplectic form and $i_* TN$ is a Lagrangian distribution on $N \subset M$. In the case $M = T^*Q$, there is a natural symplectic form ω, and an induced symplectic form $\dot{\omega}$ on TT^*Q given in local coordinates by

$$\dot{\omega} = \sum_i d\dot{q}_i \wedge dp_i + \sum_i dq_i \wedge d\dot{p}_i.$$

If X_H is a Hamiltonian vector field on T^*Q, then set

$$N = \{(x, X_H(x)) \mid x \in T^*Q\} \subset TT^*Q.$$

It turns out that N is a Lagrangian submanifold of TT^*Q (with symplectic form $\dot{\omega}$), and every Lagrangian submanifold of TT^*Q is locally parametrized by a Hamiltonian vector field on T^*Q. Thus Hamiltonian vector fields on T^*Q are equivalently defined by the Lagrangian submanifolds of TT^*Q that are parametrized by a single Hamiltonian function on T^*Q. See Weinstein [1971], Abraham and Marsden [1978], van der Schaft [1982, 1983], and, for the Poisson case, Sanchez [1986] for details. We note that this is also the setting for the generalized Legendre transformation theory of Tulczyjew [1977].

Invariance under coordinate changes and Rayleigh Dissipation. An advantage of Lagrangian models of mechanical system dynamics is their manifest invariance with respect to coordinate changes. One can extend also these models to include dissipation by defining a *dissipation function* $D(q, \dot{q})$ such that

$$\dot{q}^T D\dot{q} = \textit{rate of dissipation of energy per second.}$$

We generally assume that dissipation functions are quadratic, symmetric, and positive definite with respect to the generalized velocity variables $\dot{q}$. Letting $L(q, \dot{q})$ be the Lagrangian of the system of interest, the dissipative equations of motion are given locally by

$$\frac{d}{dt}\frac{\partial L}{\partial \dot{q}} - \frac{\partial L}{\partial q} + \frac{\partial D}{\partial \dot{q}} = 0. \tag{3.3.9}$$

We have the following results which are easy to check:

3.3.4 Theorem. *If $E(q, \dot{q})$ is the total energy of the system, then*

$$\frac{d}{dt}E(q, \dot{q}) = -\dot{q}^T\frac{\partial D}{\partial \dot{q}}.$$

3.3.5 Theorem. *The dissipative Lagrangian system is invariant under a change of coordinates $q = Q(q)$. In particular, if the system dynamics is given by a Lagrangian $L(q, \dot{q})$ and dissipation function $D(q, \dot{q})$, with corresponding equation of motion (3.3.9), then the same system dynamics is prescribed in terms of Q-coordinates by a Lagrangian $\mathcal{L}(Q, \dot{Q})$, dissipation function $\mathcal{D}(Q, \dot{Q})$, and equations of motion*

$$\frac{d}{dt}\frac{\partial \mathcal{L}}{\partial \dot{Q}} - \frac{\partial \mathcal{L}}{\partial Q} + \frac{\partial \mathcal{D}}{\partial \dot{Q}} = 0.$$

This type of rate-dependent dissipation is often called **Rayleigh dissipation**.

For a discussion of Rayleigh dissipation on manifolds see Bloch, Krishnaprasad, Marsden, and Ratiu [1996].

3.4 Mechanical Systems with External Forces

Most mechanical systems interact with the environment through (generalized) forces. The basic mechanisms that describe these interactions are Newton's laws and the Lagrange–d'Alembert principle. We describe below very briefly the Lagrange–d'Alembert principle and its relationship with constrained dynamics. We will return to this topic in Chapter 5.

Newton's Laws. To describe a version of Newton's second law consistent with our configuration space modeled by the manifold Q, we introduce axiomatically a bundle isomorphism

$$\mathcal{P} : TQ \rightarrow T^*Q, \tag{3.4.1}$$

where $\mathcal{P}_q : T_qQ \rightarrow T_q^*Q$ defines the relationship between the phase velocity $\dot{q} \in T_qQ$ and the momentum $p \in T_q^*Q$:

$$p = \mathcal{P}_q(\dot{q}). \tag{3.4.2}$$

For example, for a single particle on a line with mass m this isomorphism is simply given by the map $(q, v) \mapsto (q, mv)$, and more generally, it is given by the Legendre transformation.

We first need a notion of uniform motion given by Newton's first law. We model this as the flow on TQ induced by a second-order vector field X_0 on TQ, which we take to be the geodesic flow with respect to a suitable metric on Q (see Chapter 2). Now, $\mathcal{P}_* : TTQ \to TT^*Q$ is also a bundle isomorphism, so $\mathcal{P}_* X_0$ defines a vector field on T^*Q.

Our abstract notion of a **force field** will be a time-varying one-form $F(t)$ on Q. Thus, $F(t) \in \Gamma(T^*Q)$ (a section of the bundle T^*Q for each t).

If Z is a vector field on Q, then we may lift Z to a function P_Z on T^*Q by setting

$$P_Z(q, p) = p(Z_q), \quad p \in T_q^* Q.$$

We note that P_Z is the **momentum function** associated with Z (see Abraham and Marsden [1978] and Section 3.7).

If η is a one-form on Q, then we define the **vertical lift** of η, denoted by η^v, which will be a vector field on T^*Q, by setting

$$\eta^v(P_Z) = \eta(Z),$$

for all vector fields Z on Q. Locally, if $\eta = \sum_i \alpha_i dq^i$, then $\eta^v = \sum_i \alpha_i(\partial/\partial p_i)$.

Our definition of a **Newtonian vector field** X_F corresponding to an external force field F is the vector field on T^*Q given by

$$X_F = \mathcal{P}_* X_0 + F^v. \tag{3.4.3}$$

Since X_0 is of second order, in local coordinates (q, v) for TQ we may write it in the form

$$\dot{q} = v, \qquad \dot{v} = a_v(q, v).$$

The dynamical system $\dot{\alpha} = X_F(\alpha)$, $\alpha \in T^*Q$ can therefore be expressed in local coordinates (q, p) for T^*Q in the form

$$\dot{p} = a_p(q, p) + F, \tag{3.4.4}$$
$$\dot{q} = \mathcal{P}_q^{-1}(p),$$

where $a_v(q, p) = \frac{d}{dt}\mathcal{P}_q(v)$, $\dot{v} = a_v(q, v)$, $\dot{q} = v = \mathcal{P}_q^{-1}(p)$.

The representation (3.4.4) is useful in distinguishing two important components of the Newtonian system on T^*Q. The kinematics are the system on Q defined locally by

$$\dot{q} = \mathcal{P}_q^{-1}(p),$$

which describes the evolution of q with $p \in T_q M$ as an input. The dynamics are the system on T^*Q defined locally by

$$\dot{p} = a_p(q, p) + F,$$

which describes the evolution of p with $F \in T_q^* M$ as an input. Note that our input space in each of the examples above is a vector bundle.

For an interesting related discussion of forces as vector fields on classical spacetime, see the book Marsden and Hughes [1994] and references therein. In particular, see the original article on classical spacetime, Cartan [1923], as well as Misner, Thorne, and Wheeler [1973]. For further details on the approach discussed here see Bloch and Crouch [1998b].

Lagrange–d'Alembert Principle   The Lagrange–d'Alembert principle gives an alternative means of describing motion subject to an external force field F. Given a Lagrangian function L, the motion is governed by solutions of the variational system

$$\delta \int_a^b L(q, \dot{q})dt + \int_a^b F(\delta q)dt = 0, \quad q(t) \in Q, \tag{3.4.5}$$

subject to the condition that q is a C^1 curve satisfying $q(a) = q_a$ and $q(b) = q_b$.

Solutions of this "variational" system are flows of the vector field X_L on TQ satisfying

$$\mathbf{i}_{X_L}\Omega_L - \mathbf{d}H_L + \pi_Q^* F(t) \equiv 0. \tag{3.4.6}$$

As in the previous section, the formulation (3.4.6) includes holonomic constraints. To incorporate nonholonomic constraints, additional "forces" must be included to ensure that the constraints are satisfied. We apply the Lagrange–d'Alembert principle to this situation later on.

3.5   Lie–Poisson Brackets and the Rigid Body

An important Poisson structure that occurs in a number of basic examples, such as rigid bodies and fluids, is the Lie–Poisson bracket defined in this section. This class of examples also provides a rich class of Poisson manifolds that are *not* symplectic.

The Lie–Poisson Bracket.   Let G be a Lie group and $\mathfrak{g} = T_e G$ its Lie algebra with $[\, ,] : \mathfrak{g} \times \mathfrak{g} \to \mathfrak{g}$ the associated Lie bracket.

3.5.1 Proposition.   *The dual space $\mathfrak{g}^*$ is a Poisson manifold with either of the two brackets*

$$\{f, k\}_{\pm}(\mu) = \pm \left\langle \mu, \left[\frac{\delta f}{\delta \mu}, \frac{\delta k}{\delta \mu} \right] \right\rangle. \tag{3.5.1}$$

Here $\mathfrak{g}$ is identified with $\mathfrak{g}^{**}$ in the sense that $\delta f/\delta\mu \in \mathfrak{g}$ is defined by $\langle \nu, \delta f/\delta\mu \rangle = \mathbf{D}f(\mu) \cdot \nu$ for $\nu \in \mathfrak{g}^*$, where $\mathbf{D}$ denotes the derivative.[2] The notation $\delta f/\delta\mu$ is used to conform to the functional derivative notation in classical field theory. Assuming that $\mathfrak{g}$ is finite-dimensional and choosing coordinates $(\xi^1, \ldots, \xi^m)$ on $\mathfrak{g}$ and corresponding dual coordinates $(\mu_1, \ldots, \mu_m)$ on $\mathfrak{g}^*$, the **Lie–Poisson bracket** (3.5.1) is

$$\{f, k\}_\pm(\mu) = \pm\mu_a C^a_{bc} \frac{\partial f}{\partial \mu_b} \frac{\partial k}{\partial \mu_c}; \tag{3.5.2}$$

here C^a_{bc} are the **structure constants** of $\mathfrak{g}$ defined by $[e_a, e_b] = C^c_{ab} e_c$, where $(e_1, \ldots, e_m)$ is the coordinate basis of $\mathfrak{g}$ and where for $\xi \in \mathfrak{g}$ we write $\xi = \zeta^a e_a$, and for $\mu \in \mathfrak{g}^*, \mu - \mu_a e^a$, where $(e^1, \ldots, e^m)$ is the dual basis. Formula (3.5.2) appears explicitly in Lie [1890], Section 75.

Lie–Poisson Reduction.   Which sign to take in (3.5.2) is determined by understanding **Lie–Poisson reduction**, which can be summarized as follows. Let the **left and right translation maps to the identity** be defined as follows:

$$\lambda : T^*G \to \mathfrak{g}^* \quad \text{is defined by} \quad p_g \mapsto (T_e L_g)^* p_g \in T_e^* G \cong \mathfrak{g}^*; \tag{3.5.3}$$

$$\rho : T^*G \to \mathfrak{g}^* \quad \text{is defined by} \quad p_g \mapsto (T_e R_g)^* p_g \in T_e^* G \cong \mathfrak{g}^*. \tag{3.5.4}$$

Then λ *is a Poisson map if one takes the* $-$ *Lie–Poisson structure on* $\mathfrak{g}^*$, *and* ρ *is a Poisson map if one takes the* $+$ *Lie–Poisson structure on* $\mathfrak{g}^*$.[3]

Every left-invariant Hamiltonian and Hamiltonian vector field on T^*G is mapped by λ to a Hamiltonian and Hamiltonian vector field on $\mathfrak{g}^*$. There is a similar statement for right-invariant systems on T^*G. One says that the original system on T^*G has been **reduced** to $\mathfrak{g}^*$. The reason λ and ρ are both Poisson maps is perhaps best understood by observing that they are both equivariant momentum maps (see Section 3.7) generated by the the lift to T^*G of the action of G on itself by right and left translations, respectively.

Euler Equations.   The classical **Euler equations** of motion for rigid body dynamics are given by

$$\dot{\Pi} = \Pi \times \Omega, \tag{3.5.5}$$

where $\Pi = \mathbb{I}\Omega$ is the body angular momentum, Ω is the body angular velocity, and $\mathbb{I}$ is the moment of inertia tensor. Euler's equations are Hamiltonian relative to a Lie–Poisson structure. To see this, take $G = SO(3)$ to be the

---

[2] In the infinite-dimensional case one needs to worry about the existence of $\delta f/\delta\mu$; in this context, methods like the Hahn–Banach theorem are not always appropriate!

[3] This follows from the fact that λ and ρ are momentum maps; see Marsden and Ratiu [1999].

configuration space. Then $\mathfrak{g} \cong (\mathbb{R}^3, \times)$, and we make the identification $\mathfrak{g} \cong \mathfrak{g}^*$. The corresponding Lie–Poisson structure on $\mathbb{R}^3$ is given by

$$\{f, k\}(\Pi) = -\Pi \cdot (\nabla f \times \nabla k). \tag{3.5.6}$$

For the rigid body one chooses the minus sign in the Lie–Poisson bracket. This is because the rigid body Lagrangian (and hence Hamiltonian) is left-invariant, and so its dynamics push to $\mathfrak{g}^*$ by the map λ in (3.5.3).

Starting with the kinetic energy Hamiltonian, we directly obtain the formula $H(\Pi) = \frac{1}{2}\Pi \cdot (\mathbb{I}^{-1}\Pi)$, the kinetic energy of the rigid body. One verifies the following result from the chain rule and properties of the triple product:

3.5.2 Proposition. *Euler's equations are equivalent to the following equation for all $f \in \mathcal{F}(\mathbb{R}^3)$*

$$\dot{f} = \{f, H\}. \tag{3.5.7}$$

Casimir Functions. Some conserved quantities can be captured in the basic notion of a Casimir function.

3.5.3 Definition. *Let $(P, \{\,,\})$ be a Poisson manifold. A function $C \in \mathcal{F}(P)$ satisfying*

$$\{C, f\} = 0 \quad \text{for all} \quad f \in \mathcal{F}(P) \tag{3.5.8}$$

*is called a **Casimir function**.*

A crucial difference between symplectic manifolds and Poisson manifolds is this: On symplectic manifolds, the only Casimir functions are the constant functions (assuming that P is connected). On the other hand, on Poisson manifolds there is often a large supply of Casimir functions. In the case of the rigid body, every function $C : \mathbb{R}^3 \to \mathbb{R}$ of the form

$$C(\Pi) = \Phi(\|\Pi\|^2), \tag{3.5.9}$$

where $\Phi : \mathbb{R} \to \mathbb{R}$ is a differentiable function, is a Casimir function, as is readily checked.

Casimir functions are constants of the motion for *any* Hamiltonian, since $\dot{C} = \{C, H\} = 0$ for any H. In particular, for the rigid body, $\|\Pi\|^2$ is a constant of the motion; this is the invariant momentum sphere of rigid body dynamics.

Reduction of Dynamics. The maps λ and ρ induce Poisson isomorphisms between $(T^*G)/G$ and $\mathfrak{g}^*$ (with the $-$ and $+$ brackets, respectively), and this is a special instance of Poisson reduction. The following result is one useful way of formulating the general relation between T^*G and $\mathfrak{g}^*$. We treat the left-invariant case to be specific. Of course, the right-invariant case is similar.

3.5.4 Theorem. *Let G be a Lie group and $H : T^*G \to \mathbb{R}$ a left-invariant Hamiltonian. Let $h : \mathfrak{g}^* \to \mathbb{R}$ be the restriction of H to the identity. For a curve $p(t) \in T^*_{g(t)}G$, let $\mu(t) = (T^*_{g(t)}L) \cdot p(t) = \lambda(p(t))$ be the induced curve in $\mathfrak{g}^*$. Assume that $\dot{g} = \partial H / \partial p \in T_g G$. Then the following are equivalent:*

(i) *$p(t)$ is an integral curve of X_H; i.e., Hamilton's equations on T^*G hold.*

(ii) *For any $F \in \mathcal{F}(T^*G), \dot{F} = \{F, H\}$, where $\{\,,\}$ is the canonical bracket on T^*G.*

(iii) *$\mu(t)$ satisfies the **Lie–Poisson equations***

$$\frac{d\mu}{dt} = \mathrm{ad}^*_{\delta h/\delta \mu} \mu, \tag{3.5.10}$$

where $\mathrm{ad}_\xi : \mathfrak{g} \to \mathfrak{g}$ is defined by $\mathrm{ad}_\xi \eta = [\xi, \eta]$ and ad^*_ξ is its dual, i.e.

$$\dot{\mu}_a = C^d_{ba} \frac{\delta h}{\delta \mu_b} \mu_d. \tag{3.5.11}$$

(iv) *For any $f \in \mathcal{F}(\mathfrak{g}^*)$, we have*

$$\dot{f} = \{f, h\}_-, \tag{3.5.12}$$

where $\{\,,\}_-$ is the minus Lie–Poisson bracket.

We now make some remarks about the proof. First of all, the equivalence of (i) and (ii) is general for any cotangent bundle, as we have already noted. Next, the equivalence of (ii) and (iv) follows directly from the fact that λ is a Poisson map and $H = h \circ \lambda$. Finally, we establish the equivalence of (iii) and (iv). Indeed, $\dot{f} = \{f, h\}_-$ means

$$\left\langle \dot{\mu}, \frac{\delta f}{\delta \mu} \right\rangle = -\left\langle \mu, \left[\frac{\delta f}{\delta \mu}, \frac{\delta h}{\delta \mu} \right] \right\rangle = \left\langle \mu, \mathrm{ad}_{\delta h/\delta \mu} \frac{\delta f}{\delta \mu} \right\rangle = \left\langle \mathrm{ad}^*_{\delta h/\delta \mu} \mu, \frac{\delta f}{\delta \mu} \right\rangle.$$

Since f is arbitrary, this is equivalent to (iii).

Exercises

⋄ **3.5-1.** Prove Proposition 3.5.2

⋄ **3.5-2.** Prove that the functions C given in (3.5.9) are Casimir functions for the rigid body bracket.

3.6   The Euler–Poincaré Equations

For the rigid body there is an analogue of the above theorem on SO(3) and
$so(3)$ using the Euler–Lagrange equations and the variational principle as a
starting point. We now generalize this to an arbitrary Lie group and make
the direct link with the Lie–Poisson equations.

3.6.1 Theorem.   *Let G be a Lie group and $L : TG \to \mathbb{R}$ a left-invariant
Lagrangian. Let $l : \mathfrak{g} \to \mathbb{R}$ be its restriction to the tangent space to G at the
identity. For a curve $g(t) \in G$, let*

$$\xi(t) = g(t)^{-1} \cdot \dot{g}(t); \quad i.e., \quad \xi(t) = T_{g(t)} L_{g(t)^{-1}} \dot{g}(t).$$

Then the following are equivalent:

(i) *$g(t)$ satisfies the Euler–Lagrange equations for L on G.*

(ii) *The variational principle*

$$\delta \int L(g(t), \dot{g}(t)) dt = 0 \tag{3.6.1}$$

holds, for variations with fixed endpoints.

(iii) *The **Euler–Poincaré equations** hold:*

$$\frac{d}{dt} \frac{\delta l}{\delta \xi} = \mathrm{ad}^*_\xi \frac{\delta l}{\delta \xi}. \tag{3.6.2}$$

(iv) *The variational principle*

$$\delta \int l(\xi(t)) dt = 0 \tag{3.6.3}$$

holds on $\mathfrak{g}$, using variations of the form

$$\delta \xi = \dot{\eta} + [\xi, \eta], \tag{3.6.4}$$

where η vanishes at the endpoints.

Proof.   We will just give the main idea of the proof. First of all, the
equivalence of (i) and (ii) holds on the tangent bundle of any configuration
manifold Q, as we have seen; secondly, (ii) and (iv) are equivalent. To see
this, one needs to compute the variations $\delta \xi$ induced on $\xi = g^{-1} \dot{g} = TL_{g^{-1}} \dot{g}$
by a variation of g. To calculate this, we need to differentiate $g^{-1} \dot{g}$ in the
direction of a variation δg. If $\delta g = dg/d\epsilon$ at $\epsilon = 0$, where g is extended to
a curve g_ϵ, then

$$\delta \xi = \frac{d}{d\epsilon} \left(g^{-1} \frac{d}{dt} g \right) \bigg|_{\epsilon = 0},$$

while if $\eta = g^{-1}\delta g$, then

$$\dot{\eta} = \frac{d}{dt}\left(g^{-1}\frac{d}{d\epsilon}g\right)\Bigg|_{\epsilon=0}.$$

The difference $\delta\xi - \dot{\eta}$ is the commutator $[\xi, \eta]$. This argument is fine for matrix groups, but takes a little more work to make precise for general Lie groups. See, for example, Bloch, Krishnaprasad, Marsden, and Ratiu [1996] for the general case. Thus, (ii) and (iv) are equivalent.

To complete the proof, we show the equivalence of (iii) and (iv). Indeed, using the definitions and integrating by parts,

$$\delta \int l(\xi)dt = \int \frac{\delta l}{\delta\xi}\delta\xi\, dt = \int \frac{\delta l}{\delta\xi}(\dot{\eta} + \mathrm{ad}_\xi\,\eta)dt$$
$$= \int \left[-\frac{d}{dt}\left(\frac{\delta l}{\delta\xi}\right) + \mathrm{ad}_\xi^* \frac{\delta l}{\delta\xi}\right]\eta\, dt\,,$$

so the result follows.   ∎

Generalizing what we saw directly in the rigid body, one can check directly from the Euler–Poincaré equations that conservation of spatial angular momentum holds:

$$\frac{d}{dt}\pi = 0\,, \tag{3.6.5}$$

where π is defined by

$$\pi = \mathrm{Ad}_g^* \frac{\partial l}{\partial\xi}. \tag{3.6.6}$$

Since the Euler–Lagrange and Hamilton equations on TQ and T^*Q are equivalent, it follows that the Lie–Poisson and Euler–Poincaré equations are also equivalent. To see this directly, we make the following Legendre transformation from $\mathfrak{g}$ to $\mathfrak{g}^*$:

$$\mu = \frac{\delta l}{\delta\xi}, \quad h(\mu) = \langle\mu, \xi\rangle - l(\xi).$$

Note that

$$\frac{\delta h}{\delta\mu} = \xi + \left\langle\mu, \frac{\delta\xi}{\delta\mu}\right\rangle - \left\langle\frac{\delta l}{\delta\xi}, \frac{\delta\xi}{\delta\mu}\right\rangle = \xi\,,$$

and so it is now clear that (3.5.10) and (3.6.2) are equivalent.

An important generalization of this theorem to the case in which one has systems whose Lagrangian is parametrized, including the rigid body in a gravitational field (the heavy top), and compressible fluid mechanics that involve semidirect product theory is given in Holm, Marsden, and Ratiu [1998].

3.7 Momentum Maps

Momentum maps capture in a geometric way conserved quantities associated with symmetries, such as linear and angular momentum, that are associated with translational and rotational invariance.

Definition of Momentum Maps. Let G be a Lie group and P a Poisson manifold such that G acts on P by Poisson maps (in this case the action is called a **Poisson action**). Denote the corresponding infinitesimal action of $\mathfrak{g}$ on P by $\xi \mapsto \xi_P$, a map of $\mathfrak{g}$ to $\mathfrak{X}(P)$, the space of vector fields on P. We write the action of $g \in G$ on $z \in P$ as simply gz; the vector field ξ_P is obtained at z by differentiating gz with respect to g in the direction ξ at $g = e$. Explicitly,

$$\xi_P(z) = \frac{d}{d\epsilon}[\exp(\epsilon\xi) \cdot z]\Big|_{\epsilon=0}.$$

3.7.1 Definition. *A map* $\mathbf{J} : P \to \mathfrak{g}^*$ *is called a* **momentum map** *if* $X_{\langle \mathbf{J}, \xi \rangle} = \xi_P$ *for each* $\xi \in \mathfrak{g}$, *where* $\langle \mathbf{J}, \xi \rangle(z) = \langle \mathbf{J}(z), \xi \rangle$.

3.7.2 Theorem. *(Noether's Theorem) If H is a G-invariant Hamiltonian on P, then $\mathbf{J}$ is conserved on the trajectories of the Hamiltonian vector field X_H.*

Proof. Differentiating the invariance condition $H(gz) = H(z)$ with respect to $g \in G$ for fixed $z \in P$, we get $\mathbf{d}H(z) \cdot \xi_P(z) = 0$, and so $\{H, \langle \mathbf{J}, \xi \rangle\} = 0$, which by antisymmetry gives $\mathbf{d}\langle \mathbf{J}, \xi \rangle \cdot X_H = 0$, and so $\langle \mathbf{J}, \xi \rangle$ is conserved on the trajectories of X_H for every ξ in G. ∎

The Construction of Momentum Maps. Let Q be a manifold and let G act on Q. This action induces an action of G on T^*Q by cotangent lifting; that is, we take the transpose inverse of the tangent lift. The action of G on T^*Q is always symplectic and therefore Poisson.

3.7.3 Theorem. *A momentum map for a cotangent lifted action is given by*

$$\mathbf{J} : T^*Q \to \mathfrak{g}^* \quad \text{defined by} \quad \langle \mathbf{J}, \xi \rangle(p_q) = \langle p_q, \xi_Q(q) \rangle. \qquad (3.7.1)$$

In canonical coordinates we write $p_q = (q^i, p_j)$ and define the **action functions** $K^i{}_a$ by $(\xi_Q)^i = K^i{}_a(q)\xi^a$. Then

$$\langle \mathbf{J}, \xi \rangle(p_q) = p_i K^i{}_a(q)\xi^a, \qquad (3.7.2)$$

and therefore

$$J_a = p_i K^i{}_a(q). \qquad (3.7.3)$$

Equivariance. Recall that by differentiating the conjugation operation $h \mapsto ghg^{-1}$ at the identity, one gets the **adjoint action** of G on $\mathfrak{g}$. Taking its dual produces the **coadjoint action** of G on $\mathfrak{g}^*$. The cotangent momentum map can be checked to have the following equivariance property.

3.7.4 Proposition. *The momentum map for cotangent lifted actions is equivariant, i.e., the diagram in Figure 3.7.1 commutes.*

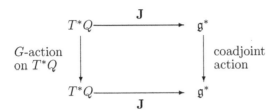

FIGURE 3.7.1. Equivariance of the momentum map.

Differentiating the equivariance relation, one gets the following:

3.7.5 Proposition. *Equivariance of the momentum map implies infinitesimal equivariance, which can be stated as the **classical commutation relations***

$$\{\langle \mathbf{J}, \xi \rangle, \langle \mathbf{J}, \eta \rangle\} = \langle \mathbf{J}, [\xi, \eta] \rangle.$$

Using this, one finds that, remarkably, momentum maps that are equivariant are always Poisson maps.

3.7.6 Proposition. *If $\mathbf{J}$ is infinitesimally equivariant, then $\mathbf{J} : P \to \mathfrak{g}^*$ is a Poisson map. If $\mathbf{J}$ is generated by a left (respectively right) action, then we use the $+$ (respectively $-$) Lie–Poisson structure on $\mathfrak{g}^*$.*

The Lagrangian Side. The above development concerns momentum maps using the Hamiltonian point of view. However, one can also consider them from the Lagrangian point of view. In this context we consider a Lie group G acting on a configuration manifold Q and lift this action to the tangent bundle TQ using the tangent operation. Given a G-invariant Lagrangian $L : TQ \to \mathbb{R}$, the corresponding momentum map is obtained by replacing the momentum p_q in (3.7.1) with the fiber derivative $\mathbb{F}L(v_q)$. Thus, $\mathbf{J} : TQ \to \mathfrak{g}^*$ is given by

$$\langle \mathbf{J}(v_q), \xi \rangle = \langle \mathbb{F}L(v_q), \xi_Q(q) \rangle , \tag{3.7.4}$$

or, in coordinates,

$$J_a = \frac{\partial L}{\partial \dot{q}^i} K_a^i, \tag{3.7.5}$$

where the action coefficients K_a^i are defined as before by writing $\xi_Q(q) = K_a^i \xi^a \partial / \partial q^i$. We have Noether's Theorem:

3.7.7 Proposition.   *For a solution of the Euler–Lagrange equations (even if the Lagrangian is degenerate),* **J** *is constant in time.*

Proof.   In the case that L is a hyperregular Lagrangian, this follows from its Hamiltonian counterpart. To include the degenerate case and to give a proof that is purely Lagrangian we use Hamilton's principle (which is the way it was originally done by Noether). To do this, choose any function $\phi(t, s)$ of two variables such that the conditions $\phi(a, s) = \phi(b, s) = \phi(t, 0) = 0$ hold. Since L is G-invariant, for each Lie algebra element $\xi \in \mathfrak{g}$, the expression

$$\int_a^b L(\exp(\phi(t, s)\xi)q, \exp(\phi(t, s)\xi)\dot{q})) \, dt \qquad (3.7.6)$$

is independent of s. Differentiating this expression with respect to s at $s = 0$ and setting $\phi' = \partial\phi/\partial s$ taken at $s = 0$ gives

$$0 = \int_a^b \left(\frac{\partial L}{\partial q^i} \xi_Q^i \phi' + \frac{\partial L}{\partial \dot{q}^i} (T\xi_Q \cdot \dot{q})^i \phi' \right) dt. \qquad (3.7.7)$$

Now we consider the variation $q(t, s) = \exp(\phi(t, s)\xi) \cdot q(t)$. The corresponding infinitesimal variation is given by

$$\delta q(t) = \phi'(t)\xi_Q(q(t)).$$

By Hamilton's principle, we have

$$0 = \int_a^b \left(\frac{\partial L}{\partial q^i} \delta q^i + \frac{\partial L}{\partial \dot{q}^i} \dot{\delta q}^i \right) dt. \qquad (3.7.8)$$

Note that

$$\dot{\delta q} = \dot{\phi}'\xi_Q + \phi'(T\xi_Q \cdot \dot{q})$$

and subtract (3.7.8) from (3.7.7) to give

$$0 = \int_a^b \frac{\partial L}{\partial \dot{q}^i} (\xi_Q)^i \dot{\phi}' \, dt = \int_a^b \frac{d}{dt}\left(\frac{\partial L}{\partial \dot{q}^i} \xi_Q^i \right) \phi' \, dt. \qquad (3.7.9)$$

Since ϕ' is arbitrary, except for endpoint conditions, it follows that the integrand vanishes, and so the time derivative of the momentum map is zero, and so the proposition is proved.   ■

Exercises

⬦ **3.7-1.**   Consider the action of $\mathbb{R}$ on $\mathbb{R}^3$ given by $x \cdot q^i \mapsto q^i + x$. Show that the action lifted to $T^*\mathbb{R}^3$ is given by $x \cdot (q^i, p_i) \mapsto (q_i + x, p_i)$. Hence show that the infinitesimal generator of this lifted action is the vector field $\xi_Q = (\xi, \xi, \xi, 0, 0, 0)$.

Thus write down the action coefficients of the lifted action and the momentum map. This gives an expression for the linear momentum of a particle in $\mathbb{R}^3$.

Finally, show this momentum is conserved along the flow of a free particle: Noether's theorem in this case.

⋄ **3.7-2.**   Consider again the configuration space $\mathbb{R}^3$, this time with the action of SO(3) given by $A \cdot \mathbf{q} \mapsto A\mathbf{q}$. Show that the action lifted to $T^*\mathbb{R}^3$ is given by $A \cdot (\mathbf{q}, \mathbf{p}) \mapsto (A\mathbf{q}, A\mathbf{p})$. Write $A\mathbf{q} = \omega \times q$ for a suitable vector ω and then show that the momentum map is given by $\mathbf{J}(\mathbf{q}, \mathbf{p}) = \mathbf{q} \times \mathbf{p}$.

3.8 Symplectic and Poisson Reduction

We have already seen how to use variational principles to reduce the Euler–Lagrange equations when the configuration manifold is a group. We will use this same method to perform reduction of nonholonomic systems in Chapter 5.

On the Hamiltonian side, there are three levels of reduction of decreasing generality: those of Poisson reduction, symplectic reduction, and cotangent bundle reduction. (For the last see Marsden [1992]). Let us first consider Poisson reduction.

Poisson Reduction.   For this situation, we start with a Poisson manifold P and let the Lie group G act on P by Poisson maps. Assuming that P/G is a smooth manifold, endow it with the unique Poisson structure such that the canonical projection $\pi : P \to P/G$ is a Poisson map. We can specify the Poisson structure on P/G explicitly as follows. For f and $k : P/G \to \mathbb{R}$, let $F = f \circ \pi$ and $K = k \circ \pi$, so F and K are f and k thought of as G-invariant functions on P. Then $\{f, k\}_{P/G}$ is defined by

$$\{f, k\}_{P/G} \circ \pi = \{F, K\}_P. \tag{3.8.1}$$

To show that $\{f, k\}_{P/G}$ is well-defined, one has to prove that $\{F, K\}_P$ is G-invariant. This follows from the fact that F and K are G-invariant and the group action of G on P consists of Poisson maps.

Lie–Poisson Reduction.   For $P = T^*G$ we get a very important special case.

3.8.1 Theorem.   **(Lie–Poisson Reduction)** *Let $P = T^*G$ and assume that G acts on P by the cotangent lift of left translations. If one endows $\mathfrak{g}^*$ with the minus Lie–Poisson bracket, then $P/G \cong \mathfrak{g}^*$.*

Symplectic Reduction.   In this case we begin with a symplectic manifold (P, Ω). Let G be a Lie group acting by symplectic maps on P; in this case the action is called a ***symplectic action***. Let $\mathbf{J}$ be an equivariant momentum map for this action and H a G-invariant Hamiltonian on P. Let

$G_\mu = \{g \in G \mid g \cdot \mu = \mu\}$ be the isotropy subgroup (symmetry subgroup) at $\mu \in \mathfrak{g}^*$. As a consequence of equivariance, G_μ leaves $\mathbf{J}^{-1}(\mu)$ invariant. Assume for simplicity that μ is a regular value of $\mathbf{J}$, so that $\mathbf{J}^{-1}(\mu)$ is a smooth manifold and G_μ acts freely and properly on $\mathbf{J}^{-1}(\mu)$, so that $\mathbf{J}^{-1}(\mu)/G_\mu =: P_\mu$ is a smooth manifold. Let $i_\mu : \mathbf{J}^{-1}(\mu) \to P$ denote the inclusion map and let $\pi_\mu : \mathbf{J}^{-1}(\mu) \to P_\mu$ denote the projection. Note that

$$\dim P_\mu = \dim P - \dim G - \dim G_\mu. \tag{3.8.2}$$

Building on classical work of Jacobi, Liouville, Arnold, and Smale, we have the following basic result of Marsden and Weinstein [1974] (see also Meyer [1973]).

3.8.2 Theorem.   (Symplectic Reduction Theorem) *There is a unique symplectic structure Ω_μ on P_μ satisfying*

$$i_\mu^* \Omega = \pi_\mu^* \Omega_\mu. \tag{3.8.3}$$

Given a G-invariant Hamiltonian H on P, define the reduced Hamiltonian $H_\mu : P_\mu \to \mathbb{R}$ by $H = H_\mu \circ \pi_\mu$. Then the trajectories of X_H project to those of X_{H_μ}. An important problem is how to reconstruct trajectories of X_H from trajectories of X_{H_μ}. Schematically, we have the situation in Figure 3.8.1.

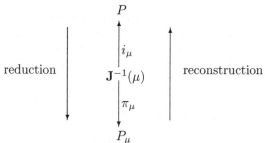

FIGURE 3.8.1. Reduction to P_μ and reconstruction back to P.

As already hinted at in Chapter 1, the reconstruction process is where the holonomy and "geometric phase" ideas enter.

Coadjoint Orbits.   Let $\mathcal{O}_\mu$ denote the coadjoint orbit through μ. As a special case of the symplectic reduction theorem, we get the following corollary.

3.8.3 Corollary. $(T^*G)_\mu \cong \mathcal{O}_\mu$.

The symplectic structure inherited on $\mathcal{O}_\mu$ is called the (*Lie–Kostant–Kirillov*) *orbit symplectic structure*. This structure is compatible with the Lie–Poisson structure on $\mathfrak{g}^*$ in the sense that the bracket of two functions on $\mathcal{O}_\mu$ equals that obtained by extending them arbitrarily to $\mathfrak{g}^*$, taking the Lie–Poisson bracket on $\mathfrak{g}^*$ and then restricting to $\mathcal{O}_\mu$.

3.8.4 Examples.

A. Rotational Coadjoint Orbits. $G = \mathrm{SO}(3), \mathfrak{g}^* = \mathfrak{so}(3)^* \cong \mathbb{R}^3$. In this case the coadjoint action is the usual action of $\mathrm{SO}(3)$ on $\mathbb{R}^3$. This is because of the orthogonality of the elements of G. The set of orbits consists of spheres and a single point. The reduction process confirms that all orbits are symplectic manifolds. One calculates that the symplectic structure on the spheres is a multiple of the area element.

B. Jacobi–Liouville Theorem. Let $G = \mathbb{T}^k$ be the k-torus and assume that G acts on a symplectic manifold P. In this case the components of $\mathbf{J}$ are in involution and $\dim P_\mu = \dim P - 2k$, so $2k$ variables are eliminated. As we shall see, reconstruction allows one to reassemble the solution trajectories on P by quadratures in this abelian case.

C. Jacobi–Deprit Elimination of the Node. Let $G = \mathrm{SO}(3)$ act on P. In the classical case of Jacobi, $P = T^*\mathbb{R}^3$, and in the generalization of Deprit [1983] one considers the phase space of n particles in $\mathbb{R}^3$. We just point out here that the reduced space P_μ has dimension $\dim P - 3 - 1 = \dim P - 4$, since $G_\mu = S^1$ (if $\mu \neq 0$) in this case.   ◆

Orbit Reduction Theorem.   This result of Marle [1976] and Kazhdan, Kostant, and Sternberg [1978] states that P_μ may be alternatively constructed as

$$P_{\mathcal{O}} = \mathbf{J}^{-1}(\mathcal{O})/G, \tag{3.8.4}$$

where $\mathcal{O} \subset \mathfrak{g}^*$ is the coadjoint orbit through μ. As above we assume that we are away from singular points. The spaces P_μ and $P_{\mathcal{O}}$ are shown to be isomorphic by using the inclusion map $l_\mu : \mathbf{J}^{-1}(\mu) \to \mathbf{J}^{-1}(\mathcal{O})$ and taking equivalence classes to induce a symplectic isomorphism $L_\mu : P_\mu \to P_{\mathcal{O}}$. The symplectic structure $\Omega_{\mathcal{O}}$ on $P_{\mathcal{O}}$ is uniquely determined by

$$j_{\mathcal{O}}^* \Omega = \pi_{\mathcal{O}}^* \Omega_{\mathcal{O}} + \mathbf{J}_{\mathcal{O}}^* \omega_{\mathcal{O}}, \tag{3.8.5}$$

where $j_{\mathcal{O}} : \mathbf{J}^{-1}(\mathcal{O}) \to P$ is the inclusion, $\pi_{\mathcal{O}} : \mathbf{J}^{-1}(\mathcal{O}) \to P_{\mathcal{O}}$ is the projection, and where $\mathbf{J}_{\mathcal{O}} = \mathbf{J}|\mathbf{J}^{-1}(\mathcal{O}) : \mathbf{J}^{-1}(\mathcal{O}) \to \mathcal{O}$ and $\omega_{\mathcal{O}}$ is the orbit symplectic form. In terms of the Poisson structure, $\mathbf{J}^{-1}(\mathcal{O})/G$ has the bracket structure inherited from P/G; in fact, $\mathbf{J}^{-1}(\mathcal{O})/G$ *is a symplectic leaf* in P/G. Thus, we get the picture in Figure 3.8.2.

Kirillov has shown that *every* Poisson manifold P is the union of symplectic leaves, although the preceding construction explicitly realizes these symplectic leaves in this case by the reduction construction. A special case is the foliation of the dual $\mathfrak{g}^*$ of any Lie algebra $\mathfrak{g}$ into its symplectic leaves, namely the coadjoint orbits. For example, $\mathfrak{so}(3)$ is the union of spheres plus the origin, each of which is a symplectic manifold. Notice that the drop in dimension from $T^*\mathrm{SO}(3)$ to $\mathcal{O}$ is from 6 to 2, a drop of 4, as in general $\mathrm{SO}(3)$ reduction. An exception is the singular point, the origin, where the drop in dimension is larger.

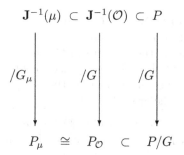

FIGURE 3.8.2. Orbit reduction gives another realization of P_μ.

Exercises

⋄ **3.8-1.** Compute the dimension of the generic coadjoint orbits of SO(n). These are important for proving integrability of the generalized (n-dimensional) rigid body equations.

3.9 A Particle in a Magnetic Field

In cotangent bundle reduction theory, one adds terms to the symplectic form called "magnetic terms" that may be identified with curvatures of certain connections that we explain shortly. To explain this terminology, we consider a particle in a magnetic field.

Let B be a closed two-form on $\mathbb{R}^3$ and $\mathbf{B} = B_x\mathbf{i}+B_y\mathbf{j}+B_z\mathbf{k}$ the associated divergence-free vector field; i.e., $\mathbf{i_B}(dx \wedge dy \wedge dz) = B$, or

$$B = B_x dy \wedge dz - B_y dx \wedge dz + B_z dx \wedge dy.$$

Thinking of $\mathbf{B}$ as a magnetic field, the equations of motion for a particle with charge e and mass m are given by the **Lorentz force law**:

$$m\frac{d\mathbf{v}}{dt} = \frac{e}{c}\mathbf{v} \times \mathbf{B},\tag{3.9.1}$$

where $\mathbf{v} = (\dot{x}, \dot{y}, \dot{z})$. On $\mathbb{R}^3 \times \mathbb{R}^3$, i.e., on $(\mathbf{x}, \mathbf{v})$-space, consider the symplectic form

$$\Omega_B = m(dx \wedge d\dot{x} + dy \wedge d\dot{y} + dz \wedge d\dot{z}) - \frac{e}{c}B.\tag{3.9.2}$$

For the Hamiltonian, take the kinetic energy:

$$H = \frac{m}{2}(\dot{x}^2 + \dot{y}^2 + \dot{z}^2).\tag{3.9.3}$$

Writing $X_H(u, v, w) = (u, v, w, (\dot{u}, \dot{v}, \dot{w}))$, the condition defining X_H, namely $\mathbf{i}_{X_H} \Omega_B = \mathbf{d}H$, is

$$m(u\,d\dot{x} - \dot{u}\,dx + v\,d\dot{y} - \dot{v}\,dy + w\,d\dot{z} - \dot{w}\,dz)$$
$$- \frac{e}{c}[B_x v\,dz - B_x w\,dy - B_y u\,dz + B_y w\,dx + B_z u\,dy - B_z v\,dx]$$
$$= m(\dot{x}\,d\dot{x} + \dot{y}\,d\dot{y} + \dot{z}\,d\dot{z}), \tag{3.9.4}$$

which is equivalent to $u = \dot{x}, v = \dot{y}, w = \dot{z}, m\dot{u} = e(B_z v - B_y w)/c, m\dot{v} = e(B_x w - B_z u)/c$, and $m\dot{w} = e(B_y u - B_x v)/c$, i.e., to

$$m\ddot{x} = \frac{e}{c}(B_z \dot{y} - B_y \dot{z}),$$
$$m\ddot{y} = \frac{e}{c}(B_x \dot{z} - B_z \dot{x}), \tag{3.9.5}$$
$$m\ddot{z} = \frac{e}{c}(B_y \dot{x} - B_x \dot{y}),$$

which is the same as (3.9.1). Thus the *equations of motion for a particle in a magnetic field are Hamiltonian, with energy equal to the kinetic energy and with the symplectic form* Ω_B.

If $B = \mathbf{d}A$, i.e., $\mathbf{B} = \nabla \times \mathbf{A}$, where A is a one-form and $\mathbf{A}$ is the associated vector field, then the map $(\mathbf{x}, \mathbf{v}) \mapsto (\mathbf{x}, \mathbf{p})$, where $\mathbf{p} = m\mathbf{v} + e\mathbf{A}/c$, pulls back the *canonical* form to Ω_B, as is easily checked. *Thus, equations* (3.9.1) *are also Hamiltonian relative to the canonical bracket on* $(\mathbf{x}, \mathbf{p})$-*space with the Hamiltonian*

$$H_{\mathbf{A}} = \frac{1}{2m} \left\| \mathbf{p} - \frac{e}{c}\mathbf{A} \right\|^2. \tag{3.9.6}$$

Even in Euclidean space, not every magnetic field can be written as $\mathbf{B} = \nabla \times \mathbf{A}$. For example, the field of a magnetic monopole of strength $g \neq 0$, namely

$$\mathbf{B}(\mathbf{r}) = g \frac{\mathbf{r}}{\|\mathbf{r}\|^3}, \tag{3.9.7}$$

cannot be written this way, since the flux of $\mathbf{B}$ through the unit sphere is $4\pi g$, yet Stokes's theorem applied to the two hemispheres would give zero. Thus, one might think that the Hamiltonian formulation involving only B (i.e., using Ω_B and H) is preferable. However, one can recover the magnetic potential A by regarding A as a connection on a nontrivial bundle over $\mathbb{R}^3 \backslash \{0\}$. The bundle over the sphere S^2 is in fact the **Hopf fibration** $S^3 \to S^2$. This same construction can be carried out using reduction. For a readable account of some aspects of this situation, see Yang [1980]. For an interesting example of Weinstein in which this monopole comes up, see Marsden [1981], p. 34.

When one studies the motion of a colored (rather than a charged) particle in a Yang–Mills field, one finds a beautiful generalization of this construction and related ideas using the theory of principal bundles; see Sternberg

[1977], Weinstein [1978b], and Montgomery [1984]. In the study of centrifugal and Coriolis forces one discovers some structures analogous to those here (see Marsden and Ratiu [1999] for more information).

3.10   The Mechanical Connection

An important connection for analyzing the dynamics and control of mechanical systems is the mechanical connection, a notion that goes back to Smale [1970].

Locked Inertia Tensor.   Assume that we have a configuration manifold Q and Lagrangian $L : TQ \to \mathbb{R}$. Let G be a Lie group with Lie algebra $\mathfrak{g}$, and assume that G acts on Q, and lift the action to TQ via the tangent mapping. We assume also that G acts freely and properly on Q. We assume also that we have a metric $\langle\!\langle\,,\rangle\!\rangle$ on Q that is invariant under the group action.

For each $q \in Q$ define the **locked inertia tensor** to be the map $\mathbb{I} : \mathfrak{g} \to \mathfrak{g}^*$ defined by

$$\langle \mathbb{I}\eta, \zeta \rangle = \langle\!\langle \eta_Q(q), \zeta_Q(q) \rangle\!\rangle . \tag{3.10.1}$$

Recall that ξ_Q denotes the infinitesimal generator of the action of G on Q. Locally, if

$$[\xi_Q(q)]^i = K_a^i(q)\xi^a \tag{3.10.2}$$

relative to the coordinates q^i of Q and a basis $e_a, a = 1, \ldots, m$, of $\mathfrak{g}$ (K_a^i are called the action coefficients), then

$$\mathbb{I}_{ab} = g_{ij} K_a^i K_b^j . \tag{3.10.3}$$

Recall also that the momentum map in this context, $\mathbf{J} : TM \to \mathfrak{g}^*$, is given by

$$\langle \mathbf{J}(q, v), \xi \rangle = \langle\!\langle \xi_Q(q), v \rangle\!\rangle . \tag{3.10.4}$$

3.10.1 Definition.   *We define the **mechanical connection** on the principal bundle $Q \to Q/G$ to be the map $\mathcal{A} : TQ \to \mathfrak{g}$ given by*

$$\mathcal{A}(q, v) = \mathbb{I}(q)^{-1} (\mathbf{J}(q, v)) ; \tag{3.10.5}$$

that is, $\mathcal{A}$ is the map that assigns to each (q, v) the corresponding angular velocity of the locked system. In coordinates,

$$A^a = \mathbb{I}^{ab} g_{ij} K_b^i v^j . \tag{3.10.6}$$

One can check that $\mathcal{A}$ is G-equivariant and $\mathcal{A}(\xi_Q(q)) = \xi$.
The horizontal space of the connection is given by

$$\mathrm{hor}_q = \{(q, v) \mid \mathbf{J} = 0\} \subset T_q Q . \tag{3.10.7}$$

The vertical space of vectors that are mapped to zero under the projection $Q \to S = Q/G$ is given by

$$\text{ver}_q = \{\xi_Q(q)|\xi \in \mathfrak{g}\}. \tag{3.10.8}$$

The horizontal–vertical decomposition of a vector $(q, v) \in T_q Q$ is given by

$$v = \text{hor}_q v + \text{ver}_q v, \tag{3.10.9}$$

where

$$\text{ver}_q v = [\mathcal{A}(q, v)]_Q(q) \quad \text{and} \quad \text{hor}_q v = v - \text{ver}_q v. \tag{3.10.10}$$

3.10.2 Example (Pendulum on a Cart). An example that illustrates the mechanical connection is the uncontrolled inverted pendulum on a cart.
Recall that the Lagrangian may be written

$$L = \frac{1}{2}\left(\alpha \dot{\theta}^2 - 2\beta \cos\theta \dot{\theta}\dot{s} + \gamma \dot{s}^2 + D\cos\theta\right). \tag{3.10.11}$$

In this case the Lagrangian is cyclic in the variable s; that is, L is invariant under the linear $\mathbb{R}^1$ action $s \mapsto s + a$.
The infinitesimal generator of this action is thus given by

$$\xi_Q = \frac{d}{dt}\Big|_{t=0}(s + at, \theta) = (a, 0). \tag{3.10.12}$$

In this case using the mechanical metric induced by the Lagrangian we have

$$J(q, v) = \langle \mathbf{J}(q, v), a \rangle = \langle \mathbb{F}L(q, v), (a, 0) \rangle \tag{3.10.13}$$

and hence

$$J = \frac{\partial L}{\partial \dot{s}} = \gamma \dot{s} - \beta \cos\theta \dot{\theta}. \tag{3.10.14}$$

The locked inertia tensor $\mathbb{I}(q)$ is given by

$$\langle \mathbb{I}(q)a, b \rangle = \langle\!\langle (a, 0), (b, 0) \rangle\!\rangle, \tag{3.10.15}$$

and hence

$$\mathbb{I}(q) = \begin{bmatrix} 1 & 0 \end{bmatrix} \begin{bmatrix} \gamma & -\beta\cos\theta \\ -\beta\cos\theta & \alpha \end{bmatrix} \begin{bmatrix} 1 \\ 0 \end{bmatrix} = \gamma. \tag{3.10.16}$$

Notice that this is indeed just the "locked" inertia "$\gamma = m + M$," the sum of pendulum and cart mass!
Hence the action of the mechanical connection on a tangent vector v at the point q is given by

$$\mathcal{A}(q, v) = \frac{1}{\gamma}\left(\gamma\dot{s} - \beta\cos\theta\dot{\theta}\right) = \left(\dot{s} - \frac{\beta}{\gamma}\cos\theta\dot{\theta}\right). \tag{3.10.17}$$

The vertical–horizontal decomposition of a vector v is given by

$$\text{ver}_q v = [\mathcal{A}(q,v)]_Q(q) = \left(\dot{s} - \frac{\beta}{\gamma}\cos\theta\dot{\theta}, 0\right),$$

$$\text{hor}_q v = v - \text{ver}_q v = \left(\frac{\beta}{\gamma}\cos\theta\dot{\theta}, \dot{\theta},\right). \tag{3.10.18}$$

Note that the horizontal component of v is clearly zero under the action of $\mathcal{A}$.   ♦

3.11   The Lagrange–Poincaré Equations

To describe the reduced Lagrange equations, we make use of a connection on the principal G-bundle $Q \to Q/G$; for the Euler–Poincaré-equations, in which $Q = G$, the group structure automatically provides such a connection. For a more general choice of Q one can choose the mechanical connection as defined in the previous section. (See Marsden and Scheurle [1993b].)

Thus, assume that the bundle $Q \to Q/G$ has a given (principal) connection $\mathcal{A}$. Divide variations into horizontal and vertical parts; this breaks up the Euler–Lagrange equations on Q into two sets of equations that we now describe. Let x^α be coordinates on the shape space Q/G and let Ω^a be coordinates for vertical vectors in a local bundle chart. Drop L to TQ/G to obtain a reduced Lagrangian $l : TQ/G \to \mathbb{R}$ in which the group coordinates are eliminated. We can represent this reduced Lagrangian in a couple of ways. First, if we choose a local trivialization as we have described earlier, we obtain l as a function of the variables $(r^\alpha, \dot{r}^\alpha, \xi^a)$. However, it will be more convenient and intrinsic to change variables from ξ^a to the local version of the locked angular velocity, which has the physical interpretation of the *body* angular velocity, namely $\Omega = \xi + \mathcal{A}_{\text{loc}}\dot{r}$, or in coordinates,

$$\Omega^a = \xi^a + A^a_\alpha(r)\dot{r}^\alpha.$$

We will write $l(r^\alpha, \dot{r}^\alpha, \Omega^a)$ for the local representation of l in these variables.

3.11.1 Theorem.   *A curve* $(q^i, \dot{q}^i) \in TQ$, *satisfies the Euler–Lagrange equations if and only if the induced curve in* TQ/G *with coordinates given in a local trivialization by* $(r^\alpha, \dot{r}^\alpha, \Omega^a)$ *satisfies the* **Lagrange–Poincaré equations**

$$\frac{d}{dt}\frac{\partial l}{\partial \dot{r}^\alpha} - \frac{\partial l}{\partial r^\alpha} = \frac{\partial l}{\partial \Omega^a}\left(-\mathcal{B}^a_{\alpha\beta}\dot{r}^\beta + \mathcal{E}^a_{\alpha d}\Omega^d\right), \tag{3.11.1}$$

$$\frac{d}{dt}\frac{\partial l}{\partial \Omega^b} = \frac{\partial l}{\partial \Omega^a}(-\mathcal{E}^a_{\alpha b}\dot{r}^\alpha + C^a_{db}\Omega^d), \tag{3.11.2}$$

where $\mathcal{B}^a_{\alpha\beta}$ *are the coordinates of the curvature* $\mathcal{B}$ *of* $\mathcal{A}$, *and* $\mathcal{E}^a_{\alpha d} = C^a_{bd}A^b_\alpha$.

The first of these equations is similar to the equations for a nonholonomic system written in terms of the constrained Lagrangian, and the second is similar to the momentum equation (see Chapter 5). It is useful to note that the first set of equations results from the variational principle of Hamilton by restricting the variations to be horizontal relative to the given connection. As we shall see, this is very similar to what one has in systems with nonholonomic constraints with the principle of Lagrange–d'Alembert.

If one uses as variables $(r^\alpha, \dot{r}^\alpha, p_a)$, where p is the body angular momentum, so that $p = \mathbb{I}_{\text{loc}}(r)\Omega = \partial l/\partial\Omega$, then the equations become (using the same letter l for the reduced Lagrangian, an admitted abuse of notation)

$$\frac{d}{dt}\frac{\partial l}{\partial \dot{r}^\alpha} - \frac{\partial l}{\partial r^\alpha} = p_a\left(-\mathcal{B}^a_{\alpha\beta}\dot{r}^\beta + \mathcal{E}^a_{\alpha d}I^{de}p_e\right) - p_d\frac{\partial I^{de}}{\partial r^\alpha}p_e\,, \qquad (3.11.3)$$

$$\frac{d}{dt}p_b = p_a(-\mathcal{E}^a_{\alpha b}\dot{r}^\alpha + C^a_{db}I^{de}p_e)\,, \qquad (3.11.4)$$

where I^{de} denotes the inverse of the matrix I_{ab}.

The intrinsic geometry of these equations is systematically developed in Cendra, Marsden, and Ratiu [2001a], and their nonholonomic counterpart is developed in Bloch, Krishnaprasad, Marsden, and Murray [1996] and Cendra, Marsden, and Ratiu [2001b].

Explicit Form of the Reduced Lagrangian. It is possible to write down a useful explicit form of the reduced Lagrangian for mechanical systems (see, e.g., Murray [1995] and Ostrowski [1995, 1998]). We shall also see a related, even simpler version of this in the next section in the context of Routh reduction.

Suppose we have a Lagrangian on L on TQ that is invariant under the action of G on Q. We define the reduced Lagrangian $l : TQ/G \to \mathbb{R}$ to be

$$l(r, \xi, \dot{r}) = L(r, g^{-1}g, \dot{r}, g^{-1}\dot{g})\,, \qquad (3.11.5)$$

where $(\xi = g^{-1}\dot{g}, r, \dot{r})$ are local coordinates on TQ/G. Here the ξ are referred to as **body** velocities or velocities with respect to the body frame, while $\xi^s = \dot{g}g^{-1} = \text{Ad}_g\xi$ are referred to as **spatial** velocities. See Marsden and Ratiu [1999] for further details.

For a mechanical system the Lagrangian has the form

$$L(q, v_q) = \frac{1}{2}\langle\langle v_q, v_q\rangle\rangle - V(q)\,. \qquad (3.11.6)$$

Then we have the following result.

3.11.2 Proposition. *For a G-invariant mechanical Lagrangian the reduced Lagrangian may be written in the form*

$$l(r, \dot{r}, \xi) = \frac{1}{2}(\xi^T, \dot{r}^T)\begin{pmatrix} I & IA \\ A^T I & m(r) \end{pmatrix}\begin{pmatrix} \xi \\ \dot{r} \end{pmatrix} - V(r)\,, \qquad (3.11.7)$$

where here I is the local form of the locked inertia tensor and A is the local form of the mechanical connection.

Proof.   Write L as

$$L(g,r,\dot{g},\dot{r}) = \frac{1}{2}(\dot{g}^T,\dot{r}^T)\begin{pmatrix} g_{11} & g_{12} \\ g_{12}^T & g_{22} \end{pmatrix}\begin{pmatrix} \dot{g} \\ \dot{r} \end{pmatrix} - V(g,r). \qquad (3.11.8)$$

We now consider each term of the kinetic energy separately: Firstly,

$$\dot{g}^T g_{11}\dot{g} = \langle\!\langle(\dot{g},0),(\dot{g},0)\rangle\!\rangle = \langle J(\dot{g},0),\xi^s\rangle = \langle \mathbb{I}A(\dot{g},0),\xi^s\rangle = \langle \mathbb{I}\xi^s,\xi^s\rangle \qquad (3.11.9)$$

using (2.9.12) for the local expression of the connection and the definition of the mechanical connection (3.10.5).

Similarly,

$$\dot{r}^T g_{21}\dot{g} = \langle\!\langle(0,\dot{r}),(\dot{g},0)\rangle\!\rangle = \langle J(0,\dot{r}),\xi^s\rangle$$
$$= \langle \mathbb{I}A(0,\dot{r}),\xi^s\rangle = \langle \mathbb{I}\mathrm{Ad}_g A(r)\dot{r},\xi^s\rangle, \qquad (3.11.10)$$

again using (2.9.12). Now set $m(r) = g_{22}(e,r)$ and $V(r) = V(e,r)$. Then the reduced Lagrangian in the spatial frame is

$$l^s(r,\dot{r},\xi^s) = \frac{1}{2}((\xi^s)^T,\dot{r}^T)\begin{pmatrix} \mathbb{I} & \mathbb{I}\mathrm{Ad}_g A \\ A^T\mathrm{Ad}_g^*\mathbb{I} & m(r) \end{pmatrix}\begin{pmatrix} \xi^s \\ \dot{r} \end{pmatrix} - V(r). \qquad (3.11.11)$$

Using $\xi^s = \mathrm{Ad}_g \xi$ and defining $I(r) = \mathbb{I}(e,r) = \mathrm{Ad}_g^*\mathbb{I}\mathrm{Ad}_g$ gives the body representation

$$l = l^b(r,\dot{r},\xi) = \frac{1}{2}(\xi^T,\dot{r}^T)\begin{pmatrix} I & IA \\ A^T I & m \end{pmatrix}\begin{pmatrix} \xi \\ \dot{r} \end{pmatrix} - V(r). \qquad (3.11.12)$$

∎

Setting $\Omega = \xi + A\dot{r}$ we obtain the block diagonal form

$$l = l^\Omega(r,\dot{r},\Omega) = \frac{1}{2}(\Omega^T,\dot{r}^T)\begin{pmatrix} I & 0 \\ 0 & m - A^T I A \end{pmatrix}\begin{pmatrix} \Omega \\ \dot{r} \end{pmatrix} - V(r). \qquad (3.11.13)$$

We can then compute the equations of motion in these reduced coordinates. See Marsden and Scheurle [1993a] and Ostrowski [1995].

Define the generalized momentum

$$p = \frac{\partial l}{\partial \xi} = I\xi + IA\dot{r}. \qquad (3.11.14)$$

Suppose also that there is a G-invariant forcing $F = (F_\alpha, F_a)$, the components corresponding to base and fiber directions, respectively. Then the reduced equations of motion are

$$g^{-1}\dot{g} = \xi = -A\dot{r} + I^{-1}p, \qquad (3.11.15)$$

$$\dot{p} = \text{ad}_\xi^* p + \overline{F}, \qquad (3.11.16)$$

$$\tilde{M}\ddot{r} + \dot{r}^T\tilde{C}(r)\dot{r} + \tilde{N} + \frac{\partial V}{\partial r} = T(r)F, \qquad (3.11.17)$$

where $\tilde{M}(r) = m(r) - A^T(r)I(r)A(r)$, and $\tilde{C}(r)$ represents reduced Coriolis and centrifugal terms,

$$\dot{r}^T\tilde{C}(r)\dot{r} = C_{\alpha\beta\gamma}\dot{r}^\alpha\dot{r}^\beta = \frac{1}{2}\left(\frac{\partial\tilde{M}_{\alpha\beta}}{\partial r^\gamma} + \frac{\partial\tilde{M}_{\alpha\gamma}}{\partial r^\beta} - \frac{\partial\tilde{M}_{\gamma\beta}}{\partial r^\alpha}\right)\dot{r}^\beta\dot{r}^\alpha, \quad (3.11.18)$$

while

$$\left\langle\tilde{N},\delta r\right\rangle = \left\langle p, dA(\dot{r},\delta r) - [A(\dot{r}, A(\delta r)] + \left[I^{-1}p, A(\delta r)\right] + \frac{1}{2}\frac{\partial I^{-1}p}{\partial r}(\delta r)\right\rangle$$
$$(3.11.19)$$

and $(T(r)F)_\alpha = F_\alpha - F_a A_\alpha^a$, $\overline{F}_a = F_b g_a^b$, with g_a^b denoting the lifted action of the group.

The reduced part of these equations is of course equivalent to the reduced Euler–Lagrange equations discussed above, but this is a useful way to look at the full set of equations.

3.11.3 Example (Pendulum on a Cart). One can see this structure quite explicitly in the pendulum on a cart discussed above in Section 3.10. The off-diagonal entries of the kinetic energy metric are $-\beta\cos\theta = -\gamma(\beta/\gamma\cos\theta$, just IA in the notation above. ◆

3.11.4 Example (Rigid Body (Satellite) with Rotor). Computationally, these equations are often quite straightforward to obtain by a direct analysis. Consider, for example, the satellite with rotor that we briefly analyzed in Chapter 1. (See also Ostrowski [1995].)

Recall from Chapter 1 that the Lagrangian for this system in the body frame is

$$l = \frac{1}{2}(\lambda_1\Omega_1^2 + \lambda_2\Omega_2^2 + I_3\Omega_3^2 + J_3(\Omega_3 + \dot{\alpha})^2). \qquad (3.11.20)$$

Hence comparing with equations (3.11.12) we see that the local form of the locked inertia tensor is the matrix

$$I = \text{diag}(\lambda_1, \lambda_2, \lambda_3), \qquad (3.11.21)$$

$$IA = [0, 0, J_3]^T, \qquad (3.11.22)$$

and the local form of the mechanical connection is

$$A = I^{-1}(IA) = [0, 0, J_3/(I_3 + J_3)]^T. \qquad (3.11.23)$$

Here
$$p = \Pi = (\lambda_1\Omega_1, \lambda_2\Omega_2, \lambda_3\Omega_3 + J_3\dot\alpha), \qquad (3.11.24)$$
and the equation
$$\xi = -A\dot r + I^{-1}p \qquad (3.11.25)$$
becomes
$$\xi = -A\dot\alpha + I^{-1}p, \qquad (3.11.26)$$
that is,
$$\Omega = -\frac{J_3}{I_3 + J_3}[0,0,\dot\alpha]^T + I^{-1}p \qquad (3.11.27)$$
with I given by (3.11.21).

Finally, identifying the algebra with its dual, the equations
$$\dot p = \mathrm{ad}_\xi^* p = \mathrm{ad}_{\hat\Omega} p = p \times \Omega \qquad (3.11.28)$$
give the equations we saw in Chapter 1:
$$\dot\Pi_1 = \left(\frac{1}{I_3} - \frac{1}{\lambda_2}\right)\Pi_2\Pi_3 - \frac{l_3\Pi_2}{I_3},$$
$$\dot\Pi_2 = \left(\frac{1}{\lambda_1} - \frac{1}{I_3}\right)\Pi_1\Pi_3 + \frac{l_3\Pi_1}{I_3},$$
$$\dot\Pi_3 = \left(\frac{1}{\lambda_2} - \frac{1}{\lambda_1}\right)\Pi_1\Pi_2 \doteq a_3\Pi_1\Pi_2.$$

The base space equation is just
$$\dot l_3 = u \qquad (3.11.29)$$
as before.   ♦

Exercises

◇ **3.11-1.** Repeat the above analysis for the case of a satellite with three rotors, one about each principal axis.

3.12   The Energy-Momentum Method

A key notion in dynamics and control is that of stability of a point or set. In this context one normally thinks either of asymptotic or nonlinear stability, the former meaning essentially that all nearby trajectories tend to the set and the latter meaning that all trajectories starting nearby the set remain near the set. Most often one considers stability of an equilibrium point of a given system. Also of interest are *relative* equilibria—equilibria modulo the action of a symmetry group. Examples of such equilibria are the steady motions of rigid body—uniform rotations about one of the principal axes. In the theory of controlled systems, the problem is often that of achieving nonlinear or asymptotic stability of an initially unstable equilibrium point or set.

In this section we sketch a few key ideas in the stability theory for relative equilibria of holonomic mechanical systems. We will not be exhaustive or completely up to date; our purpose is just to give enough background so that the corresponding methods for nonholonomic systems can be treated later. The reader can consult the references below for further details. The key idea is the use of a combination of energy and another conserved quantity, such as momentum, to provide a Lyapunov function for the system. In Chapter 8 we extend this notion to nonholonomic systems, while in Chapter 9 we will see the method applied to stabilization problems in control.

The Energy–Momentum Method for Holonomic Systems. The energy-momentum method has a long and distinguished history going back to Routh, Riemann, Poincaré, Lyapunov, Arnold, Smale, and many others. The main new feature provided in the more recent work of Simo, Lewis, and Marsden [1991] (see Marsden [1992] for an exposition) is to obtain the powerful block diagonalization structure of the second variation of the augmented Hamiltonian as well as the normal form for the symplectic structure. This formulation also allowed for the proof of a *converse* of the energy-momentum method in the context of *dissipation-induced instabilities* due to Bloch, Krishnaprasad, Marsden, and Ratiu [1994, 1996].

As we saw in Theorem 2.3.3, there is a standard procedure for determining the stability of equilibria of an ordinary differential equation

$$\dot{x} = f(x), \tag{3.12.1}$$

where $x = (x^1, \ldots, x^n)$ and f is smooth. Recall that equilibria are points x_e such that $f(x_e) = 0$; that is, points that are fixed in time under the dynamics. By **stability** of the fixed point x_e we mean that *any solution to $\dot{x} = f(x)$ that starts near x_e remains close to x_e for all future time.* A traditional method of ascertaining the stability of x_e is to examine the first variation equation

$$\dot{\xi} = \mathbf{d}f(x_e)\xi \tag{3.12.2}$$

where $\mathbf{d}f(x_e)$ is the Jacobian of f at x_e, defined to be the matrix of partial derivatives

$$\mathbf{d}f(x_e) = \left[\frac{\partial f^i}{\partial x^j}\right]_{x=x_e}. \tag{3.12.3}$$

3.12.1 Theorem (Lyapunov's theorem). *If all the eigenvalues of $\mathbf{d}f(x_e)$ lie in the strict left half plane, then the fixed point x_e is stable. If any of the eigenvalues lie in the right half plane, then the fixed point is unstable.*

For Hamiltonian systems, the eigenvalues come in quartets that are symmetric about the origin, and so they cannot all lie in the strict left half plane. (See, for example, Marsden and Ratiu [1999] for the proof of this assertion.) Thus, *the above form of Lyapunov's theorem is not appropriate to deduce whether or not a fixed point of a Hamiltonian system is stable.*

When the Hamiltonian is in canonical form, one can use a stability test for fixed points due to Lagrange and Dirichlet. This method starts with the observation that for a fixed point (q_e, p_e) of such a system,

$$\frac{\partial H}{\partial q}(q_e, p_e) = \frac{\partial H}{\partial p}(q_e, p_e) = 0.$$

Hence the *fixed point occurs at a critical point of the Hamiltonian.*

Lagrange–Dirichlet Criterion.   *If the $2n \times 2n$ matrix $\delta^2 H$ consisting of second partial derivatives (the second variation) is either positive or negative definite at (q_e, p_e) then it is a stable fixed point.*

The proof is very simple. Consider the positive definite case. Since H has a nondegenerate minimum at $z_e = (q_e, p_e)$, Taylor's theorem with remainder shows that its level sets near z_e are bounded inside and outside by spheres of arbitrarily small radius. Since energy is conserved, solutions stay on level surfaces of H, so a solution starting near the minimum has to stay near the minimum.

For a Hamiltonian of the form kinetic plus potential V, critical points occur when $p_e = 0$ and q_e is a critical point of the potential of V. The Lagrange–Dirichlet Criterion then reduces to asking for a non-degenerate minimum of V.

In fact, this criterion was used in one of the classical problems of the 19th century: the problem of rotating gravitating fluid masses. This problem was studied by Newton, MacLaurin, Jacobi, Riemann, Poincaré and others. The motivation for its study was in the conjectured birth of two planets by the splitting of a large mass of solidifying rotating fluid. Riemann [1860], Routh [1877] and Poincaré [1885, 1892]; Poincare [1901] were major contributors to the study of this type of phenomenon and used the potential energy and angular momentum to deduce the stability and bifurcation.

The Lagrange–Dirichlet method was adapted by Arnold [1966b] into what has become known as the **energy-Casimir or Arnold method.** Arnold analyzed the stability of stationary flows of perfect fluids and arrived at an explicit stability criterion when the configuration space Q for the Hamiltonian of this system is the symmetry group G of the mechanical system.

A **Casimir function** C is one that Poisson commutes with any function F defined on the phase space of the Hamiltonian system, that is,

$$\{C, F\} = 0. \tag{3.12.4}$$

Large classes of Casimirs can occur when the reduction procedure is performed, resulting in systems with non-canonical Poisson brackets. For example, in the case of the rigid body discussed previously, if Φ is a function of one variable and μ is the angular momentum vector in the inertial coordinate system, then

$$C(\mu) = \Phi(\|\mu\|^2) \qquad (3.12.5)$$

is readily checked to be a Casimir for the rigid body bracket.

Energy–Casimir method. *Choose C such that $H + C$ has a critical point at an equilibrium z_e and compute the second variation $\delta^2(H+C)(z_e)$. If this matrix is positive or negative definite, then the equilibrium z_e is stable.*

When the phase space is obtained by reduction, the equilibrium z_e is called a ***relative equilibrium*** of the original Hamiltonian system.

The energy-Casimir method has been applied to a variety of problems including problems in fluids and plasmas (see, for instanceHolm, Marsden, Ratiu, and Weinstein [1985]) and rigid bodies with flexible attachments (Krishnaprasad and Marsden [1987]). If applicable, the energy-Casimir method may permit an explicit determination of the stability of the relative equilibria. It is important to remember, however, that these techniques give stability information only. As such one cannot use them to infer instability without further investigation.

The energy-Casimir method is restricted to certain types of systems, since its implementation relies on an abundant supply of Casimir functions. In some important examples, such as the dynamics of geometrically exact flexible rods, Casimirs have not been found and may not even exist. A method developed to overcome this difficulty is known as the ***energy-momentum method***, which is closely linked to the method of reduction. It uses conserved quantities, namely the energy and momentum map, that are readily available, rather than Casimirs.

The energy momentum method (Simo, Posbergh, and Marsden [1990], Simo, Posbergh, and Marsden [1991], Simo, Lewis, and Marsden [1991], and Lewis and Simo [1990]) involves the ***augmented Hamiltonian*** defined by

$$H_\xi(q,p) = H(q,p) - \xi \cdot \mathbf{J}(q,p) \qquad (3.12.6)$$

where $\mathbf{J}$ is the momentum map and ξ may be thought of as a Lagrange multiplier. When the symmetry group is the rotation group, $\mathbf{J}$ is the angular momentum and ξ is the angular velocity of the relative equilibrium. It is a theorem that one sets the first variation of H_ξ equal to zero to obtain the relative equilibria. To ascertain stability, the second variation $\delta^2 H_\xi$ is calculated. One is then interested in determining the definiteness of the second variation.

Definiteness in this context has to be properly interpreted to take into account the conservation of the momentum map $\mathbf{J}$ and the fact that $\mathbf{d}^2 H_\xi$

may have zero eigenvalues due to its invariance under a subgroup of the symmetry group. The variations of p and q must satisfy the linearized angular momentum constraint $(\delta q, \delta p) \in \ker[\mathbf{DJ}(q_e, p_e)]$, and must not lie in symmetry directions; only these variations are used to calculate the second variation of the augmented Hamiltonian H_ξ. These define the space of **admissible variations** $\mathcal{V}$.

The energy momentum method has been applied to the stability of relative equilibria of among others, geometrically exact rods and coupled rigid bodies (Patrick [1989] and Simo, Posbergh, and Marsden [1990, 1991]). It has also undergone continued development and maturity, as in, for example, Patrick [1992, 1995], Lewis [1995] and Leonard and Marsden [1997].

A cornerstone in the development of the energy-momentum method was laid by Routh [1877] and Smale [1970] who studied the stability of relative equilibria of simple mechanical systems. Simple mechanical systems are those whose Hamiltonian may be written as the sum of the potential and kinetic energies. Part of Smale's work may be viewed as saying that there is a naturally occurring connection called the **mechanical connection** on the reduction bundle that plays an important role. A connection can be thought of as a generalization of the electromagnetic vector potential.

The **amended potential** V_μ is the potential energy of the system plus a generalization of the potential energy of the centrifugal forces in stationary rotation:

$$V_\mu(q) = V(q) + \frac{1}{2}\mu \cdot \mathbb{I}^{-1}(q)\mu \tag{3.12.7}$$

where $\mathbb{I}$ is the **locked inertia tensor** (see Section 3.10). Smale showed that relative equilibria are critical points of the amended potential V_μ. The corresponding momentum P need not be zero since the system is typically in motion.

The second variation $\delta^2 V_\mu$ of V_μ directly yields the stability of the relative equilibria. However, an interesting phenomenon occurs if the space $\mathcal{V}$ of admissible variations is split into two specially chosen subspaces $\mathcal{V}_{\mathrm{RIG}}$ and $\mathcal{V}_{\mathrm{INT}}$. In this case the second variation *block diagonalizes*:

$$\delta^2 V_\mu \mid \mathcal{V} \times \mathcal{V} = \begin{bmatrix} D^2 V_\mu \mid \mathcal{V}_{\mathrm{RIG}} \times \mathcal{V}_{\mathrm{RIG}} & 0 \\ 0 & D^2 V_\mu \mid \mathcal{V}_{\mathrm{INT}} \times \mathcal{V}_{\mathrm{INT}} \end{bmatrix} \tag{3.12.8}$$

The space $\mathcal{V}_{\mathrm{RIG}}$ (**rigid variations**) is generated by the symmetry group, and $\mathcal{V}_{\mathrm{INT}}$ are the **internal** or **shape variations**. In addition, the whole matrix $\delta^2 H_\xi$ block diagonalizes in a very efficient manner. This often allows the stability conditions associated with $\delta^2 V_\mu \mid \mathcal{V} \times \mathcal{V}$ to be recast in terms of a standard eigenvalue problem for the second variation of the amended potential.

This splitting, that is, block diagonalization, has more miracles associated with it. In fact, *the second variation $\delta^2 H_\xi$ and the symplectic structure (and therefore the equations of motion) can be explicitly brought into normal form simultaneously.*

This result has several interesting implications. In the case of pseudo-rigid bodies (Lewis and Simo [1990]), it reduces the stability problem from an unwieldy 14×14 matrix to a relatively simple 3×3 subblock on the diagonal. The block diagonalization procedure enabled Lewis and Simo to solve their problem analytically, whereas without it, a substantial numerical computation would have been necessary.

The presence of discrete symmetries gives further, or refined, subblocking properties in the second variation of $\delta^2 H_\xi$ and $\delta^2 V_\mu$ and the symplectic form.

In general, this diagonalization explicitly separates the rotational and internal modes, a result which is important not only in rotating and elastic fluid systems, but also in molecular dynamics and robotics. Similar simplifications are expected in the analysis of other problems to be tackled using the energy momentum method.

The Routhian. As we have indicated, the energy-momentum method for mechanical systems with symmetry has its historical roots in what is called the *Routh method* for stability. We now review the Routh construction in a concrete way to bring out this link. One may regard this discussion as a simple, more explicit version of what is going on in the preceding section.

An abelian version of Lagrangian reduction was known to Routh by around 1860. A modern account was given in Arnold, Kozlov, and Neishtadt [1988], and motivated by that, Marsden and Scheurle [1993a] gave a geometrization and a generalization of the Routh procedure to the nonabelian case. We now give an elementary classical description of the Routh procedure so that one can see how it involves, in a concrete way, the amended potential when the group is abelian.

Assume that Q is a product of a manifold S and a number, say k, of copies of the circle S^1, namely $Q = S \times (S^1 \times \cdots \times S^1)$. The factor S, called **shape space**, has coordinates denoted by $x^1, \ldots, x^m$, and coordinates on the other factors are written $\theta^1, \ldots, \theta^k$. Some or all of the factors of S^1 can be replaced by $\mathbb{R}$ if desired, with little change. Given a Lagrangian on TQ, we assume that the variables θ^a, $a = 1, \ldots, k$, are **cyclic**, that is, they do not appear explicitly in the Lagrangian, although their velocities do.

Invariance of L under the action of the abelian group $G = S^1 \times \cdots \times S^1$ is another way to express that fact that θ^a are cyclic variables.

A basic class of examples are those for which the Lagrangian L has the form kinetic minus potential energy:

$$L(x, \dot{x}, \dot{\theta}) = \frac{1}{2} g_{\alpha\beta}(x) \dot{x}^\alpha \dot{x}^\beta + g_{a\alpha}(x) \dot{x}^\alpha \dot{\theta}^a + \frac{1}{2} g_{ab}(x) \dot{\theta}^a \dot{\theta}^b - V(x), \quad (3.12.9)$$

where there is a sum over α, β from 1 to m and over a, b from 1 to k. Even in simple examples, such as the double spherical pendulum or the simple pendulum on a cart, the matrices $g_{\alpha\beta}$, $g_{a\alpha}$, g_{ab} can depend on x.

Because θ^a are cyclic, the corresponding conjugate momenta

$$p_a = \frac{\partial L}{\partial \dot{\theta}^a} \qquad (3.12.10)$$

are conserved quantities, as is seen directly from the Euler–Lagrange equations. In the case of the Lagrangian (3.12.9), these momenta are given by

$$p_a = g_{a\alpha}\dot{x}^\alpha + g_{ab}\dot{\theta}^b.$$

3.12.2 Definition. *The **classical Routhian** is defined by setting $p_a = \mu_a = $ constant and performing a partial Legendre transformation in the variables θ^a :*

$$R^\mu(x, \dot{x}) = \left[L(x, \dot{x}, \dot{\theta}) - \mu_a \dot{\theta}^a \right]\Big|_{p_a = \mu_a}, \qquad (3.12.11)$$

where it is understood that the variable $\dot{\theta}^a$ is eliminated using the equation $p_a = \mu_a$ and μ_a is regarded as a constant.

Now consider the Euler–Lagrange equations

$$\frac{d}{dt}\frac{\partial L}{\partial \dot{x}^a} - \frac{\partial L}{\partial x^a} = 0; \qquad (3.12.12)$$

we attempt to write these as Euler–Lagrange equations for a function from which $\dot{\theta}^a$ has been eliminated. We claim that the Routhian R^μ does the job. To see this, we compute the Euler–Lagrange expression for R^μ using the chain rule:

$$\frac{d}{dt}\left(\frac{\partial R^\mu}{\partial \dot{x}^\alpha} \right) - \frac{\partial R^\mu}{\partial x^\alpha} = \frac{d}{dt}\left(\frac{\partial L}{\partial \dot{x}^\alpha} + \frac{\partial L}{\partial \dot{\theta}^a}\frac{\partial \dot{\theta}^a}{\partial \dot{x}^\alpha} \right)$$

$$- \left(\frac{\partial L}{\partial x^\alpha} + \frac{\partial L}{\partial \dot{\theta}^a}\frac{\partial \dot{\theta}^a}{\partial x^\alpha} \right) - \frac{d}{dt}\left(\mu_a \frac{\partial \dot{\theta}^a}{\partial \dot{x}^\alpha} \right) + \mu_a \frac{\partial \dot{\theta}^a}{\partial x^\alpha}.$$

The first and third terms vanish by (3.12.12), and the remaining terms vanish using $\mu_a = p_a$. Thus, we have proved the following result.

3.12.3 Proposition. *The Euler–Lagrange equations (3.12.12) for the Lagrangian $L(x, \dot{x}, \dot{\theta})$ together with the conservation laws $p_a = \mu_a$ are equivalent to the Euler–Lagrange equations for the Routhian $R^\mu(x, \dot{x})$ together with $p_a = \mu_a$.*

The Euler–Lagrange equations for R^μ are called the **reduced Euler–Lagrange equations**, since the configuration space Q with variables (x^a, θ^a) has been *reduced* to the configuration space S with variables x^α.

Let g^{ab} denote the entries of the inverse matrix of the $m \times m$ matrix $[g_{ab}]$, and similarly, $g^{\alpha\beta}$ denote the entries of the inverse of the $k \times k$ matrix $[g_{\alpha\beta}]$. We will not use the entries of the inverse of the whole matrix tensor on Q, so there is no danger of confusion.

3.12.4 Proposition. *For L given by (3.12.9) we have*

$$R^\mu(x, \dot{x}) = g_{a\alpha} g^{ac} \mu_c \dot{x}^\alpha + \frac{1}{2}\left(g_{\alpha\beta} - g_{a\alpha} g^{ac} g_{c\beta}\right) \dot{x}^\alpha \dot{x}^\beta - V_\mu(x), \quad (3.12.13)$$

where

$$V_\mu(x) = V(x) + \frac{1}{2}g^{ab}\mu_a\mu_b$$

is the **amended potential.**

Proof. We have $\mu_a = g_{a\alpha}\dot{x}^\alpha + g_{ab}\dot{\theta}^b$, so

$$\dot{\theta}^a = g^{ab}\mu_b - g^{ab} g_{b\alpha}\dot{x}^\alpha. \quad (3.12.14)$$

Substituting this in the definition of R^μ gives

$$R^\mu(x, \dot{x}) = \frac{1}{2}g_{\alpha\beta}\dot{x}^\alpha\dot{x}^\beta + (g_{a\alpha}\dot{x}^\alpha)\left(g^{ac}\mu_c - g^{ac}g_{c\beta}\dot{x}^\beta\right)$$
$$+ \frac{1}{2}g_{ab}\left(g^{ac}\mu_c - g^{ac}g_{c\beta}\dot{x}^\beta\right)\left(g^{bd}\mu_d - g^{bd}g_{d\gamma}\dot{x}^\gamma\right)$$
$$- \mu_a\left(g^{ac}\mu_c - g^{ac}g_{c\beta}\dot{x}^\beta\right) - V(x).$$

The terms linear in $\dot{x}$ are

$$g_{a\alpha}g^{ac}\mu_c\dot{x}^\alpha - g_{ab}g^{ac}\mu_c g^{bd}g_{d\gamma}\dot{x}^\gamma + \mu_a g^{ac}g_{c\beta}\dot{x}^\beta = g_{a\alpha}g^{ac}\mu_c\dot{x}^\alpha,$$

while the terms quadratic in $\dot{x}$ are

$$\frac{1}{2}(g_{\alpha\beta} - g_{a\alpha}g^{ac}g_{c\beta})\dot{x}^\alpha\dot{x}^\beta,$$

and the terms dependent only on x are $-V_\mu(x)$, as required. ∎

Note that R^μ has picked up a term linear in the velocity, and the potential as well as the kinetic energy matrix (the **mass matrix**) have both been modified.

The term linear in the velocities has the form $A_\alpha^a \mu_a \dot{x}^\alpha$, where $A_\alpha^a = g^{ab}g_{b\alpha}$. The Euler–Lagrange expression for this term can be written

$$\frac{d}{dt}A_\alpha^a\mu_a - \frac{\partial}{\partial x^\alpha}A_\beta^a\mu_a\dot{x}^\beta = \left(\frac{\partial A_\alpha^a}{\partial x^\beta} - \frac{\partial A_\beta^a}{\partial x^\alpha}\right)\mu_a\dot{x}^\beta,$$

which is denoted by $B_{\alpha\beta}^a\mu_a\dot{x}^\beta$. If we think of the one-form $A_\alpha^a dx^\alpha$, then $B_{\alpha\beta}^a$ is its exterior derivative. The quantities A_α^a are called **connection coefficients**, and $B_{\alpha\beta}^a$ are called the **curvature coefficients**.

Introducing the modified (simpler) Routhian, obtained by deleting the terms linear in $\dot{x}$,

$$\tilde{R}^\mu = \frac{1}{2}\left(g_{\alpha\beta} - g_{a\alpha}g^{ab}g_{b\beta}\right)\dot{x}^\alpha\dot{x}^\beta - V_\mu(x),$$

the equations take the form

$$\frac{d}{dt}\frac{\partial \tilde{R}^\mu}{\partial \dot{x}^\alpha} - \frac{\partial \tilde{R}^\mu}{\partial x^\alpha} = -B^a_{\alpha\beta}\mu_a\dot{x}^\beta, \tag{3.12.15}$$

which is the form that makes intrinsic sense and generalizes to the case of nonabelian groups. The extra terms have the structure of magnetic, or Coriolis terms.

The above gives a hint of the large amount of geometry hidden behind the apparently simple process of Routh reduction. In particular, *connections* A^a_α and their *curvatures* $B^a_{\alpha\beta}$ play an important role in more general theories, such as those involving nonablelian symmetry groups (like the rotation group).

Another link with the more general theories in the preceding section is that the kinetic term in (3.12.13) can be written in the following way:

$$\frac{1}{2}(\dot{x}^\alpha, -A^a_\delta\dot{x}^\delta)\begin{pmatrix} g_{\alpha\beta} & g_{\alpha b} \\ g_{a\beta} & g_{ab} \end{pmatrix}\begin{pmatrix} \dot{x}^\beta \\ -A^b_\gamma\dot{x}^\gamma \end{pmatrix},$$

which also exhibits its positive definite nature.

Routh himself (in the mid 1800s) was very interested in rotating mechanical systems, such as those possessing an angular momentum conservation law. In this context, Routh used the term "steady motion" for dynamic motions that were uniform rotations about a fixed axis. *We may identify these with equilibria of the reduced Euler–Lagrange equations and with relative equilibria of the system before reduction.* The change of terminology from steady motions to **relative equilibria** is due to Poincaré around 1890.

Since the Coriolis term does not affect conservation of energy, we can apply the Lagrange–Dirichlet test to reach the following conclusion:

3.12.5 Proposition (Routh's Stability Criterion). *Steady motions correspond to critical points x_e of the amended potential V_μ. If $d^2 V_\mu(x_e)$ is positive definite, then the steady motion x_e is stable.*

Thus, we can see that the Routh stability criterion is a special case of the energy-momentum method.

3.13   Coupled Planar Rigid Bodies

In this section we discuss the dynamics of coupled planar rigid bodies and their reduction. This follows work of Sreenath, Oh, Krishnaprasad, and

Marsden [1988]. This is useful as a model of coupled mechanical systems and for understanding reduction, as well as good background for understanding coupled nonholonomic systems. In addition it provides a nice application of the energy-momentum method for stability. We begin with two bodies and discuss the situation for three bodies in the Internet Supplement. See Sreenath, Oh, Krishnaprasad, and Marsden [1988], Oh [1987], Grossman, Krishnaprasad, and Marsden [1988], and Patrick [1989] for further details and for the three dimensional case. Related references are Krishnaprasad [1985], Krishnaprasad and Marsden [1987], and Bloch, Krishnaprasad, Marsden, and Alvarez [1992].

Summary of Results.   Refer to Figure 3.13.1 and define the following quantities, for $i = 1, 2$:

d_i distance from the hinge to the center of mass of body i;

ω_i angular velocity of body i;

θ joint angle from body 1 to body 2;

$\lambda(\theta) \;=\; d_1 d_2 \cos\theta$;

m_i mass of body i;

$\varepsilon \;=\; m_1 m_2 / (m_1 + m_2)$ = reduced mass;

I_i moment of inertia of body i about its center of mass;

$\tilde{I}_1 \;=\; I_1 + \varepsilon d_1^2$; $\tilde{I}_2 = I_2 + \varepsilon d_2^2$ = augmented moments of inertia;

$\gamma \;=\; \varepsilon\lambda'/(\tilde{I}_1\tilde{I}_2 - \varepsilon^2\lambda^2),\quad ' = d/d\theta.$

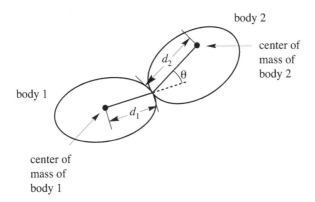

FIGURE 3.13.1. Two planar linked rigid bodies free to rotate and translate in the plane.

As we shall see, the dynamics of the system is given by the Euler–Lagrange equations for θ, ω_1, and ω_2:

$$\dot{\theta} = \omega_2 - \omega_1,$$
$$\dot{\omega}_1 = -\gamma(\tilde{I}_2\omega_2^2 + \varepsilon\lambda\omega_1^2),$$
$$\dot{\omega}_2 = \gamma(\tilde{I}_1\omega_1^2 + \varepsilon\lambda\omega_2^2). \tag{3.13.1}$$

For the Hamiltonian structure it is convenient to introduce the momenta

$$\mu_1 = \tilde{I}_1 \omega_1 + \varepsilon \lambda \omega_2, \quad \mu_2 = \tilde{I}_2 \omega_2 + \varepsilon \lambda \omega_1, \tag{3.13.2}$$

that is,

$$\begin{pmatrix} \mu_1 \\ \mu_2 \end{pmatrix} = \mathbf{J} \begin{pmatrix} \omega_1 \\ \omega_2 \end{pmatrix}, \quad \text{where} \quad \mathbf{J} = \begin{pmatrix} \tilde{I}_1 & \varepsilon \lambda \\ \varepsilon \lambda & \tilde{I}_2 \end{pmatrix} \tag{3.13.3}$$

(these may be obtained via the Legendre transform). The evolution equations for μ_i are obtained by solving for ω_1, ω_2 from (3.13.3) and substituting into (3.13.1). The Hamiltonian is

$$H = \frac{1}{2}(\omega_1, \omega_2)\mathbf{J} \begin{pmatrix} \omega_1 \\ \omega_2 \end{pmatrix}, \tag{3.13.4}$$

that is,

$$H = \frac{1}{2}(\mu_1, \mu_2)\mathbf{J}^{-1} \begin{pmatrix} \mu_1 \\ \mu_2 \end{pmatrix}, \tag{3.13.5}$$

which is the total kinetic energy for the two bodies. The Poisson structure on the (θ, μ_1, μ_2)-space is

$$\{F, H\} = \{F, H\}_2 - \{F, H\}_1, \tag{3.13.6}$$

where

$$\{F, H\}_i = \frac{\partial F}{\partial \theta} \frac{\partial H}{\partial \mu_i} - \frac{\partial H}{\partial \theta} \frac{\partial F}{\partial \mu_i}.$$

The evolution equations (3.13.1) are equivalent to Hamilton's equations $\dot{F} = \{F, H\}$. Casimirs for this bracket are readily checked to be

$$C = \Phi(\mu_1 + \mu_2) \tag{3.13.7}$$

for Φ any smooth function of one variable; that is, $\{F, C\} = 0$ for any F. One can also verify directly that, correspondingly, $(d\mu/dt) = 0$, where $\mu = \mu_1 + \mu_2$ is the total system angular momentum.

The symplectic leaves of the bracket are described by the variables $\nu = (\mu_2 - \mu_1)/2, \theta$, which parametrize a cylinder. The bracket in terms of (θ, ν) is the canonical one on T^*S^1:

$$\{F, H\} = \frac{\partial F}{\partial \theta} \frac{\partial H}{\partial \nu} - \frac{\partial H}{\partial \theta} \frac{\partial F}{\partial \nu}. \tag{3.13.8}$$

We now set up the phase space for the dynamics of our problem. Refer to Figure 3.13.2 and define the following quantities:

$\mathbf{d}_{12}$ the vector from the center of mass of body 1 to the hinge point in a fixed reference configuration;

$\mathbf{d}_{21}$ the vector from the center of mass of body 2 to the hinge point in a fixed reference configuration;

$R(\theta_i) \;=\; \begin{pmatrix} \cos\theta_i & -\sin\theta_i \\ \sin\theta_i & \cos\theta_i \end{pmatrix}$ the rotation through angle θ_i giving the current orientation of body i (written as a matrix relative to the fixed standard inertial frame);

$\mathbf{r}_i$ current position of the center of mass of body i;

$\mathbf{r}$ current position of the system center of mass;

$\mathbf{r}_i^0$ the vector from the system center of mass to the center of mass of body i;

$\theta \;=\; \theta_2 - \theta_1$ joint angle;

$R(\theta) \;=\; R(\theta_2)\cdot R(-\theta_1)$ joint rotation;

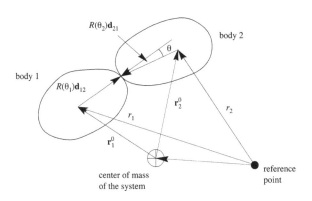

FIGURE 3.13.2. Quantities needed for setting up the dynamics of two coupled rigid bodies in the plane.

The configuration space we start with is Q, the subset of $SE(2) \times SE(2)$ (two copies of the special Euclidean group of the plane) consisting of pairs $(R(\theta_1), \mathbf{r}_1), (R(\theta_2), \mathbf{r}_2))$ satisfying the **hinge constraint**

$$\mathbf{r}_2 = \mathbf{r}_1 + R(\theta_1)\mathbf{d}_{12} - R(\theta_2)\mathbf{d}_{21}. \qquad (3.13.9)$$

Notice that Q is of dimension 4 and is parametrized by θ_1, θ_2, and, say, $\mathbf{r}_1$; that is, $Q \approx S^1 \times S^1 \times \mathbb{R}^2$. We form the velocity phase space TQ and momentum phase space T^*Q.

The Lagrangian on TQ is the kinetic energy (relative to the inertial frame) given by summing the kinetic energies of each body. To spell this out, let $\mathbf{X}_i$ denote a position vector in body 1 relative to the center of mass of body 1, and let $\rho_1(\mathbf{X}_1)$ denote the mass density of body 1. Then the current position of the point with material label $\mathbf{X}_1$ is

$$\mathbf{x}_1 = R(\theta_1)\mathbf{X}_1 + \mathbf{r}_1. \qquad (3.13.10)$$

Thus, $\dot{\mathbf{x}}_1 = \dot{R}(\theta_1)\mathbf{X}_1 + \dot{\mathbf{r}}_1$, and so the kinetic energy of body 1 is

$$
\begin{aligned}
K_1 &= \frac{1}{2} \int \rho_1(\mathbf{X}_1)\|\dot{\mathbf{x}}_1\|^2 d^2\mathbf{X}_1 \\
&= \frac{1}{2} \int \rho_1(\mathbf{X}_1) \left\langle \dot{R}\mathbf{X}_1 + \dot{\mathbf{r}}_1, \dot{R}\mathbf{X}_1 + \dot{\mathbf{r}}_1 \right\rangle d^2\mathbf{X}_1 \\
&= \frac{1}{2} \int \rho_1(\mathbf{X}_1) \left[\left\langle \dot{R}\mathbf{X}_1, \dot{R}\mathbf{X}_1 \right\rangle + 2\left\langle \dot{R}\mathbf{X}_1, \dot{\mathbf{r}}_1 \right\rangle + \|\dot{\mathbf{r}}_1\|^2 \right] d^2\mathbf{X}_1.
\end{aligned}
\qquad (3.13.11)
$$

But

$$\left\langle \dot{R}\mathbf{X}_1, \dot{R}\mathbf{X}_1 \right\rangle = \operatorname{tr}(\dot{R}\mathbf{X}_1, \dot{R}\mathbf{X}_1)^T = \operatorname{tr}(\dot{R}\mathbf{X}_1^T\mathbf{X}_1\dot{R}^T) \qquad (3.13.12)$$

and

$$\int \rho_1(\mathbf{X}_1) \left\langle \dot{R}\mathbf{X}_1, \dot{\mathbf{r}}_1 \right\rangle d^2\mathbf{X}_1 = \left\langle \dot{R} \int \rho_1(\mathbf{X}_1)\mathbf{X}_1 d^2\mathbf{X}_1, \dot{\mathbf{r}}_1 \right\rangle = 0, \quad (3.13.13)$$

since $\mathbf{X}_1$ is the vector relative to the center of mass of body 1. Substituting (3.13.12) and (3.13.13) into (3.13.11) and defining the matrix

$$\mathbf{I}^1 = \int \rho(\mathbf{X}_1)\mathbf{X}_1\mathbf{X}_1^T d^2\mathbf{X}_1 \qquad (3.13.14)$$

we get

$$K_1 = \frac{1}{2}\operatorname{tr}(\dot{R}(\theta_1)\mathbf{I}^1(\dot{R}(\theta_1)^T)) + \frac{1}{2}m_1\|\dot{\mathbf{r}}_1\|^2, \qquad (3.13.15)$$

with a similar expression for $\mathbf{K}_2$. Now let

$$L : TQ \to \mathbb{R} \quad \text{be defined by} \quad L = K_1 + K_2. \qquad (3.13.16)$$

The equations of motion then are the Euler–Lagrange equations for this L on TQ. Equivalently, they are Hamilton's equations for the corresponding Hamiltonian.

For later convenience, we shall rewrite the energy in terms of $\omega_1 = \dot{\theta}_1$, $\omega_2 = \dot{\theta}_2$, $\mathbf{r}_1^0$, and $\mathbf{r}_2^0$. To do this note that by definition,

$$m\mathbf{r} = m_1\mathbf{r}_1 + m_2\mathbf{r}_2, \qquad (3.13.17)$$

where $m = m_1 + m_2$, and so, using $\mathbf{r}_1 = \mathbf{r} + \mathbf{r}_2^0$,

$$0 = m_1\mathbf{r}_1^0 + m_2\mathbf{r}_2^0 \qquad (3.13.18)$$

and, subtracting $\mathbf{r}$ from both sides of (3.13.9),

$$\mathbf{r}_2^0 = \mathbf{r}_1^0 + R(\theta_1)\mathbf{d}_{12} - R(\theta_2)\mathbf{d}_{21}. \tag{3.13.19}$$

From (3.13.18) and (3.13.19) we find that

$$\mathbf{r}_2^0 = \frac{m_1}{m}(R(\theta_1)\mathbf{d}_{12} - R(\theta_2)\mathbf{d}_{21}) \tag{3.13.20}$$

and

$$\mathbf{r}_1^0 = -\frac{m_2}{m}(R(\theta_1)\mathbf{d}_{12} - R(\theta_2)\mathbf{d}_{21}). \tag{3.13.21}$$

Now we substitute

$$\mathbf{r}_1 = \mathbf{r} + \mathbf{r}_1^0, \quad \text{so} \quad \dot{\mathbf{r}}_1 = \dot{\mathbf{r}} + \dot{\mathbf{r}}_1^0, \tag{3.13.22}$$

and

$$\mathbf{r}_2 = \mathbf{r} + \mathbf{r}_2^0, \quad \text{so} \quad \dot{\mathbf{r}}_2 = \dot{\mathbf{r}} + \dot{\mathbf{r}}_2^0, \tag{3.13.23}$$

into our expression for the Lagrangian to give

$$L = \frac{1}{2}\operatorname{tr}\left(\dot{R}(\theta_1)\mathbf{I}^1\dot{R}(\theta_1)^T + \dot{R}(\theta_2)\mathbf{I}^2\dot{R}(\theta_2)^T\right)$$
$$+ \frac{1}{2}[m_1\|\dot{\mathbf{r}} + \dot{\mathbf{r}}_1^0\|^2 + m_2\|\dot{\mathbf{r}} + \dot{\mathbf{r}}_2^0\|^2]. \tag{3.13.24}$$

But $m_1\langle\dot{\mathbf{r}}, \dot{\mathbf{r}}_1^0\rangle + m_2\langle\dot{\mathbf{r}}, \dot{\mathbf{r}}_2^0\rangle = 0$, since $m_1\dot{\mathbf{r}}_1^0 + m_2\dot{\mathbf{r}}_2^0 = 0$. Thus L simplifies to

$$L = \frac{1}{2}\operatorname{tr}\left(\dot{R}(\theta_1)\mathbf{I}^1\dot{R}(\theta_1)^T + \dot{R}(\theta_2)\mathbf{I}^2\dot{R}(\theta_2)^T\right)$$
$$+ \left(\frac{p^2}{2m}\right) + \frac{1}{2}m_1\|\dot{\mathbf{r}}_1^0\|^2 + \frac{1}{2}m_2\|\dot{\mathbf{r}}_2^0\|^2, \tag{3.13.25}$$

where $p = m\|\dot{\mathbf{r}}\|$ is the magnitude of the system momentum.

Now write

$$\dot{R}(\theta_1) = \frac{d}{dt}\begin{pmatrix} \cos\theta_1 & -\sin\theta_1 \\ \sin\theta_1 & \cos\theta_1 \end{pmatrix} = \begin{pmatrix} -\sin\theta_1 & -\cos\theta_1 \\ \cos\theta_1 & -\sin\theta_1 \end{pmatrix}\omega_1$$
$$:= R(\theta_1)\begin{pmatrix} 0 & -\omega_1 \\ \omega_1 & 0 \end{pmatrix} := R(\theta_1)\hat{\omega}_1, \tag{3.13.26}$$

so that

$$\dot{\mathbf{r}}_2^0 = \frac{m_1}{m}(R(\theta_1)\hat{\omega}_1\mathbf{d}_{12} - (R(\theta_2)\hat{\omega}_2\mathbf{d}_{21}), \tag{3.13.27}$$

$$\dot{\mathbf{r}}_1^0 = -\frac{m_2}{m}(R(\theta_1)\hat{\omega}_1\mathbf{d}_{12} - (R(\theta_2)\hat{\omega}_2\mathbf{d}_{21}). \tag{3.13.28}$$

Thus we obtain

$$L = \frac{1}{2} \operatorname{tr}((\hat{\omega}_1 \mathbf{I}^1 \hat{\omega}_1^T) + \hat{\omega}_1 \mathbf{I}^2 \hat{\omega}_2^T)) + \frac{p^2}{2m} + \frac{m_1 m_2}{m} \|\hat{\omega}_1 \mathbf{d}_{12} - R(\theta_2 - \theta_1)\hat{\omega}_2 \mathbf{d}_{21}\|^2.$$
(3.13.29)

Finally, we note that

$$\operatorname{tr}(\hat{\omega}_1 \mathbf{I}^1 \hat{\omega}_1^T) = \operatorname{tr}(\hat{\omega}_1^T \hat{\omega}_1 \mathbf{I}^1) = \operatorname{tr}\left(\begin{pmatrix} \omega_1^2 & 0 \\ 0 & \omega_1^2 \end{pmatrix} \mathbf{I}^1\right) = \omega_1^2 \operatorname{tr} \mathbf{I}^1 := \omega_1^2 I_1,$$
(3.13.30)

where

$$I_1 = \int \rho(X_1, Y_1)(X_1^2 + Y_1^2) dX_1 dY_1$$

is the moment of inertia of body 1 about its center of mass. One similarly derives an expression where 1 is replaced by 2 throughout. We note also that

$$\|\hat{\omega}_1 \mathbf{d}_{12} - R(\theta)\hat{\omega}_2 \mathbf{d}_{21}\|^2 = \|\hat{\omega}_1 \mathbf{d}_{12}\|^2 - 2\langle \hat{\omega}_1 \mathbf{d}_{12}, R(\theta)\hat{\omega}_2 \mathbf{d}_{21}\rangle + \|\hat{\omega}_2 \mathbf{d}_{21}\|^2$$
$$= \omega_1^2 d_1^2 + \omega_2^2 d_2^2 - 2\langle \hat{\omega}_1 \mathbf{d}_{12}, \hat{\omega}_2 R(\theta) \mathbf{d}_{21}\rangle$$
$$= \omega_1^2 d_1^2 + \omega_2^2 d_2^2 - 2\omega_1 \omega_2 \langle \mathbf{d}_{12}, R(\theta)\mathbf{d}_{21}\rangle. \quad (3.13.31)$$

Finally, we get

$$L = \frac{1}{2}\left[(\omega_1^2 \tilde{I}_1 + \omega_2^2 \tilde{I}_2 + 2\omega_1 \omega_2 \varepsilon \lambda(\theta)\right] + \frac{p^2}{2m}, \quad (3.13.32)$$

where

$$\lambda(\theta) = -\langle \mathbf{d}_{12}, R(\theta)\mathbf{d}_{21}\rangle = -[\mathbf{d}_{12} \cdot \mathbf{d}_{21} \cos\theta - (\mathbf{d}_{12} \times \mathbf{d}_{21}) \cdot \mathbf{k} \sin\theta]. \quad (3.13.33)$$

Remarks.

(i) If $\mathbf{d}_{12}$ and $\mathbf{d}_{21}$ are parallel (that is, the reference configuration is chosen with $\mathbf{d}_{12}$ and $\mathbf{d}_{21}$ aligned), then $\lambda(\theta) = d_1 d_2 \cos\theta$.

(ii) The quantities $\tilde{I}_1, \tilde{I}_2$ are the moments of inertia of "augmented" bodies; for example, $\tilde{I}_1$ is the moment of inertia of body 1 augmented by putting a mass ε at the hinge point.

Reduction to the Center of Mass Frame. Now we reduce the dynamics by the action of the translation group $\mathbb{R}^2$. This group acts on the original configuration space Q by

$$\mathbf{v} \cdot ((R(\theta_1), \mathbf{r}_1), (R(\theta_2), \mathbf{r}_2)) = ((R(\theta_1), \mathbf{r}_1 + \mathbf{v}), (R(\theta_2), \mathbf{r}_2 + \mathbf{v})). \quad (3.13.34)$$

This is well-defined, since the hinge constraint is preserved by this action. The induced momentum map on TQ is calculated by the standard formula

$$J_\xi = \frac{\partial L}{\partial \dot{q}_i} \xi_Q^i(q), \quad (3.13.35)$$

or on T^*Q by

$$J_\xi = p_i \xi^i_Q(q), \tag{3.13.36}$$

where ξ^i_Q is the infinitesimal generator of the action on Q. To compute
(3.13.36) we parametrize Q by θ_1, θ_2, and $\mathbf{r}$. The momentum conjugate to
$\mathbf{r}$ is

$$\mathbf{p} = \frac{\partial L}{\partial \dot{\mathbf{r}}} = m\dot{\mathbf{r}}, \tag{3.13.37}$$

and so (3.13.36) gives

$$J_\xi = \langle \mathbf{p}, \xi \rangle, \quad \xi \in \mathbb{R}^2. \tag{3.13.38}$$

Thus $J = \mathbf{p}$ is conserved, since H is cyclic in $\mathbf{r}$, and so H is translation-
invariant. The corresponding reduced space is obtained by fixing $\mathbf{p} = \mathbf{p}_0$
and letting

$$P_{\mathbf{p}_0} = \mathbf{J}^{-1}(\mathbf{p}_0)/\mathbb{R}^2.$$

But $P_{\mathbf{p}_0}$ is isomorphic to $T^*(S^1 \times S^1)$, that is, to the space of θ_1, θ_2 and their
conjugate momenta. The reduced Hamiltonian is simply the Hamiltonian
corresponding to (3.13.32) with $\mathbf{p}_0$ regarded as a constant.

In this case the reduced symplectic manifold is a cotangent bundle, and
the reduced phase space has the canonical symplectic form: One can also
check this directly.

In (3.13.32) we can adjust L by a constant and thus assume that $\mathbf{p}_0 = 0$;
this obviously does not affect the equations of motion.

The reduced system is given by geodesic flow on $S^1 \times S^1$, since (3.13.32)
is quadratic in the velocities. Indeed, the metric tensor is just the matrix
$\mathbf{J}$ given by (3.13.3), so the conjugate momenta are μ_1 and μ_2 given by
(3.13.3).

Reduction by Rotations.   To complete the reduction, we reduce by the
diagonal action of S^1 on the configuration space $S^1 \times S^1$ that was obtained
above. The momentum map for this action is

$$J((\theta_1, \mu_1), (\theta_2, \mu_2)) = \mu_1 + \mu_2. \tag{3.13.39}$$

To facilitate stability calculations, form the Poisson reduced space

$$P := T^*(S^1 \times S^1)/S^1 \tag{3.13.40}$$

whose symplectic leaves are the reduced symplectic manifolds

$$P_\mu = J^{-1}(\mu)/S^1 \subset P.$$

We coordinatize P by $\theta = \theta_2 - \theta_1$, μ_1, and μ_2; topologically, $P = S^1 \times \mathbb{R}^2$.
The Poisson structure on P is computed as follows: Take two functions
$F(\theta, \mu_1, \mu_2)$ and $H(\theta, \mu_1, \mu_2)$. Regard them as functions of $\theta_1, \theta_2, \mu_1, \mu_2$
by substituting $\theta = \theta_2 - \theta_1$ and compute the canonical bracket. The as-
serted bracket (3.13.6) is what results. The Casimirs on P are obtained by

composing J with Casimirs on the dual of the Lie algebra of S^1, that is, with arbitrary functions of one variable; thus (3.13.7) results. This can be checked directly.

If we parametrize P_μ by θ and set $\nu = (\mu_2 - \mu_1)/2$, then the Poisson bracket on P_μ becomes the canonical one.

The realization of P_μ as $T^* S^1$ is not unique. For example, we can parametrize P_μ by (θ_2, μ_2) or by (θ_1, μ_1), each of which also gives the canonical bracket. The reduced bracket on $T^*(S^1 \times S^1)/S^1$ can also be obtained from the general formula for the bracket on $(P \times T^*G)/G \cong P \times \mathbf{3}^*$ (see Krishnaprasad and Marsden [1987]): It produces one of the variants above, depending on whether we take G to be parametrized by θ_1, θ_2, or $\theta_2 - \theta_1$.

The reduced Hamiltonian on P is (3.13.5) regarded as a function of μ_1, μ_2, and θ:

$$H = \frac{1}{2\Delta}(\mu_1, \mu_2) \begin{pmatrix} \tilde{I}_2 & -\varepsilon\lambda \\ -\varepsilon\lambda & \tilde{I}_1 \end{pmatrix} \begin{pmatrix} \mu_1 \\ \mu_2 \end{pmatrix}, \tag{3.13.41}$$

where $\Delta = \tilde{I}_1 \tilde{I}_2 - \varepsilon^2 \lambda^2$. Substituting $\mu_1 = (\mu/2) - \nu$ and $\mu_2 = \nu + (\mu/2)$ gives

$$H = \frac{1}{2\Delta}(\tilde{I}_1 + \tilde{I}_2 + 2\varepsilon\lambda)\nu^2 + \frac{1}{2\Delta}\left[\left(\tilde{I}_1 - \tilde{I}_2\right)\mu\right]\nu$$
$$+ \frac{1}{2\Delta}\left(\frac{1}{4}\mu^2\left(\tilde{I}_1 + \tilde{I}_2 - 2\varepsilon\lambda\right)\right). \tag{3.13.42}$$

The potential piece here is the *amended potential*.

We summarize as follows:

3.13.1 Theorem.   *The reduced phase space for two coupled planar rigid bodies is the three-dimensional Poisson manifold $P = S^1 \times \mathbb{R}^2$ with the bracket (3.13.6); its symplectic leaves are the cylinders with canonical variables (θ, ν). Casimirs are given by (3.13.7).*

The reduced dynamics are given by $\dot{F} = \{F, H\}$, or equivalently,

$$\dot{\theta} = \frac{\partial H}{\partial \mu_2} - \frac{\partial H}{\partial \mu_1}, \quad \dot{\mu}_1 = \frac{\partial H}{\partial \theta}, \quad \dot{\mu}_2 = -\frac{\partial H}{\partial \theta}, \tag{3.13.43}$$

where H is given by (3.13.5). The equivalent dynamics on the leaves are given by

$$\frac{\partial \theta}{\partial t} = \frac{\partial H}{\partial \nu}, \quad \frac{\partial \nu}{\partial t} = -\frac{\partial H}{\partial \theta}, \tag{3.13.44}$$

where H is given by (3.13.42).

We shall continue this discussion and illustrate the energy-momentum and energy-Casimir (or Arnold) method in the Internet Supplement.

3.14   Phases and Holonomy, the Planar Skater

In this section we give a brief introduction to phases and holonomy, emphasizing aspects that apply to control theory. This section is an abbreviated version of Marsden and Ostrowski [1998]. The main point is that we can view motion generation as a question of relating internal *shape changes* to net changes in position via a coupling mechanism, most often either interaction with the environment or via some type of conservation law.

Perhaps the best-known example of the generation of rotational motion using internal shape changes is that of the falling cat (discussed in more detail later). Released from rest with its feet above its head, the cat is able to execute a 180° reorientation and land safely on its feet. One observes that the cat achieves this net change in orientation by wriggling to create changes in its internal shape or configuration. On the surface, this provides a seeming contradiction—since the cat is dropped from rest, it has zero angular momentum at the beginning of the fall and hence, by conservation of angular momentum, throughout the duration of the fall. The cat has effectively changed its angular position while at the same time having zero angular momentum.

To understand how the reorientation works we need to keep in mind that for a rotating articulated structure, the angular momentum is the sum of the angular momenta of its rigid parts. Each part has its own angular velocity. At a deeper level this has to do with the changing locked inertial tensor due to shape changes. In fact, these ideas actually define a **principal connection** as explained in Chapter 2. While the study of this problem has a long history, e.g., Kane and Scher [1969], new and interesting insights have recently been obtained using these geometric methods; see Enos [1993], Montgomery [1990], and references therein.

One can say that this effect is an example of *the holonomy of a natural connection on the principal bundle associated with symplectic reduction.*

Another example to help visualize this effect is to consider astronauts who wish to reorient themselves in a free space environment. This motion can again be achieved using internal gyrations, or shape changes. For example, consider a motion where the arms are held out forward to lie in a horizontal plane that goes through the shoulders, parallel to the floor. If the astronaut carries out one complete cycle of arm movement, the body undergoes a net rotation in the opposite direction of arm motion. When the desired orientation is achieved, the astronaut need merely stop the arm motion in order to come to rest. One often refers to the extra motion that is achieved by the name **geometric phase**.

Historical Perspective on Geometric Phases.   The history of geometric phases and its applications is a long and complex story. We shall only mention a few highlights. The shift in the plane of the swing in the Foucault pendulum (commonly seen in science centers) as Earth rotates around its axis is certainly one of the earliest known examples of this phenomenon. Anomalous spectral shifts in rotating molecules is another. Phase formulas for special problems such as rigid body motion, elastic rods, and polarized light in helical fibers were understood already by the early 1950s, although the geometric roots to these problems go back to MacCullagh [1840] and Thomson and Tait [1879], Sections 123–126. See Berry [1990] and Marsden and Ratiu [1999] for additional historical information.

More recently this subject has become better understood, through the work of Berry[1984, 1985] and Simon [1983], whose papers first brought into clear focus the ubiquity of, and the geometry behind, all these phenomena. It was quickly realized that the phenomenon occurs in essentially the same way in both classical and quantum mechanics (see Hannay [1985]). It was also realized by Shapere and Wilczek [1987] that these ideas for classical systems were of great importance in the understanding of locomotion of microorganisms.

That geometric phases can be linked in a fundamental way with the reduction theory for mechanical systems with symmetry was realized by Gozzi and Thacker [1987] and Montgomery [1988], and developed extensively in terms of reconstruction theory for mechanical systems with symmetry (not necessarily abelian) by Marsden, Montgomery, and Ratiu [1990]. This relation with reduction has played an important role in developing an understanding of the geometric nature of many general forms of locomotion. Other relations with symplectic geometry were found by Weinstein [1990].

The theory of geometric phases has an interesting link with non-Euclidean geometry. A simple way to explain this link is as follows. Hold your hand at arms length, but allow rotation in your shoulder joint. Move your hand along three great circles, forming a triangle on the sphere, and during the motion, keep your thumb "parallel"; that is, forming a *fixed* angle with the direction of motion. After completing the circuit around the triangle, your thumb will return rotated through an angle relative to its starting position. See Figure 3.14.1. In fact, this angle (in radians) is given by $\Theta = \Delta - \pi$ where Δ is the sum of the angles of the triangle. The fact that $\Theta \neq 0$ is of course one of the basic facts of non-Euclidean geometry: In curved spaces, the sum of the angles of a triangle is not necessarily π. This angle is also related to the *area* A enclosed by the triangle through the relation $\Theta = A/r^2$, where r is the radius of the sphere.

The Role of Geometry in Control Theory.   Taking a control-theoretic perspective, this motion of the thumb above is closely related to nonholonomic control systems: We have two allowable input motions—motion along the sphere in two directions—and we generate a third, indirectly controlled, motion, rotation of our thumb. This is similar in spirit to what happens in the Heisenberg system or rolling wheel. In each case, cyclic motion in

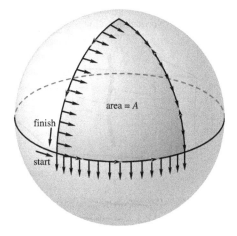

FIGURE 3.14.1. A parallel movement of your thumb around a spherical triangle produces a phase shift.

one set of variables (usually called the *internal*, *base*, or *shape* variables) produces motion in another set of variables (the *group* or *fiber* variables). In fact, the generation of net translational motion via the mechanism of internal shape changes is common to a wide range of biological and robotic and mechanical systems. See, e.g., Shapere and Wilczek [1987], Ostrowski and Burdick [1996], and Brockett [1989], and the large body of literature devoted to controlling the reorientation of satellites using only internal rotors (see later). Central to these ideas is the role of connections, which we discussed in the previous chapter.

Connections and Bundles.   As we have seen, in the general theory, connections are associated with bundle mappings, which project larger spaces onto smaller ones, as in Figure 3.14.2. The larger space is the *bundle*, and the smaller space is the *base*. Directions in the larger space that project to zero are *vertical* directions. The *connection* is a specification of a set of directions, the *horizontal* directions, at each point, which complements the space of vertical directions.

  In the example of moving one's thumb around the sphere, the larger space is the space of all tangent vectors to the sphere, and this space maps down to the sphere itself by projecting a vector to its point of attachment on the sphere. The horizontal directions are the directions with zero acceleration within the intrinsic geometry of the sphere; that is, the directions determined by great circles. When the thumb moves along a great circle at a fixed angle to its path, it is said to be *parallel transported*. Equivalently, this motion corresponds to moving in horizontal directions with respect to the connection. The rotational shift that the thumb undergoes during the course of its journey is directly related to the curvature of the sphere (and

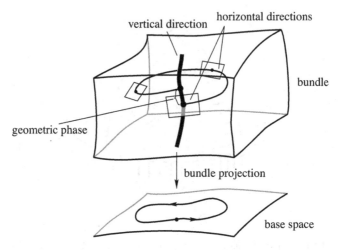

FIGURE 3.14.2. A connection divides the space into vertical and horizontal directions.

hence the curvature of the connection) and to the area enclosed by the path that is traced out.

In general, we can expect that for a horizontal motion in the bundle corresponding to a cyclic motion in the base, the vertical motion will undergo a shift, called a **phase shift**, between the beginning and the end of its path. The magnitude of the shift will depend on the curvature of the connection and the area that is enclosed by the path in the base space. This shift in the vertical element is often given by an element of a group, such as a rotation or translation group, and is called the *geometric phase*. In many of the examples discussed so far, the base space is the *control space* in the sense that the path in the base space can be chosen by suitable control inputs to the system, e.g., changes in internal shape.

In the locomotion setting, the base space describes the internal shape of the object, and cyclic paths in the shape space correspond to the movements that lead to translational and rotational motion of the body.

This setting of connections provides a framework in which one can understand the phrase "when one variable in a system moves in a periodic fashion, motion of the whole object can result." In mechanics, the basic connection is the mechanical connection that we discussed in Section 3.10.

Connections from Constraints: Momentum and Rolling. In the control and mechanics literature one confusing point is that "constraints" are often taken to be either mechanical externally imposed constraints (such as rolling) or those arising from a conservation law, such as angular momentum conservation. While these are quite different in principle, from the point of view of phases they may often be treated in the same fashion. Momentum constraints are typified by the constraint of zero angular momentum for the falling cat. This law of conservation of angular momentum exactly defines the horizontal space of the *mechanical connection*. (This connection was defined in a somewhat different, but equivalent, way in Section 3.10 above. Recall that it was discovered through the combined efforts of Smale [1970], Abraham and Marsden [1978], and Kummer [1981]. These ideas were further developed by many people such as Guichardet [1984] and Iwai [1987].)

We note that the generation of geometric phases is closely linked with the reconstruction problem discussed in Section 3.8.

One of the simplest systems in which one can see these phenomena is called the *planar skater*. This device consists of three coupled rigid bodies lying in the plane: the three-body version of the system we analyzed above. (See also the Internet Supplement for details of the three-body system).

These bodies are free to rotate and translate in the plane, somewhat like three linked ice hockey pucks. This has been a useful model example for a number of investigations, and was studied fairly extensively in Oh, Sreenath, Krishnaprasad, and Marsden [1989], Krishnaprasad [1989], and references therein (see Section 3.13).

The only forces acting on the three bodies are the forces they exert on each other as they move. Because of their translational and rotational invariance, the total linear and angular momentum remain constant as the bodies move. This holds true even if one applies controls to the joints of the device. See Figure 3.14.3.

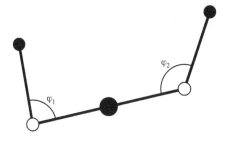

FIGURE 3.14.3. The planar skater consists of three interconnected bodies that are free to rotate about their joints.

The planar skater illustrates well some of the basic ideas of geometric phases. If the device starts with zero angular momentum and it moves its arms in a periodic fashion, then the whole assemblage can rotate, keeping, of course, zero angular momentum. The definition of angular momentum allows one to reconstruct the overall attitude of the device in terms of the motion of the joints. Doing so, one gets a motion generated in the overall attitude of the skater. This is indeed a geometric phase, dependent only on the path followed and not on the speed at which it is traversed or the overall energy of the system.

In the case of the planar skater (and in fact, for all mechanical systems with Lie group symmetries), we can write down the information encoded by the connection in a very simple form. Denoting as usual by g the group position (the vertical direction in the fiber bundle) and by r the internal shape configuration, we know from Chapter 2 that any horizontal motion, that is, any motion compatible with the given connection, must satisfy an equation of the form

$$g^{-1}\dot{g} = -A(r)\dot{r}.$$

In the event that the angular momentum of the planar skater is not zero, the system experiences a steady drift in addition to the motions caused by the internal shape changes. If we were to fix the shape variables (ϕ_1, ϕ_2), this drift would manifest itself as a steady angular rotation of the body with speed proportional to the momentum. More generally, the reorientation of the planar skater can always be decomposed into two components: the **geometric phase**, determined by the shape of the path and the area enclosed by it, and the **dynamic phase**, driven by the internal kinetic energy of the system characterized by the momentum. Figure 3.14.4 shows a schematic representation of this decomposition for general rigid body motion, which also serves to illustrate the motion of the planar skater. In this figure the sphere represents the reduced space, with a loop in the shape space shown as a circular path on the sphere. The closed circle above the sphere represents the fiber of this bundle attached to the indicated point. Given any path in the reduced (shape) space, there is an associated path, called the *horizontal lift*, that is independent of the time parametrization of the path and of the initial vertical position of the system. Following the lifted path along a loop in the shape space leads to a net change in vertical position along the fiber. This net change is just the geometric phase. On top of that, but decoupled from it, there is the motion of the system driven by the momentum, which leads to the dynamic phase. Combining these two provides the actual trajectory of the system.

Elroy's Beanie. We now describe an elementary example of geometric phases—Elroy's beanie—which still illustrates many of the interesting features of more complicated examples. This is used as a key example in the monograph Marsden, Montgomery, and Ratiu [1990]. The main point here is that conservation of angular momentum implies that a motion by one part of a system results in a corresponding shift in coordinates of another part of the system. This shift turns out to be important in control applications, such as reorientation of a system of bodies.

Consider two planar rigid bodies joined by a pin joint *at their center of masses*. Let I_1 and I_2 be their moments of inertia, and θ_1 and θ_2 the angles they make with a fixed inertial direction, as in Figure 3.14.5.

Conservation of angular momentum states that $I_1\dot{\theta}_1 + I_2\dot{\theta}_2 = \mu = $ constant in time, where the overdot means time derivative. The **shape space** of a system is the space whose points give the shape of the system. In

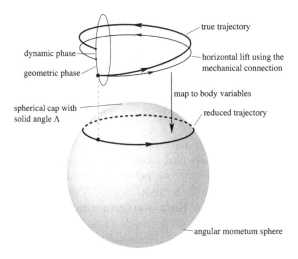

FIGURE 3.14.4. The geometric phase formula for rigid body motion; motion in the body angular momentum sphere can be periodic (lower portion of the figure), while the corresponding motion in the space of attitudes and their conjugate momenta, which carries the extra attitude information, is aperiodic (upper portion of the figure). The vertical arrow represents the map from the material representation to body representation.

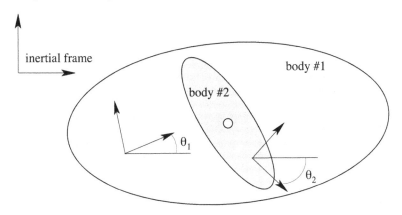

FIGURE 3.14.5. Elroy and his beanie.

this case, shape space is the circle S^1 parametrized by the **hinge angle** $\psi = \theta_2 - \theta_1$. We parametrize the configuration space of the system not by θ_1 and θ_2 but by $\theta = \theta_1$ and ψ. Conservation of angular momentum reads

$$I_1\dot{\theta} + I_2(\dot{\theta} + \dot{\psi}) = \mu; \quad \text{that is,} \quad d\theta + \frac{I_2}{I_1 + I_2}d\psi = \frac{\mu}{I_1 + I_2}\,dt. \quad (3.14.1)$$

The left-hand side of the second equation is the **mechanical connection**. Suppose that the beanie (body #2) goes through one full revolution, so that ψ increases from 0 to 2π. Suppose, moreover, that the total angular momentum is zero: $\mu = 0$. Then we see that the entire configuration undergoes a net rotation of

$$\Delta\theta = -\frac{I_2}{I_1 + I_2} \int_0^{2\pi} d\psi = -\left(\frac{I_2}{I_1 + I_2}\right) 2\pi. \qquad (3.14.2)$$

This is the amount by which Elroy rotates each time his beanie goes around once.

Notice that this result is independent of the detailed dynamics and depends only on the fact that angular momentum is conserved and the beanie goes around once. In particular, we get the same answer even if there is a "hinge potential" hindering the motion or if there is a control present in the joint. Also note that if Elroy wants to rotate by $(-2\pi k)I_1/I_1 + I_2$ radians, where k is an integer, all he needs to do is spin his beanie around k times, then reach up and stop it. By conservation of angular momentum, he will stay in that orientation after stopping the beanie.

Here is a geometric interpretation of this calculation: The connection we used is $A_{\text{mech}} = d\theta + I_2/(I_1 + I_2)d\psi$. This is a flat connection for the trivial principal S^1-bundle $\pi : S^1 \times S^1 \to S^1$ given by $\pi(\theta, \psi) = \psi$. Formula (1.1.2) is the holonomy of this connection when we traverse the base circle through an angle of $0 \leq \psi \leq 2\pi$.

4

An Introduction to Aspects of Geometric Control Theory

4.1 Nonlinear Control Systems

There are many texts on linear control theory, and a number of introductions to nonlinear control theory and in particular its differential geometric formulation, which is important for this book. We do not pretend here to give a comprehensive introduction to this subject; we just touch on a few aspects that are important to our major topics. We mention the books by Jurdjevic [1997], Isidori [1995], Nijmeijer and van der Schaft [1990], Sontag [1990], and Brockett [1970b]. The first three deal with the differential-geometric approach, Sontag's book gives a mathematical treatment of both linear theory and various parts of the nonlinear theory, while Brockett's book gives an approach to linear theory. We refer the reader to these books for a detailed treatment of nonlinear control theory as well as for a more exhaustive list of references. We also mention the important articles of Sussmann [1987], Sussmann and Jurdjevic [1972], Hermann and Krener [1977], and Brockett [1970a,b]. Two recent books that are relevant are van der Schaft [2000] and Ortega, Loria, Nicklasson and Sira-Ramirez [1998]. There are, of course, many other good sources as well, and we shall refer to them as needed.

Nonlinear Control Systems. We begin with the general notion of a control system.

4.1.1 Definition. *A finite-dimensional nonlinear control system on a smooth n-manifold M is a differential equation of the form*

$$\dot{x} = f(x, u), \tag{4.1.1}$$

*where $x \in M$, $u(t)$ is a time-dependent map from the nonnegative reals $\mathbb{R}^+$ to a constraint set $\Omega \subset \mathbb{R}^m$, and f is taken to be C^∞ (smooth) or C^ω (analytic) from $M \times \mathbb{R}^m$ into TM such that for each fixed u, f is a vector field on M. The map u is assumed to be piecewise smooth or piecewise analytic. Such a map u is said to be **admissible**. The manifold M is said to be the **state space** or the **phase space** of the system.*

It is possible to generalize this definition further; for example, Brockett [1973b] and Bloch and Crouch [1998b] deal with control systems defined on bundles. It is also possible to specialize the definition to include special structures; for example, van der Schaft [1982], Crouch and van der Schaft [1987], and Chang, Bloch, Leonard, Marsden, and Woolsey [2002] deal with Hamiltonian and Lagrangian control systems.

However, Definition 4.1.1 will be sufficient for the purposes of this chapter. The assumptions on f and u may be weakened; see, for example, Sussmann [1973], and Sussmann [1987]. The general definition of a control system is quite subtle, as reflected in the "behavioral" approach of Willems [1986].

An enormous amount can be said about control systems. We restrict ourselves here to a development of the concepts of **accessibility, controllability, stabilizability,** and **feedback linearizability.** We also restrict ourselves primarily to **affine nonlinear control systems**, which have the form

$$\dot{x} = f(x) + \sum_{i=1}^{m} g_i(x)u_i, \tag{4.1.2}$$

where f and the g_i, $i = 1, \ldots, m$, are smooth vector fields on M. We usually suppose that the constraint set Ω contains an open set of the origin in $\mathbb{R}^m$. Thus $u \equiv 0$ is an admissible control resulting in trajectories generated by the vector field f. Hence the vector field f is usually called the **drift vector field**, and the g_i are called the **control vector fields**.

General Remarks on Control Systems.   Before we go over some of these concepts in detail, we make a few general remarks about the key concepts with which one is concerned in analyzing controlled dynamical systems. In one sense, the basic goals can be divided into two parts: One goal is being able to drive the given system from one part of the state space (the phase space for a mechanical system) to another, and a second goal is being able to stabilize a given system about a given equilibrium or equilibrium manifold.

For the first goal, one is interested in the first instance in **controllability,** the question of whether one can drive the system from one point to another with the given class of controls, but one does not concern oneself with the path taken. On the other hand, one might want to prescribe a given path. This is the problem of **path planning**. Or one might want to choose a path that is optimal in some sense: the problem of **optimal control**.

Similarly, for the second goal, in analyzing the stabilizability of the given system, one can consider different definitions of stability. We concern ourselves in this book for the most part with **nonlinear stability**, but one may ask many other questions, regarding, for example, the change in system output in response to a system input. This is the question of **input–output stability**, a subject in which there has been much interest recently; see, for example, Georgiou, Praly, Sontag, and Teel [1995], Sontag and Wang [1996], Sontag and Wang [1997], and Teel [1996]. One also might wish the output to remain close to a prescribed one: the problem of **output regulation**.

There are many other related problems of varying importance; for example, a key issue is that of system **robustness**; once one has stabilized the system, is it robust to system uncertainty and external perturbations (noise)? In many cases one does not know the parameters of the system one is studying, and one can then ask for the best method of **identification** of the system at hand. System parameters may also be changing with time, and one might wish the system behavior to adapt to the changing parameters. This is the problem of **adaptive control**. Also of interest is the problem of classifying systems; for example, when is there an underlying variational or Hamiltonian structure (see, for example, Crouch and van der Schaft [1987])?

Needless to say, many of the above questions are much easier to analyze in the linear than the nonlinear setting.

Our concern in this book is mainly with analyzing stability, controllability, and optimal control in the nonlinear setting, particularly for nonholonomic mechanical systems. We discuss briefly certain classical aspects of these ideas in this chapter, and return to them in much more detail in subsequent chapters. This chapter is intended merely as minimal background.

4.2   Controllability and Accessibility

Controllability.   We begin by making precise the general notion of controllability that was discussed informally in the previous section. We assume in this section that the set of admissible controls contains the set of piecewise constant controls with values in Ω.

4.2.1 Definition.   *The system (4.1.2) is said to be **controllable** if for any two points x_0 and x_f in M there exists an admissible control $u(t)$ defined on some time interval $[0, T]$ such that the system (4.1.2) with initial condition x_0 reaches the point x_f in time T.*

Controllability is a basic concept in control theory, and much work has been done on deriving useful sufficient conditions to ensure it. However, proving that a given system is controllable is not easy in general. A related property, called (local) *accessibility*, is often much easier to prove.

Accessibility.   To define accessibility we first need the notion of a reachable set. This notion will depend on the choice of a positive time T. The reachable set from a given point at time T will be defined to be, essentially, the set of points that may be reached by the system by traveling on trajectories from the initial point in a time at most T. In particular, if $q \in M$ is of the form $x(t)$ for some trajectory with initial condition $x(0) = p$ and for some t with $0 \leq t \leq T$, then q will be said to be reachable from p in time T. More precisely:

4.2.2 Definition.   *Given $x_0 \in M$ we define $R(x_0, t)$ to be the set of all $x \in M$ for which there exists an admissible control u such that there is a trajectory of* (4.1.2) *with $x(0) = x_0$, $x(t) = x$. The **reachable set from** x_0 **at time** T is defined to be*

$$R_T(x_0) = \bigcup_{0 \leq t \leq T} R(x_0, t). \tag{4.2.1}$$

4.2.3 Definition.   *The **accessibility algebra** $\mathcal{C}$ of* (4.1.2) *is the smallest Lie algebra of vector fields on M that contains the vector fields f and $g_1, \dots, g_m$.*

Note that the accessibility algebra is just the span of all possible Lie brackets of f and the g_i.

4.2.4 Definition.   *We define the **accessibility distribution** C of* (4.1.2) *to be the distribution generated by the vector fields in $\mathcal{C}$; i.e., $C(x)$ is the span of the vector fields X in $\mathcal{C}$ at x.*

4.2.5 Definition.   *The system* (4.1.2) *on M is said to be **accessible** from $p \in M$ if for every $T > 0$, $R_T(p)$ contains a nonempty open set.*

Roughly speaking, this means that there is *some* point q (not necessarily even close to a desired objective point) that is reachable from p in time no more than T and that points close to q are also reachable from p in time no more than T.

Accessibility, while relatively easy to prove, is far from proving controllability (see the exercise at the end of this section for an example).

4.2.6 Theorem.   *Consider the system* (4.1.2) *and assume that the vector fields are C^∞. If $\dim C(x_0) = n$ (i.e., the accessibility algebra spans the tangent space to M at x_0), then for any $T > 0$, the set $R_T(x_0)$ has a nonempty interior; i.e., the system has the accessibility property from x_0.*

When the hypotheses of this theorem, namely $\dim C(x_0) = n$, hold, we say that the **accessibility rank condition** holds at x_0.

Note that while this spanning condition is an intuitively reasonable condition, the resulting theorem is quite weak, since it is far from implying controllability. The problem is that one cannot move "backward" along the drift vector field f. If f is absent, this is a strong condition; see below.

We remark also that under the conditions of Theorem 4.2.6 not only do we have a local Frobenius theorem as in Chapter 2, but the manifold M is partitioned into orbits of the generators of C. This result is a refinement of the result of Chow [1939] and is elegantly proved, for example, in Krener [1974]. Its role in control theory was developed in various papers, including those mentioned at the beginning of Section 4.1. For further references and history, see, e.g., Nijmeijer and van der Schaft [1990].

We remark that if the system and the underlying manifold is C^ω, then the converse of the above theorem holds: Accessibility implies the accessibility rank condition. Further refinements of these results may be found in Lobry [1972], Hermann [1963], Sussmann and Jurdjevic [1972], and Krener [1974].

Accessibility and Controllability. In certain special cases the accessibility rank condition *does* imply controllability, however. (We assume here that all vector fields are real analytic; the nonanalytic case can present difficulties. See, e.g., Jurdjevic [1997].)

4.2.7 Theorem. *Suppose the system* (4.1.2) *is analytic. If* $\dim C(x) = n$ *everywhere on* M *and either*

1. $f = 0$, *or*

2. f *is divergence-free and* M *is compact and Riemannian,*

then (4.1.2) *is controllable.*

The idea behind this result is that one cannot move "backward" along the drift directions, and hence a spanning condition involving the drift vector field does not guarantee controllability. A particular case of item 2 above is that in which f is Hamiltonian. This ensures a drift "backward" eventually.

4.2.8 Example (The Heisenberg System). Recall from Chapter 1 the Heisenberg system

$$\dot{x} = u,$$
$$\dot{y} = v, \qquad\qquad (4.2.2)$$
$$\dot{z} = vx - uy,$$

which may be written as

$$\dot{q} = u_1 g_1 + u_2 g_2, \qquad\qquad (4.2.3)$$

where $g_1 = (1, 0, -y)^T$ and $g_2 = (0, 1, x)^T$. As observed in Chapter 1, it is easy to compute that $[g_1, g_2] = 2g_3$, where $g_3 = (0, 0, 1)$. The three vector fields g_1, g_2, g_3 span all of $\mathbb{R}^3$, and hence this system is accessible everywhere and hence controllable everywhere in $\mathbb{R}^3$, since there is no drift vector field.   ♦

Controllability Rank Condition.   Another case of interest where accessibility implies controllability is a linear system of the form

$$\dot{x} = Ax + \sum_{i=1}^{m} b_i u_i = Ax + Bu, \qquad (4.2.4)$$

where $x \in \mathbb{R}^n$, and $A \in \mathbb{R}^n \times \mathbb{R}^n$ and $B \in \mathbb{R}^n \times \mathbb{R}^m$ are constant matrices, b_i being the columns of B.

The Lie bracket of the drift vector field Ax with b_i is readily checked to be the constant vector field $-Ab_i$. Bracketing the latter field with Ax and so on tells us that $\mathcal{C}$ is spanned by $Ax, b_i, Ab_i, \ldots, A^{n-1}b_i$, $i = 1, \ldots, m$. Thus, the accessibility rank condition at the origin is equivalent to the classical controllability rank condition

$$\text{rank}[B, AB, \ldots, A^{n-1}B] = n. \qquad (4.2.5)$$

In fact, the following theorem holds.

4.2.9 Theorem.   *The system* (4.2.4) *is controllable if and only if the controllability rank condition holds.*

Note that in this case accessibility is equivalent to controllability but that the drift vector field is involved. See, e.g., Sontag [1990] and references therein for the proof.

In this linear setting if the system is not controllable, the **reachable subspace** $\mathcal{R}$ of the system (the space reachable from the origin) is given by the range of $[B, AB, \ldots, A^{n-1}B]$.

Strong Accessibility.   In the preceding discussion, note that the term span$\{Ax\}$ is not present in the controllability rank condition. This motivates a slightly stronger definition of accessibility in the nonlinear setting, where the g_i (over which we have direct control) play a more prominent role in the rank condition:

4.2.10 Definition.   *The system* (4.1.2) *is said to be* **strongly accessible** *from x_0 if the set $R(x_0, T)$ contains a nonempty open set for any T sufficiently small.*

Thus this means that points that we can reach in *exactly* time t contain a nonempty open set.

4.2.11 Definition.   *Let $\mathcal{C}$ be the accessibility algebra of* (4.1.2). *Define $\mathcal{C}_0$ to be the smallest subalgebra containing $g_1, \ldots, g_n$ and such that $[f, X] \in \mathcal{C}_0$ for all $X \in \mathcal{C}_0$.*

4.2.12 Definition.   *We define the* **strong accessibility distribution** *C_0 of* (4.1.2) *to be the distribution generated by the vector fields in $\mathcal{C}_0$.*

If $\dim C_0(x_0) = n$, then the system (4.1.2) can be shown to be strongly accessible at x_0. The proof may be found in Nijmeijer and van der Schaft [1990]; it relies on augmenting the original system by the equation $\dot{t} = 1$.

Strong accessibility can be nicely illustrated by the Euler equations for the rigid body with external torques, as discussed in analyzed in Crouch [1984] and discussed in Nijmeijer and van der Schaft [1990]:

4.2.13 Example. Recall the Euler equations for the rigid body from Chapter 1. With two controls about the first two principal axes one has the equations

$$\dot{\Omega}_1 = \frac{I_2 - I_3}{I_1}\Omega_2\Omega_3 + a_1u_1,$$

$$\dot{\Omega}_2 = \frac{I_3 - I_1}{I_2}\Omega_3\Omega_1 + a_2u_2, \qquad (4.2.6)$$

$$\dot{\Omega}_3 = \frac{I_1 - I_2}{I_3}\Omega_2\Omega_3,$$

where the a_i are nonzero constants. Writing the system in the standard affine form

$$\dot{\Omega} = f(\Omega) + u_1g_1(\Omega) + u_2g_2(\Omega), \qquad (4.2.7)$$

we find that $[g_2, [g_1, f]](\Omega)=[0, 0, a_1a_2(I_1 - I_2)/I_3]^T$. The vectors $[a_1, 0, 0]^T$, $[a_2, 0, 0]^T$, and $[0, 0, a_1a_2(I_1 - I_2)/I_3]^T$ are contained in $C_0(0)$, and hence if $I_1 \neq I_2$, the system is strongly accessible from $\Omega = 0$. Further, $I_1 \neq I_2$ is also necessary, for if $I_1 = I_2$, then $\dot{\Omega}_3 = 0$. ◆

Small-Time Local Controllability. An important controllability condition is that of *small-time local controllability* (see again Sussmann [1987]).

4.2.14 Definition. *We say the system* (4.1.2) *is **small-time locally controllable from** x_0 if x_0 is an interior point of $R_T(x_0)$ for any $T > 0$.*

This is a genuine local controllability result as opposed to accessibility and is an important notion for control of dynamic nonholonomic control systems (which have drift), as we shall see in Chapter 6, where more details are given. We remark that small-time local controllability can often be proved for the restricted set of admissible controls that are piecewise constant (see Sussmann [1987]).

Exercises

◇ **4.2-1.** Show that the dynamic vertical penny system

$$(I + mR^2)\ddot{\theta} = u_1,$$
$$J\ddot{\varphi} = u_2,$$
$$\dot{x} = R(\cos\varphi)\dot{\theta},$$
$$\dot{y} = R(\sin\varphi)\dot{\theta},$$

is accessible at the origin. Does this imply controllability at the origin? Write down the corresponding kinematic control system and answer the same questions.

◇ **4.2-2.** Consider the system on $\mathbb{R}^2$ (see Nijmeijer and van der Schaft [1990])

$$
\begin{aligned}
\dot{x} &= y^2, \\
\dot{y} &= u.
\end{aligned}
\tag{4.2.8}
$$

Show that this system is accessible everywhere but not controllable anywhere.

◇ **4.2-3.** Show that the controllability rank condition fails for the linearized Heisenberg system.

◇ **4.2-4.** Compute the equations for the linearized cart on a pendulum equations (linearized about the unstable vertical equilibrium of the pendulum) and show that for this system the controllability rank condition holds.

◇ **4.2-5.** Consider the symmetric rigid body equations with $I_1 = I_2$ and with a single control that is not aligned along a principal axis (see Crouch [1984] and Nijmeijer and van der Schaft [1990]):

$$
\begin{aligned}
\dot{\Omega}_1 &= \frac{I_1 - I_3}{I_1}\Omega_2\Omega_3 + b_1 u, \\
\dot{\Omega}_2 &= \frac{I_3 - I_1}{I_1}\Omega_3\Omega_1 + b_2 u, \\
\dot{\Omega}_3 &= b_3 u,
\end{aligned}
\tag{4.2.9}
$$

where the b_i are constants. Show that if $I_1 \neq I_3$, $b_3 \neq 0$, and at least one of b_1 and b_2 is not zero, the system is strongly accessible from the origin.

4.3 Representation of System Trajectories

The previous section considered the general problem of controllability but did not consider the problem of constructing specific trajectories and corresponding controls. In order to attack this latter problem, we will begin below by considering useful techniques for representing system trajectories. In the following section we will apply these results to systems with periodically time-varying control functions, as in the work of Gurvitz [1992], Sussmann and Liu [1991], and Leonard and Krishnaprasad [1993, 1995]. This and the following section are quite technical and can be omitted without loss of continuity.

We employ two different coordinate systems to explore these mechanisms: coordinates of the first and second kind.

The Wei–Norman Representation—Coordinates of the Second Kind   In this section we restrict ourselves to analytic affine systems of the form (4.1.2), and in addition, we assume that the accessibility rank conditions holds everywhere on M. Assume that $X_1, \ldots, X_n$ is a set of vector fields on M that spans the accessibility distribution C in a neighborhood of a given point $x_0 \in M$.

We have the following result (see Crouch [1977, 1981]).

4.3.1 Lemma.   *Let $x(t)$ be a solution of equation (4.1.2) with $x(0) = x_0$. Then there exists a $t_0 > 0$ and functions F and G on $\mathbb{R}^n$ with components that are real analytic functions in a neighborhood of $0 \in \mathbb{R}^n$, such that for $|t| < t_0$,*

$$x(t) = e^{h_1(t)X_1} e^{h_2(t)X_2} \cdots e^{h_n(t)X_n} x_0, \tag{4.3.1}$$

where the time-dependent parameters $h = (h_1, \cdots, h_n)^T$ satisfy

$$\dot{h} = F(h) + G(h)u, \quad h(0) = 0. \tag{4.3.2}$$

If the accessibility algebra C is solvable, the spanning set $X_1, \ldots, X_n$ can be chosen such that (4.3.1) and (4.3.2) hold globally.

We call $h = (h_1, \ldots, h_n)^T$ the coordinates of x *of the second kind.* In the case that $M = G$ is a Lie group, this result is due to Wei and Norman [1964], and we call (4.3.2) the corresponding Wei–Norman system associated with the system (4.1.2).

We sketch a proof of Lemma 4.3.1, since it is instructive and will be useful later. To do this we substitute the expression (4.3.1) into the system (4.1.2). From the identity

$$T_{e^{-hZ}x} e^{hZ} Y \left(e^{-hZ} x \right) = \exp{-h \operatorname{ad}_Z}(Y)(x),$$

where X, Y, and Z are vector fields on M, we obtain

$$\dot{x} = \dot{h}_1 X_1 + \dot{h}_2 \exp{-h_1 \operatorname{ad}_{X_1}}(X_2) + \cdots \tag{4.3.3}$$
$$+ \dot{h}_n \exp{-h_1 \operatorname{ad}_{X_1}} \cdots \exp{-h_{n-1} \operatorname{ad}_{X_{n-1}}}(X_n)$$

$$= \sum_i f_i X_i + \sum_{j=1}^m g_{ij} X_i u_j$$

$$= f(x) + \sum_{j=1}^m u_j g_j(x),$$

and $f_i(x)$, $g_{ij}(x)$ are analytic functions on M in a neighborhood of x_0.

By expanding the exponential series, which are analytic, since the data is analytic, we may rewrite this expression in the form

$$\sum_{i=1}^n \dot{h}_i e_{ik}(h_1, \ldots, h_{i-1}, x) X_k = \sum_{i=1}^n \left(f_k(x) + \sum_{j=1}^m g_{kj}(x) u_j \right) X_k,$$

where the e_{ik} are the coefficients in the expansion of (4.3.3). Now for $h_1 = h_2 = \cdots = h_n = 0$, we have

$$e_{ik}(0, \ldots, 0, x) \equiv \delta_{ik},$$

so for h in a neighborhood of $0 \in \mathbb{R}^n$ we may invert the matrix $[e_{ik}]$, and once more substitute for $x(t)$ using the formula (4.3.1) to obtain the expression (4.3.2). Note that if $f \equiv 0$, then we find that $F \equiv 0$ also. Also, if we can choose $g_i = X_i$, $i = 1, \ldots, m$, then the resulting system (4.3.2), written as

$$\dot{h}_i = \sum_{j=1}^{m} G_{ij}(h) u_j, \tag{4.3.4}$$

has the property $G_{ij}(0) = \delta_{ij}$, $1 \leq i, j \leq m$, and $G_{ij}(0) = 0$ for $i > m$.

The case studied by Wei and Norman, where $M = G$ is a Lie group, with Lie algebra $\mathfrak{g}$, is already interesting. We study this case, where we additionally assume that we have no drift in the dynamics, so that $f \equiv 0$ in (4.1.2). We write the system in this case as

$$\dot{x} = \sum_{i=1}^{m} u_i X_i(x), \quad x(0) = x_0. \tag{4.3.5}$$

The case where $m < n$ is of special interest for nonholonomic systems (such as a wheeled robot), and for coupled and articulated systems such as spacecraft. We may also assume that in this case

$$X_1, \ldots, X_m, X_{m+1}, \ldots, X_n$$

is a basis for the Lie algebra $\mathfrak{g}$ of G.

4.3.2 Example (The Unicycle). In this example we consider the kinematic representation of a unicycle system with steering speed $\dot{\theta} = u_1$ and rolling speed $v = u_2$ as controls.

The system takes the form

$$\begin{aligned} \dot{x} &= u_2 \cos \theta, \\ \dot{y} &= u_2 \sin \theta, \\ \dot{\theta} &= u_1. \end{aligned} \tag{4.3.6}$$

We consider the Lie algebra $\mathfrak{se}(2)$ spanned by the matrix elements

$$\begin{pmatrix} 0 & 1 & 0 \\ -1 & 0 & 0 \\ 0 & 0 & 0 \end{pmatrix} = X_1, \quad \begin{pmatrix} 0 & 0 & 1 \\ 0 & 0 & 0 \\ 0 & 0 & 0 \end{pmatrix} = X_2, \quad \begin{pmatrix} 0 & 0 & 0 \\ 0 & 0 & 1 \\ 0 & 0 & 0 \end{pmatrix} = X_3.$$

We have

$$\begin{aligned} [X_1, X_2] &= -X_3, \\ [X_1, X_3] &= X_2, \\ [X_2, X_3] &= 0. \end{aligned}$$

Associated with $\mathfrak{se}(2)$ we have a system evolving on the corresponding Lie group $SE(2) \subset GL(3)$ by

$$\dot{x} = u_1 X_1 x + u_2 X_2 x, \quad x(0) = \text{id}. \tag{4.3.7}$$

We may restrict attention to those $x \in SE(2)$ parametrized in the form

$$x = \begin{pmatrix} \cos\phi & \sin\phi & a \\ -\sin\phi & \cos\phi & b \\ 0 & 0 & 1 \end{pmatrix}, \quad a, b, \phi \in \mathbf{R}.$$

In terms of these coordinates the system (4.3.7) may be expressed in the form

$$\begin{aligned} \dot{\phi} &= u_1, \\ \dot{a} &= bu_1 + u_2, \\ \dot{b} &= -au_1. \end{aligned} \tag{4.3.8}$$

It is easily seen that the system (4.3.8) is related to the unicycle system (4.3.6) via the expression

$$\begin{pmatrix} a \\ b \end{pmatrix} = \begin{pmatrix} \cos\theta & \sin\theta \\ -\sin\theta & \cos\theta \end{pmatrix} \begin{pmatrix} x \\ y \end{pmatrix}, \quad \theta = \phi.$$

Thus the group equations (4.3.7) represent the unicycle equations in a frame that is fixed in the unicycle.

We may also represent x in system (4.3.7) in a Wei–Norman expansion

$$x = e^{X_1 h_1} e^{X_2 h_2} e^{X_3 h_3}, \tag{4.3.9}$$

where we assume that $h_1(0) = h_2(0) = h_3(0) = 0$, so that $x(0) = \text{id}$. Using the values for X_1, X_2, and X_3 we obtain

$$\begin{aligned} x(t) &= \begin{pmatrix} \cos h_1 & \sinh_1 & 0 \\ -\sin h_1 & \cos h_1 & 0 \\ 0 & 0 & 1 \end{pmatrix} \begin{pmatrix} 1 & 0 & h_2 \\ 0 & 1 & 0 \\ 0 & 0 & 1 \end{pmatrix} \begin{pmatrix} 1 & 0 & 0 \\ 0 & 1 & h_3 \\ 0 & 0 & 1 \end{pmatrix} \\ &= \begin{pmatrix} \cos h_1 & \sin h_1 & h_2 \cosh_1 + h_3 \sin h_1 \\ -\sinh_1 & \cos h_1 & -h_2 \sin h_1 + h_3 \cos h_1 \\ 0 & 0 & 1 \end{pmatrix}. \end{aligned}$$

Substituting this expression for $x(t)$ into the system (4.3.7) to obtain the Wei–Norman equations, we obtain

$$\dot{h}_1 = u_1, \quad \dot{h}_2 = \cos h_1 u_2, \quad \dot{h}_3 = \sin h_1 u_2. \tag{4.3.10}$$

These equations are, of course, just the unicycle equations (4.3.6) under the identification

$$(h_1, h_2, h_3) = (\theta, x, y).$$

We have shown that the equations (4.3.8) on the Lie group SE(2), generate a corresponding associated Wei–Norman system that is simply the usual unicycle system (4.3.6). As we have seen, this is simply a change of reference frame in which the unicycle system is written.   ♦

Approximations   As discussed above, if the system is accessible, we know that there is a control that will drive the system from any initial state to any other desired state. However, except in the linear case, there is no immediate procedure for explicitly constructing the desired controls. However, through the representation (4.3.1) one can obtain better information on the effect of controls. In particular, we study approximations of our trajectories obtained by truncating perturbation expansions. In the next section we shall relate this to averaging. We examine these approximations by introducing a parameter ϵ into the system (4.3.5):

$$\dot{x} = \epsilon \sum_{i=1}^{m} u_i X_i(x), \quad x(0) = x_0. \tag{4.3.11}$$

Applying the Wei–Norman expansion (4.3.1) and the analysis of the corresponding Wei–Normann system (4.3.2), we obtain from the expression (4.3.3)

$$\sum_{i=1}^{n} \dot{h}_i X_i - \sum_{j \le k} \dot{h}_k [X_j, X_k] h_j = \epsilon \sum_{i=1}^{m} u_i X_i + O\left(h^2\right) \dot{h}.$$

We let γ_{ij}^k denote the structure constants of the Lie algebra $\mathfrak{g}$, so that

$$[X_i, X_j] = \sum_{k=1}^{n} \gamma_{ij}^k X_k, \quad \gamma_{ij}^k = -\gamma_{ji}^k.$$

Setting $u_i = 0$ for $i > m$ we may therefore reduce the equation to the form

$$\dot{h}_i - \sum_{j < k} \gamma_{jk}^i \dot{h}_k h_j = \epsilon u_i + O\left(h^2\right) \dot{h}.$$

Setting $h_i = \epsilon \bar{u}_i + \epsilon^2 r_i + O\left(\epsilon^3\right)$, where

$$\bar{u}_i = \int_0^t u_i(\sigma) d\sigma, \tag{4.3.12}$$

we obtain

$$\epsilon^2 \dot{r}_i - \epsilon^2 \sum_{j < k} \gamma_{jk}^i u_k \bar{u}_j = O\left(\epsilon^3\right),$$

or

$$\dot{r}_i = \sum_{j < k} \gamma_{jk}^i \bar{u}_j u_k. \tag{4.3.13}$$

Since $h_i(0) = \bar{u}_i(0) = 0$, we have $r_i(0) = 0$ for $1 \le i \le n$. Thus the Wei–Norman system corresponding to the system (4.3.11) has a solution

$$h_i = \epsilon \bar{u}_i + \epsilon^2 \sum_{j<k} \gamma^i_{jk} \int_0^t \bar{u}_j u_k d\sigma + O\left(\epsilon^3\right). \tag{4.3.14}$$

Setting

$$h'_i = \epsilon \bar{u}_i + \epsilon^2 \sum_{j<k} \gamma^i_{jk} \int_0^t \bar{u}_j u_k d\sigma \tag{4.3.15}$$

we may define an approximation $x'(t)$ to the trajectory $x(t)$ of system (4.3.11) as

$$x'(t) = e^{h'_1(t)X_1} e^{h'_2(t)X_2} \cdots e^{h'_n(t)X_n} x_0. \tag{4.3.16}$$

In the case of matrix Lie groups it is easy to see that $x'(t) = x(t) + O\left(\epsilon^3\right)$. We make this approximation more precise in the next section.

4.3.3 Example (Unicycle Continued). We may write the second-order approximation to the bilinear system (4.3.7) using the expansion (4.3.9) and the expression (4.3.15) for the second-order approximation to the solution of the Wei–Norman system (4.3.10). Noting the Lie algebra structure constants are $\bar{\gamma}^3_{12} = 1$, $\bar{\gamma}^2_{13} = -1$, and all others are zero, we see that

$$h'_1(t, \epsilon) = \epsilon \bar{u}_1(t), \quad h'_2(t, \epsilon) = \epsilon \bar{u}_2(t),$$

$$h'_3(t, \epsilon) = -\epsilon^2 \frac{t}{T} \int_0^T \bar{u}_1(t) u_2(t) \, dt.$$

◆

The Magnus Representation—Coordinates of the First Kind.
The Magnus representation of the solution of a system of differential equations applies to time-varying matrix systems of the form

$$\dot{x} = A(t)x, \quad x \in \mathrm{GL}(n), \quad x(0) = x_0. \tag{4.3.17}$$

As we explain below, the solution to this system of equations may be expressed in the form

$$x(t) = e^{Z(t)}, \quad Z(0) = 0, \tag{4.3.18}$$

and a series expression for $Z(t)$ developed, as investigated by Magnus [1954] (see also, e.g., Iserles [2002] and Iserles, Nørsett, and Rasmussen [2001] for further background). In fact, defining

$$\{A, Z^n\} = [\ldots [A, Z], Z] \ldots Z],$$

where we have n repeated commutators $[A, Z] = AZ - ZA$ of matrices, Magnus showed that formally

$$\frac{dZ}{dt}(t) = \left\{A, Z\left(I - e^{-Z}\right)^{-1}\right\} = \sum_{m=0}^{\infty} \beta_m \{A, Z^m\},$$

where $\beta_{2m} = (-1)^m B_{2m}/(2m)!$, where B_{2m} (for $m = 1, 2, 3, \ldots$) are the Bernoulli numbers and β_m vanish for $m = 3, 5, 7, \ldots$.

Thus,

$$\frac{dZ(t)}{dt} = A(t) + \frac{1}{2}[A(t), Z(t)] + \frac{1}{12}[[A(t), Z(t)], Z(t)] + \cdots.$$

From a process of series expansion one finds an expression for $Z(t)$, which up to terms involving three brackets, is given by

$$Z(t) = \int_0^t A(\sigma_1)d\sigma_1 + \frac{1}{2}\int_0^t \left[A(\sigma_1), \int_0^{\sigma_1} A(\sigma_2)d\sigma_2\right] d\sigma_1 \qquad (4.3.19)$$

$$+ \frac{1}{12}\int_0^t \left[\left[A(\sigma_1), \int_0^{\sigma_1} A(\sigma_2)d\sigma_2\right], \int_0^{\sigma_1} A(\sigma_3)d\sigma_3\right] d\sigma_1$$

$$+ \frac{1}{4}\int_0^t \left[A(\sigma_1), \int_0^{\sigma_1} \left[A(\sigma_2), \int_0^{\sigma_2} A(\sigma_3)d\sigma_3\right]\right] d\sigma_2 d\sigma_1$$

$$+ \cdots.$$

Letting $\bar{A}(t) = \int_0^t A(\sigma)d\sigma$, we may write this in the form

$$Z(t) = \bar{A}(t) + \frac{1}{2}\int_0^t \left[\dot{\bar{A}}(\sigma_1), \bar{A}(\sigma_1)\right] d\sigma_1 \qquad (4.3.20)$$

$$+ \frac{1}{12}\int_0^t \left[\left[\dot{\bar{A}}(\sigma_1), \bar{A}(\sigma_1)\right], \bar{A}(\sigma_1)\right] d\sigma_1$$

$$+ \frac{1}{4}\int_0^t \int_0^{\sigma_1} \left[\dot{\bar{A}}(\sigma_1), \left[\dot{\bar{A}}(\sigma_2), \bar{A}(\sigma_2)\right]\right] d\sigma_2 d\sigma_1$$

$$+ \cdots.$$

To determine the convergence properties of this series, we refer the reader to a generalization of this result in the case of vector fields, where (4.3.17) is replaced by a system

$$\dot{x} = f(t, x), \quad x \in M, \quad x(0) = x_0, \qquad (4.3.21)$$

where f is a smooth, time-varying vector field on a manifold M. The solution may be expressed in the form

$$x(t) = (\exp Z(t))(x_0), \quad Z(0) = 0, \qquad (4.3.22)$$

where $Z(t)$ is another vector field on M, and $z(t) = (\exp Z(t))(x_0)$ is the solution of the autonomous differential equation

$$\frac{dz(\sigma)}{d\sigma} = Z(t)(z(\sigma)), \quad z(0) = x_0,$$

on M. The explicit form and convergence of $Z(t)$ were studied in Strichartz [1987].

We call (4.3.18), or more generally (4.3.22), expressions of the solutions of the corresponding differential equations (4.3.17) and (4.3.21) in *coordinates of the first kind*.

We now apply the expansion (4.3.20) to the system (4.3.11), where we now assume that the group G is a subgroup of the matrix group $GL(n)$, and $X_i(x) = X_i x$ are right-invariant vector fields. Hence

$$A(t) = \epsilon \sum_{i=1}^{M} u_i X_i.$$

Applying this to (4.3.20) we obtain the following expression for $Z(t)$:

$$
\begin{aligned}
Z(t) = {} & \epsilon \sum_i \bar{u}_i X_i + \epsilon^2 \frac{1}{2} \sum_{j,k} \int_0^t \dot{\bar{u}}_j(\sigma_1) \bar{u}_k(\sigma_1) [X_j, X_k] d\sigma_1 \\
& + \frac{\epsilon^3}{12} \int_0^t \sum_{j,k,m} \dot{\bar{u}}_j(\sigma_1) \bar{u}_k(\sigma_1) \bar{u}_m(\sigma_1) [[X_j, X_k], X_m] d\sigma_1 \\
& + \frac{\epsilon^3}{12} \int_0^t \int_0^{\sigma_1} \sum_{j,k,m} \dot{\bar{u}}_j(\sigma_1) \dot{\bar{u}}_k(\sigma_2) \bar{u}_m(\sigma_2) [X_j, [X_k, X_m]] d\sigma_2 d\sigma_1 \\
& + O\left(\epsilon^4\right).
\end{aligned}
$$

We shall concentrate on the expansion to second order. Since $[X_j, X_k] = -[X_k, X_j]$, we may rewrite this expression in the form

$$Z(t) = \epsilon \sum_i \bar{u}_i X_i + \epsilon^2 \sum_{j<k} A_{jk}(t) [X_j, X_k] + O\left(\epsilon^3\right), \tag{4.3.23}$$

where

$$A_{jk}(t) = \frac{1}{2} \int_0^t (\dot{\bar{u}}_j(\sigma_1) \bar{u}_k(\sigma_1) - \dot{\bar{u}}_k(\sigma_1) \bar{u}_j(\sigma_1)) d\sigma_1. \tag{4.3.24}$$

It is interesting to compare the two results of expanding solutions of (4.3.11) in coordinates of the first kind and coordinates of the second kind. We apply the expression (4.3.1) for coordinates of the second kind in the case where X_i are now matrices and the coordinates h_i are given by the expression (4.3.14). (Note: By insisting that the $X_i(x)$ be right-invariant vector fields $X_i x$, we in effect, change the structure constants of the Lie algebra of $\bar{\gamma}_{j,k}^i = -\gamma_{j,k}^i$, where as matrices, $[X_i, X_j] = \sum_k X_k \bar{\gamma}_{ij}^k$.) Thus

$$h_i = \epsilon \bar{u}_i - \epsilon^2 \sum_{j<k} \bar{\gamma}_{jk}^i \int_0^t \bar{u}_j u_k d\sigma + O\left(\epsilon^3\right). \tag{4.3.25}$$

To reduce the expression (4.3.1) to a single exponential to compare with our expression in coordinates of the first kind we need the Baker–Campbell–Hausdorff formula for matrix exponentials (see Abraham, Marsden, and

Ratiu [1988])

$$e^{\epsilon x} e^{\epsilon y} = e^{\epsilon x + \epsilon y + \frac{1}{2}\epsilon^2 [x,y] + O(\epsilon^3)}.$$

Applying this repeatedly to the expression (4.3.1) we see that $x(t) = e^{Z'(t)} x_0$, where

$$Z'(t) = \sum_{i=1}^{n} h_i X_i + \frac{1}{2} \sum_{i<j} h_i h_j [X_i, X_j] + \text{ h.o.t.}$$

Substituting the formula (4.3.25) we obtain

$$Z'(t) = \epsilon \sum_{i=1}^{n} \bar{u}_i X_i - \epsilon^2 \sum_{j<k} X_i \bar{\gamma}^i_{jk} \int_0^t \bar{u}_j \dot{\bar{u}}_k d\sigma + \frac{\epsilon^2}{2} \sum_{j<k} \bar{u}_j \bar{u}_k [X_j, X_k] + O\left(\epsilon^3\right).$$

Now, $\bar{u}_j \bar{u}_k = \int_0^t (\dot{\bar{u}}_j \bar{u}_k + \dot{\bar{u}}_k \bar{u}_j) d\sigma$. Thus we may write

$$Z'(t) = \epsilon \sum_{i=1}^{n} \bar{u}_i X_i + \epsilon^2 \sum_{j<k} X_i \bar{\gamma}^i_{jk} \int_0^t \left(\frac{\dot{\bar{u}}_j \bar{u}_k + \dot{\bar{u}}_k \bar{u}_j}{2} - \bar{u}_j \dot{\bar{u}}_k \right) d\sigma + O\left(\epsilon^3\right)$$

$$= \epsilon \sum_{i=1}^{n} \bar{u}_i X_i + \epsilon^2 \sum_{j<k} X_i \bar{\gamma}^i_{jk} A_{jk}(t) + O\left(\epsilon^3\right). \tag{4.3.26}$$

But this expression coincides with the expression (4.3.23) for the coordinate representation of the first kind.

4.4   Averaging and Trajectory Planning

In this section we apply the results of the previous section to periodically time-varying control functions. The results of applying these control functions on the trajectory generated by them can be assessed by applying averaging techniques and the expansion techniques of the previous section. See, for example, Gurvitz [1992], Sussmann and Liu [1991], and Leonard and Krishnaprasad [1993, 1995]. We loosely follow the work in the latter references. Another recent reference of interest is Bullo [2001]. A basic exposition on averaging may be found in Guckenheimer and Holmes [1983] and references therein.

The Averaging Theorem.   A classical averaging result as contained in Khalil [1992] relates a system of the form

$$\dot{x} = \epsilon f(t, x, \epsilon), \quad x \in \mathbb{R}^n, \tag{4.4.1}$$

where f is a smooth mapping, periodic in t, with period T, to the average system

$$\dot{\bar{x}} = \epsilon f_{\text{av}}(\bar{x}), \quad x \in \mathbb{R}^n, \tag{4.4.2}$$

where $f_{av} = \frac{1}{T} \int_0^T f(t, x, 0) dt$.

Theorem 7.4 of Khalil [1992] states that if $\|x(0, \epsilon) - \bar{x}(0, \epsilon)\| = O(\epsilon)$, then $\|x(t, \epsilon) - \bar{x}(t, \epsilon)\| = O(\epsilon)$, where $x(t, \epsilon)$ and $\bar{x}(t, \epsilon)$ are solutions of (4.4.1) and (4.4.2), respectively, and $t \leq O(1/\epsilon)$.

Of course, a special case is presented by the system

$$\dot{x} = \epsilon h(t), \quad x(0) = 0,$$

where $h(t)$ is periodic of period T. The averaged system is

$$\dot{\bar{x}} = \frac{\epsilon}{T} \int_0^T h(t) dt, \quad \bar{x}(0) = 0.$$

Since $\frac{t}{T} \int_0^T h(\sigma) d\sigma - \int_0^t h(\sigma) d\sigma$ is also periodic of period T, we have

$$\|x(t, \epsilon) - \bar{x}(t, \epsilon)\| = O(\epsilon) \tag{4.4.3}$$

for all $t > 0$.

Systems on Lie Groups. We first consider a system on a Lie group G governed by the equations (4.3.11), where we apply periodic controls $u_i(t)$ with period T. To apply the above averaging result on $\mathbb{R}^n$ above to this system on a Lie group we proceed as follows.

We apply the result to the **Wei–Norman system** defined by

$$\dot{h} = \epsilon G(h) u, \quad h(0) = 0, \quad h \in \mathbb{R}^n, \tag{4.4.4}$$

corresponding to the expansion (4.3.1).

Let

$$u_T = \frac{1}{T} \int_0^T u(\sigma) d\sigma.$$

Then the corresponding average system is given by

$$\dot{\bar{h}} = \epsilon G(\bar{h}) u_T, \quad \bar{h}(0) = 0.$$

By (4.4.3) $\|h(t, \epsilon) - \bar{h}(t, \epsilon)\| = O(\epsilon)$ for $t \leq O(1/\epsilon)$. The trajectory may be substituted into the expansion (4.3.1) to obtain an expression for the averaged trajectory G.

We now consider second-order effects. We assume that $u(t)$ is periodic as before but with the additional constraint that $u_T = 0$. We let

$$h(t) = \epsilon \bar{u}_0(t) + r(t)$$

in (4.4.4) and note as in (4.3.4) that $G(0)u = u_0$, where $u_0 = (u_1, \ldots, u_m, 0, \ldots, 0)^T$, to obtain

$$\dot{r} = \epsilon \bar{G}(r + \epsilon \bar{u}_0) u, \quad r(0) = 0, \tag{4.4.5}$$

where $G(h) = G(0) + \overline{G}(h) = G(0) + G_L(h) + G_N(h)$ and $G_L(h) = \overline{G}(h) - G_N(h)$ is linear in h. We may therefore rewrite (4.4.5) in the form

$$\dot{r} = \epsilon^2 G_L(\bar{u}_0) u + (\epsilon G_L(r) u + \epsilon G_N(r + \epsilon \bar{u}_0) u).$$

4.4.1 Lemma (Leonard and Krishnaprasad [1993]). *If*

$$\dot{\bar{r}} = \frac{\epsilon^2}{T} \int_0^T G_L(\bar{u}_0)u \, dt,$$

with $\bar{r}(0) = 0$, *then*

$$\|r(t,\epsilon) - \bar{r}(t,\epsilon)\| = O\left(\epsilon^3\right)$$

for $t \leq O(1/\epsilon)$.

The critical element in this result is noting that since u is periodic with zero average $u_T = 0$, $G_L(\bar{u}_0)u$ is also periodic so that

$$\left(\frac{t}{T} \int_0^T G_L(\bar{u}_0)u \, dt - \int_0^t G_L(\bar{u}_0)u \, dt \right) = O(1)$$

for all t. Now combining this result with the decomposition of h, we obtain the following theorem:

4.4.3 Theorem (Leonard and Krishnaprasad [1993]). *If u is a periodic control with $u_T = 0$, then the Wei–Norman system (4.4.4) has a solution $h(t, \epsilon)$ satisfying*

$$\left\| h(t,\epsilon) - \left(\epsilon \bar{u}_0(t) + \frac{\epsilon^2}{T} t \int_0^T G_L(\bar{u}_0)u \, dt \right) \right\| = O\left(\epsilon^3\right) \qquad (4.4.6)$$

for $t \leq O(1/\epsilon)$.

By comparing this result with the analysis of the Wei–Norman decomposition in the previous section (see 4.3.15) we see that

$$[G_L(\bar{u}_0)u]_i = \sum_{1 \leq j < k \leq m} \gamma_{jk}^i \bar{u}_j u_k. \qquad (4.4.7)$$

Similarly, we can obtain the approximation in coordinates of the first kind, using the approximation obtained in the previous section, by employing the Magnus representation in the case of a matrix Lie group. From (4.3.26) we obtain the following result:

4.4.4 Theorem (Leonard and Krishnaprasad [1993]). *For the system*

$$\dot{x} = \epsilon \sum_{i=1}^m u_i X_i x,$$

with $x(0) = x_0$ and $x \in \mathrm{GL}(n)$, where u is a periodic control with $u_T = 0$, $x(t) = e^{Z(t)} x_0$ with

$$\left\| Z(t) - \left(\epsilon \sum_{i=1}^m \bar{u}_i(t) X_i + \epsilon^2 t \sum_{\substack{1 \leq j < k \leq m \\ i=1}}^{i=n} X_i \bar{\gamma}_{jk}^i A_{jk}(T) \right) \right\| = O\left(\epsilon^3\right) \qquad (4.4.8)$$

for $t \leq O(1/\epsilon)$.

Both of the results in Theorems 4.4.3 and 4.4.4 represent the second-order approximations to the solutions of the bilinear system (4.3.11), expressed in coordinates of the second and first kind, respectively. They both consist of an $O(\epsilon)$ periodic term and an $O(\epsilon^2)$ secular term (linear in t). In the case of representation by coordinates of the first kind, the secular term is proportional to area terms that describe the area bounded by the closed curves defined by $\bar{u}_i$ and $\bar{u}_j$ over one period, namely, $A_{ij}(T)$.

Motion Planning.   We now briefly describe the application of Theorems 4.4.3 and 4.4.4 to trajectory planning. For simplicity we consider only the result in Theorem 4.4.4 and apply it to the system described there, consisting of a system on a matrix Lie group G with right-invariant vector fields. Let us suppose that in system (4.3.11) the Lie algebra $\mathfrak{g}$ is spanned by

$$X_1, \ldots, X_m, \{[X_i, X_j]; \quad 1 \leq i < j \leq m\}. \tag{4.4.9}$$

Let us also suppose we wish to drive the system from x_0 to x_f, where $x_0, x_f \in G$, so we seek $Z(t_f)$ with

$$x_f = e^{Z(t_f)} x_0,$$

where we suppose that $\log(x_f x_0^{-1})$ is defined on G. From the fact that the set in (4.4.9) spans $\mathfrak{g}$, there exist constants c_i, c_{ij} such that

$$\log(x_f x_0^{-1}) = \sum_{i=1}^m c_i X_i + \sum_{j<k} c_{jk}[X_j, X_k] \tag{4.4.10}$$

$$= \sum_{i=1}^m c_i X_i + \sum_{i=1}^m \sum_{j<k} c_{jk} \bar{\gamma}_{jk}^i X_i.$$

With reference to the expression (4.4.8) we see that by choosing t_f and the controls u_i that satisfy the relations

$$\epsilon t_f = 1,$$
$$\epsilon \bar{u}_i(t_f) = c_i, \tag{4.4.11}$$
$$\epsilon^2 t_f A_{jk}(T) = c_{jk},$$

we have

$$\|Z(t_f) - \log(x_f x_0^{-1})\| = O(\epsilon^2). \tag{4.4.12}$$

Clearly, equations (4.4.11) can be satisfied by periodic controls u, but more importantly, they can be satisfied by suitably chosen sinusoids, with variable parameters of amplitude, frequency, and phase. This makes the process of satisfying the system (4.4.11) very mechanistic.

That we meet the objective of reaching the terminal value x_f only to $O(\epsilon^2)$ is clearly a drawback. This can be mitigated by iterating this process. The first reference to such a process (even in the case where third-order approximations are required) is described in Crouch [1984], with respect to the attitude control of spacecraft.

Exercises

◇ **4.4-1.** Write down the Wei–Norman equations for the kinematic controlled rigid-body system

$$\dot{X} = X\hat{u}, \qquad (4.4.13)$$

where $X \in \mathrm{SO}(3)$, $u = (u_1, u_2, u_3)$ is the vector of angular velocities viewed as controls, and

$$\hat{u} \equiv \begin{pmatrix} 0 & -u_3 & u_2 \\ u_3 & 0 & -u_1 \\ -u_2 & u_1 & 0 \end{pmatrix}.$$

Hint: Use the Euler angles; see Leonard and Krishnaprasad [1993].

4.5   Stabilization

4.5.1 Definition. *Let x_0 be an equilibrium of the control system $\dot{x} = f(x, u)$. The system is said to be **nonlinearly (asymptotically) stabilizable** at x_0 if a feedback control $u(x)$ can be found that renders the system nonlinearly (asymptotically) stable.*

In much of the control literature stabilizability is taken to mean asymptotic stabilizability. However, in this book we will distinguish between the two.

The classical linear case is of interest: In this case the system is stabilizable (in the sense of asymptotic stability) if and only if it is controllable. More than that, in this case one can show that one can find a feedback $u = Fx$ such that the closed loop system $\dot{x} = (A + BF)x$ has an arbitrarily assigned set of eigenvalues.

Stabilization Techniques. Now we discuss some stabilization techniques in the literature that are of relevance to the material in this book. Good references for classical material on this subject are Sontag [1990] and Nijmeijer and van der Schaft [1990]. See also Brockett [1983].

The simplest result on stabilization for nonlinear systems is based on linearization. Consider the nonlinear control system

$$\dot{x} = f(x, u), \qquad (4.5.1)$$

with $x \in \mathbb{R}^n$ (for simplicity) and $u \in \mathbb{R}^m$, and its linearization about (x_0, u_0),

$$\dot{x} = Ax + Bu, \qquad (4.5.2)$$

where

$$A = \frac{\partial f}{\partial x}(x_0, u_0) \quad \text{and} \quad B = \frac{\partial f}{\partial u}(x_0, u_0).$$

Then if $\mathcal{R}$ is the reachable subspace of the linearized system (see Section 4.2), we may find a change of basis such that the system takes the form

$$
\frac{d}{dt}\begin{bmatrix} x_u \\ x_l \end{bmatrix} = \begin{bmatrix} A_{11} & A_{12} \\ 0 & A_{22} \end{bmatrix}\begin{bmatrix} x_u \\ x_l \end{bmatrix} + \begin{bmatrix} B_u \\ 0 \end{bmatrix} u , \tag{4.5.3}
$$

where x_u has dimension equal to that of Range $[B, AB, \ldots, A^{n-1}B]$.

If A_{22} has all eigenvalues in the left half-plane, a feedback may be found that stabilizes the system asymptotically to the origin. (For further details see, e.g., Nijmeijer and van der Schaft [1990]).

Brockett's Necessary Conditions. A beautiful general theorem on necessary conditions for feedback stabilization of nonlinear systems was given by Brockett [1983].

4.5.2 Theorem (Brockett).   *Consider the nonlinear system $\dot{x} = f(x, u)$ with $f(x_0, 0) = 0$ and $f(\cdot, \cdot)$ continuously differentiable in a neighborhood of $(x_0, 0)$. Necessary conditions for the existence of a continuously differentiable control law for asymptotically stabilizing $(x_0, 0)$ are:*

(*i*) *The linearized system has no uncontrollable modes associated with eigenvalues with positive real part.*

(*ii*) *There exists a neighborhood N of $(x_0, 0)$ such that for each $\xi \in N$ there exists a control $u_\xi(t)$ defined for all $t > 0$ that drives the solution of $\dot{x} = f(x, u_\xi)$ from the point $x = \xi$ at $t = 0$ to $x = x_0$ at $t = \infty$.*

(*iii*) *The mapping $\gamma : N \times \mathbb{R}^m \to \mathbb{R}^n$, N a neighborhood of the origin, defined by $\gamma : (x, u) \to f(x, u)$ should be onto an open set of the origin.*

Proof.   Part (i) is clear from equation (4.5.3). Part (ii) holds, since if a system is asymptotically stabilizable at the origin, solutions near the origin must approach it.

To prove (iii) consider the closed-loop system

$$
\dot{x} = f(x, u(x)) \equiv a(x) \tag{4.5.4}
$$

and suppose that x_0 is locally asymptotically stable. Now compute the index of x_0: For sufficiently small r the map from the ball of radius r about x_0 into S^{n-1} given by

$$
x \mapsto \frac{a(x)}{\|a(x)\|} \tag{4.5.5}
$$

has degree $(-1)^n$ (see the arguments in Chapter 2 and Milnor [1965]). Since this degree is nonzero, the map is actually onto. Hence for any small α and β we can solve

$$
\frac{a(x)}{\|a(x)\|} = \frac{\alpha}{\|\alpha\|} , \qquad \|a(x)\| = \beta, \tag{4.5.6}
$$

for $a(x)$. Hence $a(x) = f(x, u(x)) = \alpha$ is solvable for f.   ∎

In fact, the necessary conditions above hold for continuous control laws (see Coron [1990]).

Lyapunov Methods for Asymptotic Stability. Another case of interest is that where the free system is Lyapunov stable and the controls may be used to make the system asymptotically stable. Issues of this sort are of key importance in the recent works Bloch, Leonard, and Marsden [1997, 1998]; Chang, Bloch, Leonard, Marsden, and Woolsey [2002], for example, and we shall return to them later. In that case the Hamiltonian structure of the system is used for energy shaping, and then a suitable dissipation is added.

For the moment, we just summarize the kind of general argument that one can find, for example, in Nijmeijer and van der Schaft [1990]. Consider the affine control system

$$\dot{x} = f(x) + \sum_{i=1}^{m} u_i g_i(x) \,. \tag{4.5.7}$$

Suppose that the free system $\dot{x} = f(x)$ has an equilibrium x_0 and suppose there exists a Lyapunov function $V(x)$ for the free system about x_0 in some neighborhood U of x_0.

Consider a feedback of the following form:

$$u_i(x) = -\mathcal{L}_{g_i} V(x) \,. \tag{4.5.8}$$

Then x_0 is an equilibrium point for the closed-loop system and V remains a Lyapunov function for the the closed-loop system. The question is now whether the system is in addition asymptotically stable. For this we can use a LaSalle invariance principle argument (see Chapter 2).

Consider the set

$$W = \{x \in U \mid \mathcal{L}_f V(x) = 0, \, \mathcal{L}_{g_i} V(x) = 0, \, i = 1, \ldots, m\} \,. \tag{4.5.9}$$

Let W_0 be the largest invariant set in W under the closed-loop dynamics. We observe that x_0 is in W, since V is a Lyapunov function. If W_0 is identically equal to $\{x_0\}$, then x_0 is locally asymptotically stable equilibrium point by LaSalle. Notice also that any trajectory of the closed-loop system in W is also a trajectory of the free dynamics. Hence by the LaSalle theorem, the system is locally asymptotically stable about x_0 if the largest invariant subset of the free dynamics in W equals x_0. If in addition $dV(x) \neq 0$ for $x \in U \backslash \{x_0\}$, this condition is also necessary, since by the definition of W, along any trajectory in W of the closed-loop system and hence of the free system the Lyapunov function is constant, and a nontrivial trajectory can therefore not approach x_0.

Of course, one would like to apply this reasoning without knowing the free dynamics. Some conditions for this are discussed in Nijmeijer and van der Schaft [1990]. The intuitive idea, of course, is that one would like W to be as small as possible. Locally, a sufficient condition for stability is simply that the distribution

$$\text{span}\{f(x), \text{ad}_f^k g_i(x),\ i = 1, \ldots, m,\ \text{for all } k \geq 0\} \tag{4.5.10}$$

should have rank n at x_0. But this is equivalent to the linearized system being stabilizable.

The Center Manifold. Another important tool in the analysis of the stabilization of nonlinear control systems is center manifold theory. Given that there exists a center manifold for the system and the remaining dynamics are stable, one can concentrate on stabilizing the center manifold dynamics. A nice application of this technique to the rigid-body dynamics is given in the work of Aeyels [1989] and the discussion in Nijmeijer and van der Schaft [1990]. We discussed the general theory of the center manifold earlier (see Section 2.4) and will return to it when we consider nonholonomic systems.

Feedback Linearization. An important technique for analyzing nonlinear systems is feedback linearization.

The basic question is to determine when a nonlinear control system of the form

$$\dot{x} = f(x, u) \tag{4.5.11}$$

satisfying $f(x_0, u_0) = 0$ can be transformed by nonlinear feedback $u = \alpha(x, v)$, $v \in \mathbb{R}^m$, into the form of a standard linear system

$$\dot{z} = Az + Bv. \tag{4.5.12}$$

More specifically, one can ask when an affine system of the form

$$\dot{x} = f(x) + \sum_{j=1}^{m} u_j g_j(x) \tag{4.5.13}$$

can be transformed by suitable feedback of the form

$$u_i = \alpha_i(x) + \sum_j \beta_{ij}(x) v_j \tag{4.5.14}$$

to a linear system.

This question is interesting for many reasons. In particular, it provides a method of classifying systems. For example, when are they equivalent to linear systems? Secondly, once the system is in linear form, one can apply standard linear techniques to it, for example, for acheiving stabilization.

It turns out one can give conditions for feedback linearizability in terms of a nested sequence of vector fields associated with the problem. We will not give any details here, since we shall not make use of the technique, but refer the reader to Isidori [1995], Nijmeijer and van der Schaft [1990], and Krener [1999] and references therein.

Exercises

◇ **4.5-1.** Consider again the linearized cart/pendulum equations (linearized about the unstable equilibrium of the pendulum). Find a linear feedback control that stabilizes the system.

◇ **4.5-2.** Show that the necessary condition (iii) of Brockett's theorem fails for the Heisenberg system.

◇ **4.5-3.** Show that the necessary condition (iii) of Brockett's theorem fails for the dynamic vertical penny system.

4.6 Hamiltonian and Lagrangian Control Systems

The extension of the notion of Hamiltonian and Lagrangian systems to the setting of control was formally proposed in Brockett [1976b] and extended and formalized by Willems [1979], van der Schaft [1983, 1986], and others. The book Nijmeijer and van der Schaft [1990] gives a nice summary of many of the ideas. Also see Ortega, Loria, Nicklasson and Sira-Ramirez [1998].

Lagrangian Control Systems.   We begin with the Lagrangian side. The simplest form of Lagrangian control system is a Lagrangian system with external forces: in local coordinates we have

$$\frac{d}{dt}\left(\frac{\partial L}{\partial \dot{q}_i}\right) - \frac{\partial L}{\partial q_i} = u_i, \ i = 1, \ldots, m,$$

$$\frac{d}{dt}\left(\frac{\partial L}{\partial \dot{q}_i}\right) - \frac{\partial L}{\partial q_i} = 0, \ i = m+1, \ldots, n. \tag{4.6.1}$$

More generally, we have the system

$$\frac{d}{dt}\left(\frac{\partial L(q, \dot{q}, u)}{\partial \dot{q}_i}\right) - \frac{\partial L(q, \dot{q}, u)}{\partial q_i} = 0 \tag{4.6.2}$$

for $q \in \mathbb{R}^n$ and $u \in \mathbb{R}^m$.

The latter system is a nontrivial generalization of the former, for example, if the control input is a velocity rather than a force (see Brockett [1976a], Nijmeijer and van der Schaft [1990]). The system (4.6.1) is said to be *underactuated* if $m < n$.

Similarly, one can define a Hamiltonian control system. In local coordinates, these have the form

$$\dot{q}_i = \frac{\partial H(q, p, u)}{\partial p_i},$$

$$\dot{p}_i = -\frac{\partial H(q, p, u)}{\partial q_i}, \tag{4.6.3}$$

for $q, p \in \mathbb{R}^n$ and $u \in \mathbb{R}^m$. One can readily generalize this notion to a Hamiltonian control system on a symplectic or Poisson manifold. We turn to this now.

Affine Hamiltonian Control Systems.   Let P be a Poisson manifold and let $H_0, H_1, \ldots, H_m$ be smooth functions on P. Then an **affine Hamiltonian control system** on P is given by

$$\dot{x} = X_{H_0}(x) + \sum_{j=1}^{m} X_{H_j}(x)u_i, \tag{4.6.4}$$

where $x \in M$, and as usual, X_{H_j} is the Hamiltonian vector field corresponding to H_j.

This relates to the basic question asked by Brockett [1976a]: How can one take a generic nonlinear control system of the form

$$\dot{x} = f(x, u), \tag{4.6.5}$$
$$y = h(x, u),$$

where $x \in M$, $u \in N$, $y \in P$ are smooth manifolds, $f : M \times N \to TM$ is a parametrized smooth vector field on M, and $h : M \times N \to P$ is an output mapping, and introduce appropriate additional structure so that one may systematically study problems in mechanics with external forces. The output mapping h and manifold P are important, not least in the context of examining the (output) feedback control $u(t) = k(y(t))$ for mappings $k :$ $P \to N$, but also in the context of modeling the duality between input and output spaces and interpreting in this generality such concepts as work and dissipation. The duality is especially important in modeling generalizations of conservative mechanical systems. Obvious candidates for M, P, and N for mechanical systems are $M = TQ$, $N = V$, $P = V^*$, where Q models the generalized configuration space, V is a vector bundle over Q, and V^* is the dual bundle to V. Thus if $x = (q, v) \in TQ$, with $v \in T_q Q$, the tangent space to Q at $q \in Q$, the triple (x, u, y) in (4.6.5) would have the form $u \in V_q$, $y \in V_q^*$, where $V_q(V_q^*)$ are the fibers above q in $V(V^*)$. The cases where $N = TQ$ or T^*Q arise in this book, while other choices are also used.

Additional structure needs to be added to the model (4.6.5) to reflect the Hamiltonian or Lagrangian structure, at least for the uncontrolled system, nominally modeled by choosing u to be the zero section of the bundle V, and also to reflect the manner in which controls (external forces or torques) interact with the uncontrolled system. Understanding this additional structure and appropriate specializations that reflect specific classes of mechanical systems with external forces has been a topic of much research in the past two decades, building upon both the existing control theories and analytical mechanics, but also prompting further work in both areas separately.

If a system is Hamiltonian, one can use the Hamiltonian structure of the problem to analyze basic concepts such as accessibility and stabilizability.

Key ideas in this direction may be found in Brockett [1976a], and these were developed in various ways by van der Schaft; see, for example, Nijmeijer and van der Schaft [1990] and references therein. Suppose, for example, that M is a Poisson manifold of dimension $2n$ with nondegenerate Poisson bracket; let $H_0, H_1, \ldots, H_m$ be smooth functions on M, and consider the Hamiltonian control system

$$\dot{x} = X_{H_0} + \sum_{j=1}^{m} u_j X_{H_j}(x) \,. \tag{4.6.6}$$

Consider the linear space $\mathcal{C}$ of functions spanned by all repeated Poisson brackets of the form

$$\{F_1, \{F_2, \{\cdots \{F_k, H_j\}\} \cdots \}\} \tag{4.6.7}$$

for F_i in the set $\{H_0, H_1, \ldots, H_m\}$.

Then if dim $d\mathcal{C}(x_0) = 2n$, the system (4.6.6) is strongly accessible at x_0.

To consider stabilization one wants more structure in the system. In particular, we want to consider mechanical control systems on the tangent bundle T^*Q of a symplectic manifold Q, with local coordinates (q^i, p_j) and where the Hamiltonian H_0 is of the form

$$H_0(q, p) = \frac{1}{2} p_i g^{ij}(q) p_j + V(q) \tag{4.6.8}$$

for g a Riemannian metric on Q and

$$H_j(q, p) = H_j(q) \,. \tag{4.6.9}$$

Potential and Kinetic Shaping. If $(q, p) = (q, p_0)$ is an equilibrium and $V(q) - V(q_0)$ positive definite, the equilibrium is Lyapunov stable, but not asymptotically stable. One problem of interest is then to choose a feedback control $u = K(x)$ such that the system becomes asymptotically stable. On the other hand, if the potential energy is not definite, one can think of methods of "shaping" potential energy by feedback so as to stabilize the equilibrium. There has been much interesting work on this by van der Schaft and others. For instance, a generalization of this approach to systems with symmetry was given in Jalnapurkar and Marsden [1999, 2000].

The simplest situation occurs when $m = n$ and $H_1, \ldots, H_n$ are independent. Then, setting the system $H_i = y_i$ (viewing H_i as the system outputs), any feedback of the form

$$u_i = -k_i y_i - c_i \dot{y}_i \tag{4.6.10}$$

with $k_i, c_i > 0$ and k_i sufficiently large will asymptotically stabilize the system. Indeed, this may be viewed as adding in a shaping potential $\frac{1}{2}\sum_{i=1}^{m} k_i y_i^2$ and a Rayleigh dissipation function $R(\dot{y}) = \frac{1}{2}\sum_{i=1}^{m} c_i \dot{y}_i^2$. (It is also interesting to consider the notion of the Rayleigh dissipation function on a manifold; this is analyzed in the paper Bloch, Krishnaprasad, Marsden, and Ratiu [1996].)

Of course, the interesting case is $m < n$. In van der Schaft [1986] and Jalnapurkar and Marsden [1999] conditions are given for the achievement of asymptotic stabilization in this case. We describe a version of this briefly here.

Before doing this, however, we note that there are many underactuated systems for which the method of potential shaping just cannot work (as the preceding references show). To overcome this, recent works of Bloch, Leonard, and Marsden [1997, 1998, 1999a,b, 2000, 2001]; Bloch, Chang, Leonard, Marsden, and Woolsey [2000]; Bloch, Chang, Leonard, and Marsden [2001] have considered a method of shaping *both* the potential and kinetic energy by the so-called matching method. We discuss some of this briefly in Chapter 9 as it pertains to nonholonomic systems. (See also recent work by Ortega, Loria, Nicklasson and Sira-Ramirez [1998], Ortega, van der Schaft, Mashcke and Escobar [1999], Blankenstein, Ortega, and van der Schaft [2002], Hamberg [1999], Auckly, Kapitanski, and White [2000], Shiriaev and Fradkov [2000], Polushin, Fradkov, and Khill [2000], and Chang, Bloch, Leonard, Marsden, and Woolsey [2002].)

Lagrangian Setting.   It is convenient to consider the Lagrangian setting: We assume that the given mechanical system has configuration space Q and that a Lie group G acts freely and properly on Q. It is useful to keep in mind the case in which $Q = S \times G$ with G acting only on the second factor by acting on the left by group multiplication. For example, for the inverted planar pendulum on a cart, we have $Q = S^1 \times \mathbb{R}$ with $G = \mathbb{R}$, the group of real numbers under addition (corresponding to translations of the cart). We are interested in the underactuated problem in which the controls act directly only on the variables lying in G, but that all variables in the state space are to be controlled. We suppose that G is a symmetry group for the kinetic energy of the system, but the potential energy V need not be G invariant.

Let θ^a be coordinates for G, and let x^α be coordinates for Q/G. Let the metric tensor $g(\cdot, \cdot)$ define the kinetic energy $\frac{1}{2}g(\dot{q}, \dot{q})$ and let $L : TQ \longrightarrow \mathbb{R}$ be the original Lagrangian given by the kinetic minus potential energy:

$$
\begin{aligned}
L(x^\alpha, \theta^a, \dot{x}^\alpha, \dot{\theta}^a) &= K(x^\alpha, \theta^a, \dot{x}^\alpha, \dot{\theta}^a) - V(x^\alpha, \theta^a) \\
&= \frac{1}{2}g_{\alpha\beta}\dot{x}^\alpha \dot{x}^\beta + g_{\alpha a}\dot{x}^\alpha \dot{\theta}^a + \frac{1}{2}g_{ab}\dot{\theta}^a \dot{\theta}^b - V(x^\alpha, \theta^a).
\end{aligned}
$$
(4.6.11)

Then $(x_e, \theta_e, 0, 0) \in TQ$ is the equilibrium of interest, where (x_e, θ_e) is a critical point of the original potential V. Furthermore, $D^2K(x_e, \theta_e, 0, 0)$, the second derivative of the kinetic energy K with respect to $(\dot{x}^\alpha, \dot{\theta})$ at the origin in (x, θ)-space, is taken to be a positive definite matrix.

Now assume that the following definiteness condition holds:

$$\frac{\partial^2 V}{\partial x^\alpha \partial x^\beta}(x_e, \theta_e) > 0; \tag{4.6.12}$$

i.e., the equilibrium is a minimum of the original potential energy in the x^α variables. In the case that (4.6.12) holds, the Lyapunov stabilization of the equilibrium $(x_e, \theta_e, 0, 0)$ can be achieved by just introducing a potential shaping. Indeed, choose any function $V_\epsilon : G \longrightarrow \mathbb{R}$ that has a minimum at θ_e. Let the control input u be of the form

$$u_a = -\frac{\partial V_\epsilon}{\partial \theta^a} + \tilde{u}_a. \tag{4.6.13}$$

Then, we can check that the Euler–Lagrange equations of the given Lagrangian L with the force u is equal to those of the new Lagrangian $\tilde{L}$ defined by

$$\tilde{L}(x^\alpha, \theta, \dot{x}^\alpha, \dot{\theta}) = L(x^\alpha, \theta, \dot{x}^\alpha, \dot{\theta}) - V_\epsilon(\theta^a) = K(x^\alpha, \theta^a, \dot{x}^\alpha, \dot{\theta}^a) - \tilde{V}(x^\alpha, \theta^a)$$

with the force $\tilde{u}$, where $\tilde{V} = V + V_\epsilon$. Let $\tilde{E}$ be the energy from the Lagrangian $\tilde{L}$ defined by

$$\tilde{E} = K + \tilde{V}.$$

By the choice of V_ϵ, we can see that $(x_e, \theta_e, 0, 0)$ is a critical point of $\tilde{E}$. The second derivative of $\tilde{E}$ at $z_e = (x_e, \theta_e, 0, 0)$ is

$$D^2\tilde{E}(z_e) = \begin{pmatrix} D^2\tilde{V} & 0 \\ 0 & D^2K \end{pmatrix}\bigg|_{z=z_e}, \tag{4.6.14}$$

where $D^2\tilde{V}$ is given by

$$D^2\tilde{V} = \begin{pmatrix} \dfrac{\partial^2 V}{\partial x^\alpha \partial x^\beta} & \dfrac{\partial^2 V}{\partial x^\alpha \partial \theta^a} \\ \dfrac{\partial^2 V}{\partial \theta^a \partial x^\alpha} & \dfrac{\partial^2 V}{\partial \theta^a \partial \theta^b} + \dfrac{\partial^2 V_\epsilon}{\partial \theta^a \partial \theta^b} \end{pmatrix},$$

and as above, D^2K denotes the second derivative of the kinetic energy K with respect to $(\dot{x}^\alpha, \dot{\theta})$. We already know that $D^2K(x_e, \theta_e, 0, 0)$ is a positive definite matrix. By simple linear algebra and (4.6.12), we can make $D^2\tilde{V}(x_e, \theta_e)$ positive definite by choosing V_ϵ such that its second derivative at (x_e, θ_e) is positive definite and the magnitude of its eigenvalues is very big. Thus, $\tilde{E}$ has a minimum at $(x_e, \theta_e, 0, 0)$, and we can use $\tilde{E}$ as a Lyapunov function. We introduce the input term

$$\tilde{u}_a = c_a^b g_{bc} \dot{\theta}^c, \tag{4.6.15}$$

where c_a^b is a negative definite matrix with respect to the g_{ab} metric. Then, we have

$$\frac{d}{dt}\tilde{E} = c_a^b g_{bc} \dot{\theta}^a \dot{\theta}^c \le 0. \tag{4.6.16}$$

Hence, $(x_e, \theta_e, 0, 0)$ is still an equilibrium of the closed-loop system and becomes Lyapunov stable.

To prove the asymptotic stability of the equilibrium, we will use the LaSalle invariance principle discussed in Chapter 2.

By (4.6.16) and the fact that $\tilde{E}$ has a minimum at $(x_e, \theta_e, 0, 0)$, there exists $c \in \mathbb{R}$ such that $\Omega_c := \{z = (x^\alpha, \theta^a, \dot{x}^\alpha, \dot{\theta}^a) \in TQ \,|\, \tilde{E}(z) \le c\}$ becomes a nonempty, compact, and invariant set. Define

$$\mathcal{E} := \left\{ z = (x^\alpha, \theta^a, \dot{x}^\alpha, \dot{\theta}^a) \in \Omega_c \,\left|\, \frac{d}{dt}\tilde{E}(z) = 0 \right.\right\}$$
$$= \{z = (x^\alpha, \theta^a, \dot{x}^\alpha, \dot{\theta}^a) \in \Omega_c \,|\, \dot{\theta}^a = 0\}.$$

Let $\mathcal{M}$ be the largest invariant subset of $\mathcal{E}$. Instead of directly looking into the dynamics on $\mathcal{M}$, we follow the approach given in van der Schaft [1986] and Jalnapurkar and Marsden [1999]. Let $\mathbb{F}\tilde{L} : TQ \longrightarrow T^*Q$ be the Legendre transform induced from the Lagrangian $\tilde{L}$ (see Chapter 3 for the definition). Since the Lagrangian $\tilde{L}$ is regular, we can define $\tilde{H} : T^*Q \longrightarrow \mathbb{R}$ by

$$\tilde{H} = \tilde{E} \circ \mathbb{F}\tilde{L}^{-1}.$$

Define $G_b : TQ \longrightarrow \mathbb{R}$ by

$$G_b(x^\alpha, \theta^a, \dot{x}^\alpha, \dot{\theta}^a) = \dot{\theta}^b.$$

Define $F_b : T^*Q \longrightarrow \mathbb{R}$ by

$$F_b = G_b \circ \mathbb{F}\tilde{L}^{-1}.$$

Let $(q(t), \dot{q}(t)) \in TQ$ be a trajectory of the closed-loop Lagrangian system with the force $\tilde{u}$. Then it is well known that the curve $(q(t), p(t)) \in T^*Q$ defined by

$$(q(t), p(t)) = \mathbb{F}\tilde{L}(q(t), \dot{q}(t))$$

satisfies the following Hamiltonian equations:

$$\dot{q}^i = \frac{\partial \tilde{H}}{\partial p_i},$$
$$\dot{p}_i = -\frac{\partial \tilde{H}}{\partial q^i} + \tilde{u}_i, \tag{4.6.17}$$

where $\tilde{u}_i$ is short for $\tilde{u}_i \circ \mathbb{F}\tilde{L}^{-1}$. Notice that $\mathbb{F}\tilde{L}(x_e, \theta_e, 0, 0)$ becomes an equilibrium of the system (4.6.17) and that Ω_c, $\mathcal{E}$, and $\mathcal{M}$ are diffeomorphically mapped into T^*Q via $\mathbb{F}\tilde{L}$, if necessary after shrinking Ω_c into the domain of $\mathbb{F}\tilde{L}$. Let $\{\,,\}$ be the Poisson bracket on T^*Q induced from the standard symplectic form on T^*Q (see Chapter 3 for the definition). Consider the set of functions defined by

$$\mathcal{C} = \mathrm{span}\{F_b, \{\tilde{H}, F_b\}, \{\tilde{H}, \{\tilde{H}, F_b\}\}, \ldots\}, \quad b = 1, \ldots, \dim G,$$

where by span we mean the collection of all linear combinations with real coefficients. Define the codistribution $d\mathcal{C}$ as follows:

$$d\mathcal{C} = \mathrm{span}\{\, dg \mid g \in \mathcal{C}\,\}.$$

Notice that the equilibrium $(x_e, \theta_e, 0, 0)$ is an isolated equilibrium by the Morse lemma, since every equilibrium is a critical point of the energy $\tilde{E}$ and $(x_e, \theta_e, 0, 0)$ is a nondegenerate critical point. Now one can easily see (see Jalnapurkar and Marsden [1999]) that when the dimension of $d\mathcal{C}$ is $2n$ in a neighborhood of $\mathbb{F}\tilde{L}(x_e, \theta_e, 0, 0)$, the only flow in $\mathcal{M}$ is the equilibrium. (Here n is the dimension of the configuration space Q.) This follows simply by observing that since F_b and all the brackets $\{H, F_b\}$ are conserved along the flow of a trajectory $z(t)$, the vector $\dot{z}$ lies in the annihilator of $d\mathcal{C}$. But since $d\mathcal{C}$ spans the whole cotangent space, $\dot{z}(t)$ must be zero, and thus $z(t)$ must be an equilibrium.

Now consider a more general case, that there is a subcodistribution of $d\mathcal{C}$ whose locally constant dimension is $(2n - 1)$ around the equilibrium. The subcodistribution defines a one-dimensional (regular) submanifold of T^*Q, which contains the invariant set $\mathbb{F}\tilde{L}(\mathcal{M})$ as well as the equilibrium. Since the equilibrium is stable and isolated, the flow in the one-dimensional submanifold should converge to the equilibrium if necessary after shrinking the domain. Thus the *(bi-)invariant* set $\mathcal{M}$ is the equilibrium itself. By the LaSalle invariance principle, the equilibrium is asymptotically stable. We have thus proved the following:

4.6.1 Theorem.   *If (4.6.12) holds, then $(x_e, \theta_e, 0, 0)$ is Lyapunov stabilizable. If in addition to (4.6.12), the dimension of the codistribution $d\mathcal{C}$ is greater than or equal to $(2n - 1)$ in a neighborhood of $\mathbb{F}\tilde{L}(x_e, \theta_e, 0, 0)$, where n is the dimension of the configuration space Q, then $(x_e, \theta_e, 0, 0)$ becomes an asymptotically stable equilibrium of the closed-loop system with the input u given by (4.6.13) and (4.6.15).*

The work of Bloch, Leonard, and Marsden cited above gives a combination of kinetic energy and potential energy shaping techniques for achieving stabilization in the case $m < n$ in the Lagrangian setting. We discuss this briefly in the final section of this book.

We repeat the obvious here: This section barely touches on the vast and exciting subject of nonlinear control theory. We have merely introduced some key ideas related to the subject of this book.

Exercises

◇ **4.6-1.** Formulate the cart–pendulum control system as a Hamiltonian control system. Can you achieve stabilization of the inverted pendulum via potential shaping?

◇ **4.6-2.** Show that a controlled Lagrangian system on a Riemannian manifold may be locally written in the form

$$M(y)\ddot{y} + \hat{\Gamma}(y, \dot{y}) + \frac{\partial V}{\partial y} = \begin{pmatrix} u \\ 0 \end{pmatrix}, \qquad (4.6.18)$$

where

$$\hat{\Gamma}(y, \dot{y}) = \left(\ \hat{\Gamma}_1(y, \dot{y}), \quad \ldots, \quad \hat{\Gamma}_n(y, \dot{y}) \ \right)^T, \qquad \hat{\Gamma}_k(y, \dot{y}) = \sum_{i,j} \Gamma_{ijk} \dot{y}_i \dot{y}_j,$$

and

$$\Gamma_{ijk} = \frac{1}{2} \left(\frac{\partial m_{ki}}{\partial y_j} + \frac{\partial m_{kj}}{\partial y_i} - \frac{\partial m_{ij}}{\partial y_k} \right).$$

5
Nonholonomic Mechanics

Nonholonomic systems provide an important class of mechanical control systems. One reason for this importance is that nonintegrability is essential to both the mechanics and the control: Nonintegrable constraint distributions are the essence of nonholonomic systems, while a nonintegrable distribution of control vector fields is the key to controllability of nonlinear systems. We will learn how these two different types of nonintegrability work together when we study control of nonholonomic mechanical systems.

Nonholonomic mechanical systems—systems with constraints on the velocity that are not derivable from position constraints—arise in mechanical systems such as rolling contact (wheels) or sliding contact (a skate). However, such constraints also occur in less obvious ways. For example, one may view angular momentum constraints, which are really integrals of motion and are integrable constraints on the phase space (functions of position and momentum) as nonintegrable constraints on the configuration space. This point of view is sometimes helpful for controllability. Further, many first-order control systems may be simply viewed as controlled distributions lying in the kernel of a nonintegrable constraint. A classic example of the latter is the Heisenberg system.

Some History. The history of nonholonomic mechanical systems is long and complex. There has been recurring confusion over the very equations of motion as well as the deeper questions associated with the geometry and analysis of these equations.

First of all, in terms of the equations of motion themselves, the confusion has mainly centered on *whether or not the equations can be derived from a variational principle in the usual sense*; that is, is there an action function defined on a space of curves that is being extremized? An important issue is whether the constraints are to be imposed before or after taking variations. When one first imposes the constraints on the class of curves being considered, one gets equations that are variational in the usual sense just explained. This type of approach is *certainly appropriate for optimal control problems*, as we shall see. However, for mechanics, this is *not what one should do*, as we have already seen in Chapter 1 and shall remark on further below. Namely, what is correct is that one imposes the constraints *after* taking variations. This is the **Lagrange–d'Alembert principle**.

It is a fundamental fact that the Lagrange–d'Alembert principle can be derived from balance of forces, or if you like, $F = ma$, together with balance of torques. We have seen this already in some of the basic examples in Chapter 1, and it can be proven the same way in more general contexts, such as for general rolling rigid bodies (see, for instance, Jalnapurkar [1994]). Thus, anyone who doubts that the Lagrange–d'Alembert principle is correct for the dynamical equations of nonholonomic mechanics should take up the issue with Mr. Newton. In summary, *there is no doubt that the correct equations of motion for nonholonomic mechanical systems are given by the Lagrange–d'Alembert principle.*

This issue of whether the equations of nonholonomic mechanics are variational or not was discussed extensively and "put to rest" by Korteweg [1899] in favor of the Lagrange–d'Alembert approach. Despite Korteweg's work, the issue concerning the variational nature of the equations curiously resurfaces from time to time, and the confusion is surprisingly persistent, perhaps because specific examples can exhibit nonintuitive behavior,[1] and so one is tempted to question the validity of the very equations of motion.

One such recurrence is a different set of equations of motion for nonholonomically constrained mechanical systems proposed by Kozlov [1983]. Recall that the Euler–Lagrange equations for the motion of an unconstrained mechanical system can be derived from Hamilton's principle: The trajectory followed by the system is the trajectory that is a critical point of the action integral. Similarly, in Kozlov's proposed formulation the trajectory of the constrained system is the trajectory that is a critical point of the action integral *restricted to the space of curves satisfying the constraints*. See also Arnold, Kozlov, and Neishtadt [1988] (and references therein) and Chetaev [1959], Rumiantsev [1978, 1979, 1982a], Griffiths [1983], and Bryant and Griffiths [1983]. Kozlov called the resulting equations the *vakonomic* formulation, a term we *strongly discourage* using, preferring to use the term **variational nonholonomic systems**. The proper place of these equa-

---

[1]An example of nonintuitive behavior is the motion of a golf ball as it enters a hole, drops, circles around a few times, and then pops out again. This example is discussed further at the end of this section.

tions is in the setting of optimal control, which we discuss in Chapter 7.

Besides the equations of motion themselves, the geometric foundations of nonholonomic mechanics have proved to be very useful in the analysis of the equations, including the finding of specific solutions, stability analysis, bifurcations, and control. In the developments in these directions there have been many important fundamental contributions, both in books and papers, such as Chaplygin [1897a,b, 1903, 1911, 1949, 1954], Vranceneau [1936], Cartan [1952], Neimark and Fufaev [1972], Rosenberg [1977], Weber [1986], Koiller [1992], Bloch and Crouch [1995], Karapetyan [1980], Yang [1992], Yang, Krishnaprasad, and Dayawansa [1993], Bates and Sniatycki [1993], Cushman, Kemppainen, Sniatycki, and Bates [1995], Marle [1998], van der Schaft and Maschke [1994], Bloch, Krishnaprasad, Marsden, and Murray [1996], Koon and Marsden [1997a,b, 1998], Cortés and de León [1999], Cantrijn, Cortés, de León, de Diego [2002],Cortés, de León, Martín de Diego, and Martínez [2001], and Cortés [2002] and, of course, literally thousands of very interesting papers on applications.

So-called quasicoordinates or quasivelocities can be a useful tool for analyzing nonholonomic systems, see for example, Neimark and Fufaev [1972], Greenwood [2003] and Bloch, Marsden and Zenkov [2007]. Various other forms of the nonholonomic equations such as Boltzman-Hamel equations are also discussed in these references.

Dynamic Nonholonomic vs. Kinematic Nonholonomic.   Nonholonomic systems come in two varieties. First of all, there are those with *dynamic nonholonomic constraints*, i.e., constraints preserved by the basic Euler–Lagrange or Hamilton equations, such as angular momentum, or more generally momentum maps. Of course, these "constraints" are not externally imposed on the system, but rather are consequences of the equations of motion, and so it is sometimes convenient to treat them as conservation laws rather than constraints per se. On the other hand, *kinematic nonholonomic constraints* are those imposed by the kinematics, such as rolling constraints, which are constraints linear in the velocity.

Examples of Nonholonomic Systems.   There have, of course, been many classical examples of nonholonomic systems studied, and we have mentioned many of these in Chapter 1. For example, Routh [1860] showed that a uniform sphere rolling on a surface of revolution is an integrable system (in the classical sense). For more modern treatments of Routh's problem see Zenkov [1995] and Hermans [1995]. Another example is the rolling disk (not necessarily vertical), which was treated in Vierkandt [1892]; this paper shows that the solutions of the equations on what we would call the reduced space (denoted by $\mathcal{D}/G$ later in this chapter) are all periodic.[2]

---

[2]This example is also treated in Appel [1900] and Korteweg [1899]. For this example from a more modern point of view, see, for example, O'Reilly [1996], and Getz and Marsden [1994].

A related example is the bicycle; see Getz and Marsden [1995] and Koon and Marsden [1997b]. The work of Chaplygin [1897a] is a very interesting study of the rolling of a solid of revolution on a horizontal plane. In this case, it is also true that the orbits are periodic on the reduced space.[3] One should note that a limiting case of this result (when the body of revolution limits to a disk) is that of Vierkandt. Chaplygin[1897b, 1903] also studied the case of a rolling sphere on a horizontal plane that additionally allowed for the possibility of spheres with an inhomogeneous mass distribution. For some recent work on this system see Schneider [2000, 2002] and Tai [2001].

Another classical example is the wobblestone (sometimes called the Celtic stone or rattleback), studied in a variety of papers and books such as Walker [1896], Crabtree [1909], Bondi [1986], Pascal [1983], Karapetyan [1983], and Markeev [1983]. See Hermans [1995] and Burdick, Goodwine and Ostrowski [1994] for additional information and references. In particular, the paper of Walker establishes important stability properties of relative equilibria by a spectral analysis; it shows, under rather general conditions (including the crucial one that the axes of principal curvature do not align with the inertia axes), that rotation in one direction is spectrally stable. This implies linear stability and hence nonlinear asymptotic stablility by use of the Lyapunov–Malkin theorem, as described below in Chapter 8 and in Karapetyan [1983] and Zenkov, Bloch, and Marsden [1998]. Karaptyan was the first to prove nonlinear stability of the rattleback.

By time reversibility, rotation in the other direction is unstable. On the other hand, one can have a relative equilibrium with eigenvalues in both halfplanes, so that rotations in opposite senses about it can *both* be unstable, as Walker has shown. Presumably this is consistent with the fact that some wobblestones execute multiple reversals (see Markeev [1983]). However, the global geometry of this mechanism is still not fully understood analytically.

The vertical rolling disk and the spherical ball rolling on a rotating table may be used as examples of systems with *both* dynamic and kinematic nonholonomic constraints. In either case, the angular momentum about the vertical axis is conserved; see Bloch, Reyhanoglu, and McClamroch [1992], Bloch and Crouch [1995], Brockett and Dai [1992], and Yang [1992].

A related modern example is the snakeboard (see Lewis, Ostrowski, Murray, and Burdick [1994] and Bloch, Krishnaprasad, Marsden, and Murray [1996] and Section 5.8), which shares some of the features of these examples but which has a crucial difference as well. This example, like many of the others, has the symmetry group SE(2) of Euclidean motions of the plane, but *the corresponding momentum is not conserved*. However, the *equation* satisfied by the momentum associated with the symmetry is use-

---

[3]This is proved by a nice technique of Birkhoff utilizing the reversible symmetry in Hermans [1995].

ful for understanding the dynamics of the problem and how group motion and locomotion can be generated. The nonconservation of momentum occurs even with no forces applied (besides the forces of constraint) and yet is consistent with the conservation of energy for these systems. In fact, this nonconservation is crucial to the generation of movement in a control-theoretic context.

5.1 Equations of Motion

Lagrange's Equations.   The basic equations for holonomic mechanics, normally derived from $F = ma$ and possibly torque balance as well, were rewritten by Lagrange in 1760 in a way that is *covariant*, that is, valid in general coordinate systems. This achievement is one of the most important advances in the modeling of mechanical systems in the last two hundred years. We now recall some of the essential features from Chapters 1 and 3.

To write down the equations of motion for a mechanical system, we first identify the configuration manifold of the system, which is the set of all possible configurations. We then choose local coordinates $q = (q^1, \ldots, q^n)$ on the configuration manifold, which are called **generalized coordinates**. Next, we express the kinetic energy K as a function of q and its time derivative $\dot{q}$, and the potential energy V as a function of q. The **Lagrangian** L is, for most systems, defined to be $K - V$.[4] Lagrange's equations are

$$\frac{d}{dt}\frac{\partial L}{\partial \dot{q}^i}(q(t), \dot{q}(t)) - \frac{\partial L}{\partial q^i}(q(t), \dot{q}(t)) = F_i(t), \quad i = 1, \ldots, n, \qquad (5.1.1)$$

or in short,

$$\frac{d}{dt}\frac{\partial L}{\partial \dot{q}} - \frac{\partial L}{\partial q} = F, \qquad (5.1.2)$$

where

$$\frac{\partial L}{\partial \dot{q}} = \left[\frac{\partial L}{\partial \dot{q}^1}, \cdots, \frac{\partial L}{\partial \dot{q}^n}\right], \quad \frac{\partial L}{\partial q} = \left[\frac{\partial L}{\partial q^1}, \cdots, \frac{\partial L}{\partial q^n}\right],$$

and where $F = [F_1, \cdots, F_n,]$. Both sides of (5.1.2) are *row* vectors, and the geometric object they represent is a cotangent vector to the configuration manifold at the point $q(t)$. The quantity F is called the **generalized force vector**, and is related to the nonconservative external forces on the system. The ith component F_i of F is given by

$$F_i = \left\langle F, \frac{\partial}{\partial q^i} \right\rangle,$$

---

[4]For some systems, such as a particle in a magnetic field, one needs to choose an appropriate Lagrangian, which is *not* simply the kinetic minus potential energy.

where $\partial/\partial q^i$ is the ith element of the standard tangent space basis that we get from the coordinates $(q^1, \ldots, q^n)$. These equations provide a simple and systematic way of modeling complicated systems, such as a system of interconnected rigid bodies.

In contrast, an alternative approach (which, for example, has been extensively used in the textbook by Smart [1951]), involves drawing "free body diagrams" for each body in the system and then using for each body the laws of balance of linear and angular momentum. In addition, in specific examples there may be special geometrical features that one can exploit to simplify the equations obtained (see, e.g., Kane and Levinson [1980]). This process is generally regarded as equivalent to but less efficient than Lagrange's approach.

Our goal in this chapter will be to take a closer look at Lagrange's equations, especially the situation where the generalized forces arise from constraints on the system.

Constrained Systems.   Consider a mechanical system, subject to a linear velocity constraint that in generalized coordinates can be expressed as

$$[a_1(q) \cdots a_n(q)]\dot{q} = 0,$$

where $\dot{q}$ is regarded as a column vector. The constraint is said to be **holonomic** or **integrable** if (locally) there is a real-valued function h of q such that the constraint can be rewritten as $h(q) = $ constant, or in differentiated form, $(\partial h/\partial q^i)\dot{q}^i = 0$. Thus the configuration of the system is actually constrained to be on a submanifold of the configuration manifold. This condition of being holonomic is equivalent by the Frobenius theorem to the integrability of the corresponding distribution (see Chapter 2). We can then write Lagrange's equations using coordinates on this submanifold and thus get a system of equations with fewer variables.

If no such function h exists, the constraint is said to be **nonintegrable** or **nonholonomic**. As we saw already in Chapter 1, nonholonomic constraints arise, for example, when one body is constrained to roll without slipping on another body or surface.

The Nonholonomic Principle.   The extension of Lagrange's equations for modeling systems subject to nonholonomic constraints was developed by Ferrers [1871], Neumann [1888], and Vierkandt [1892]. For the simple case of linear, time-independent constraints, the argument, which can be found in standard textbooks on mechanics (see, for example, Whittaker [1988], Pars [1965], Rosenberg [1977], and Neimark and Fufaev [1972]), is as follows: Suppose that our system is subject to m velocity constraints, represented by the equation

$$A(q)\dot{q} = 0.$$

Here $A(q)$ is an $m \times n$ matrix and $\dot{q}$ is a column vector.[5] At any configuration q, the set of all possible **virtual displacements** is defined to be the subspace of the tangent space to the configuration manifold at q consisting of vectors δq that satisfy the constraints, i.e., the subspace $\mathcal{D}_q$ defined by

$$\mathcal{D}_q = \{\delta q \in T_q Q \mid A(q) \cdot \delta q = 0\}.$$

The (generalized) constraint force, which is regarded as a cotangent vector at q, is assumed to lie in the annihilator of the space of virtual displacements. Thus, F has to be a linear combination of the rows of $A(q)$:

$$F = \lambda A(q).$$

We shall call this assumption the **nonholonomic principle**. Here F is regarded as a row vector, and λ is a row vector whose elements are called "Lagrange multipliers."[6]

The equations we obtain are thus

$$\frac{d}{dt}\left(\frac{\partial L}{\partial \dot{q}}\right) - \frac{\partial L}{\partial q} = \lambda A(q), \quad A(q)\dot{q} = 0. \tag{5.1.3}$$

Note that we have n second-order differential equations (which can be rewritten as $2n$ first-order equations) and m constraint equations, which we need to solve for the $2n + m$ unknowns $q, \dot{q}, \lambda$. As in Chapter 1, we call the above equations the **nonholonomic equations** or the **Lagrange-d'Alembert equations** for a mechanical system with velocity constraints.

Comments on the Derivation. A problem with the above classical derivation of the Lagrange–d'Alembert equations is that no adequate justification is given for the nonholonomic principle, i.e., the assertion that the vector of generalized forces always has to annihilate *all* possible virtual displacements (in the case of "ideal" constraints, which do no work). With this assumption, the total energy of the system is conserved, and conservation of energy indeed holds for many systems with nonholonomic constraints—for example, systems involving constraints of rolling without slipping. The rate of change of the total energy of the system is equal to the rate of work done by the generalized forces, which is $\langle F, \dot{q} \rangle$.

Therefore, conservation of energy requires only that the work done by the generalized forces at each instant be zero, i.e., that $\langle F, \dot{q} \rangle = 0$. The

---

[5]We will see how to view $A(q)$ intrinsically in the next section. Here, we do not consider nonholonomic constraints which are nonlinear in the velocities. Discussion of such constraints may be found, for example, in Appell [1911], Marle [1996, 1998], and Terra and Kobayashi [2002] and references therein.

[6]The name "Lagrange multipliers" is somewhat inappropriate here, since λ has, at the moment, nothing to do with the Lagrange multipliers of the Lagrange multiplier theorem for constrained optimization found in textbooks on multivariable calculus. But we will continue to use it for historical reasons.

constraints ensure that the vector $\dot{q}$ does lie in the space of all possible virtual displacements at q, but, and here is the rub, *conservation of energy in itself does not explain why the generalized force vector should annihilate* **all** *the possible virtual displacements.*

It has long been the general consensus in the mechanics community that the Lagrange–d'Alembert equations do indeed provide an accurate model of the observed behavior of constrained physical systems. However, what the confusion over the equations mentioned above did do was, quite properly, to *highlight the inadequacies in the classical derivation of the Lagrange–d'Alembert equations.* We shall comment on this further in the next paragraph.

The final resolution of this situation is quite simple, as we have already remarked: One can indeed derive the Lagrange–d'Alembert principle from $F = ma$ along with Newton's third law and the assumption that the constraints do no work.

Variational Problems. As we have mentioned, the correct equations for nonholonmic mechanical systems are not literally variational. However, the variational approach is very appropriate for *optimal control problems* where one by definition wants to optimize some function (usually called the *cost function*). Much more will be said about the associated *variational equations of motion* in Chapter 7 in conjunction with our study of optimal control. In fact, the optimal control setting is the only setting known to us in which these equations are useful for physical applications.

To elaborate on what it means to have a variational problem with nonholonomic contraints, let $q_0 : [a, b] \to \mathbb{R}^n$ be a trajectory of the constrained system, with $q_0(a) = q_a$, and $q_0(b) = q_b$. Let $C(q_a, q_b, [a, b])$ denote the space of smooth curves on the interval $[a, b]$, taking values in $\mathbb{R}^n$ and with endpoints q_a and q_b. Let $\mathfrak{S} : C(q_a, q_b, [a, b]) \to \mathbb{R}$ be the action integral, defined as

$$\mathfrak{S}(q) = \int_a^b L(q(t), \dot{q}(t))\, dt.$$

Let C' be the subset of $C(q_a, q_b, [a, b])$ consisting of curves that satisfy the constraints. Then the variational equations require that the curve q_0 be a critical point of $\mathfrak{S}|C'$.

This formulation, in general, *leads to equations that are different from the Lagrange–d'Alembert equations* (5.1.3), though in the case of holonomic constraints, both formulations obviously yield the same equations.

It has long been known (Carathéodory [1933], Neimark and Fufaev [1972]) that the Lagrange–d'Alembert equations can be obtained by starting with an unconstrained system subject to appropriately chosen dissipative forces, and then letting these forces go to infinity in an appropriate manner.

Kozlov showed that the variational equations too can be obtained as the result of another limiting process: He added a parameter-dependent "inertial term" to the Lagrangian of the constrained system, and then showed that the unconstrained equations approach the variational equations as the parameter approaches infinity. We will not present the details here; Arnold, Kozlov, and Neishtadt [1988] gives a nice exposition.

The issue of to what extent the variational formulation is applicable to actual physical systems has been a matter of some controversy, although, as we mentioned, it was settled already by Korteweg [1899] (see also Kharlamov [1992]). Kozlov [1992], replying to this paper of Kharmalov, points out that for the Lagrange–d'Alembert equations, "the origin of the initial axioms [(i.e. the requirement that the generalized force annihilate all possible virtual displacements] remains unknown," and that "the question of applicability of the nonholonomic model ... cannot, in any specific situation, be solved within the framework of an axiomatic scheme without recourse to experimental results." In fact, there *are* experimental results that do confirm that the Lagrange–d'Alembert principle is correct. See, for example, Lewis and Murray [1995], which studies a rolling ball on a rotating turntable. They showed that if the ball is sufficiently rigid and if the slippage is minimized, then the nonholonomic formulation best approximates the observed behavior.

5.1.1 Example (A Counterintuitive Example). We now discuss an interesting example from Smart [1951] that illustrates in how counterintuitive a manner systems with rolling constraints can sometimes behave. The system consists of a spherical body rolling on the inner surface of a vertical pipe. We assume that the sphere is rolling fast enough around the pipe to prevent slipping. (See Figure 5.1.1.) Smart actually obtains the equation

FIGURE 5.1.1. Sphere rolling inside a cylinder

of motion the hard way—using balance of momentum laws and a moving coordinate frame attached to the body (i.e., *not* using generalized coordinates). The body initially rolls downwards, as one would expect. But then it changes direction and moves back up! In fact, the height of the body oscillates in a simple harmonic manner. In fact, Routh [1860] considered the more general problem of a sphere rolling in a surface of revolution and showed that it is integrable (in the sense that one could in principle write down the solutions in terms of integrals); see Zenkov [1995].   ◆

5.2   The Lagrange–d'Alembert Principle

With the above considerations behind us, let us "restart" from a somewhat more general point of view. Namely, we start with a configuration space Q and a distribution $\mathcal{D}$ that describes the kinematic constraints of interest. Thus, $\mathcal{D}$ is a collection of linear subspaces denoted by $\mathcal{D}_q \subset T_q Q$, one for each $q \in Q$. A curve $q(t) \in Q$ will be said to **satisfy the constraints** if $\dot{q}(t) \in \mathcal{D}_{q(t)}$ for all t. This distribution will, in general, be nonintegrable in the sense of Frobenius's theorem; i.e., the constraints are, in general, nonholonomic.

The above setup describes *linear* constraints; for *affine* constraints, for example, a ball on a rotating turntable (where the rotational velocity of the turntable represents the affine part of the constraints), one way to describe them is to assume that there is a given vector field V_0 on Q and then write the constraints as $\dot{q}(t) - V_0(q(t)) \in \mathcal{D}_{q(t)}$.

The Lagrange–d'Alembert Principle.   Consider a given Lagrangian $L : TQ \to \mathbb{R}$. In (generalized) coordinates, q^i, $i = 1, \ldots, n$, on Q with induced coordinates $(q^i, \dot{q}^i)$ for the tangent bundle, we write, as before, $L(q^i, \dot{q}^i)$. Following the discussion in the preceding section we assume the following principle of Lagrange–d'Alembert.

5.2.1 Definition.   *The **Lagrange–d'Alembert equations of motion** for the system are those determined by*

$$\delta \int_a^b L(q^i, \dot{q}^i) \, dt = 0, \tag{5.2.1}$$

where we choose variations $\delta q(t)$ of the curve $q(t)$ that satisfy $\delta q(t) \in \mathcal{D}_q(t)$ for each t, $a \le t \le b$, and $\delta q(a) = \delta q(b) = 0$.

This principle is supplemented by the condition that the curve $q(t)$ itself satisfy the constraints.

As explained before, in such a principle we take the variation δq *before* imposing the constraints; that is, we *do not* impose the constraints on the family of curves defining the variation. The usual arguments in the calculus of variations show that this constrained variational principle is equivalent to the equations

$$-\delta L = \left(\frac{d}{dt} \frac{\partial L}{\partial \dot{q}^i} - \frac{\partial L}{\partial q^i} \right) \delta q^i = 0, \tag{5.2.2}$$

for all variations δq such that $\delta q \in \mathcal{D}_q$ at each point of the underlying curve $q(t)$.

Structure of the Equations of Motion. To explore the structure of the equations determined by (5.2.2) in more detail, let $\{\omega^a\}$ be a set of m independent one-forms whose vanishing describes the constraints on the system; that is, the constraints on $\delta q \in TQ$ are defined by the m conditions $\omega^a \cdot v = 0$, $a = 1, \ldots, m$. Using the fact that these m one-forms are independent, we leave it as an exercise for the reader to show that one can choose local coordinates such that the one-forms ω^a have the form

$$\omega^a(q) = ds^a + A_\alpha^a(r, s)dr^\alpha, \qquad a = 1, \ldots, m, \tag{5.2.3}$$

where $q = (r, s) \in \mathbb{R}^{n-m} \times \mathbb{R}^m$. With this choice, the constraints on $\delta q = (\delta r, \delta s)$ are given by the conditions

$$\delta s^a + A_\alpha^a \delta r^\alpha = 0. \tag{5.2.4}$$

The equations of motion for the system are given by (5.2.2), where we choose variations $\delta q(t)$ that satisfy the constraints. Substituting (5.2.4) into (5.2.2) and using the fact that δr is arbitrary gives

$$\left(\frac{d}{dt} \frac{\partial L}{\partial \dot{r}^\alpha} - \frac{\partial L}{\partial r^\alpha} \right) = A_\alpha^a \left(\frac{d}{dt} \frac{\partial L}{\partial \dot{s}^a} - \frac{\partial L}{\partial s^a} \right), \qquad \alpha = 1, \ldots, n - m. \tag{5.2.5}$$

The equations (5.2.5) combined with the constraint equations

$$\dot{s}^a = -A_\alpha^a \dot{r}^\alpha, \qquad a = 1, \ldots, m, \tag{5.2.6}$$

give a complete description of the *equations of motion* of the system. Notice that they consist of $n - m$ second-order equations and m first-order equations.

The Constrained Lagrangian. We now define the "constrained" Lagrangian by substituting the constraints (5.2.6) into the Lagrangian:

$$L_c(r^\alpha, s^a, \dot{r}^\alpha) = L(r^\alpha, s^a, \dot{r}^\alpha, -A_\alpha^a(r, s)\dot{r}^\alpha).$$

The equations of motion (5.2.5) can be written in terms of the constrained Lagrangian in the following way, as a direct coordinate calculation shows:[7]

$$\frac{d}{dt} \frac{\partial L_c}{\partial \dot{r}^\alpha} - \frac{\partial L_c}{\partial r^\alpha} + A_\alpha^a \frac{\partial L_c}{\partial s^a} = -\frac{\partial L}{\partial \dot{s}^b} B_{\alpha\beta}^b \dot{r}^\beta, \tag{5.2.7}$$

where

$$B_{\alpha\beta}^b = \left(\frac{\partial A_\alpha^b}{\partial r^\beta} - \frac{\partial A_\beta^b}{\partial r^\alpha} + A_\alpha^a \frac{\partial A_\beta^b}{\partial s^a} - A_\beta^a \frac{\partial A_\alpha^b}{\partial s^a} \right). \tag{5.2.8}$$

---

[7]See the Internet Supplement for some details of this calculation

Letting $\mathbf{d}\omega^b$ be the exterior derivative of ω^b, another straightforward computation using properties of differential forms shows that

$$\mathbf{d}\omega^b(\dot{q}, \cdot) = B^b_{\alpha\beta}\dot{r}^\beta dr^\alpha,$$

and hence the equations of motion have the form

$$-\delta L_c = \left(\frac{d}{dt}\frac{\partial L_c}{\partial \dot{r}^\alpha} - \frac{\partial L_c}{\partial r^\alpha} + A^a_\alpha \frac{\partial L_c}{\partial s^a} \right) \delta r^\alpha = -\frac{\partial L}{\partial \dot{s}^b}\mathbf{d}\omega^b(\dot{q}, \delta r).$$

This form of the equations isolates the effects of the constraints, and shows that if the constraints are integrable (which is equivalent to $d\omega^b = 0$, that is, to $B^b_{\alpha\beta} = 0$), then the correct equations of motion are obtained by substituting the constraints into the Lagrangian and setting the variation of L_c to zero. However in the nonintegrable case, which is the typical case for nonholonomic systems, the constraints generate extra forces that must be taken into account.

Intrinsic Formulation of the Equations. We can now rephrase our coordinate computations in the language of the Ehresmann connections that we discussed in Chapter 2. We shall do this first for systems with homogeneous constraints and then treat the affine case.

Suppose that we have chosen a bundle and an Ehresmann connection A on that bundle such that the constraint distribution $\mathcal{D}$ is given by the horizontal subbundle associated with A. In other words, we assume that the connection A is chosen such that the constraints are written as $A \cdot \dot{q} = 0$. Note that this is an intrinsic way of writing the constraints and a way of thinking of the collection of one-forms that we used in the coordinate description. In those coordinates one can choose the bundle to be that given in coordinates by $(s, r) \mapsto r$, and the connection is, in this choice of bundle, defined by the contstraints. It is clear that this choice of bundle is not unique; sometimes this sort of ambiguity is removed for systems with symmetry, as we shall see later.

In the language of connections, the ***constrained Lagrangian*** can be written as

$$L_c(q, \dot{q}) = L(q, \text{hor}\,\dot{q}),$$

and we have the following theorem.

5.2.2 Theorem. *The Lagrange–d'Alembert equations may be written as the equations*

$$\delta L_c = \langle \mathbb{F}L, B(\dot{q}, \delta q)\rangle,$$

where $\langle,\rangle$ denotes the pairing between a vector and a dual vector and where

$$\delta L_c = \left\langle \delta q^\alpha, \frac{\partial L_c}{\partial q^\alpha} - \frac{d}{dt}\frac{\partial L_c}{\partial \dot{q}^\alpha} \right\rangle,$$

in which δq is a horizontal variation (i.e., it takes values in the horizontal space) and B is the curvature regarded as a vertical-valued two-form, in addition to the constraint equations

$$A(q) \cdot \dot{q} = 0.$$

Affine Constraints. We next consider the modifications necessary to allow affine constraints of the form

$$A(q) \cdot \dot{q} = \gamma(q, t),$$

where A is an Ehresmann connection as described above and $\gamma(q, t)$ is a vertical-valued (possibly time-dependent) vector field on Q. The expression γ here is related to the vector field V_0 given in the introduction to this section by $\gamma(q) = A(q) \cdot V_0(q)$. Affine constraints arise, for example, in studying the motion of a ball on a spinning turntable. Since the turntable is moving underneath the ball, the velocity in the constraint directions is not zero, but is instead determined by the position of the ball on the turntable and the angular velocity of the turntable.

Since $\gamma(q, t)$ is vertical, we can define the **covariant derivative** of γ along X as

$$D\gamma(X) = \text{ver}[\text{hor}\, X, \gamma]$$

(see Chapter 2 and Marsden, Montgomery, and Ratiu [1990]). Relative to bundle coordinates $q = (r, s)$, we write γ as

$$\gamma(q, t) = \gamma^a(q, t) \frac{\partial}{\partial s^a},$$

and the covariant derivative along a *horizontal* vector field

$$X = X^\alpha \left(\frac{\partial}{\partial r^\alpha} - A^a_\alpha \frac{\partial}{\partial s^a} \right)$$

is given by

$$D\gamma(X) = X^\alpha \left(\frac{\partial \gamma^a}{\partial r^\alpha} - A^b_\alpha \frac{\partial \gamma^a}{\partial s^b} + \gamma^b \frac{\partial A^a_\alpha}{\partial s^b} \right) \frac{\partial}{\partial s^a} =: \gamma^a_\alpha X^\alpha \frac{\partial}{\partial s^a},$$

which defines the symbols γ^a_α.

In this affine case, we define the **constrained Lagrangian** as

$$L_c(q, \dot{q}, t) = L\left(q, \text{hor}\, \dot{q} + \gamma(q, t)\right).$$

A long calculation, similar to that for linear (homogeneous) constraints, shows that the dynamics have the form

$$\delta L_c = \langle \mathbb{F}L, B(\dot{q}, \delta q) \rangle + \langle \mathbb{F}L, D\gamma(\delta q) \rangle,$$

$$\tag{5.2.9}$$

$$A(q) \cdot \dot{q} = \gamma(q, t),$$

where the δq are, as in the homogeneous case, restricted to satisfy $A(q) \cdot \delta q = 0$. In coordinates, the first of these equations reads as

$$\frac{d}{dt}\frac{\partial L_c}{\partial \dot{r}^\alpha} - \frac{\partial L_c}{\partial r^\alpha} + A^a_\alpha \frac{\partial L_c}{\partial s^a} = -\frac{\partial L}{\partial \dot{s}^b} B^b_{\alpha\beta}\dot{r}^\beta - \frac{\partial L}{\partial \dot{s}^a}\gamma^a_\alpha, \qquad (5.2.10)$$

while the second reads as $\dot{s}^a + A^a_\alpha \dot{r}^\alpha = \gamma^a$. Notice that these equations show how, in the affine case, the covariant derivative of the affine part γ enters into the description of the system; in particular, note that the covariant derivative in (5.2.9) is with respect to the configuration variables and not with respect to the time.

Nonholonomic Equations of Motion with Lagrange Multipliers.
We can obtain the nonholonomic equations of motion with Lagrange multipliers from the Lagrange–d'Alembert principle as follows (see, e.g., Neimark and Fufaev [1972]):

Recall that the Lagrange–d'Alembert principle gives us

$$\left(\frac{d}{dt}\frac{\partial}{\partial \dot{q}^i}L - \frac{\partial}{\partial q^i}L\right)\delta q^i = 0, \quad i = 1, \ldots, n, \qquad (5.2.11)$$

for variations $\delta q^i \in \mathcal{D}$, i.e., for variations in the constraint distribution. Using the notation in Chapter 1 (see equation (1.3.1)) for the constraints, we write the constraints on δq^i as

$$\sum_{i=1}^n a^j_i \delta q^i = 0, \qquad j = 1, \ldots, m, \qquad (5.2.12)$$

where a^j_i constitute an $m \times n$ matrix, whose entries depend on q. This implies, in a trivial way, that for any constants λ_j,

$$\sum_{j=1}^m \sum_{i=1}^n \lambda_j a^j_i \delta q^i = 0. \qquad (5.2.13)$$

Hence we can append this to the Lagrange–d'Alembert equations to obtain

$$\sum_{i=1}^n \left(\frac{d}{dt}\frac{\partial}{\partial \dot{q}^i}L - \frac{\partial}{\partial q^i}L - \sum_{j=1}^m \lambda_j a^j_i\right)\delta q^i = 0. \qquad (5.2.14)$$

Assuming that our constraints are independent implies that one of the $m \times m$ minors of the $m \times n$ matrix a^j_i must be nonzero. Let us assume it to be that given by letting the indices run from 1 to m. We can then assume that the variations $\delta q^{m+1}, \ldots, \delta q^n$ are arbitrary, since the constraint equations (5.2.12) may then be satisfied by a choice of resulting definite values for variations $\delta q^1, \ldots, \delta q^m$.

We can choose values of the Lagrange multipliers such that the expression in brackets in (5.2.14) vanishes for each dependent variation $\delta q^1, \ldots, \delta q^m$. This entails solving a linear systems of algebraic equations in the λ_i, which is solvable by virtue of our assumptions on the a_i^j.

Once this is done, equation (5.2.14) becomes

$$\sum_{i=m+1}^{n} \left(\frac{d}{dt} \frac{\partial L}{\partial \dot{q}^i} - \frac{\partial L}{\partial q^i} - \sum_{j=1}^{m} \lambda_j a_i^j \right) \delta q^i = 0, \qquad (5.2.15)$$

where the variations are independent. Hence each term in parentheses must also vanish independently. Putting the observations for the dependent and independent variables together gives us the set of n equations

$$\frac{d}{dt} \frac{\partial L}{\partial \dot{q}^i} - \frac{\partial L}{\partial q^i} = \sum_{j=1}^{m} \lambda_j a_i^j . \qquad (5.2.16)$$

We have seen various applications of this approach, for example, the Chaplygin sleigh and the rolling ball discussed in the first chapter.

Exercises

⋄ **5.2-1.** Use the above formalism to compute the nonholonomic equations of motion for the the Chaplygin sleigh of Chapter 1.

⋄ **5.2-2.** Show that if one has a conserved quantity for a holonomic mechanical system (treat either the specific case of angular momentum or the general case of a momentum map associated with a symmetry) and one treats it as a nonholonomic constraint using the Lagrange–d'Alembert principle, then one gets the correct holonomic equations restricted to a surface of constant momentum.

5.3 Projected Connections and Newton's Law

In this section we discuss the formulation of the equations of motion of nonholonomic systems with forces or controls and indicate how to write such systems as second-order forced systems (i.e., satisfying Newton's law) on the constraint subbundle. This follows an approach due to Vershik and Faddeev [1981]; see also Montgomery [1990], Yang [1992], and Bloch and Crouch [1995, 1997].

Let us rewrite the equations of a forced nonholonomic system (see equation (1.3.6)) on a Riemannian manifold Q. We assume also in this section that the Lagrangian is pure kinetic energy and any potential forces are part of the external force F:

$$\frac{D^2q}{dt^2} = \sum_{a=1}^{m} \lambda_a W_a + F, \tag{5.3.1}$$

where F is an arbitrary external force field and the system is subject to the independent constraints

$$\omega^a(\dot{q}) = \langle W_a, \dot{q} \rangle = 0, \qquad 1 \leq a \leq m, \tag{5.3.2}$$

where the W_a are vector fields on M.

Locally, these constraints may be written as in (5.2.3), but we do not use this local notation here.

The constraint distribution N is given by

$$N_p = \{X_p \mid \omega^a(X) = 0, 1 \leq a \leq m\} \tag{5.3.3}$$

for X a vector field on Q and p a point on Q.

We now use the independence of the one-forms ω^a to eliminate the multipliers: Let $a_{ij} = \omega^i(W_j)$, $1 \leq i, j \leq m$. Then the matrix with entries a_{ij} is invertible on Q. Differentiating the constraints gives

$$\frac{D\omega^i}{dt}(\dot{q}) + \omega^i \left(\sum_j \lambda_j W_j + F \right) = 0, \ 1 \leq i \leq m. \tag{5.3.4}$$

Hence the multipliers have the explicit form

$$\lambda_k = - \sum_j a_{kj}^{-1} \left(\frac{D\omega^j}{dt}(\dot{q}) + \omega^j(F) \right), \ 1 \leq k \leq m. \tag{5.3.5}$$

Hence the equations of motion may be rewritten as

$$\frac{D^2q}{dt^2} + \sum_{k,j} W_k a_{kj}^{-1} \frac{D\omega^j}{dt} = F - \sum_{k,j} W_k a_{kj}^{-1} \omega^j(F),$$

$$\omega^i(\dot{q}) = 0, \quad 1 \leq i \leq m. \tag{5.3.6}$$

5.3.1 Definition.   *For X a vector field on Q, let*

$$\pi_N(X) = X - \sum_{ki} W_k a_{ki}^{-1} \omega_i(X). \tag{5.3.7}$$

Notice that $\pi_N(X)$ is a projection of X onto the constraint distribution N. Now apply this projection to D^2q/dt^2 and observe that differentiating the constraints gives

$$\omega^i \left(\frac{D^2q}{dt^2} \right) + \frac{D\omega^i}{dt}(\dot{q}) = 0.$$

Thus,

$$\pi_N\left(\frac{D^2q}{dt^2}\right) = \frac{D^2q}{dt^2} + \sum_{i,k} W_k a_{ki}^{-1} \frac{D\omega^i}{dt}(\dot{q}). \qquad (5.3.8)$$

Hence, equation (5.3.6) may be written

$$\pi_N\left(\frac{D^2q}{dt^2}\right) = \pi_N(F), \ \dot{q} \in N. \qquad (5.3.9)$$

We now make the following natural definition:

5.3.2 Definition.   *If $\overline{\nabla}$ is any connection on Q with corresponding covariant derivative $\overline{D}/dt$, then the second-order system*

$$\frac{\overline{D}^2q}{dt^2} = \overline{F}, \ q \in Q, \qquad (5.3.10)$$

*is said to be a system on Q satisfying **Newton's law with forces $\overline{F}$**.*

Thus our nonholonomic system is a projection of the Newton law system $D^2q/dt^2 = F$ to N.

However, as Vershik and Faddeev [1981] pointed out, by redefining the connection, the system may be written as a system obeying Newton's law on N directly. To see this, define a new connection on Q,

$$\nabla'_X Y = \nabla_X Y + \sum_{i=1}^m W_i a_{ik}^{-1} (\nabla_X \omega^k)(Y), \qquad (5.3.11)$$

for X, Y vector fields on Q. This in turn defines a covariant derivative D'/dt. From (5.3.11) we note that

$$\omega^i(\nabla'_X Y) = \omega^i(\nabla_X Y) + (\nabla_X \omega^i)(Y) = X(\omega^i(Y)). \qquad (5.3.12)$$

Thus since X and Y are vector fields on N, so are $\nabla'_X Y$, and thus $\nabla'|_N$ defines a connection on the subbundle N of TQ, and the nonholonomic system (5.3.1) may be viewed as a system of Newton law type on N:

$$\frac{D'^2q}{dt^2} = \pi_N(F), \ \dot{q} \in N. \qquad (5.3.13)$$

Thus the nonholonomic system is a system of Newton law type with respect to a modified connection. This connection is not metric, however. In Bloch and Crouch [1997] we explore extensions of this system to a nonmetric Newton law system on the whole of M. See also the work of Lewis [1998].

5.4   Systems with Symmetry

We now add symmetry to our nonholonomic system. We will begin with some general remarks about symmetry.

Group Actions and Invariance. We remind the reader of a few of the concepts discussed in Chapter 3. Assume that we are given a Lie group G and an action of G on Q. This action will be denoted by $q \mapsto gq = \Phi_g(q)$. The group orbit through a point q, which is an (immersed) submanifold, is denoted by

$$\mathrm{Orb}(q) := \{gq \mid g \in G\}.$$

When there is danger of confusion about which group is meant, we write the orbit as $\mathrm{Orb}_G(q)$.

Let $\mathfrak{g}$ denote the Lie algebra of the Lie group G. For an element $\xi \in \mathfrak{g}$, we write ξ_Q, a vector field on Q for the corresponding infinitesimal generator; recall that this is obtained by differentiating the flow $\Phi_{\exp(t\xi)}$ with respect to t at $t = 0$. The tangent space to the group orbit through a point q is given by the set of infinitesimal generators at that point:

$$T_q(\mathrm{Orb}(q)) = \{\xi_Q(q) \mid \xi \in \mathfrak{g}\}.$$

For simplicity, we make the assumption that the action of G on Q is free (none of the maps Φ_g has any fixed points) and proper (the map $(q, g) \mapsto gq$ is proper; that is, the inverse images of compact sets are compact). The case of nonfree actions is very important, and the investigation of the associated singularities needs to be carried out, but that topic is not the subject of the present book.[8]

The quotient space $M = Q/G$, whose points are the group orbits, is called **shape space**. It is known that if the group action is free and proper, then shape space is a smooth manifold and the projection map $\pi : Q \to Q/G$ is a smooth surjective map with a surjective derivative $T_q\pi$ at each point. We will denote the projection map by $\pi_{Q,G}$ if there is any danger of confusion. The kernel of the linear map $T_q\pi$ is the set of infinitesimal generators of the group action at the point q, i.e.,

$$\ker T_q\pi = \{\xi_Q(q) \mid \xi \in \mathfrak{g}\},$$

so these are also the tangent spaces to the group orbits.

Invariance Properties We now introduce some assumptions concerning the relations among the given group action, the Lagrangian, and the constraint distribution.

5.4.1 Definition.

(L1) *We say that the Lagrangian is **invariant** under the group action if L is invariant under the induced action of G on TQ.*

---

[8]This would take us into the subject of singular nonholonomic reduction; see, for example, Bates [1998].

(L2) *We say that the Lagrangian is **infinitesimally invariant** if for any Lie algebra element $\xi \in \mathfrak{g}$ we have $dL \circ \dot{\xi}_Q = 0$ where, for a vector field X on Q, $\dot{X}$ denotes the vector field on TQ naturally induced by it (if F_t is the flow of X then the flow of $\dot{X}$ is TF_t).*

(S1) *We say that the distribution $\mathcal{D}$ is **invariant** if the subspace $\mathcal{D}_q \subset T_qQ$ is mapped by the tangent of the group action to the subspace $\mathcal{D}_{gq} \subset T_{gq}Q$.*

(S2) *An Ehresmann connection A on Q (that has $\mathcal{D}$ as its horizontal distribution) is **invariant** under G if the group action preserves the bundle structure associated with the connection (in particular, it maps vertical spaces to vertical spaces) and if, as a map from TQ to the vertical bundle, A is G-equivariant.*

(S3) *A Lie algebra element ξ is said to act **horizontally** if $\xi_Q(q) \in \mathcal{D}_q$ for all $q \in Q$.*

Some relationships among these conditions are as follows: Condition (L1) implies (L2), as is obtained by differentiating the invariance condition. It is also clear that condition (S2) implies condition (S1), since the invariance of the connection A implies that the group action maps its kernel to itself. Condition (S1) may be stated as follows:

$$T_q\Phi_g \cdot \mathcal{D}_q = \mathcal{D}_{gq}. \tag{5.4.1}$$

In the case of affine constraints, one needs, where appropriate, the assumption that the vector field γ is invariant under the action.

To help explain condition (S1), we will rewrite it in infinitesimal form. Let $\mathfrak{X}_\mathcal{D}$ be the space of sections X of the distribution $\mathcal{D}$; that is, the space of vector fields X that take values in $\mathcal{D}$. The condition (S1) implies that for each $X \in \mathfrak{X}_\mathcal{D}$, we have $\Phi_g^* X \in \mathfrak{X}_\mathcal{D}$. Here, $\Phi_g^* X$ denotes the pullback of the vector field X under the map Φ_g. Differentiation of this condition with respect to g proves the following result.

5.4.2 Proposition. *Assume that condition (S1) holds and let X be a section of $\mathcal{D}$. Then for each Lie algebra element ξ, we have*

$$[\xi_Q, X] \in \mathfrak{X}_\mathcal{D}, \tag{5.4.2}$$

which we write as $[\xi_Q, \mathfrak{X}_\mathcal{D}] \subset \mathfrak{X}_\mathcal{D}$.

We now have the following result.

5.4.3 Proposition. *Under assumptions (L1) and (S1), we can form the **reduced velocity phase space** TQ/G and the **constrained reduced velocity phase space** $\mathcal{D}/G$. The Lagrangian L induces well-defined functions, the **reduced Lagrangian***

$$l : TQ/G \to \mathbb{R}$$

satisfying $L = l \circ \pi_{TQ}$, *where* $\pi_{TQ} : TQ \to TQ/G$ *is the projection, and the* **constrained reduced Lagrangian**

$$l_c : \mathcal{D}/G \to \mathbb{R},$$

which satisfies $L|\mathcal{D} = l_c \circ \pi_{\mathcal{D}}$, *where* $\pi_{\mathcal{D}} : \mathcal{D} \to \mathcal{D}/G$ *is the projection. In addition, the Lagrange–d'Alembert equations induce well-defined* **reduced Lagrange–d'Alembert equations** *on* $\mathcal{D}/G$. *That is, the vector field on the manifold* $\mathcal{D}$ *determined by the Lagrange–d'Alembert equations (including the constraints) is G-invariant, and so defines a reduced vector field on the quotient manifold* $\mathcal{D}/G$.

This proposition follows from general symmetry considerations. For example, to get the constrained reduced Lagrangian l_c we restrict the given Lagrangian to the distribution $\mathcal{D}$ and then use its invariance to pass to the quotient. The problem of constrained Lagrangian reduction is the detailed determination of these reduced structures and will be dealt with later.

The Principal or Purely Kinematic Case.   To illustrate how symmetries affect the equations of motion, we will start with one of the simplest cases, in which *the group orbits exactly complement the constraints*, which we call the **principal or the purely kinematic case**, sometimes called the **Chaplygin** or the nonabelian Chaplygin case. This case goes back to Chaplygin [1897a], Hamel [1904], and was put into a geometric context by Koiller [1992]. See also Bloch, Reyhanoglu, and McClamroch [1992], Yang, Krishnaprasad, and Dayawansa [1993], and Cantrijn, Cortés, de León, de Diego [2002].

An example of the purely kinematic case is the vertical rolling disk discussed in the examples section below. However, in other examples, such as the snakeboard, this condition is not valid, and its failure is crucial to understanding the dynamic behavior of this system. Because of this, we will consider the more general case below in Section 5.7.

5.4.4 Definition.   *The* **principal kinematic case** *is the case in which* (L1) *and* (S1) *hold and where at each point* $q \in Q$, *the tangent space* $T_q Q$ *is the direct sum of the tangent to the group orbit and to the constraint distribution; that is, we require that at each point,* $\mathcal{S}_q = \{0\}$ *and that*

$$T_q Q = T_q \operatorname{Orb}(q) \oplus \mathcal{D}_q =: V_q \oplus \mathcal{D}_q.$$

In other words, we require that the group directions provide a vertical space for the Ehresmann connection introduced earlier; thus, in this situation there is a preferred vertical space, and so there is no freedom in choosing the associated Ehresmann connection whose horizontal space is the given constraint distribution. In other words, the nonholonomic kinematic constraints provide a connection on the principal bundle $\pi : Q \to Q/G$, so that we can choose this bundle to be coincident with the bundle

$\pi_{Q,R} : Q \to R$ introduced earlier. If the Lagrangian and the constraints are invariant with respect to the group action (assumptions (L1) and (S1)), then as we explained above, the equations of motion in Theorem 5.2.2 drop to the reduced space $\mathcal{D}/G$. As we shall see, in the principal kinematic case these reduced equations may be regarded as second-order equations on Q/G together with the constraint equations. The connection that describes the constraints provides the information necessary to reconstruct the trajectory on the full space. In essence, the constraints provide a connection that replaces the mechanical connection that is used in the reduction theory of unconstrained systems with symmetry. The general case, described in Section 5.7, requires a synthesis of the two approaches.

Since a principal connection is uniquely determined by the specification of its horizontal spaces as an invariant complement to the group orbits, we get the following.

5.4.5 Proposition. *In the principal kinematic case, there is a unique principal connection on $Q \to Q/G$ whose horizontal space is the given distribution $\mathcal{D}$.*

We now make these considerations more explicit. The vertical space for the principal bundle $\pi : Q \to Q/G$ is $V_q = \ker T_q\pi$, which is the tangent space to the group orbit through q. Thus, each vertical fiber at a point q is isomorphic to the Lie algebra $\mathfrak{g}$ by means of the map $\xi \in \mathfrak{g} \mapsto \xi_Q(q)$. In the principal kinematic case, the splitting of the tangent space to Q given in the preceding definition defines a projection onto the vertical space and hence defines an Ehresmann connection that, as before, we denote by A. If condition (S1) holds, then $A : TQ \to V$ will be group-invariant (assumption (S2)), and there exists a Lie-algebra–valued one-form $\mathcal{A} : TQ \to \mathfrak{g}$ such that

$$A(q) \cdot \dot{q} = (\mathcal{A}(q) \cdot \dot{q})_Q (q), \quad \text{or} \quad A = \mathcal{A}_Q.$$

Thus on a principal bundle we can express our results in terms of $\mathcal{A}$ instead of A. In bundle coordinates, $\mathcal{A}$ can be written as

$$\mathcal{A}(r, g) \cdot (\dot{r}, \dot{g}) = \mathrm{Ad}_g(g^{-1}\dot{g} + \mathcal{A}_{\mathrm{loc}}(r)\dot{r}),$$

as in equation (5.2.3).

We now turn to the expression in a local trivialization for the *constrained reduced Lagrangian* l_c. This is obtained by substituting the constraints $\mathcal{A}(q) \cdot \dot{q} = 0$ into the reduced Lagrangian. Thus $l_c : T(Q/G) \to \mathbb{R}$ is given by

$$l_c(r, \dot{r}) = l(r, \dot{r}, -\mathcal{A}_{\mathrm{loc}}(r)\dot{r}). \tag{5.4.3}$$

Alternatively, note that we can write

$$l_c(r, \dot{r}) = L(q, \mathrm{hor}\,\dot{q}),$$

where $r = \pi(q)$ and $\dot{r} = T_q\pi(\dot{q})$.

Using this notation, the equations of motion can be read off of Theorem 5.2.2 to give the following theorem.

5.4.6 Theorem. *In the principal kinematic case, the equations of motion may be written*

$$\delta l_c = \left\langle \frac{\partial l}{\partial \xi}, \mathcal{B}_{\mathrm{loc}}(\dot{r}, \delta r) \right\rangle,$$

$$\dot{g} = -g \mathcal{A}_{\mathrm{loc}}(r)\dot{r}, \tag{5.4.4}$$

where $\delta r \in T(Q/G)$ *and* $\mathcal{B}_{\mathrm{loc}}$ *is the curvature of* $\mathcal{A}_{\mathrm{loc}}$ *and where* $\xi = -\mathcal{A}_{\mathrm{loc}}(r)\dot{r}$.

This theorem goes back to the work of Chaplygin [1897a] (see the references) for the abelian principal case and was extended to the nonabelian case by Koiller [1992]. This result is also a consequence of the results of Marsden and Scheurle [1993a, 1993b]; indeed, they show that the first of these equations is a consequence of the horizontal variations in the action (i.e., the Lagrange–d'Alembert principle) and that in this calculation one can choose any connection, in particular the principal kinematic connection, in this case. Of course, the second of the equations is just the condition of horizontality, that is, the kinematic constraints themselves.

We see in local coordinates that the dynamics of the system can be completely written in terms of the dynamics in base coordinates $r \in Q/G$, and the full dynamics are given by reconstruction of $\dot{g}$ using the constraints. Thus, in the purely kinematic case, we recover the process of reduction and reconstruction with the kinematic connection $\mathcal{A}$ replacing the mechanical connection. We stress, in particular, that in the principal kinematic case, something special happens, namely there is no dynamic equation for $\xi = g^{-1}\dot{g}$, but rather ξ can be expressed directly in terms of r and $\dot{r}$ using the constraints, and when this is substituted into the first of equations (5.4.4), we obtain second-order equations for r. Thus, in this case, the equations actually reduce from equations on $\mathcal{D}/G$ to equations on Q/G. The dynamics of g itself are then recovered by the constraint equation, which may be regarded as the *reconstruction problem*, which is also encountered in the problem of calculating holonomy, as in Marsden, Montgomery, and Ratiu [1990]. In particular, for abelian groups, the dynamics of g can be written in terms of those of r by an explicit quadrature.

The purely kinematic case can easily be extended to allow affine constraints.

5.5   The Momentum Equation

In this section we use the Lagrange–d'Alembert principle to derive an equation for a generalized momentum as a consequence of the symmetries. Under the hypothesis that the action of some Lie algebra element is horizontal (that is, the infinitesimal generator is automatically in the constraint distribution), this yields a conservation law in the usual sense. We refer the reader to Chapter 3 for the classical momentum map.

As we shall see, the momentum equation does not directly involve the choice of an Ehresmann connection to describe the distribution $\mathcal{D}$, but the choice of such a connection will be useful for the coordinate versions.

We have already mentioned that simple physical systems that have symmetries do not have associated conservation laws, namely the wobblestone and the snakeboard. It is also easy to see why this is not generally the case from the equations of motion. The simplest situation would be the case of cyclic variables. Recall that the equations of motion have the form

$$\frac{d}{dt}\frac{\partial L_c}{\partial \dot{r}^\alpha} - \frac{\partial L_c}{\partial r^\alpha} + A^a_\alpha \frac{\partial L_c}{\partial s^a} = -\frac{\partial L}{\partial \dot{s}^b} B^b_{\alpha\beta} \dot{r}^\beta.$$

If this had a cyclic variable, say r^1, then all the quantities $L_c, L, B^b_{\alpha\beta}$ would be independent of r^1. This is equivalent to saying that there is a translational symmetry in the r^1 direction. Let us also suppose, as is often the case, that the s variables are also cyclic. Then the above equation for the momentum $p_1 = \partial L_c/\partial \dot{r}^1$ becomes

$$\frac{d}{dt}p_1 = -\frac{\partial L}{\partial \dot{s}^b} B^b_{1\beta} \dot{r}^\beta.$$

This fails to be a conservation law in general. Note that the right-hand side is linear in $\dot{r}$ (the first term is linear in p_r), and the equation does not depend on r^1 itself. This is a very special case of the momentum equation that we shall develop in this chapter. Even for systems like the snakeboard, the symmetry group is not abelian, so the above analysis for cyclic variables fails to capture the full story. In particular, the momentum equation is not of the preceding form in that example, and thus it must be generalized.

The Derivation of the Momentum Equation.   We now derive a generalized momentum map for nonholonomic systems. The number of equations obtained will equal the dimension of the intersection of the orbit with the given constraints. As we will see, this result will give conservation laws as a particular case.

To formulate this result, some additional ideas and notation will be useful. As the examples show, in general the tangent space to the group orbit through q intersects the constraint distribution at q nontrivially. It will be helpful to give this intersection a name.

5.5.1 Definition.   *The intersection of the tangent space to the group orbit through the point $q \in Q$ and the constraint distribution at this point is denoted by $\mathcal{S}_q$, as in Figure 5.5.1, and we let the union of these spaces over $q \in Q$ be denoted by $\mathcal{S}$. Thus,*

$$\mathcal{S}_q = \mathcal{D}_q \cap T_q(\mathrm{Orb}(q)).$$

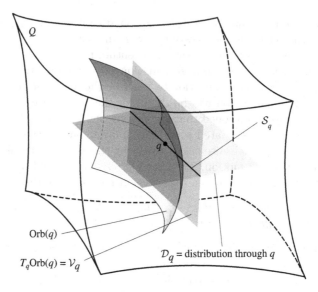

FIGURE 5.5.1. The intersection of the tangent space to the group orbit with the constraint distribution; here the tangent spaces are superimposed on the spaces themselves.

5.5.2 Definition.   *Define, for each $q \in Q$, the vector subspace $\mathfrak{g}^q$ to be the set of Lie algebra elements in $\mathfrak{g}$ whose infinitesimal generators evaluated at q lie in S_q:*

$$\mathfrak{g}^q = \{\xi \in \mathfrak{g} : \xi_Q(q) \in S_q\}.$$

The corresponding bundle over Q whose fiber at the point q is given by $\mathfrak{g}^q$ is denoted by $\mathfrak{g}^{\mathcal{D}}$.

Consider a section of the vector bundle S over Q, i.e., a mapping that takes q to an element of $S_q = \mathcal{D}_q \cap T_q(\mathrm{Orb}(q))$. Assuming that the action is free, a section of S can be uniquely represented as ξ_Q^q, and it defines a section ξ^q of the bundle $\mathfrak{g}^{\mathcal{D}}$. For example, one can construct the section by orthogonally projecting (using the kinetic energy metric) $\xi_Q(q)$ to the subspace S_q. However, as we shall see in later examples, it is often easy to choose a section by inspection.

Next, we choose the variation analogously to what we chose in the case of the standard Noether theorem in the proof of Proposition 3.7.7, given in Section 3.7, namely, $q(t, s) = \exp\left(\phi(t, s)\xi^{q(t)}\right) \cdot q(t)$. The corresponding infinitesimal variation is given by

$$\delta q(t) = \phi'(t)\xi_Q^q(q(t)).$$

Letting $\partial\xi^q$ denote the derivative of ξ^q with respect to q, we have

$$\dot{\delta q} = \dot{\phi}'\xi_Q^{q(t)} + \phi'\left[(T\xi_Q^{q(t)} \cdot \dot{q}) + (\partial\xi^{q(t)} \cdot \dot{q})_Q\right].$$

In this equation the term $T\xi_Q^{q(t)}$ is computed by taking the derivative of the vector field $\xi_Q^{q(t)}$ with $q(t)$ held fixed. By construction, the variation δq satisfies the constraints, and the curve $q(t)$ satisfies the Lagrange–d'Alembert equations, so that the following variational equation holds:

$$0 = \int_a^b \left(\frac{\partial L}{\partial q^i}\delta q^i + \frac{\partial L}{\partial \dot{q}^i}\dot{\delta q}^i\right) dt. \tag{5.5.1}$$

In addition, the invariance identity of (3.7.7) holds using ξ^q:

$$0 = \int_a^b \left(\frac{\partial L}{\partial q^i}\left(\xi_Q^{q(t)}\right)^i \phi' + \frac{\partial L}{\partial \dot{q}^i}\left(T\xi_Q^{q(t)} \cdot \dot{q}\right)^i \phi'\right) dt. \tag{5.5.2}$$

Subtracting equations (5.5.1) and (5.5.2) and using the arbitrariness of ϕ' and integration by parts shows that

$$\frac{d}{dt}\frac{\partial L}{\partial \dot{q}^i}(\xi^{q(t)})_Q^i = \frac{\partial L}{\partial \dot{q}^i}\left[\frac{d}{dt}(\xi^{q(t)})\right]_Q^i.$$

The quantity whose rate of change is involved here is the nonholonomic version of the momentum map in geometric mechanics.

5.5.3 Definition.   *The **nonholonomic momentum map** J^{nhc} is the bundle map taking TQ to the bundle $(\mathfrak{g}^{\mathcal{D}})^*$ whose fiber over the point q is the dual of the vector space $\mathfrak{g}^q$ that is defined by*

$$\left\langle J^{\mathrm{nhc}}(v_q), \xi\right\rangle = \frac{\partial L}{\partial \dot{q}^i}(\xi_Q)^i,$$

where $\xi \in \mathfrak{g}^q$. Intrinsically, this reads

$$\left\langle J^{\mathrm{nhc}}(v_q), \xi\right\rangle = \left\langle \mathbb{F}L(v_q), \xi_Q\right\rangle,$$

where $\mathbb{F}L$ is the fiber derivative of L and where $\xi \in \mathfrak{g}^q$. For notational convenience, especially when the variable v_q is suppressed, we will often write the left-hand side of this equation as $J^{\mathrm{nhc}}(\xi)$.

Notice that the nonholonomic momentum map may be viewed as giving just some of the components of the ordinary momentum map, namely along those symmetry directions that are consistent with the constraints.

We summarize these results in the following theorem.

5.5.4 Theorem.   *Assume that condition (L2) of definition 5.4.1 holds (which is implied by (L1)) and that ξ^q is a section of the bundle $\mathfrak{g}^{\mathcal{D}}$. Then any solution of the Lagrange–d'Alembert equations for a nonholonomic system must satisfy, in addition to the given kinematic constraints, the **momentum equation***

$$\frac{d}{dt}\left(J^{\text{nhc}}(\xi^{q(t)})\right) = \frac{\partial L}{\partial \dot{q}^i}\left[\frac{d}{dt}(\xi^{q(t)})\right]_Q^i. \tag{5.5.3}$$

When the momentum map is paired with a section in this way, we will just refer to it as the momentum. The following is a direct corollary of this result.

5.5.5 Corollary.   *If ξ is a horizontal symmetry (see (S3) above), then the following conservation law holds:*

$$\frac{d}{dt}J^{\text{nhc}}(\xi) = 0. \tag{5.5.4}$$

A somewhat restricted version of the momentum equation was given by Kozlov and Kolesnikov [1978], and the corollary was given by Arnold, Kozlov, and Neishtadt [1988], page 82 (see Bloch and Crouch [1992, 1995] for the controlled case).

Remarks.

1. The right-hand side of the momentum equation (5.5.3) can be written in more intrinsic notation as

$$\left\langle \mathbb{F}L(\dot{q}(t)), \left(\frac{d}{dt}\xi^{q(t)}\right)_Q \right\rangle.$$

2. In the theorem and the corollary we do not need to assume that the distribution itself is G-invariant; that is, we do not need to assume condition (S1). In particular, as we shall see in the examples, one can get conservation laws in some cases in which the distribution is not invariant.

3. The validity of the form of the momentum equation is not affected by any "internal forces," that is, any control forces on shape space. Indeed, such forces would be invariant under the action of the Lie group G and so would be annihilated by the variations taken to prove the above result.

4. The momentum equation still holds in the presence of affine constraints. We do *not* need to assume that the affine vector field defining the affine constraints is invariant under the group. However, this

vector field may appear in the final momentum equation (or conservation law) because the constraints may be used to rewrite the resulting equation. We will see this explicitly in the example of a ball on a rotating table.

5. Assuming that the distribution is invariant (hypothesis (S1)), the nonholonomic momentum map as a bundle map is equivariant with respect to the action of the group G on the tangent bundle TQ and on the bundle $(\mathfrak{g}^{\mathcal{D}})^*$. In fact, since the distribution is invariant, using the general identity $(\mathrm{Ad}_g\,\xi)_Q = \Phi^*_{g^{-1}}\xi_Q$, valid for any group action, we see that the space $\mathfrak{g}^g$ is mapped to $\mathfrak{g}^{gq}$ by the map Ad_g, and so in this sense, the adjoint action acts in a well-defined manner on the bundle $\mathfrak{g}^{\mathcal{D}}$. By taking its dual, we see that the coadjoint action is well-defined on $(\mathfrak{g}^{\mathcal{D}})^*$. In this setting, equivariance of the nonholonomic momentum map follows as in the usual proof (see, for example, Marsden and Ratiu [1999], Chapter 11).

6. One can find an *invariant* momentum if the section is chosen such that

$$(\mathrm{Ad}_{g^{-1}}\,\xi^{g\cdot q})_Q = \xi^q_Q.$$

This can always be done in the case of trivial bundles; one chooses any ξ^q at the identity in the group variable and translates it around by using the action to get a ξ^q at all points. This direction of reasoning (initiated by remarks of Ostrowski, Lewis, Burdick, and Murray; see, e.g., Lewis, Ostrowski, Murray, and Burdick [1994]) is discussed in the paragraph "The Momentum Equation in Body Representation" below. As we will see later, this point of view is useful in the case of the snakeboard.

7. The form of the momentum equation in this section is valid for any curve $q(t)$ that satisfies the Lagrange–d'Alembert principle; we do *not* require that the constraints be satisfied for this curve. The version of the momentum equation given below will explicitly require that the constraints be satisfied. Of course, in examples we always will impose the constraints, so this is really a comment about the logical structure of the various versions of the equation.

8. In some interesting cases one can get conservation laws *without* having horizontal symmetries, as required in the preceding corollary. These are cases in which, for reasons other than horizontality, the right-hand side of the momentum equation vanishes. This may be an important observation for the investigation of completely integrable nonholonomic systems. A specific case in which this occurs is the vertical rolling disk discussed below. There are also interesting implicit conservations laws that do not arise from symmetry; see Chapter 8.

9. Another derivation of the momentum equation and a deeper explo-
ration of the associated geometry can be found in Cendra, Marsden,
and Ratiu [2001b].   ♦

The Momentum Equation in a Moving Basis. There are several
ways of rewriting the momentum equation that are useful; the examples
will show that each of them can reveal interesting aspects of the system
under consideration. This subsection develops the first of these coordinate
formulas, which is in some sense the most naive, but also the most direct. The next subsection will develop a form that is suitable for a local
trivialization of the bundle $Q \to Q/G$. Later on, when the nonholonomic
connection is introduced, we shall come back to both of these forms and
rewrite them in a more sophisticated but also more revealing way.

Introduce coordinates $q^1, \ldots, q^n$ in the neighborhood of a given point q_0
in Q. At the point q_0, introduce a basis

$$\{e_1, e_2, \ldots, e_m, e_{m+1}, \ldots, e_k\}$$

of the Lie algebra such that the first m elements form a basis of $\mathfrak{g}^{q_0}$. Thus,
$k = \dim \mathfrak{g}$ and $m = \dim \mathfrak{g}^q$, which, by assumption, is locally constant. We
can introduce a similar basis

$$\{e_1(q), e_2(q), \ldots, e_m(q), e_{m+1}(q), \ldots, e_k(q)\}$$

at neighboring points q. For example, one can choose an orthonormal basis
(in either the locked inertia metric or relative to a Killing form) that varies
smoothly with q. We introduce a change of basis matrix by writing

$$e_b(q) = \sum_{a=1}^{k} \psi_b^a(q) e_a$$

for $b = 1, \ldots, k$. Here, the change of basis matrix $\psi_b^a(q)$ is an invertible
$k \times k$ matrix. Relative to the dual basis, we write the components of the
nonholonomic momentum map as J_b. By definition,

$$J_b = \sum_{i=1}^{n} \frac{\partial L}{\partial \dot{q}^i} [e_b(q)]_Q^i.$$

Using this notation, the momentum equation, with the choice of section
given by

$$\xi^{q(t)} = e_b(q(t)), \quad 1 \leq b \leq m,$$

reads as follows:

$$\frac{d}{dt} J_b = \sum_{i=1}^{n} \left(\frac{\partial L}{\partial \dot{q}^i} \left[\frac{d}{dt} e_b(q(t)) \right]_Q^i \right). \tag{5.5.5}$$

Next, we define Christoffel-like symbols by

$$\Gamma^c_{bl} = \sum_{a=1}^{k} (\psi^{-1})^c_a \frac{\partial \psi^a_b}{\partial q^l}, \tag{5.5.6}$$

where the matrix $(\psi^{-1})^d_a$ denotes the inverse of the matrix ψ^a_b. Observe that

$$\frac{d}{dt} e_b(q(t)) = \sum_{c=1}^{k} \sum_{l=1}^{n} \Gamma^c_{bl} \dot{q}^l e_c(q(t)), \tag{5.5.7}$$

which implies that

$$\left[\frac{d}{dt} e_b(q(t)) \right]^i_Q = \sum_{c=1}^{k} \sum_{l=1}^{n} \Gamma^c_{bl} \dot{q}^l [e_c(q(t))]^i_Q. \tag{5.5.8}$$

Thus, we can write the momentum equation as

$$\frac{d}{dt} J_b = \sum_{c=1}^{k} \sum_{i,l=1}^{n} \frac{\partial L}{\partial \dot{q}^i} \Gamma^c_{bl} \dot{q}^l [e_c(q(t))]^i_Q. \tag{5.5.9}$$

Introducing the shorthand notation $e^i_c := [e_c(q(t))]^i_Q$, the momentum equation reads

$$\frac{d}{dt} J_b = \sum_{c=1}^{k} \sum_{i,l=1}^{n} \frac{\partial L}{\partial \dot{q}^i} \Gamma^c_{bl} \dot{q}^l e^i_c, \tag{5.5.10}$$

Breaking the summation over c into two ranges and using the definition

$$J_c = \frac{\partial L}{\partial \dot{q}^i} e^i_c, \qquad 1 \leq c \leq m,$$

gives the following form of the momentum equation.

5.5.6 Proposition (Momentum equation in a moving basis). *The momentum equation in the above coordinate notation reads*

$$\frac{d}{dt} J_b = \sum_{c=1}^{m} \sum_{l=1}^{n} \Gamma^c_{bl} J_c \dot{q}^l + \sum_{c=m+1}^{k} \sum_{i,l=1}^{n} \frac{\partial L}{\partial \dot{q}^i} \Gamma^c_{bl} \dot{q}^l e^i_c. \tag{5.5.11}$$

Assuming that the Lagrangian is of the form kinetic minus potential energy, the second term on the right-hand side of this equation vanishes if the orbit and the constraint distribution are orthogonal (in the kinetic energy metric), that is, if we can choose the basis so that the vectors $[e_c(q(t))]_Q$ for $c \geq m+1$ are orthogonal to the constraint distribution. In this case, the momentum equation has the form of an equation of parallel transport along the curve $q(t)$. The connection involved is the natural one associated with the bundle $(\mathfrak{g}^{\mathcal{D}})^*$ over Q, using a chosen decomposition of $\mathfrak{g}$, such as the orthogonal one. In the general case, the momentum equation is an equality between the covariant derivative of the nonholonomic momentum and the last term on the right-hand side of the preceding equation. Below we shall write the momentum equation in a body frame, which will be important for understanding how to decouple the momentum equation from the group variables.

The Momentum Equation in Body Representation.   Next, we develop an alternative coordinate formula for the momentum equation that is adapted to a choice of local trivialization. Thus, let a local trivialization be chosen on the principal bundle $\pi : Q \to Q/G$, with the local representation having coordinates denoted by (r, g). Let r have components denoted by r^{α} as before, being coordinates on the base Q/G, and let g be group variables for the fiber G. In such a representation, the action of G is the left action of G on the second factor. We calculate the nonholonomic momentum map using well-known ideas (see, for example, Marsden and Ratiu [1999], Chapter 12), as follows. Let $v_q = (r, g, \dot{r}, \dot{g})$ be a tangent vector at the point $q = (r, g)$, $\eta \in \mathfrak{g}^q$, and let $\xi = g^{-1}\dot{g}$, i.e., $\xi = T_g L_{g^{-1}} \dot{g}$. Since L is G-invariant, we can define a new function l by writing

$$L(r, g, \dot{r}, \dot{g}) = l(r, \dot{r}, \xi).$$

Use of the chain rule shows that

$$\frac{\partial L}{\partial \dot{g}} = T_g^* L_{g^{-1}} \frac{\partial l}{\partial \xi},$$

and so

$$\left\langle J^{\mathrm{nhc}}(v_q), \eta \right\rangle = \left\langle \mathbb{F}L(r, g, \dot{r}, \dot{g}), \eta_Q(r, g) \right\rangle = \left\langle \frac{\partial L}{\partial \dot{g}}, (0, TR_g \cdot \eta) \right\rangle$$

$$= \left\langle \frac{\partial l}{\partial \xi}, \mathrm{Ad}_{g^{-1}} \eta \right\rangle.$$

The preceding equation shows that we can write the momentum map in a local trivialization by making use of the Ad mapping in much the same way as we did with the connection and the local formulas in the principal kinematic case. We define $J_{\mathrm{loc}}^{\mathrm{nhc}} : TQ/G \to (\mathfrak{g}^{\mathcal{D}})^*$ in a local trivialization by

$$\left\langle J_{\mathrm{loc}}^{\mathrm{nhc}}(r, \dot{r}, \xi), \eta \right\rangle = \left\langle \frac{\partial l}{\partial \xi}, \eta \right\rangle.$$

Thus, as with the previous local forms, J^{nhc} and its version in a local trivialization are related by the Ad map; precisely,

$$J^{\mathrm{nhc}}(r, g, \dot{r}, \dot{g}) = \mathrm{Ad}_{g^{-1}}^* J_{\mathrm{loc}}^{\mathrm{nhc}}(r, \dot{r}, \xi).$$

Secondly, choose a q-dependent basis $e_a(q)$ for the Lie algebra such that the first m elements span the subspace $\mathfrak{g}^q$. In a local trivialization, this is done in a very simple way. First, one chooses, for each r, such a basis at the identity element $g = \mathrm{Id}$, say

$$e_1(r), e_2(r), \ldots, e_m(r), e_{m+1}(r), \ldots, e_k(r).$$

For example, this could be a basis such that the corresponding generators are orthonormal in the kinetic energy metric. (Keep in mind that the subspaces $\mathcal{D}_q$ and $T_q\,\mathrm{Orb}$ need not be orthogonal, but here we are choosing a basis corresponding only to the subspace $T_q\,\mathrm{Orb}$.) Define the **body fixed basis** by

$$e_a(r, g) = \mathrm{Ad}_g\, e_a(r);$$

then the first m elements will indeed span the subspace $\mathfrak{g}^q$, provided the distribution is invariant (condition (S1)). Thus, in this basis we have

$$\left\langle J^{\mathrm{nhc}}(r, g, \dot{r}, \dot{g}), e_b(r, g) \right\rangle = \left\langle \frac{\partial l}{\partial \xi}, e_b(r) \right\rangle := p_b, \tag{5.5.12}$$

which defines p_b, a function of r, $\dot{r}$, and ξ. We are deliberately introducing the new notation p for the momentum in body representation to signal its special role. Note that in this body representation, the functions p_b are *invariant* rather than equivariant, as is usually the case with the momentum map. The time derivative of p_b may be evaluated using the momentum equation (5.5.3). This gives

$$\frac{d}{dt} p_b = \frac{\partial L}{\partial \dot{q}^i} \left[\frac{d}{dt} e_b(r, g) \right]_Q^i = \left\langle (T_g L_{g^{-1}})^* \frac{\partial l}{\partial \xi}, \left[\frac{d}{dt} \left(\mathrm{Ad}_g \cdot e_b(r) \right) \right]_Q \right\rangle$$

$$= \left\langle \frac{\partial l}{\partial \xi}, [\xi, e_b] + \frac{\partial e_b}{\partial r^\alpha} \dot{r}^\alpha \right\rangle.$$

We summarize the conclusion drawn from this calculation as follows.

5.5.7 Proposition (Momentum equation in body representation). *The momentum equation in body representation on the principal bundle $Q \to Q/G$ is given by*

$$\frac{d}{dt} p_b = \left\langle \frac{\partial l}{\partial \xi}, [\xi, e_b] + \frac{\partial e_b}{\partial r^\alpha} \dot{r}^\alpha \right\rangle. \tag{5.5.13}$$

Moreover, the momentum equation in this representation is independent of, that is, decouples from, the group variables g.

In this representation the variable ξ is related to the group variable g by $\xi = g^{-1}\dot{g}$. In particular, in this representation, reconstruction of the group variable g can be done by means of the equation

$$\dot{g} = g\xi. \tag{5.5.14}$$

On the other hand, this variable $\xi = g^{-1}\dot{g}$, as in the case of the reduced Euler–Poincaré equations, *is not the vertical part of the velocity vector $\dot{q}$ relative to the nonholonomic connection to be constructed below.* The vertical part is related to the variable ξ by a velocity shift, and this velocity shift will make the reconstruction equation look affine, as in the case of the snakeboard (see Lewis, Ostrowski, Murray, and Burdick [1994]). In that example, the decoupling of the momentum equation from the group variables played a useful role. We also recall (as in the example of the rigid body with rotors) that it is often the shifted velocity and not ξ that diagonalizes the kinetic energy, so this shift is fundamental for a number of reasons. As we shall see later, the same ideas in this section, combined with the calculations of Marsden and Scheurle [1993b], will show how to calculate the fully reduced equations.

In the above local trivialization form of the momentum equation, we may write the terms $(\partial e_b/\partial r^\alpha)\dot{r}^\alpha$ in terms of a connection, as we did in deriving the momentum equation in a moving basis.

Other noteworthy features of this form of the momentum equation are the following direct consequences of the preceding proposition.

5.5.8 Corollary.

1. *If e_b, $b = 1, \ldots, m$, are independent of r, then the momentum equation in body representation is equivalent to the Euler–Poincaré equations projected to the subspace $\mathfrak{g}^q$.*

2. *If $\mathfrak{g}$ is abelian, then the momentum equations reduce to*

$$\frac{d}{dt}p_b = \left\langle \frac{\partial l}{\partial \xi}, \frac{\partial e_b}{\partial r^\alpha}\dot{r}^\alpha \right\rangle. \tag{5.5.15}$$

3. *If $\mathfrak{g}$ is abelian, or more generally, if the bracket of an element of $\mathfrak{g}^q$ with one in $\mathfrak{g}$ is annihilated by $\partial l/\partial \xi$, and if $e_b, b = 1, \ldots, m$, are independent of r, then the quantities $p_b, b = 1, \ldots, m$, are constants of motion.*

Regarding the first item, see Marsden and Ratiu [1999] for a discussion of the Euler–Poincaré equations; see Chapter 3. In this case, the spatial form of the momentum is conserved, just as in the case of systems with holonomic constraints. The last case occurs for the vertical rolling penny.

5.6   Examples of the Nonholonomic Momentum Map

5.6.1 Example (The Vertical Rolling Disk). We begin by developing the equations of motion using the Ehresmann connection given by the constraints and deriving the reduced Lagrangian, thus illustrating the material of Section 5.2. The equations are then written explicitly in terms of the reduced Lagrangian and the curvature of the connection. We then discuss the momentum equation of Section 5.5. Using different subgroups of the full symmetry group we show how one gets conservation laws from both horizontal and nonhorizontal symmetries. The different forms of the conservation laws are also illustrated.

Thus consider as in Chapter 1 a vertical disk free to roll on the xy-plane and to rotate about its vertical axis. Let x and y denote the position of contact of the disk in the xy-plane. The remaining variables are θ and φ, denoting the orientation of a chosen material point P with respect to the vertical and the "heading angle" of the disk.

Thus, the unconstrained configuration space for the vertical rolling disk is $Q = \mathbb{R}^2 \times S^1 \times S^1$. The velocities associated with the coordinates x, y, θ, φ are denoted by $\dot{x}, \dot{y}, \dot{\theta}$, and $\dot{\varphi}$, which provide the remaining coordinates for the velocity phase space TQ. The Lagrangian for the problem is taken to be the kinetic energy

$$L\left(x, y, \theta, \phi, \dot{x}, \dot{y}, \dot{\theta}, \dot{\phi}\right) = \frac{1}{2}m\left(\dot{x}^2 + \dot{y}^2\right) + \frac{1}{2}I\dot{\theta}^2 + \frac{1}{2}J\dot{\varphi}^2, \tag{5.6.1}$$

where m is the mass of the disk, and I and J are its moments of inertia. Note that so far, we use the full configuration space, ignoring the constraints, and that the Lagrangian is the standard "free" Lagrangian.

The rolling constraints (assuming that the disk has radius R) may be written as

$$\begin{aligned} \dot{x} &= R(\cos\varphi)\dot{\theta}, \\ \dot{y} &= R(\sin\varphi)\dot{\theta}. \end{aligned} \tag{5.6.2}$$

At first one can close one's eyes to the symmetry of the problem and just think of the constraints as the horizontal space of an Ehresmann connection. To do this, one must choose a bundle $Q \to R$. Given the nature of the constraints and the fact that one imagines that eventually controls would be added to either the θ or the φ variable, one is motivated to choose the base R to be $S^1 \times S^1$ parametrized by θ and φ with the projection to R being the naive one $(s^1, s^2, r^1, r^2) = (x, y, \theta, \varphi) \mapsto (r^1, r^2) = (\theta, \varphi)$. From the constraints one can read off the components of the Ehresmann connection:

$$\begin{aligned} A_1^1 &= -R(\cos\varphi), \\ A_1^2 &= -R(\sin\varphi), \end{aligned} \tag{5.6.3}$$

and the remaining A_α^a are zero. If we choose to regard the bundle $Q \to R$ as a principal bundle with group $G = \mathbb{R}^2$, we get an abelian purely kinematic system (see Bloch, Reyhanoglu, and McClamroch [1992] and Bloch and Crouch [1992]). Indeed, note that using the obvious action of G, we get

$$T_q \text{Orb}(q) = \text{span}\left\{\frac{\partial}{\partial x}, \frac{\partial}{\partial y}\right\}. \tag{5.6.4}$$

Notice that $\mathcal{D}_q \cap T_q(\mathrm{Orb}(q)) = \{0\}$ and that the components of A are independent of x and y.

The constrained Lagrangian L_c is given by substituting (5.6.2) into (5.6.1):

$$L_c\left(\theta, \varphi, x, y, \dot\theta, \dot\varphi\right) = \frac{1}{2}(mR^2 + I)\dot\theta^2 + \frac{1}{2}J\dot\varphi^2. \qquad (5.6.5)$$

Note that if the mass density of the disk were constant, then we would have $I = \frac{1}{2}mR^2$, and we could simplify the coefficient of $\dot\theta^2$ to $\frac{3}{4}mR^2\dot\theta^2$, but we need not make this assumption. The curvature of the connection A is computed using formula (5.2.8) to be

$$B^1_{21} = -B^1_{12} = -R\sin\varphi, \quad B^2_{12} = -B^2_{21} = -R\cos\varphi, \qquad (5.6.6)$$

with the remaining $B^a_{\alpha\beta}$ zero. The equations of motion

$$\frac{d}{dt}\left(\frac{\partial L_c}{\partial \dot r^\alpha}\right) - \frac{\partial L_c}{\partial r^\alpha} = -\left(\frac{\partial L}{\partial \dot s^a}\right)B^a_{\alpha\beta}\dot r^\beta \qquad (5.6.7)$$

become

$$\begin{aligned}
(mR^2 + I)\ddot\theta &= (mR(\cos\varphi)\dot\theta)(-R(\sin\varphi)\dot\varphi) \\
&\quad + (mR(\sin\varphi)\dot\theta)(R(\cos\varphi)\dot\varphi) = 0, \qquad (5.6.8) \\
J\ddot\varphi &= (mR(\cos\varphi)\dot\theta)(R(\sin\varphi)\dot\theta) \\
&\quad + (mR(\sin\varphi)\dot\theta)(-R\cos\varphi)\dot\theta = 0. \qquad (5.6.9)
\end{aligned}$$

Thus, $\dot\theta = \Omega$ and $\dot\varphi = \omega$ are constants, so $\theta = \Omega t + \theta_0, \varphi = \omega t + \varphi_0$, and equation (5.6.2) gives

$$\begin{aligned}
\dot x &= \Omega R\cos(\omega t + \varphi_0), \\
\dot y &= \Omega R\sin(\omega t + \varphi_0).
\end{aligned}$$

Hence,

$$x = \frac{\Omega}{\omega}R\sin(\omega t + \varphi_0) + x_0 \quad \text{and} \quad y = -\frac{\Omega}{\omega}R\cos(\omega t + \varphi_0) + y_0.$$

We now turn to the momentum equation. It is clear that in the example as presented, one has the whole group $\mathrm{SE}(2) \times S^1$ as a symmetry group. In such a case, the orbit of the group spans the entire constraint distribution. While this is certainly allowed by the theory, it is an extreme case that one does not have in general. In the presence of controls some of the symmetry will be broken, so it is appropriate to consider a smaller symmetry group, namely a subgroup of $\mathrm{SE}(2) \times S^1$, to be the group G in the general theory. We will, for illustrative purposes, make two choices, namely the subgroup

SE(2) and the *direct* product $\mathbb{R}^2 \times S^1$. To keep things clear, we will write these two choices as

$$G_1 = \text{SE}(2) \quad \text{and} \quad G_2 = \mathbb{R}^2 \times S^1.$$

It is interesting that, as we shall see, the actions of G_1 and G_2 give rise to the two conservation laws $\dot{\theta} = \Omega$ and $\dot{\varphi} = \omega$, respectively, one being induced by a horizontal symmetry, the other not.

The action of $G_1 = \text{SE}(2)$ on $\mathbb{R}^4$ is given by

$$(x, y, \theta, \varphi) \mapsto (x \cos \alpha - y \sin \alpha + a, x \sin \alpha + y \cos \alpha + b, \theta, \varphi + \alpha), \quad (5.6.10)$$

where $(a, b, \alpha) \in \text{SE}(2)$. The $\mathbb{R}^2 \times S^1$ action is given by

$$(x, y, \theta, \varphi) \mapsto (x + \lambda, y + \mu, \theta + \beta, \varphi). \quad (5.6.11)$$

The tangent space to the orbits of the SE(2) action is given by

$$T_q \text{Orb}(q) = \text{span} \left\{ \frac{\partial}{\partial x}, \frac{\partial}{\partial y}, \frac{\partial}{\partial \varphi} \right\}, \quad (5.6.12)$$

while for the $G_2 = \mathbb{R}^2 \times S^1$ action, they are

$$T_q \text{Orb}(q) = \text{span} \left\{ \frac{\partial}{\partial x}, \frac{\partial}{\partial y}, \frac{\partial}{\partial \theta} \right\}. \quad (5.6.13)$$

One checks that the Lagrangian and the constraints are invariant under each of these actions.

We now consider the momentum equations corresponding to these two actions. The preceding calculations show that the constraint distribution $\mathcal{D}_q$ is given by

$$\mathcal{D}_q = \text{span} \left\{ \frac{\partial}{\partial \varphi}, R \cos \varphi \frac{\partial}{\partial x} + R \sin \varphi \frac{\partial}{\partial y} + \frac{\partial}{\partial \theta} \right\}. \quad (5.6.14)$$

Recall that the space $\mathcal{S}_q$ is given by the intersection of the tangent space to the orbit with the constraint distribution itself. Hence, for the SE(2) action we have

$$\mathcal{S}_q = \mathcal{D}_q \cap T_q \text{Orb}_{G_1}(q) = \text{span} \left\{ \frac{\partial}{\partial \varphi} \right\}, \quad (5.6.15)$$

and for the $\mathbb{R}^2 \times S^1$ action we have

$$\mathcal{S}_q = \mathcal{D}_q \cap T_q \text{Orb}_{G_2}(q) = \text{span} \left\{ R \cos \varphi \frac{\partial}{\partial x} + R \sin \varphi \frac{\partial}{\partial y} + \frac{\partial}{\partial \theta} \right\}. \quad (5.6.16)$$

To obtain the corresponding momentum equations, we consider the bundles whose fibers are the span of the tangent vectors in the preceding two equations in the respective cases, and choose sections of these bundles. The bundles are, of course, trivial. In the case of the $G_1 = $ SE(2) action, note that the generators corresponding to the Lie algebra elements represented by the standard basis in $\mathbb{R}^3$ (with translations being the first two components and rotations the third) are given by

$$(1,0,0)_Q = \frac{\partial}{\partial x}, \qquad (0,1,0)_Q = \frac{\partial}{\partial y}, \qquad (0,0,1)_Q = -y\frac{\partial}{\partial x} + x\frac{\partial}{\partial y} + \frac{\partial}{\partial \varphi}.$$

To obtain the section of $\mathcal{S}_q$ given by vector field

$$\xi_Q^q = \frac{\partial}{\partial \varphi}, \tag{5.6.17}$$

we thus choose the Lie algebra element

$$\xi^q = (y, -x, 1), \tag{5.6.18}$$

while for the $G_2 = \mathbb{R}^2 \times S^1$ action we take the section to be the vector field

$$\xi_Q^q = R\cos\varphi\frac{\partial}{\partial x} + R\sin\varphi\frac{\partial}{\partial y} + \frac{\partial}{\partial \theta} \tag{5.6.19}$$

with corresponding Lie algebra element

$$\xi^q = (R\cos\varphi, R\sin\varphi, 1). \tag{5.6.20}$$

For the SE(2) action, the nonholonomic momentum map is

$$J^{\mathrm{nhc}}(\xi^q) = \frac{\partial L}{\partial \dot{q}^i}(\xi_Q^q)^i = J\dot{\varphi}, \tag{5.6.21}$$

and hence the momentum equation becomes

$$\frac{d}{dt}J^{\mathrm{nhc}}(\xi^q) = \frac{d}{dt}(J\dot{\varphi}) = \frac{\partial L}{\partial \dot{q}^i}\left[\frac{d}{dt}(\xi^q)\right]_Q^i = m\dot{x}(\dot{y}) + m\dot{y}(-\dot{x}) + 0 = 0. \tag{5.6.22}$$

This is, of course, an ordinary conservation law (conservation of vertical component of angular momentum) and is one of the equations of motion.

Note that corresponding to this action, $\mathcal{D}_q \cap T_q(\mathrm{Orb}_{G_1}(q)) = T_q(\mathrm{Orb}_H(q))$, where $H = S^1$, and we obtain a conservation law corresponding to the horizontal action of S^1. This law can, of course, also be obtained by directly considering the S^1 action.

For the G_1 action, a straightforward calculation shows that the third part of Corollary 5.5.8 applies, and so this is one way to find the constants of motion. Rather than giving the details of this calculation, we will give them for the $G_2 = \mathbb{R}^2 \times S^1$ action.

Using the G_2 action, the nonholonomic momentum map is

$$J^{\mathrm{nhc}}(\xi^q) = \frac{\partial L}{\partial \dot{q}^i}(\xi_Q^q)^i = m\dot{x}R\cos\varphi + m\dot{y}R\sin\varphi + I\dot{\theta}, \tag{5.6.23}$$

and so the momentum equation becomes

$$\frac{d}{dt}(m\dot{x}R\cos\varphi + m\dot{y}R\sin\varphi + I\dot{\theta}) = m\dot{x}\frac{d}{dt}(R\cos\varphi) + m\dot{y}\frac{d}{dt}(R\sin\varphi),$$

$$(5.6.24)$$

i.e.,

$$R\cos\varphi\, m\ddot{x} + R\sin\varphi\, m\ddot{y} + I\ddot{\theta} = 0. \qquad (5.6.25)$$

Using the constraints to eliminate $\ddot{x}$ and $\ddot{y}$ from this equation we get

$$(mR^2 + I)\ddot{\theta} = 0, \qquad (5.6.26)$$

which we derived in (5.6.8). Alternatively, observe that after imposing the constraints, the right-hand side of equation (5.6.24) is zero and the left-hand side reduces to the left-hand side of (5.6.26). Thus the two momentum equations yield the reduced equations of motion.

We now illustrate, for the case of the $G_2 = \mathbb{R}^2 \times S^1$ action, the momentum equation in a moving basis (5.5.11) and the momentum equation in a body frame (5.5.13). The latter is equivalent to the reduced form of the momentum equation given in Theorem 5.7.3. We first treat the version (5.5.11). Choose a fixed basis for the Lie algebra of $G_2 = \mathbb{R}^2 \times S^1$, namely $(1,0,0), (0,1,0)$, and $(0,0,1)$. From $\xi^q = \xi^a e_a$ we have

$$\xi^1 = R\cos\varphi, \qquad \xi^2 = R\sin\varphi, \qquad \xi^3 = 1.$$

Choose the moving basis

$$e_1(q) = (R\cos\varphi, R\sin\varphi, 1), \qquad e_2(q) = (1,0,0), \qquad e_3(q) = (0,1,0),$$

and write $e_b(q) = \psi_b^a(q)e_a$. We find that

$$\psi_1^1 = R\cos\varphi, \qquad \psi_1^2 = R\sin\varphi, \qquad \psi_1^3 = 1, \qquad \psi_2^1 = 1, \qquad \psi_3^2 = 1,$$

and $\psi_b^a = 0$ otherwise. Writing $\xi_a^i = K_a^i \xi^a$, we find that the infinitesimal generator coefficients are given by $K_1^1 = K_2^2 = K_3^3 = 1$, and $K_a^i = 0$ otherwise. From the formula $J_b = (\partial L/\partial \dot{q}^i)K_a^i \psi_b^a$, we obtain

$$J_1 = m\dot{x}R\cos\varphi + m\dot{y}R\sin\varphi + I\dot{\theta},$$

noting that S_q is one-dimensional, so the range of the index b in the nonholonomic momentum map is simply $b = 1$. We find that

$$\Gamma_{14}^2 = (\psi^{-1})_a^2 \frac{\partial \psi_1^a}{\partial \varphi} = -R\sin\varphi,$$

$$\Gamma_{14}^3 = (\psi^{-1})_a^3 \frac{\partial \psi_1^a}{\partial \varphi} = R\cos\varphi,$$

$$\Gamma_{1k}^d = 0 \quad \text{otherwise.}$$

With $r = 1$, $n = 4$, and $k = 3$, these calculations show that the momentum equation (5.5.11) becomes

$$\frac{d}{dt}J_1 = \sum_{l=1}^{4} \Gamma^1_{1l}J_1\dot{q}^l + \sum_{i,l=1}^{4} \frac{\partial L}{\partial \dot{q}^i}\Gamma^2_{1l}\dot{q}^l[e_2(q(t))]^i_Q + \sum_{i,l=1}^{4} \frac{\partial L}{\partial \dot{q}^i}\Gamma^3_{1l}\dot{q}^l[e_3(q(t))]^i_Q.$$

(5.6.27)

The first term is zero, and the momentum equation simplifies to

$$\frac{d}{dt}J_1 = m\dot{x}\frac{d}{dt}(R\cos\varphi) + m\dot{y}\frac{d}{dt}(R\sin\varphi),$$

(5.6.28)

which is indeed the correct momentum equation. We now discuss the version (5.5.13) of the momentum equation, continuing with the G_2 action. Here the shape variable r is φ, and $\xi = (\dot{x}, \dot{y}, \dot{\theta})$, and so the reduced Lagrangian is

$$l\left(\varphi, \dot{\varphi}, \dot{x}, \dot{y}, \dot{\theta}\right) = \frac{1}{2}m\left(\dot{x}^2 + \dot{y}^2\right) + \frac{1}{2}I\dot{\theta}^2 + \frac{1}{2}J\dot{\varphi}^2.$$

We choose $e_1(\varphi) = (R\cos\varphi, R\sin\varphi, 1), e_2(\varphi) = (1, 0, 0)$, and $e_3(\varphi) = (0, 1, 0)$. Then (5.5.12) gives

$$p_1 = \left\langle\frac{\partial l}{\partial \xi}, e_1\right\rangle = m\dot{x}R\cos\varphi + m\dot{y}R\sin\varphi + I\dot{\theta},$$

which, when the constraints are substituted, gives

$$p_1 = (mR^2 + I)\dot{\theta}.$$

The momentum equation (5.5.13) now becomes, since the group is abelian,

$$\frac{d}{dt}p_1 = \left\langle\frac{\partial l}{\partial \xi}, \frac{\partial e_1}{\partial \varphi}\dot{\varphi}\right\rangle$$
$$= \left\langle(m\dot{x}, m\dot{y}, I\dot{\theta}), (-R\sin\varphi, R\cos\varphi, 0)\right\rangle$$
$$= -m\dot{x}R\sin\varphi + m\dot{y}R\cos\varphi,$$

which vanishes in view of the constraints. Thus, we recover $dp_1/dt = 0$, as before. Observe that this formulation directly gives us a conservation law even though the symmetry is *not* horizontal.   ◆

5.6.2 Example (A Nonholonomically Constrained Particle). An instructive academic example due to Rosenberg [1977] that illustrates the momentum equation is the following example of a nonholonomically constrained free particle. This example was also used to illustrate the theory in Bates and Sniatycki [1993]. We show here that the momentum equation in an orthogonal body frame is a pure parallel transport equation with respect to the nonmetric connection for the particle, a result that was observed by

Bates and Sniatycki. We thus provide a general method for deriving such a connection.

Consider a particle with the Lagrangian

$$L = \frac{1}{2}\left(\dot{x}^2 + \dot{y}^2 + \dot{z}^2\right) \tag{5.6.29}$$

and the nonholonomic constraint

$$\dot{z} = y\dot{x}. \tag{5.6.30}$$

The constraints and Lagrangian are invariant under the $\mathbb{R}^2$ action on $\mathbb{R}^3$ given by

$$(x, y, z) \mapsto (x + \lambda, y, z + \mu). \tag{5.6.31}$$

The tangent space to the orbits of this action is given by

$$T_q \mathrm{Orb}(q) = \mathrm{span}\left\{\frac{\partial}{\partial x}, \frac{\partial}{\partial z}\right\}, \tag{5.6.32}$$

and the kinematic constraint distribution is given by

$$\mathcal{D}_q = \mathrm{span}\left\{\frac{\partial}{\partial x} + y\frac{\partial}{\partial z}, \frac{\partial}{\partial y}\right\}. \tag{5.6.33}$$

and thus

$$\mathcal{D}_q \cap T_q(\mathrm{Orb}(q)) = \mathrm{span}\left\{\frac{\partial}{\partial x} + y\frac{\partial}{\partial z}\right\}. \tag{5.6.34}$$

Consider the bundle $\mathcal{S}$ with fibers the span of these tangent vectors. To obtain the momentum equations we begin by taking an arbitrary section of this bundle. The bundle is of course trivial, and for simplicity we take the section to be the vector field

$$\xi_Q^q = \frac{\partial}{\partial x} + y\frac{\partial}{\partial z}. \tag{5.6.35}$$

The corresponding Lie algebra element $\xi^q \in \mathbb{R}^2$ is

$$\xi^q = (1, y). \tag{5.6.36}$$

The nonholonomic momentum map in this case is

$$J^{\mathrm{nhc}}(\xi^q) = \frac{\partial L}{\partial \dot{q}^i}(\xi_Q^q)^i = \langle(\dot{x}, \dot{y}, \dot{z}), (1, 0, y)\rangle = \dot{x} + y\dot{z}. \tag{5.6.37}$$

Hence the momentum equation becomes

$$\frac{dJ^{\mathrm{nhc}}(\xi^q)}{dt} = \frac{d}{dt}(\dot{x} + y\dot{z}) = \frac{\partial L}{\partial \dot{q}^i}\left[\frac{d}{dt}(\xi^q)\right]_Q^i = \langle(\dot{x}, \dot{y}, \dot{z}), (0, 0, \dot{y})\rangle = \dot{z}\dot{y},$$
$$\tag{5.6.38}$$

i.e.,

$$\ddot{x} + y\ddot{z} = 0. \tag{5.6.39}$$

Using the constraint $\dot{z} = y\dot{x}$, the momentum equation may be rewritten as

$$\ddot{x} + \frac{y}{1 + y^2}\dot{x}\dot{y} = 0. \tag{5.6.40}$$

Together with the Lagrangian equation of motion $\ddot{y} = 0$, this completely specifies the motion, and in fact, these two equations are a (nonmetric) geodesic flow, as pointed out in Bates and Sniatycki [1993]. In this example we note that the momentum equation is the total derivative of a first-order conservation law:

$$\dot{x} - \frac{c}{(1 + y^2)^{\frac{1}{2}}} = 0 \tag{5.6.41}$$

for c an arbitrary constant. Note, however, that this equation, which is used in the Bates–Sniatycki reduction, is a conservation law, but is not directly a component of a conserved momentum map. In other words, the fact that the second-order momentum equation here is the derivative of a first-order conservation law is not due to considerations of symmetry.

Note also that if one chooses the right base and fiber, this system is again an abelian Chaplygin system. Here we take $\mathbb{R}^2$ with coordinates $\{x, y\}$ to be the base and $\mathbb{R}$ with coordinate z to be the fiber. Then

$$T_q(\mathrm{Orb}(q)) = \mathrm{span}\left\{\frac{\partial}{\partial z}\right\} \tag{5.6.42}$$

and $\mathcal{D}_q \cap T_q(\mathrm{Orb}(q)) = 0$.

We again illustrate both coordinate versions of the momentum equation, namely (5.5.11) and (5.5.13), first treating the version (5.5.11). We choose a fixed basis for $\mathfrak{g} = \mathbb{R}^2$, namely $e_1 = (1,0)$ and $e_2 = (0,1)$, then $\xi^q = \xi^1 e_1 + \xi^2 e_2$, where $\xi^1 = 1$, $\xi^2 = y$. As before, choose a moving basis

$$e_1(q) = (1, y), \qquad e_2(q) = (0, 1).$$

Then if

$$e_b(q) = \sum_{a=1}^{2} \psi_b^a(q)e_a,$$

clearly $\psi_1^1 = 1$, $\psi_1^2 = y$, $\psi_2^1 = 0$, $\psi_2^2 = 1$. Writing $\xi_Q^i = K_a^i \xi^a$, we obtain $K_1^1 = 1$, $K_2^3 = 1$, and $K_a^i = 0$ otherwise. Hence,

$$J_1 = \frac{\partial L}{\partial \dot{q}^i} K_a^i \psi_1^a = \dot{x} + y\dot{z},$$

where we note that $\mathcal{S}_q$ is one-dimensional, so the range of the index b in the nonholonomic momentum map is simply $b = 1$.

Next we compute the connection coefficients. We obtain

$$(\psi^{-1})_1^1 = 1, \qquad (\psi^{-1})_1^2 = -y, \qquad (\psi^{-1})_2^1 = 0, \qquad (\psi^{-1})_2^2 = 1,$$

and hence $\Gamma_{12}^2 = 1$, and $\Gamma_{bk}^1 = 0$ otherwise. These calculations with $r = 1$, $n = 3$, and $k = 2$ show that the momentum equation (5.5.11) becomes

$$\frac{d}{dt} J_1 = \sum_{l=1}^{3} \Gamma_{1l}^1 J_1 \dot{q}^l + \sum_{i,l=1}^{3} \frac{\partial L}{\partial \dot{q}^i} \Gamma_{1l}^2 \dot{q}^l [e_2(q(t))]_Q^i. \tag{5.6.43}$$

The first term is zero, and so the momentum equation simplifies to

$$\frac{d}{dt} J_1 = \dot{z}\dot{y}, \tag{5.6.44}$$

the correct momentum equation.

Now we discuss the version (5.5.13) of the momentum equation. First, we must orthogonalize the preceding moving basis. Here the shape variable r is y, and $\xi = (\dot{x}, \dot{z})$, and so the reduced Lagrangian is

$$l(r, \dot{r}, \xi) = \frac{1}{2} \left(\dot{x}^2 + \dot{y}^2 + \dot{z}^2 \right).$$

We choose $e_1(r) = (1, y)$ and $e_2(r) = (-y, 1)$. Then (5.5.12) gives $p_1 = \dot{x}(1 + y^2)$. Again the group is abelian, and so the momentum equation (5.5.13) becomes

$$\frac{d}{dt} p_1 = \left\langle \frac{\partial l}{\partial \xi}, \frac{\partial e_1}{\partial y} \dot{y} \right\rangle = \langle (\dot{x}, \dot{z}), (0, 1) \rangle = \dot{z}\dot{y},$$

as before. Writing

$$\frac{\partial e_1}{\partial y} = (0, 1) = \gamma_{11}^1 e_1 + \gamma_{11}^2 e_2,$$

we see that

$$\gamma_{11}^1 = \frac{y}{1 + y^2}, \qquad \gamma_{11}^2 = \frac{1}{1 + y^2},$$

and so the momentum equation becomes

$$\frac{d}{dt} p_1 = \frac{y\dot{y}}{1 + y^2} p_1,$$

which is in parallel transport form. Note that the connection we have just constructed using the general principles of the momentum equation is the same nonmetric connection as in Bates and Sniatycki [1993]. ◆

Remark. There is a rather beautiful mechanical interpretation of this example due to Andy Ruina (personal communication). Let the variables above be denoted by $(x, y, z) \mapsto (\theta, y, \phi)$ for convenience of exposition, where we let θ denote the angle of a rotating turntable and ϕ denote the angle of a rotating hoop with free rotating balls on it that touches the turntable at right angles at distance y from the center. This is like a nonholonomic gear mechanism!

Exercises

$\diamond$ **5.6-1.** Compute the (asymptotic) dynamics of the constrained particle system.

5.7   More General Nonholonomic Systems with Symmetries

We consider here the nonholonomic equations of motion and the momentum equation in the body representation.

The Nonholonomic Connection. Assume that the Lagrangian has the form kinetic energy minus potential energy, and that the constraints and the orbit directions span the entire tangent space to the configuration space, sometimes called the *dimension assumption*:

$$\mathcal{D}_q + T_q(\operatorname{Orb}(q)) = T_q Q. \tag{5.7.1}$$

In this case, the momentum equation can be used to augment the constraints and provide a connection on the shape space bundle $Q \mapsto Q/G$.

5.7.1 Definition. *Under the dimension assumption in equation (5.7.1), and the assumption that the Lagrangian is of the form kinetic minus potential energy, the* **nonholonomic connection** $\mathcal{A}$ *is the connection on the principal bundle* $Q \mapsto Q/G$ *whose horizontal space at the point* $q \in Q$ *is given by the orthogonal complement to the space* $\mathcal{S}_q$ *within the space* $\mathcal{D}_q$.

Let $\mathbb{I}(q) : \mathfrak{g}^{\mathcal{D}} \to (\mathfrak{g}^{\mathcal{D}})^*$ be the locked inertia tensor relative to $\mathfrak{g}^{\mathcal{D}}$, defined by

$$\langle \mathbb{I}(q)\xi, \eta \rangle = \langle\!\langle \xi_Q, \eta_Q \rangle\!\rangle, \qquad \xi, \eta \in \mathfrak{g}^q,$$

where $\langle\!\langle \cdot, \cdot \rangle\!\rangle$ is the kinetic energy metric. Define a map

$$A_q^{\text{sym}} : T_q Q \to \mathcal{S}_q = \mathcal{D}_q \cap T_q(\operatorname{Orb}(q))$$

given by

$$A_q^{\text{sym}}(v_q) = (\mathbb{I}^{-1} J^{\text{nhc}}(v_q))_Q.$$

This map is equivariant and is a projection onto $\mathcal{S}_q$. Choose $\mathcal{U}_q \subset T_q(\operatorname{Orb}(q))$ such that $T_q(\operatorname{Orb}(q)) = \mathcal{S}_q \oplus \mathcal{U}_q$. Let $A_q^{\text{kin}} : T_q Q \to \mathcal{U}_q$ be a $\mathcal{U}_q$-valued form that projects $\mathcal{U}_q$ onto itself and maps $\mathcal{D}_q$ to zero; for example, it can be given by orthogonal projection relative to the kinetic energy metric (this will be our default choice).

5.7.2 Proposition. *The nonholonomic connection regarded as an Ehresmann connection is given by*

$$A = A^{\text{kin}} + A^{\text{sym}}. \tag{5.7.2}$$

When the connection is regarded as a principal connection (i.e., takes values in the Lie algebra rather than the vertical space) we will use the symbol $\mathcal{A}$.

Given a velocity vector $\dot{q}$ that satisfies the constraints, we orthogonally decompose it into a piece in $\mathcal{S}_q$ and an orthogonal piece denoted by $\dot{r}^h$. We regard $\dot{r}^h$ as the horizontal lift of a velocity vector $\dot{r}$ on the shape space; recall that in a local trivialization, the horizontal lift to the point (r, g) is given by

$$\dot{r}^h = (\dot{r}, -\mathcal{A}_{\text{loc}}\dot{r}) = (\dot{r}^\alpha, -\mathcal{A}^a_\alpha \dot{r}^\alpha),$$

where $\mathcal{A}^a_\alpha$ are the components of the nonholonomic connection that is a principal connection in a local trivialization.

We will denote the decomposition of $\dot{q}$ by

$$\dot{q} = \Omega_Q(q) + \dot{r}^h,$$

so that for each point q, Ω is an element of the Lie algebra and represents the spatial angular velocity of the locked system. In a local trivialization, we can write, at a point (r, g),

$$\Omega = \text{Ad}_g(\Omega_{\text{loc}}),$$

so that Ω_{loc} represents the body angular velocity. Thus,

$$\Omega_{\text{loc}} = \mathcal{A}_{\text{loc}}\dot{r} + \xi,$$

and at each point q, the constraints are that Ω belongs to $\mathfrak{g}^q$, i.e.,

$$\Omega_{\text{loc}} \in \text{span}\{e_1(r), e_2(r), \ldots, e_m(r)\}.$$

The vector $\dot{r}^h$ need not be orthogonal to the whole orbit, just to the piece $\mathcal{S}_q$. Even if $\dot{q}$ does not satisfy the constraints, we can decompose it into three parts and write

$$\dot{q} = \Omega_Q(q) + \dot{r}^h = \Omega_Q^{\text{nh}}(q) + \Omega_Q^{\perp}(q) + \dot{r}^h,$$

where Ω_Q^{nh} and $\Omega_Q^{\perp}$ are orthogonal and where $\Omega_Q^{\text{nh}}(q) \in \mathcal{S}_q$. The relation $\Omega_{\text{loc}} = \mathcal{A}_{\text{loc}}\dot{r} + \xi$ is valid even if the constraints do not hold; also note that this decomposition of Ω corresponds to the decomposition of the nonholonomic connection, $A = A^{\text{kin}} + A^{\text{sym}}$, that was given in equation (5.7.2).

To avoid confusion, we will make the following index and summation conventions:

1. The first batch of indices range from 1 to m corresponding to the symmetry directions along the constraint space. These indices will be denoted by $a, b, c, d, \ldots$, and a summation from 1 to m will be understood.

2. The second batch of indices range from $m + 1$ to k corresponding to the symmetry directions not aligned with the constraints. Indices for this range or for the whole range 1 to k will be denoted by $a', b', c', \ldots$, and the summations will be given explicitly.

3. The indices $\alpha, \beta, \ldots$ on the shape variables r range from 1 to σ. Thus, σ, is the dimension of the shape space Q/G, and so $\sigma = n - k$. The summation convention for these indices will be understood.

The Reduced Nonholonomic Equations. The reduced equations of motion are given as follows.

5.7.3 Theorem. *The following **reduced nonholonomic Lagrange–d'Alembert equations** hold for each $1 \le \alpha \le \sigma$ and $1 \le b \le m$:*

$$\frac{d}{dt}\frac{\partial l_c}{\partial \dot{r}^\alpha} - \frac{\partial l_c}{\partial r^\alpha} = -\frac{\partial I^{cd}}{\partial r^\alpha}p_c p_d - \mathcal{D}_{b\alpha}^c I^{bd} p_c p_d - \mathcal{B}_{\alpha\beta}^c p_c \dot{r}^\beta$$
$$- \mathcal{D}_{\beta\alpha b}I^{bc}p_c \dot{r}^\beta - \mathcal{K}_{\alpha\beta\gamma}\dot{r}^\beta \dot{r}^\gamma,$$

$$\frac{d}{dt}p_b = C_{ab}^c I^{ad} p_c p_d + \mathcal{D}_{b\alpha}^c p_c \dot{r}^\alpha + \mathcal{D}_{\alpha\beta b}\dot{r}^\alpha \dot{r}^\beta.$$

Here $l_c(r^\alpha, \dot{r}^\alpha, p_a)$ is the constrained Lagrangian; r^α, $1 \le \alpha \le \sigma$, are coordinates in the shape space; p_a, $1 \le a \le m$, are components of the momentum map in the body representation, $p_a = \langle \partial l_c / \partial \Omega_{\mathrm{loc}}, e_a(r) \rangle$; I^{ad} are the components of the inverse locked inertia tensor; $\mathcal{B}_{\alpha\beta}^a$ are the local coordinates of the curvature $\mathcal{B}$ of the nonholonomic connection $\mathcal{A}$ corrected by certain terms (see below); and the coefficients $\mathcal{D}_{b\alpha}^c$, $\mathcal{D}_{\alpha\beta b}$, $\mathcal{K}_{\alpha\beta\gamma}$ are given by the formulae

$$\mathcal{D}_{b\alpha}^c = \sum_{a'=1}^{k} -C_{a'b}^c \mathcal{A}_\alpha^{a'} + \gamma_{b\alpha}^c + \sum_{a'=m+1}^{k} \lambda_{a'\alpha}C_{ab}^{a'} I^{ac},$$

$$\mathcal{D}_{\alpha\beta b} = \sum_{a'=m+1}^{k} \lambda_{a'\alpha}\left(\gamma_{b\beta}^{a'} - \sum_{b'=1}^{k} C_{b'b}^{a'} \mathcal{A}_\beta^{b'}\right),$$

$$\mathcal{K}_{\alpha\beta\gamma} = \sum_{a'=m+1}^{k} \lambda_{a'\gamma}\mathcal{B}_{\alpha\beta}^{a'},$$

where

$$\mathcal{B}_{\alpha\beta}^{a'} = \frac{\partial \mathcal{A}_\alpha^{a'}}{\partial r^\beta} - \frac{\partial \mathcal{A}_\beta^{a'}}{\partial r^\alpha} - C_{b'c'}^{a'}\mathcal{A}_\alpha^{b'}\mathcal{A}_\beta^{c'} + \gamma_{b'\beta}^{a'}\mathcal{A}_\alpha^{b'} - \gamma_{b'\alpha}^{a'}\mathcal{A}_\beta^{b'},$$

for $a' = 1, \ldots, k$

$$\lambda_{a'\alpha} = l_{a'\alpha} - \sum_{b'=1}^{k} l_{a'b'}\mathcal{A}_\alpha^{b'} := \frac{\partial l}{\partial \xi^{a'}\partial \dot{r}^\alpha} - \sum_{b'=1}^{k} \frac{\partial l}{\partial \xi^{a'}\partial \xi^{b'}}\mathcal{A}_\alpha^{b'}$$

for $a' = m+1, \ldots, k$. Here $C^{b'}_{a'c'}$ are the structure constants of the Lie algebra defined by $[e_{a'}, e_{c'}] = C^{b'}_{a'c'} e_{b'}$, $a', b', c' = 1, \ldots, k$, and the coefficients $\gamma^{c'}_{b\alpha}$ are defined by

$$\frac{\partial e_b}{\partial r^\alpha} = \sum_{c'=1}^{k} \gamma^{c'}_{b\alpha} e_{c'}.$$

This result is proved in Bloch, Krishnaprasad, Marsden, and Murray [1996]; see also the Internet Supplement. A statement of the reduced equations in intrinsic geometric language is given in Cendra, Marsden, and Ratiu [2001b]. They also, for good reason, call these reduced equations the **Lagrange–d'Alembert–Poincaré** equations.

Relative Equilibria.   A *relative equilibrium* is an equilibrium of the reduced equations; that is, it is a solution that is given by a one-parameter group orbit, just as in the holonomic case (see, e.g., Marsden [1992] for a discussion).

The Constrained Routhian.   This function is defined by analogy with the usual Routhian by

$$R(r^\alpha, \dot{r}^\alpha, p_a) = l_c(r^\alpha, \dot{r}^\alpha, I^{ab} p_b) - I^{ab} p_a p_b, \tag{5.7.3}$$

and in terms of it, the reduced equations of motion become

$$\frac{d}{dt} \frac{\partial R}{\partial \dot{r}^\alpha} - \frac{\partial R}{\partial r^\alpha} = -\mathcal{D}^c_{b\alpha} I^{bd} p_c p_d - \mathcal{B}^c_{\alpha\beta} p_c \dot{r}^\beta$$
$$- \mathcal{D}_{\beta\alpha b} I^{bc} p_c \dot{r}^\beta - \mathcal{K}_{\alpha\beta\gamma} \dot{r}^\beta \dot{r}^\gamma, \tag{5.7.4}$$

$$\frac{d}{dt} p_b = C^c_{ab} I^{ad} p_c p_d + \mathcal{D}^c_{b\alpha} p_c \dot{r}^\alpha + \mathcal{D}_{\alpha\beta b} \dot{r}^\alpha \dot{r}^\beta. \tag{5.7.5}$$

The Reduced Constrained Energy.   The kinetic energy in the variables r^α, $\dot{r}^\alpha$, Ω^a, $\Omega^{a'}$ equals

$$\frac{1}{2} g_{\alpha\beta} \dot{r}^\alpha \dot{r}^\beta + \frac{1}{2} I_{ac} \Omega^a \Omega^c$$
$$+ \sum_{a'=m+1}^{k} \left(l_{a'\alpha} - l_{a'c'} \mathcal{A}^{c'}_\alpha \right) \Omega^{a'} \dot{r}^\alpha + \frac{1}{2} \sum_{a',c'=m+1}^{k} l_{a'c'} \Omega^{a'} \Omega^{c'}, \tag{5.7.6}$$

where $g_{\alpha\beta}$ are coefficients of the kinetic energy metric induced on the manifold Q/G. Substituting the relations $\Omega^a = I^{ab} p_b$ and the constraint equations $\Omega^{a'} = 0$ in (5.7.6) and adding the potential energy, we define the function E by

$$E = \frac{1}{2} g_{\alpha\beta} \dot{r}^\alpha \dot{r}^\beta + U(r^\alpha, p_a), \tag{5.7.7}$$

which represents the reduced constrained energy in the coordinates r^α, $\dot{r}^\alpha$, p_a, where $U(r^\alpha, p_a)$ is the **amended potential** defined by

$$U(r^\alpha, p_a) = \frac{1}{2} I^{ab} p_a p_b + V(r^\alpha), \qquad (5.7.8)$$

and $V(r^\alpha)$ is the potential energy of the system.

Now we show that the reduced constrained energy is conserved along the solutions of (5.7.4), (5.7.5).

5.7.4 Theorem.   *The reduced constrained energy is a constant of motion.*

Proof.   One way to prove this is to note that the reduced energy is a constant of motion, because it equals the energy represented in coordinates r, $\dot{r}$, g, ξ and because the energy is conserved, since the Lagrangian and the constraints are time-invariant. Along the trajectories, the constrained energy and the energy are the same. Therefore, the reduced constrained energy is a constant of motion.

One may also prove this fact by a direct computation of the time derivative of the reduced constrained energy (5.7.7) along the vector field defined by the equations of motion.   ∎

5.7.5 Example (The Falling Rolling Disk).   Consider again the disk rolling without sliding on the xy-plane as discussed in the introduction and Zenkov, Bloch, and Marsden [1998].

Recall that we have the following: Denote the coordinates of contact of the disk in the xy-plane by x, y. Let θ, ϕ, and ψ denote the angle between the plane of the disk and the vertical axis, the "heading angle" of the disk and "self-rotation" angle of the disk, respectively, as was introduced earlier.

The Lagrangian and the constraints in these coordinates are given by

$$L = \frac{m}{2}\left[(\xi - R(\dot{\phi}\sin\theta + \dot{\psi}))^2 + \eta^2 \sin^2\theta + (\eta\cos\theta + R\dot{\theta})^2\right]$$
$$\quad + \frac{1}{2}\left[A(\dot{\theta}^2 + \dot{\phi}^2\cos^2\theta) + B(\dot{\phi}\sin\theta + \dot{\psi})^2\right] - mgR\cos\theta,$$
$$\dot{x} = -\dot{\psi}R\cos\phi,$$
$$\dot{y} = -\dot{\psi}R\sin\phi,$$

where $\xi = \dot{x}\cos\phi + \dot{y}\sin\phi + R\dot{\psi}$, $\eta = -\dot{x}\sin\phi + \dot{y}\cos\phi$. Note that the constraints may be written as $\xi = 0$, $\eta = 0$.

This system is invariant under the action of the group $G = \mathrm{SE}(2)\times\mathrm{SO}(2)$; the action by the group element (a, b, α, β) is given by

$$(\theta, \phi, \psi, x, y) \mapsto (\theta, \phi + \alpha, \psi + \beta, x\cos\alpha - y\sin\alpha + a, x\sin\alpha + y\cos\alpha + b).$$

Obviously,

$$T_q \mathrm{Orb}(q) = \mathrm{span}\left(\frac{\partial}{\partial\phi}, \frac{\partial}{\partial\psi}, \frac{\partial}{\partial x}, \frac{\partial}{\partial y}\right)$$

and

$$\mathcal{D}_q = \text{span}\left(\frac{\partial}{\partial\theta}, \frac{\partial}{\partial\phi}, R\cos\phi\frac{\partial}{\partial x} + R\sin\phi\frac{\partial}{\partial y} - \frac{\partial}{\partial\psi}\right),$$

which imply

$$\mathcal{S}_q = \mathcal{D}_q \cap T_q\,\text{Orb}(q) = \text{span}\left(\frac{\partial}{\partial\phi}, -R\cos\phi\frac{\partial}{\partial x} - R\sin\phi\frac{\partial}{\partial y} + \frac{\partial}{\partial\psi}\right).$$

Choose vectors $(1,0,0,0)$, $(0,1,0,0)$, $(0,0,1,0)$, $(0,0,0,1)$ as a basis of the Lie algebra $\mathfrak{g}$ of the group G. The corresponding generators are

$$\partial_x, \quad \partial_y, \quad -y\partial_x + x\partial_y + \partial_\phi, \quad \partial_\psi.$$

Taking into account that the generators ∂_ϕ, $-R\cos\phi\partial_x - R\sin\phi\partial_y + \partial_\psi$ correspond to the elements $(y, -x, 1, 0)$, $(-R\cos\phi, -R\sin\phi, 0, 1)$ of the Lie algebra $\mathfrak{g}$, we obtain the following momentum equations:

$$\begin{aligned} \dot{p}_1 &= mR^2\cos\theta\,\dot{\theta}\dot{\psi}, \\ \dot{p}_2 &= -mR^2\cos\theta\,\dot{\theta}\dot{\phi}, \end{aligned} \tag{5.7.9}$$

where

$$\begin{aligned} p_1 &= A\dot{\phi}\cos^2\theta + (mR^2 + B)(\dot{\phi}\sin\theta + \dot{\psi})\sin\theta, \\ p_2 &= (mR^2 + B)(\dot{\phi}\sin\theta + \dot{\psi}), \end{aligned} \tag{5.7.10}$$

into which the constraints have been substituted. One notices that

$$p_1 = \frac{\partial l_c}{\partial\dot{\phi}}, \quad p_2 = \frac{\partial l_c}{\partial\dot{\psi}}.$$

These equations completely determine the motion together with a single "shape space" equation for the variable $r = \theta$. (In virtually all examples of interest the shape space is no more than one-dimensional. For the vertical penny in this setting there is no shape space.)

The θ equation can be determined from the general formalism above or more easily directly from the unreduced constrained equations (5.2.7). ◆

Reduced constrained equations in mechanical form.   As for the unconstrained reduced equations, we can write the base space equations for the reduced nonholonomic equations in a standard mechanical form. See, e.g., Ostrowski [1995].

Let l be the Lagrangian reduced by the group action and and let p_b denote the body momenta in group directions in the constraint manifold as defined earlier.

Then we can write the constraints as

$$\xi = -A^{\text{kin}}(r)\dot{r} + \tilde{I}^{-1}p, \tag{5.7.11}$$

where $\tilde{I}$ is used here to denote the locked inertia in the group direction in the constraint manifold and A^{kin} denotes the kinematic (external) constraints.

5.7.6 Proposition.   *The reduced constrained Lagrangian, i.e., the reduced Lagrangian with the constraints substituted,*

$$l_c(r, \dot{r}, p) = l(r, \dot{r}, \xi)|_{\xi = -A^{\text{kin}}\dot{r} + \tilde{I}^{-1}p}, \tag{5.7.12}$$

takes the form

$$l_c(r, \dot{r}, p) = \frac{1}{2}\dot{r}^T \tilde{M}\dot{r} + \frac{1}{2}\left\langle p, \tilde{I}^{-1}p \right\rangle - V(r), \tag{5.7.13}$$

where

$$\tilde{M} = m - A^T I A + \left(A - A^{\text{kin}}\right)^T I \left(A - A^{\text{kin}}\right). \tag{5.7.14}$$

The proof of this follows simply from substituting the constraints (5.7.11) into the reduced Lagrangian written in the form (3.11.7).

Then the base dynamics can be written, like those in the reduced unconstrained case, as

$$\tilde{M}\ddot{r} + \dot{r}^T \tilde{C}\dot{r} + \tilde{N} + \frac{\partial V}{\partial r} = F, \tag{5.7.15}$$

where $\tilde{M}$ is now defined as in the proposition, $\tilde{C}$ is a Coriolis term defined in the same way as before, and

$$\langle \tilde{N}, \delta r \rangle = \left\langle \frac{\partial l}{\partial \xi}, dA^{\text{kin}}(\dot{r}, \delta r) - [\xi, A(\delta r)] + \frac{(\tilde{I}^{-1}p)}{\delta r}(\delta r) \right\rangle$$
$$- \frac{1}{2}\left\langle (\tilde{I}^{-1}p)^T, \frac{\partial(\tilde{I}^{-1}p)}{\partial r}(\delta r) \right\rangle,$$

where F is a possible external forcing term.

One can read off these coefficients from our general formula, but in practical calculations one just wants compute them for the specific mechanical system at hand, as for the satellite done earlier in Chapter 3 and for the nonholonomic examples in Chapter 8. See also Ostrowski [1995].

5.8   The Poisson Geometry of Nonholonomic Systems

This section consdiders in detail the Hamiltonian description of nonholonomic systems. Because of the necessary replacement of conservation laws with the momentum equation, it is natural to let the value of the momentum be a variable, and for this reason it is natural to take a Poisson viewpoint. Some of this theory was initiated in van der Schaft and Maschke [1994]. What follows builds on their work, further develops the theory of nonholonomic Poisson reduction, and ties this theory to other work in the area. We use this reduction procedure to organize nonholonomic dynamics into a reconstruction equation, a nonholonomic momentum equation, and the reduced Lagrange–d'Alembert equations in Hamiltonian form. We also show that these equations are equivalent to those given by the Lagrangian reduction methods of Bloch, Krishnaprasad, Marsden, and Murray [1996] discussed above. Because of the results of Koon and Marsden [1997b], this is also equivalent to the results of Bates and Sniatycki [1993], obtained by nonholonomic symplectic reduction.

Two interesting complications make this effort especially interesting. First of all, as we have mentioned, symmetry need not lead to conservation laws but rather to a momentum equation. Second, the natural Poisson bracket fails to satisfy the Jacobi identity. In fact, the so-called Jacobiator (the cyclic sum that vanishes when the Jacobi identity holds), or equivalently, the Schouten bracket, is an interesting expression involving the curvature of the underlying distribution describing the nonholonomic constraints. Thus in the nonholonomic setting we really have an *almost Poisson* struture as defined in Chapter 3 (see also Cannas Da Silva and Weinstein [1999], Cantrijn, de León, and de Diego [1999], and Śniatycki [2001]).

As before, the setting is a configuration space Q with a (nonintegrable) distribution $\mathcal{D} \subset TQ$ describing the constraints. For simplicity, we consider only homogeneous velocity constraints. We are given a Lagrangian L on TQ and a Lie group G acting on the configuration space that leaves the constraints and the Lagrangian invariant. In many example, the group encodes position and orientation information. For example, for the snakeboard, the group is $SE(2)$ of rotations and translations in the plane. The quotient space Q/G is called *shape space*.

As we have seen, the dynamics of such a system are described by a set of equations of the following form:

$$g^{-1}\dot{g} = -\mathcal{A}^{\mathrm{nh}}(r)\dot{r} + I^{-1}(r)p, \qquad (5.8.1)$$

$$\dot{p} = \dot{r}^T H(r)\dot{r} + \dot{r}^T K(r)p + p^T D(r)p, \qquad (5.8.2)$$

$$M(r)\ddot{r} = \delta(r,\dot{r},p) + \tau. \qquad (5.8.3)$$

The first equation is a reconstruction equation for a group element g, the second is an equation for the nonholonomic momentum p (strictly speaking, p is the *body* representation of the nonholonomic momentum map, which is not conserved in general), and the third represents equations of motion for the reduced variables r that describe the "shape" of the system. The momentum equation is bilinear in $(\dot{r}, p)$. The variable τ represents the external forces applied to the system, and is assumed to affect only the shape variables; i.e., the external forces are G-invariant. Note that the evolution of the momentum p and the shape r decouple from the group variables.

This framework has been very useful for studying controllability, gait selection, and locomotion for systems such as the snakeboard. It has also helped in the study of optimality of certain gaits, by using optimal control ideas in the context of nonholonomic mechanics (Koon and Marsden

[1997a], Ostrowski, Desai and Kumar [1997]). Hence, it is natural to explore ways to develop a similar framework on the Hamiltonian side.

Bates and Sniatycki [1993] developed the symplectic geometry on the Hamiltonian side of nonholonomic systems, while [BKMM] explored the Lagrangian side. It was not obvious how these two approaches were equivalent, especially how the momentum equation, the reduced Lagrange–d'Alembert equations, and the reconstruction equation correspond to the developments in Bates and Sniatycki [1993].

Koon and Marsden [1997b] established the specific links between these two sides and used the ideas and results of each to shed light on the other, deepening our understanding of both points of view. For example, in proving the equivalence of Lagrangian reduction and symplectic reduction, they have shown where the momentum equation lies on the Hamiltonian side and how this is related to the organization of the dynamics of nonholonomic systems with symmetry into the three parts displayed above: a reconstruction equation for the group element g, an equation for the nonholonomic momentum p, and the reduced Hamilton equations for the shape variables r, p_r. Koon and Marsden [1997b] illustrate the basic theory with the snakeboard, as well as a simplified model of the bicycle (see Getz and Marsden [1995]). The latter is an important prototype control system because it is an underactuated balance system.

Poisson Formulation.   The approach of van der Schaft and Maschke [1994] starts on the Lagrangian side with a configuration space Q and a Lagrangian L (possibly of the form kinetic energy minus potential energy, i.e.,

$$L(q, \dot{q}) = \frac{1}{2}\langle\langle \dot{q}, \dot{q}\rangle\rangle - V(q),$$

where $\langle\langle \ , \ \rangle\rangle$ is a metric on Q defining the kinetic energy and V is a potential energy function).

As above, our nonholonomic constraints are given by a distribution $\mathcal{D} \subset TQ$. We also let $\mathcal{D}^0 \subset T^*Q$ denote the annihilator of this distribution. Using a basis ω^a of the annihilator $\mathcal{D}^\circ$, we can write the constraints as

$$\omega^a(\dot{q}) = 0,$$

where $a = 1, \ldots, k$.

As above, the basic equations are given by the Lagrange–d'Alembert principle and are written as

$$\frac{d}{dt}\frac{\partial L}{\partial \dot{q}^i} - \frac{\partial L}{\partial q^i} = \lambda_a \omega_i^a,$$

where λ_a is a set of Lagrange multipliers.

The Legendre transformation $\mathbb{F}L : TQ \to T^*Q$, assuming that it is a diffeomorphism, is used to define the Hamiltonian $H : T^*Q \to \mathbb{R}$ in the standard fashion (ignoring the constraints for the moment):

$$H = \langle p, \dot{q} \rangle - L = p_i \dot{q}^i - L.$$

Here, the momentum is $p = \mathbb{F}L(v_q) = \partial L / \partial \dot{q}$. Under this change of variables, the equations of motion are written in the Hamiltonian form as

$$\dot{q}^i = \frac{\partial H}{\partial p_i}, \tag{5.8.4}$$

$$\dot{p}_i = -\frac{\partial H}{\partial q^i} + \lambda_a \omega_i^a, \tag{5.8.5}$$

where $i = 1, \ldots, n$, together with the constraint equations

$$\omega_i^a \dot{q}^i = \omega_i^a \frac{\partial H}{\partial p_i} = 0.$$

The preceding constrained Hamiltonian equations can be rewritten as

$$\begin{pmatrix} \dot{q}^i \\ \dot{p}_i \end{pmatrix} = J \begin{pmatrix} \frac{\partial H}{\partial q^j} \\ \frac{\partial H}{\partial p_j} \end{pmatrix} + \begin{pmatrix} 0 \\ \lambda_a \omega_i^a \end{pmatrix}, \quad \omega_i^a \frac{\partial H}{\partial p_i} = 0. \tag{5.8.6}$$

Recall that the cotangent bundle T^*Q is equipped with a canonical Poisson bracket and is expressed in the canonical coordinates (q, p) as

$$\{F, G\}(q, p) = \frac{\partial F}{\partial q^i} \frac{\partial G}{\partial p_i} - \frac{\partial F}{\partial p_i} \frac{\partial G}{\partial q^i} = \left(\frac{\partial F^T}{\partial q}, \frac{\partial F^T}{\partial p} \right) J \begin{pmatrix} \frac{\partial G}{\partial q} \\ \frac{\partial G}{\partial p} \end{pmatrix}.$$

Here J is the canonical Poisson tensor

$$J = \begin{pmatrix} 0_n & I_n \\ -I_n & 0_n \end{pmatrix},$$

which is intrinsically determined by the Poisson bracket $\{,\}$ as

$$J = \begin{pmatrix} \{q^i, q^j\} & \{q^i, p_j\} \\ \{p_i, q^j\} & \{p_i, p_j\} \end{pmatrix}. \tag{5.8.7}$$

On the Lagrangian side, we saw that one can get rid of the Lagrangian multipliers. On the Hamiltonian side, it is also desirable to model the Hamiltonian equations without the Lagrange multipliers by a vector field on a submanifold of T^*Q. In van der Schaft and Maschke [1994] it is done through a clever change of coordinates. We now recall how they do this.

First, a constraint phase space $\mathcal{M} = \mathbb{F}L(\mathcal{D}) \subset T^*Q$ is defined in the same way as in Bates and Sniatycki [1993] so that the constraints on the Hamiltonian side are given by $p \in \mathcal{M}$. In local coordinates,

$$\mathcal{M} = \left\{ (q, p) \in T^*Q \,\Big|\, \omega_i^a \frac{\partial H}{\partial p_i} = 0 \right\}.$$

Let $\{X_\alpha\}$ be a local basis for the constraint distribution $\mathcal{D}$ and let $\{\omega^a\}$ be a local basis for the annihilator $\mathcal{D}^0$. Let $\{\omega_a\}$ span the complementary subspace to $\mathcal{D}$ such that $\langle \omega^a, \omega_b \rangle = \delta_b^a$, where δ_b^a is the usual Kronecker delta. Here $a = 1, \ldots, k$ and $\alpha = 1, \ldots, n - k$. Define a coordinate transformation $(q, p) \mapsto (q, \tilde{p}_\alpha, \tilde{p}_a)$ by

$$\tilde{p}_\alpha = X_\alpha^i p_i, \qquad \tilde{p}_a = \omega_a^i p_i. \tag{5.8.8}$$

It is shown in van der Schaft and Maschke [1994] that in the new (generally not canonical) coordinates $(q, \tilde{p}_\alpha, \tilde{p}_a)$, the Poisson tensor becomes

$$\tilde{J}(q, \tilde{p}) = \begin{pmatrix} \{q^i, q^j\} & \{q^i, \tilde{p}_j\} \\ \{\tilde{p}_i, q^j\} & \{\tilde{p}_i, \tilde{p}_j\} \end{pmatrix}, \tag{5.8.9}$$

and the constrained Hamiltonian equations (5.8.6) transform into

$$\begin{pmatrix} \dot{q}^i \\ \dot{\tilde{p}}_\alpha \\ \dot{\tilde{p}}_a \end{pmatrix} = \tilde{J}(q, \tilde{p}) \begin{pmatrix} \frac{\partial \tilde{H}}{\partial q^j} \\ \frac{\partial \tilde{H}}{\partial \tilde{p}_\beta} \\ \frac{\partial \tilde{H}}{\partial \tilde{p}_b} \end{pmatrix} + \begin{pmatrix} 0 \\ 0 \\ \lambda_a \end{pmatrix}, \qquad \frac{\partial \tilde{H}}{\partial \tilde{p}_a}(q, \tilde{p}) = 0, \tag{5.8.10}$$

where $\tilde{H}(q, \tilde{p})$ is the Hamiltonian $H(q, p)$ expressed in the new coordinates $(q, \tilde{p})$.

Let $(\tilde{p}_\alpha, \tilde{p}_a)$ satisfy the constraint equations $\frac{\partial \tilde{H}}{\partial \tilde{p}_a}(q, \tilde{p}) = 0$. Since

$$\mathcal{M} = \left\{ (q, \tilde{p}_\alpha, \tilde{p}_a) \,\middle|\, \frac{\partial \tilde{H}}{\partial \tilde{p}_a}(q, \tilde{p}_\alpha, \tilde{p}_a) = 0 \right\},$$

van der Schaft and Maschke [1994] use $(q, \tilde{p}_\alpha)$ as induced local coordinates for $\mathcal{M}$. It is easy to show that

$$\frac{\partial \tilde{H}}{\partial q^j}(q, \tilde{p}_\alpha, \tilde{p}_a) = \frac{\partial H_\mathcal{M}}{\partial q^j}(q, \tilde{p}_\alpha),$$

$$\frac{\partial \tilde{H}}{\partial \tilde{p}_\beta}(q, \tilde{p}_\alpha, \tilde{p}_a) = \frac{\partial H_\mathcal{M}}{\partial \tilde{p}_\beta}(q, \tilde{p}_\alpha),$$

where $H_\mathcal{M}$ is the constrained Hamiltonian on $\mathcal{M}$ expressed in the induced coordinates.

Now we are ready to eliminate the Lagrange multipliers. Notice that $\frac{\partial \tilde{H}}{\partial \tilde{p}_b}(q, \tilde{p}) = 0$ on $\mathcal{M}$, and by restricting the dynamics on $\mathcal{M}$, we can disregard the last equations involving λ in equations (5.8.10). In fact, we can also truncate the Poisson tensor $\tilde{J}$ in (5.8.9) by leaving out its last k columns and last k rows and then describe the constrained dynamics on $\mathcal{M}$ expressed in the induced coordinates $(q^i, \tilde{p}_\alpha)$ as follows:

$$\begin{pmatrix} \dot{q}^i \\ \dot{\tilde{p}}_\alpha \end{pmatrix} = J_\mathcal{M}(q, \tilde{p}_\alpha) \begin{pmatrix} \frac{\partial H_\mathcal{M}}{\partial q^j}(q, \tilde{p}_\alpha) \\ \frac{\partial H_\mathcal{M}}{\partial \tilde{p}_\beta}(q, \tilde{p}_\alpha) \end{pmatrix}, \qquad \begin{pmatrix} q^i \\ \tilde{p}_\alpha \end{pmatrix} \in \mathcal{M}. \tag{5.8.11}$$

Here $J_{\mathcal{M}}$ is the $(2n-k) \times (2n-k)$ truncated matrix of $\tilde{J}$ restricted to $\mathcal{M}$ and is expressed in the induced coordinates.

The matrix $J_{\mathcal{M}}$ defines a bracket $\{,\}_{\mathcal{M}}$ on the constraint submanifold $\mathcal{M}$ as follows:

$$\{F_{\mathcal{M}}, G_{\mathcal{M}}\}_{\mathcal{M}}(q, \tilde{p}_\alpha) := \left(\frac{\partial F_{\mathcal{M}}^T}{\partial q^i} \; \frac{\partial F_{\mathcal{M}}^T}{\partial \tilde{p}_\alpha} \right) J_{\mathcal{M}}(q^i, \tilde{p}_\alpha) \left(\begin{array}{c} \frac{\partial G_{\mathcal{M}}}{\partial q^j} \\ \frac{\partial G_{\mathcal{M}}}{\partial \tilde{p}_\beta} \end{array} \right),$$

for any two smooth functions $F_{\mathcal{M}}, G_{\mathcal{M}}$ on the constraint submanifold $\mathcal{M}$. Clearly, this bracket satisfies the first two defining properties of a Poisson bracket, namely, skew symmetry and the Leibniz rule, and one can show that it satisfies the Jacobi identity if and only if the constraints are holonomic. Furthermore, the constrained Hamiltonian $H_{\mathcal{M}}$ is an integral of motion for the constrained dynamics on $\mathcal{M}$ due to the skew symmetry of the bracket.

Below we will develop a general formula for the Jacobiator (the cyclic sum that vanishes when the Jacobi identity holds), which is an interesting expression involving the curvature of the underlying distribution that describes the nonholonomic constraints. From this formula one can see clearly that the Poisson bracket defined here satisfies the Jacobi identity if and only if the constraints are holonomic.

Remarks.   Marle [1998] has shown that the bracket $\{,\}_{\mathcal{M}}$ obtained in van der Schaft and Maschke [1994] can be given an intrinsic interpretation as follows:

1. Suppose that we are given a constrained Hamiltonian system $\{T^*Q, \{,\}, H, \mathcal{M}, \mathcal{V}\}$, where the first four objects are defined as above and the last object $\mathcal{V}$ is a vector subbundle of $T_{\mathcal{M}}(T^*Q)$ defined by

$$\mathcal{V}_p = \{\mathrm{vert}_p(\eta) \mid \eta \in \mathcal{D}^0\}.$$

Here $T_{\mathcal{M}}(T^*Q)$ is the restriction of the tangent bundle of $T(T^*Q)$ to the constraint submanifold $\mathcal{M}$, and $\mathrm{vert}_p(\eta) \in T_p(T^*Q)$ is the vertical lift of $\eta \in T_q^*Q$, where $p \in T^*Q$; in coordinates, $\mathrm{vert}_{(q,p)}(q, \eta) = (q, p, 0, \eta)$. Marle [1998] uses this subbundle $\mathcal{V}$ to encode the fact that the *constraint forces* obey the Lagrange–d'Alembert principle.

It can be shown that the sum of the vector subbundles $T\mathcal{M}$ and $\mathcal{V}$ of $T_{\mathcal{M}}(T^*Q)$ is a direct sum, and

$$T\mathcal{M} \oplus \mathcal{V} = T_{\mathcal{M}}(T^*Q).$$

Moreover, the restriction $X_H|_{\mathcal{M}}$ of the Hamiltonian vector field to the constraint submanifold $\mathcal{M}$ splits into a sum

$$X_H|_{\mathcal{M}} = X_{\mathcal{M}} + X_{\mathcal{V}},$$

where $X_{\mathcal{M}}$ is the constrained Hamiltonian vector field that is tangent to $\mathcal{M}$, and $X_{\mathcal{V}}$ is a smooth section of the subbundle $\mathcal{V}$ (whose opposite is the *constraint force* field).

2. The canonical Poisson tensor Λ (associated with the Poisson bracket $\{,\}$) of T^*Q can be projected on $\mathcal{M}$, and its projection $\Lambda_{\mathcal{M}}$ is a contravariant skew-symmetric 2-tensor on $\mathcal{M}$. More precisely, let $p \in \mathcal{M}$, α and $\beta \in T_p^*\mathcal{M}$. There exists a unique pair $(\hat{\alpha}, \hat{\beta})$ of elements of $T_p^*(T^*Q)$ that vanish on the vector subspace $\mathcal{V}_p$ of $T_p(T^*Q)$, and whose restrictions to the vector subspace $T_p\mathcal{M}$ are α and β, respectively. We can therefore define $\Lambda_{\mathcal{M}}(p)$ by setting

$$\Lambda_{\mathcal{M}}(p)(\alpha, \beta) = \Lambda(p)(\hat{\alpha}, \hat{\beta}).$$

Then the vector field $X_{\mathcal{M}}$, whose integral curves are the motions of the system, is

$$X_{\mathcal{M}} = \Lambda_{\mathcal{M}}^{\sharp}(dH_{\mathcal{M}}).$$

Using the 2-tensor $\Lambda_{\mathcal{M}}$, Marle [1998] defines an intrinsic bracket $\{,\}_{\mathcal{M}}$ on the space of smooth functions on $\mathcal{M}$ by setting

$$\{f, g\}_{\mathcal{M}} = \Lambda_{\mathcal{M}}(df, dg).$$

In the local coordinates $(q, \tilde{p})$, the truncated matrix $J_{\mathcal{M}}(q, \tilde{p}_\alpha)$ obtained in van der Schaft and Maschke [1994] is exactly the matrix associated with the 2-tensor $\Lambda_{\mathcal{M}}$ and is nothing but the projection of the Poisson matrix $\tilde{J}(q, \tilde{p})$ onto the constraint submanifold $\mathcal{M}$.

The above considerations show that the nonholonomic brackets constructed by a quotient operation, are also intrinsic.

A Formula for the Constrained Hamilton Equations.   In holonomic mechanics, it is well known that the Poisson and the Lagrangian formulations are equivalent via a Legendre transform. And it is natural to ask whether the same relation holds for the nonholonomic mechanics as developed in van der Schaft and Maschke [1994] and [BKMM]. First we use the general procedures of van der Schaft and Maschke [1994] to write down a compact formula for the nonholonomic equations of motion.

5.8.1 Theorem.   *Let $q^i = (r^\alpha, s^a)$ be the local coordinates in which ω^a has the form*

$$\omega^a(q) = ds^a + A_\alpha^a(r, s)dr^\alpha, \tag{5.8.12}$$

where $A_\alpha^a(r, s)$ is the coordinate expression of the Ehresmann connection described. Then the nonholonomic constrained Hamilton equations of motion on $\mathcal{M}$ can be written as

$$\dot{s}^a = -A^a_\beta \frac{\partial H_\mathcal{M}}{\partial \tilde{p}_\beta}, \tag{5.8.13}$$

$$\dot{r}^\alpha = \frac{\partial H_\mathcal{M}}{\partial \tilde{p}_\alpha}, \tag{5.8.14}$$

$$\dot{\tilde{p}}_\alpha = -\frac{\partial H_\mathcal{M}}{\partial r^\alpha} + A^b_\alpha \frac{\partial H_\mathcal{M}}{\partial s^b} - p_b B^b_{\alpha\beta} \frac{\partial H_\mathcal{M}}{\partial \tilde{p}_\beta}, \tag{5.8.15}$$

where $B^b_{\alpha\beta}$ are the coefficients of the curvature of the Ehresmann connection. Here p_b should be understood as p_b restricted to $\mathcal{M}$ and more precisely should be denoted by $(p_b)_\mathcal{M}$.

Proof. One can always choose local coordinates in which

$$\omega^a(q) = ds^a + A^a_\alpha(r,s)dr^\alpha.$$

In this local coordinate system,

$$\mathcal{D} = \text{span}\{\partial_{r^\alpha} - A^a_\alpha \partial_{s^a}\}. \tag{5.8.16}$$

Then the new coordinates $(r^\alpha, s^a, \tilde{p}_\alpha, \tilde{p}_a)$ of van der Schaft and Maschke [1994] are defined by

$$\tilde{p}_\alpha = p_\alpha - A^a_\alpha p_a, \quad \tilde{p}_a = p_a + A^\alpha_a p_\alpha, \tag{5.8.17}$$

and we can use $(r^\alpha, s^a, \tilde{p}_\alpha)$ as the induced coordinates on $\mathcal{M}$.

Moreover, we can obtain the constrained Poisson structure matrix $J_\mathcal{M}(r^\alpha, s^a, \tilde{p}_\alpha)$ by computing $\{q^i, q^j\}, \{q^i, \tilde{p}_\alpha\}, \{\tilde{p}_\alpha, \tilde{p}_\beta\}$ and then restricting them to $\mathcal{M}$. Recall that $J_\mathcal{M}$ is constructed out of the Poisson tensor $\tilde{J}$ in equation (5.8.9) by leaving out its last k columns and last k rows and restricting its remaining elements to $\mathcal{M}$.

Clearly,

$$\{q^i, q^j\} = 0.$$

In addition, we have

$$\{r^\beta, \tilde{p}_\alpha\} = \{r^\beta, p_\alpha - A^a_\alpha p_a\} = \{r^\beta, p_\alpha\} - \{r^\beta, A^a_\alpha p_a\} = \delta^\beta_\alpha,$$

$$\{s^b, \tilde{p}_\alpha\} = \{s^b, p_\alpha - A^a_\alpha p_a\} = \{s^b, p_\alpha\} - \{s^b, A^a_\alpha p_a\} = -A^b_\alpha,$$

where δ^β_α is the usual Kronecker delta. It is also straightforward to obtain

$$\{\tilde{p}_\alpha, \tilde{p}_\beta\} = \{p_\alpha - A^a_\alpha p_a, p_\beta - A^b_\beta p_b\}$$

$$= -\{p_\alpha, A^b_\beta p_b\} - \{A^a_\alpha p_a, p_\beta\} + \{A^a_\alpha p_a, A^b_\beta p_b\}$$

$$= \frac{\partial A^b_\beta}{\partial r^\alpha} p_b - \frac{\partial A^b_\alpha}{\partial r^\beta} p_b + \frac{\partial A^a_\alpha}{\partial s^b} p_a A^b_\beta - A^a_\alpha \frac{\partial A^b_\beta}{\partial s^a} p_b$$

$$= \left(\frac{\partial A^b_\beta}{\partial r^\alpha} - \frac{\partial A^b_\alpha}{\partial r^\beta} + A^a_\beta \frac{\partial A^b_\alpha}{\partial s^a} - A^a_\alpha \frac{\partial A^b_\beta}{\partial s^a} \right) p_b$$

$$= -B^b_{\alpha\beta} p_b.$$

After restricting the above results to $\mathcal{M}$, all other terms remain the same, but the last line should be understood as $-B^b_{\alpha\beta}(p_b)_\mathcal{M}$. But for notational simplicity, we keep writing it as $-B^b_{\alpha\beta}p_b$. Putting the above computations together, we can write the nonholonomic equations of motion as

$$
\begin{pmatrix} \dot{s}^a \\ \dot{r}^\alpha \\ \dot{\tilde{p}}_\alpha \end{pmatrix} = \begin{pmatrix} 0 & 0 & -A^a_\beta \\ 0 & 0 & \delta^\alpha_\beta \\ (A^b_\alpha)^T & -\delta^\beta_\alpha & -p_c B^c_{\alpha\beta} \end{pmatrix} \begin{pmatrix} \frac{\partial H_\mathcal{M}}{\partial s^b} \\ \frac{\partial H_\mathcal{M}}{\partial r^\beta} \\ \frac{\partial H_\mathcal{M}}{\partial \tilde{p}_\beta} \end{pmatrix}, \tag{5.8.18}
$$

which is the desired result. Notice that the order of the variables r^α and s^a has been switched to make the block diagonalization of the constrained Poisson tensor more apparent. Also, it may be important to point out here that $(p_b)_\mathcal{M}$ can be expressed in terms of the induced coordinates $(q, \tilde{p}_\beta)$ on the constraint submanifold $\mathcal{M}$ by the following formula:

$$
(p_b)_\mathcal{M} = (g_{b\alpha} - g_{ba}A^a_\alpha)\dot{r}^\alpha = (g_{b\alpha} - g_{ba}A^a_\alpha)\frac{\partial H_\mathcal{M}}{\partial \tilde{p}_\alpha} = K^\beta_b \tilde{p}_\beta,
$$

where $g_{b\alpha}$ and g_{ba} are the components of the kinetic energy metric and K^β_b is defined by the last equality.

Equivalence of the Poisson and Lagrange–d'Alembert Formulations.   Now we are ready to state and prove the equivalence of the Poisson and Lagrange–d'Alembert formulations.

5.8.2 Theorem.   *The Lagrange–d'Alembert equations*

$$
\dot{s}^a = -A^a_\alpha \dot{r}^\alpha, \tag{5.8.19}
$$

$$
\frac{d}{dt}\frac{\partial L_c}{\partial \dot{r}^\alpha} - \frac{\partial L_c}{\partial r^\alpha} + A^a_\alpha \frac{\partial L_c}{\partial s^a} = -\frac{\partial L}{\partial \dot{s}^b}B^b_{\alpha\beta}\dot{r}^\beta, \tag{5.8.20}
$$

are equivalent to the constrained Hamilton equations

$$
\dot{s}^a = -A^a_\beta \frac{\partial H_\mathcal{M}}{\partial \tilde{p}_\beta}, \tag{5.8.21}
$$

$$
\dot{r}^\alpha = \frac{\partial H_\mathcal{M}}{\partial \tilde{p}_\alpha}, \tag{5.8.22}
$$

$$
\dot{\tilde{p}}_\alpha = -\frac{\partial H_\mathcal{M}}{\partial r^\alpha} + A^b_\alpha \frac{\partial H_\mathcal{M}}{\partial s^b} - p_b B^b_{\alpha\beta}\frac{\partial H_\mathcal{M}}{\partial \tilde{p}_\beta}, \tag{5.8.23}
$$

via a constrained Legendre transform that is given by

$$
\tilde{p}_\alpha = \frac{\partial L_c}{\partial \dot{r}^\alpha}, \qquad \dot{r}^\alpha = \frac{\partial H_\mathcal{M}}{\partial \tilde{p}_\alpha}. \tag{5.8.24}
$$

Proof.   Recall that

$$\mathcal{D} = \{(r, s, \dot{r}, \dot{s}) \in TQ \mid \dot{s} + A_\alpha^a \dot{r}^\alpha = 0\},$$

and we can use $(r, s, \dot{r})$ as the induced coordinates for the submanifold $\mathcal{D}$. Since the constrained Lagrangian is given by

$$L_c(r^\alpha, s^a, \dot{r}^\alpha) = L(r^\alpha, s^a, \dot{r}^\alpha, -A_\alpha^a(r, s)\dot{r}^\alpha),$$

we have

$$\frac{\partial L_c}{\partial \dot{r}^\alpha} = \frac{\partial L}{\partial \dot{r}^\alpha} - \frac{\partial L}{\partial \dot{s}^a} A_\alpha^a = p_\alpha - p_a A_\alpha^a = \tilde{p}_\alpha. \tag{5.8.25}$$

Hence, $\frac{\partial L_c}{\partial \dot{r}^\alpha} = \tilde{p}_\alpha$ does define the right constrained Legendre transform between the submanifolds $\mathcal{D}$ and $\mathcal{M}$ with the corresponding induced coordinates $(r^\alpha, s^a, \dot{r}^\alpha)$ and $(r^\alpha, s^a, \tilde{p}_\alpha)$.

Now notice that if $E = \frac{\partial L}{\partial \dot{q}^i} \dot{q}^i - L$, then restricting it to $\mathcal{D}$ we will get

$$
\begin{aligned}
E_\mathcal{D} &= \left. \left(\frac{\partial L}{\partial \dot{r}^\alpha} \dot{r}^\alpha + \frac{\partial L}{\partial \dot{s}^a} \dot{s}^a \right) \right|_\mathcal{D} - L_c \\
&= \frac{\partial L_c}{\partial \dot{r}^\alpha} \dot{r}^\alpha + A_\alpha^a \frac{\partial L}{\partial \dot{s}^a} \dot{r}^\alpha - A_\alpha^a \frac{\partial L}{\partial \dot{s}^a} \dot{r}^\alpha - L_c \\
&= \frac{\partial L_c}{\partial \dot{r}^\alpha} \dot{r}^\alpha - L_c.
\end{aligned}
$$

Hence, the constrained Hamiltonian is given by

$$H_\mathcal{M} = \tilde{p}_\alpha \dot{r}^\alpha - L_c, \tag{5.8.26}$$

and it is straightforward to show that

$$\frac{\partial H_\mathcal{M}}{\partial \tilde{p}_\alpha} = \dot{r}^\alpha + \tilde{p}_\beta \frac{\partial \dot{r}^\beta}{\partial \tilde{p}_\alpha} - \frac{\partial L_c}{\partial \dot{r}^\beta} \frac{\partial \dot{r}^\beta}{\partial \tilde{p}_\alpha} = \dot{r}^\alpha,$$

which gives the equation (5.8.22). Clearly, $\dot{s}^a = -A_\beta^a \dot{r}^\beta$ together with equation (5.8.22) gives equation (5.8.21).

Furthermore, we have

$$\frac{\partial H_\mathcal{M}}{\partial r^\beta} = \tilde{p}_\alpha \frac{\partial \dot{r}^\alpha}{\partial r^\beta} - \frac{\partial L_c}{\partial r^\beta} - \frac{\partial L_c}{\partial \dot{r}^\alpha} \frac{\partial \dot{r}^\alpha}{\partial r^\beta} = -\frac{\partial L_c}{\partial r^\beta} \tag{5.8.27}$$

and

$$\frac{\partial H_\mathcal{M}}{\partial s^b} = \tilde{p}_\alpha \frac{\partial \dot{r}^\alpha}{\partial s^b} - \frac{\partial L_c}{\partial s^b} - \frac{\partial L_c}{\partial \dot{r}^\alpha} \frac{\partial \dot{r}^\alpha}{\partial s^b} = -\frac{\partial L_c}{\partial s^b}. \tag{5.8.28}$$

Substituting the results of (5.8.27) and (5.8.28) into equation (5.8.20), we get the remaining equation (5.8.23).

A Formula for the Jacobiator. Recall that in the proof of Theorem 5.8.1 we have obtained

$$\{q^i, q^j\}_{\mathcal{M}} = 0,$$
$$\{r^\beta, \tilde{p}_\alpha\}_{\mathcal{M}} = \delta^\beta_\alpha,$$
$$\{s^b, \tilde{p}_\alpha\}_{\mathcal{M}} = -A^b_\alpha,$$
$$\{\tilde{p}_\alpha, \tilde{p}_\beta\}_{\mathcal{M}} = B^b_{\alpha\beta}(p_b)_{\mathcal{M}} = B^b_{\alpha\beta} K^\gamma_b \tilde{p}_\gamma.$$

Clearly,

$$\{\{q^i, q^j\}_{\mathcal{M}}, q_k\}_{\mathcal{M}} + \text{cyclic} = 0, \tag{5.8.29}$$

$$\{\{q^i, q^j\}_{\mathcal{M}}, \tilde{p}_\alpha\}_{\mathcal{M}} + \text{cyclic} = 0. \tag{5.8.30}$$

It is also straightforward to obtain

$$\{\{r^\gamma, \tilde{p}_\alpha\}_{\mathcal{M}}, \tilde{p}_\beta\}_{\mathcal{M}} + \text{cyclic} = K^\gamma_b B^b_{\alpha\beta}, \tag{5.8.31}$$

$$\{\{s^a, \tilde{p}_\alpha\}_{\mathcal{M}}, \tilde{p}_\beta\}_{\mathcal{M}} + \text{cyclic} = -B^a_{\alpha\beta} - A^a_\gamma K^\gamma_b B^b_{\alpha\beta}. \tag{5.8.32}$$

As for $\{\{\tilde{p}_\alpha, \tilde{p}_\beta\}_{\mathcal{M}}\}_{\mathcal{M}} + \text{cyclic}$, it takes slightly more work to obtain

$$\{\{\tilde{p}_\alpha, \tilde{p}_\beta\}_{\mathcal{M}}, \tilde{p}_\gamma\}_{\mathcal{M}} + \text{cyclic}$$
$$= \tilde{p}_\tau K^\tau_a B^a_{\delta\gamma} K^\delta_b B^b_{\alpha\beta} - \tilde{p}_\tau \left(\frac{\partial K^\tau_b}{\partial r^\gamma} - A^a_\gamma \frac{\partial K^\tau_b}{\partial s^a} \right) B^b_{\alpha\beta}$$
$$- \tilde{p}_\tau K^\tau_b \left(\frac{\partial B^b_{\alpha\beta}}{\partial r^\gamma} - A^a_\gamma \frac{\partial B^b_{\alpha\beta}}{\partial s^a} \right) + \text{cyclic}.$$

Notice that the right-hand side of the last equation involves the derivatives of the curvature. However, it can be shown that the identity

$$\frac{\partial B^b_{\alpha\beta}}{\partial r^\gamma} - A^c_\gamma \frac{\partial B^b_{\alpha\beta}}{\partial s^c} + B^a_{\alpha\beta} \frac{\partial A^b_\gamma}{\partial s^a} + \text{cyclic} = 0$$

holds, and we can apply it to rewrite the last equation using only the curvature but not its derivatives:

$$\{\{\tilde{p}_\alpha, \tilde{p}_\beta\}_{\mathcal{M}}, \tilde{p}_\gamma\}_{\mathcal{M}} + \text{cyclic}$$
$$= \tilde{p}_\tau K^\tau_a B^a_{\delta\gamma} K^\delta_b B^b_{\alpha\beta} - \tilde{p}_\tau \left(\frac{\partial K^\tau_b}{\partial r^\gamma} - A^a_\gamma \frac{\partial K^\tau_b}{\partial s^a} \right) B^b_{\alpha\beta}$$
$$+ \tilde{p}_\tau K^\tau_a \frac{\partial A^a_\gamma}{\partial s^b} B^b_{\alpha\beta} + \text{cyclic}. \tag{5.8.33}$$

Equations (5.8.29) to (5.8.33) give the Jacobiator of the Poisson bracket on $\mathcal{M}$.[9]

---

[9] One can also use the formalism of the Schouten bracket to do the computations and obtain the same results.

Notice that from the formulas for the Jacobiator, one can see clearly that if the constraints are holonomic and hence the Ehresmann connection has zero curvature, then the Jacobiator is zero and the Jacobi identity holds. Conversely, if the Jacobi identity holds, then we have

$$0 = K_b^\gamma B_{\alpha\beta}^b,$$
$$0 = -B_{\alpha\beta}^a - A_\gamma^a K_b^\gamma B_{\alpha\beta}^b.$$

Therefore, $B_{\alpha\beta}^a = 0$, and the constraints are holonomic.

Poisson Reduction. Here we focus on reduction from the Poisson point of view, that is, the Hamiltonian formulation using Poisson brackets. There is a corresponding symplectic view of noholonomic mechanics and reduction that is presented in the Internet Supplement.

As in our discussion of symmetry earlier, let G be the symmetry group of the system and assume that the quotient space $\bar{\mathcal{M}} = \mathcal{M}/G$ of the G-orbit in $\mathcal{M}$ is a smooth quotient manifold with projection map $\rho : \mathcal{M} \longrightarrow \bar{\mathcal{M}}$. Since G is a symmetry group, all intrinsically defined vector fields push down to $\bar{\mathcal{M}}$. In this section we will write the equations of motion for the reduced constrained Hamiltonian dynamics using a reduced "Poisson" bracket on the reduced constraint phase space $\bar{\mathcal{M}}$. Moreover, an explicit expression for this reduced bracket will be given.

The crucial step here is how to represent the constraint distribution $\mathcal{D}$ in a way that is both intrinsic and ready for reduction. Recall that a body fixed basis

$$e_b(g,r) = \mathrm{Ad}_g\, e_b(r)$$

has been constructed such that the infinitesimal generators $(e_i(g,r))_Q$ of its first m elements at a point q span $\mathcal{S}_q = \mathcal{D}_q \cap T_q(\mathrm{Orb}(q))$. Assume that G is a matrix group and e_i^d is the component of $e_i(r)$ with respect to a fixed basis $\{b_a\}$ of the Lie algebra $\mathfrak{g}$, where $(b_a)_Q = \partial_{g^a}$. Then

$$(e_i(g,r))_Q = g_d^a e_i^d \partial_{g^a}.$$

Since $\mathcal{D}_q$ is the direct sum of $\mathcal{S}_q$ and the horizontal space of the nonholonomic connection $\mathcal{A}$, it can be represented by

$$\mathcal{D} = \mathrm{span}\{g_d^a e_i^d \partial_{g^a}, -g_b^a A_\alpha^b \partial_{g^a} + \partial_{r^\alpha}\}. \tag{5.8.34}$$

Before we state the theorem and do some computations, we want to make sure that the reader understands the index convention used in this section:

1. The first batch of indices is denoted by $a, b, c, \ldots$ and range from 1 to k, corresponding to the symmetry direction ($k = \dim \mathfrak{g}$).

2. The second batch of indices will be denoted by $i, j, k, \ldots$ and range from 1 to m, corresponding to the symmetry direction along the constraint space (m is the number of momentum functions).

3. The indices $\alpha, \beta, \ldots$ on the shape variables r range from 1 to $n - k$ ($n - k = \dim (Q/G)$, i.e., the dimension of the shape space).

Then the induced coordinates $(g^a, r^\alpha, \tilde{p}_i, \tilde{p}_\alpha)$ for the constraint submanifold $\mathcal{M}$ are defined by

$$\tilde{p}_i = g_d^a e_i^d p_a = \mu_d e_i^d, \tag{5.8.35}$$

$$\tilde{p}_\alpha = p_\alpha - g_b^a A_\alpha^b p_a = p_\alpha - \mu_b A_\alpha^b. \tag{5.8.36}$$

Here μ is an element of the dual of the Lie algebra $\mathfrak{g}^*$, and μ_a are its coordinates with respect to a fixed dual basis. Notice that $\tilde{p}_i$ are nothing but the corresponding momentum functions on the Hamiltonian side.

We can find the constrained Poisson structure matrix $J_\mathcal{M}(g^a, r^\alpha, \tilde{p}_i, \tilde{p}_\alpha)$ by computing $\{g^a, g^b\}$, etc., and then restricting them to $\mathcal{M}$. Recall that $J_\mathcal{M}$ is constructed out of the Poisson tensor $\tilde{J}$ in (5.8.9) by leaving out its last k columns and last k rows and restricting its remaining elements to $\mathcal{M}$.

Clearly,

$$\{g^a, g^b\} = 0, \qquad \{g^a, r^\alpha\} = 0, \qquad \{r^\alpha, r^\beta\} = 0.$$

And we also have

$$\{g^a, \tilde{p}_i\} = \{g^a, g_c^b e_i^c p_b\} = g_c^a e_i^c,$$

$$\{g^a, \tilde{p}_\alpha\} = \{g^a, p_\alpha - g_b^c A_\alpha^b p_c\} = -g_b^a A_\alpha^b,$$

$$\{r^\alpha, \tilde{p}_i\} = \{r^\beta, g_c^b e_i^c p_b\} = 0,$$

$$\{r^\alpha, \tilde{p}_\beta\} = \{r^\alpha, p_\beta - g_b^c A_\beta^b p_c\} = \delta_\beta^\alpha.$$

It is also straightforward to obtain

$$\begin{aligned}
\{\tilde{p}_i, \tilde{p}_j\} &= \{g_c^a e_i^c p_a, g_d^b e_j^d p_b\} \\
&= p_b \frac{\partial g_c^b}{\partial g^\sigma} e_i^c g_d^\sigma e_j^d - p_b \frac{\partial g_d^b}{\partial g^\tau} e_i^c g_c^\tau e_j^d \\
&= p_b \left(\frac{\partial g_c^b}{\partial g^\sigma} g_d^\sigma - \frac{\partial g_d^b}{\partial g^\tau} g_c^\tau \right) e_i^c e_j^d \\
&= -p_a g_b^a C_{cd}^b e_i^c e_j^d \\
&= -\mu_a C_{cd}^a e_i^c e_j^d,
\end{aligned}$$

where C_{cd}^a are the structure constants of the Lie algebra $\mathfrak{g}$. Similarly, we have

$$\begin{aligned}
\{\tilde{p}_i, \tilde{p}_\alpha\} &= \{g_c^a e_i^c p_a, p_\alpha - g_d^b A_\alpha^b p_d\} \\
&= \{g_c^a e_i^c p_a, p_\alpha\} - \{g_c^a e_i^c p_a, g_d^b A_\alpha^b p_d\} \\
&= \mu_a \frac{\partial e_i^a}{\partial r^\alpha} + \mu_a C_{bd}^a e_i^b A_\alpha^d
\end{aligned}$$

and

$$\{\tilde{p}_\alpha, \tilde{p}_\beta\} = \{p_\alpha - g_b^a A_\alpha^b p_a, p_\beta - g_d^c A_\beta^d p_c\}$$

$$= -\{p_\alpha, g_d^c A_\beta^d p_c\} - \{g_b^a A_\alpha^b p_a, p_\beta\} + \{g_b^a A_\alpha^b p_a, g_d^c A_\beta^d p_c\}$$

$$= \mu_b \frac{\partial A_\beta^b}{\partial r^\alpha} - \mu_b \frac{\partial A_\alpha^b}{\partial r^\beta} - \mu_b C_{ac}^b A_\alpha^a A_\beta^c$$

$$= -\mu_b B_{\alpha\beta}^b,$$

where $B_{\alpha\beta}^b$ are the coefficients of the curvature of the nonholonomic connection and are given by

$$B_{\alpha\beta}^b = \frac{\partial A_\alpha^b}{\partial r^\beta} - \frac{\partial A_\beta^b}{\partial r^\alpha} + C_{ac}^b A_\alpha^a \Lambda_\beta^c.$$

Therefore, the constrained Hamilton equations can be written as

$$\begin{pmatrix} \dot{g}^a \\ \dot{\tilde{p}}_i \\ \dot{r}^\alpha \\ \dot{\tilde{p}}_\alpha \end{pmatrix} = \begin{pmatrix} 0 & g_c^a e_j^c & 0 & -g_c^a A_\beta^c \\ -(g_c^b e_i^c)^T & -\mu_a C_{bd}^a e_i^b e_j^d & 0 & \mu_a F_{i\beta}^a \\ 0 & 0 & 0 & \delta_\beta^\alpha \\ (g_c^b A_\alpha^c)^T & -(\mu_a F_{j\alpha}^a)^T & -\delta_\alpha^\beta & -\mu_a B_{\alpha\beta}^a \end{pmatrix} \begin{pmatrix} \frac{\partial H_\mathcal{M}}{\partial g^b} \\ \frac{\partial H_\mathcal{M}}{\partial \tilde{p}_j} \\ \frac{\partial H_\mathcal{M}}{\partial r^\beta} \\ \frac{\partial H_\mathcal{M}}{\partial \tilde{p}_\beta} \end{pmatrix},$$

$$(5.8.37)$$

where $F_{i\beta}^a$ is defined by

$$F_{i\beta}^a = \frac{\partial e_i^a}{\partial r^\beta} + C_{bd}^a e_i^b A_\beta^d. \qquad (5.8.38)$$

Notice that the order of the variables r^α and $\tilde{p}_i$ has been switched to make the diagonalization of the constrained Poisson tensor more apparent.

Since G is the symmetry group of the system and the Hamiltonian H is G-invariant, we have $H_\mathcal{M} = h_{\bar{\mathcal{M}}}$. Hence

$$\frac{\partial H_\mathcal{M}}{\partial g^b} = 0,$$

$$\frac{\partial H_\mathcal{M}}{\partial \tilde{p}_j} = \frac{\partial h_\mathcal{M}}{\partial \tilde{p}_j},$$

$$\frac{\partial H_\mathcal{M}}{\partial r^\beta} = \frac{\partial h_\mathcal{M}}{\partial r^\beta},$$

$$\frac{\partial H_\mathcal{M}}{\partial \tilde{p}_\beta} = \frac{\partial h_\mathcal{M}}{\partial \tilde{p}_\beta}.$$

After the reduction by the symmetry group G, we have

$$\begin{pmatrix} \xi^b \\ \dot{\tilde{p}}_i \\ \dot{r}^\alpha \\ \dot{\tilde{p}}_\alpha \end{pmatrix} = \begin{pmatrix} 0 & e_j^b & 0 & -A_\beta^b \\ -(e_i^c)^T & -\mu_a C_{bd}^a e_i^b e_j^d & 0 & \mu_a F_{i\beta}^a \\ 0 & 0 & 0 & \delta_\beta^\alpha \\ (A_\alpha^c)^T & -\mu_a(F_{j\alpha}^a)^T & -\delta_\alpha^\beta & -\mu_a B_{\alpha\beta}^a \end{pmatrix} \begin{pmatrix} c0 \\ \frac{\partial h_{\bar{\mathcal{M}}}}{\partial \tilde{p}_j} \\ \frac{\partial h_{\bar{\mathcal{M}}}}{\partial r^\beta} \\ \frac{\partial h_{\bar{\mathcal{M}}}}{\partial \tilde{p}_\beta} \end{pmatrix}, \qquad (5.8.39)$$

where $\xi^b = (g^{-1})^b_a \dot{g}^a$.

The above computations prove the following theorem.

5.8.3 Theorem.   *The momentum equation and the reduced Hamilton equations on the reduced constraint submanifold $\bar{\mathcal{M}}$ can be written as follows:*

$$\dot{\tilde{p}}_i = -\mu_a C^a_{bd} e^b_i e^d_j \frac{\partial h_{\bar{\mathcal{M}}}}{\partial \tilde{p}_j} + \mu_a F^a_{i\beta} \frac{\partial h_{\bar{\mathcal{M}}}}{\partial \tilde{p}_\beta}, \tag{5.8.40}$$

$$\dot{r}^\alpha = \frac{\partial h_{\bar{\mathcal{M}}}}{\partial \tilde{p}_\alpha}, \tag{5.8.41}$$

$$\dot{\tilde{p}}_\alpha = -\frac{\partial h_{\bar{\mathcal{M}}}}{\partial r^\alpha} - \mu_a F^a_{j\alpha} \frac{\partial h_{\bar{\mathcal{M}}}}{\partial \tilde{p}_j} - \mu_a B^a_{\alpha\beta} \frac{\partial h_{\bar{\mathcal{M}}}}{\partial \tilde{p}_\beta}. \tag{5.8.42}$$

Adding in the reconstruction equation

$$\xi^b = -A^b_\beta \frac{\partial h_{\bar{\mathcal{M}}}}{\partial \tilde{p}_\beta} + e^b_j \frac{\partial h_{\bar{\mathcal{M}}}}{\partial \tilde{p}_j}, \tag{5.8.43}$$

we recover the full dynamics of the system.

Notice that equation (5.8.40) can be considered as the momentum equation on the Hamiltonian side, which corresponds to the momentum equation. It generalizes the Lie–Poisson equation to the nonholonomic case.

Moreover, if we now truncate the reduced Poisson matrix $\tilde{J}$ in equation (5.8.39) by leaving out its first column and first row, the new matrix $J_{\bar{\mathcal{M}}}$ given by

$$J_{\bar{\mathcal{M}}} = \begin{pmatrix} -\mu_a C^a_{bd} e^b_i e^d_j & 0 & \mu_a F^a_{i\beta} \\ 0 & 0 & \delta^\alpha_\beta \\ -\mu_a (F^a_{j\alpha})^T & -\delta^\beta_\alpha & -\mu_a B^a_{\alpha\beta} \end{pmatrix} \tag{5.8.44}$$

defines a bracket $\{,\}_{\bar{\mathcal{M}}}$ on the reduced constraint submanifold $\bar{\mathcal{M}}$ as follows:

$$\{F_{\bar{\mathcal{M}}}, G_{\bar{\mathcal{M}}}\}_{\bar{\mathcal{M}}}(\tilde{p}_i, r^\alpha, \tilde{p}_\alpha)$$

$$:= \left(\frac{\partial F^T_{\bar{\mathcal{M}}}}{\partial \tilde{p}_i} \quad \frac{\partial F^T_{\bar{\mathcal{M}}}}{\partial r^\alpha} \quad \frac{\partial F^T_{\bar{\mathcal{M}}}}{\partial \tilde{p}_\alpha} \right) J_{\bar{\mathcal{M}}}(\tilde{p}_i, r^\alpha, \tilde{p}_\alpha) \begin{pmatrix} \frac{\partial G_{\bar{\mathcal{M}}}}{\partial \tilde{p}_j} \\ \frac{\partial G_{\bar{\mathcal{M}}}}{\partial r^\beta} \\ \frac{\partial G_{\bar{\mathcal{M}}}}{\partial \tilde{p}_\beta} \end{pmatrix},$$

for any two smooth functions $F_{\bar{\mathcal{M}}}, G_{\bar{\mathcal{M}}}$ on the reduced constraint submanifold $\bar{\mathcal{M}}$ whose induced coordinates are $(\tilde{p}_i, r^\alpha, \tilde{p}_\alpha)$. Clearly, this bracket satisfies the first two defining properties of a Poisson bracket, namely, skew-symmetry and the Leibniz rule. Moreover, the reduced constrained Hamiltonian $h_{\bar{\mathcal{M}}}$ is an integral of motion for the reduced Hamiltonian dynamics on $\bar{\mathcal{M}}$ due to the skew symmetry of the reduced bracket.

The Equivalence of Poisson and Lagrangian Reduction. That the Poisson reduction procedure and the Lagrangian reduction procedures give the same set of reduced equations is proved in the following theorem.

5.8.4 Theorem. *Equations (5.8.40) to (5.8.43) given by Poisson reduction are equivalent to the equations given by Lagrangian reduction,*

$$\xi^b = -A^b_\beta \dot{r}^\beta + \Gamma^{bi} p_i = -A^b_\beta \dot{r}^\beta + e^b_j \Omega^j, \qquad (5.8.45)$$

$$\dot{p}_i = \frac{\partial l}{\partial \xi^a} \left(C^a_{bd} \xi^b e^d_i + \frac{\partial e^a_i}{\partial r^\beta} \dot{r}^\beta \right), \qquad (5.8.46)$$

$$\frac{d}{dt} \left(\frac{\partial l_c}{\partial \dot{r}^\alpha} \right) - \frac{\partial l_c}{\partial r^\alpha} = -\frac{\partial l}{\partial \xi^b} (B^b_{\alpha\beta} \dot{r}^\beta + F^b_{\alpha i} \Omega^i), \qquad (5.8.47)$$

via a reduced Legendre transform

$$\tilde{p}_\alpha = \frac{\partial l_c}{\partial \dot{r}^\alpha}, \qquad \tilde{p}_i = \frac{\partial l_c}{\partial \Omega^i}.$$

Proof. Define the reduced constrained Lagrangian

$$l_c(r, \dot{r}, \Omega) = l(r, \dot{r}, -A\dot{r} + \Omega e),$$

where Ω is the body angular velocity and $e(r)$ is the body fixed basis at the identity defined earlier. Notice first that

$$\frac{\partial l}{\partial \dot{r}^\alpha} = \frac{\partial L}{\partial \dot{r}^\alpha} = p_\alpha.$$

Since

$$p_b = \frac{\partial L}{\partial \dot{g}^b} = \frac{\partial l}{\partial \xi^a} \frac{\partial \xi^a}{\partial \dot{g}^b} = \frac{\partial l}{\partial \xi^a} (g^{-1})^a_b,$$

we have

$$\frac{\partial l}{\partial \xi^a} = \mu_a.$$

Hence,

$$\frac{\partial l_c}{\partial \dot{r}^\alpha} = \frac{\partial l}{\partial \dot{r}^\alpha} + \frac{\partial l}{\partial \xi^a} \frac{\partial \xi^a}{\partial \dot{r}^\alpha} = \frac{\partial l}{\partial \dot{r}^\alpha} - \frac{\partial l}{\partial \xi^a} A^a_\alpha = p_\alpha - \mu_a A^a_\alpha = \tilde{p}_\alpha$$

and

$$\frac{\partial l_c}{\partial \Omega^i} = \frac{\partial l}{\partial \xi^a} \frac{\partial \xi^a}{\partial \Omega^i} = \frac{\partial l}{\partial \xi^a} e^a_i = \tilde{p}_i.$$

That is, $\tilde{p}_\alpha = \frac{\partial l_c}{\partial \dot{r}^\alpha}$ and $\tilde{p}_i = \frac{\partial l_c}{\partial \Omega^i}$, do define the right reduced constrained Legendre transform between the reduced constraint submanifolds $\bar{\mathcal{D}}$ and $\bar{\mathcal{M}}$ with the corresponding reduced coordinates $(r^\alpha, \dot{r}^\alpha, \Omega^i)$ and $(r^\alpha, \tilde{p}_\alpha, \tilde{p}_i)$.

To find the reduced constrained Hamiltonian $h_\mathcal{M}$, notice first that since E is G-invariant, we have

$$E = \frac{\partial L}{\partial \dot{q}^i} \dot{q}^i - L = \frac{\partial L}{\partial \dot{g}^a} \dot{g}^a + \frac{\partial L}{\partial \dot{r}^\alpha} \dot{r}^\alpha - L = \frac{\partial l}{\partial \xi^a} \xi^a + \frac{\partial l}{\partial \dot{r}^\alpha} \dot{r}^\alpha - l.$$

After restricting E to the submanifold $\mathcal{D}$, we have

$$
\begin{aligned}
E_\mathcal{D} &= \frac{\partial l}{\partial \xi^a}\left(-A_\alpha^a \dot{r}^\alpha + \Omega^i e_i^a\right) + \left(\frac{\partial l_c}{\partial \dot{r}^\alpha} + A_\alpha^a \frac{\partial l}{\partial \xi^a}\right)\dot{r}^\alpha - l_c \\
&= \frac{\partial l}{\partial \xi^a}\Omega^i e_i^a + \frac{\partial l_c}{\partial \dot{r}^\alpha}\dot{r}^\alpha - l_c \\
&= \frac{\partial l_c}{\partial \Omega^i}\Omega^i + \frac{\partial l_c}{\partial \dot{r}^\alpha}\dot{r}^\alpha - l_c \, .
\end{aligned}
$$

Therefore, we have

$$
h_{\bar{\mathcal{M}}} = \tilde{p}_i \Omega^i + \tilde{p}_\alpha \dot{r}^\alpha - l_c, \tag{5.8.48}
$$

via the Legendre transform $(r^\alpha, \dot{r}^\alpha, \Omega^i) \mapsto (r^\alpha, \tilde{p}_\alpha, \tilde{p}_i)$. Differentiate $h_{\bar{\mathcal{M}}}$ with respect to $\tilde{p}_\alpha$ and $\tilde{p}_j$ and use the Legendre transform to obtain

$$
\frac{\partial h_{\bar{\mathcal{M}}}}{\partial \tilde{p}_\alpha} = \tilde{p}_i \frac{\partial \Omega^i}{\partial \tilde{p}_\alpha} + \tilde{p}_\beta \frac{\partial \dot{r}^\beta}{\partial \tilde{p}_\alpha} + \dot{r}^\alpha - \frac{\partial l_c}{\partial \dot{r}_\beta}\frac{\partial \dot{r}^\beta}{\partial \tilde{p}_\alpha} - \frac{\partial l_c}{\partial \Omega^i}\frac{\partial \Omega^i}{\partial \tilde{p}_\alpha} = \dot{r}^\alpha,
$$

which is equation (5.8.41). Also, we have

$$
\frac{\partial h_{\bar{\mathcal{M}}}}{\partial \tilde{p}_j} = \Omega^j + \tilde{p}_i \frac{\partial \Omega^i}{\partial \tilde{p}_j} + \tilde{p}_\alpha \frac{\partial \dot{r}^\alpha}{\partial \tilde{p}_j} - \frac{\partial l_c}{\partial \dot{r}_\alpha}\frac{\partial \dot{r}^\alpha}{\partial \tilde{p}_j} - \frac{\partial l_c}{\partial \Omega^i}\frac{\partial \Omega^i}{\partial \tilde{p}_j} = \Omega^j,
$$

which, together with equation (5.8.45), gives equation (5.8.43). Moreover, since $\frac{\partial l}{\partial \xi^b} = g_b^a \frac{\partial L}{\partial \dot{g}^a} = \mu_b$ and $\tilde{p}_i = p_i$, we have

$$
\begin{aligned}
\dot{\tilde{p}}_i &= \frac{\partial l}{\partial \xi^a}\left(C_{bd}^a \xi^b e_i^d + \frac{\partial e_i^a}{\partial r^\beta}\dot{r}^\beta\right) \\
&= \mu_a\left(C_{bd}^a e_i^d\left(-A_\beta^b \dot{r}^\beta + e_j^b \Omega^j\right) + \frac{\partial e_i^a}{\partial r^\beta}\dot{r}^\beta\right) \\
&= \mu_a\left(C_{bd}^a e_i^d\left(-A_\beta^b \frac{\partial h_{\bar{\mathcal{M}}}}{\partial \tilde{p}_\beta} + e_j^b \frac{\partial h_{\bar{\mathcal{M}}}}{\partial \tilde{p}_j}\right) + \frac{\partial e_i^a}{\partial r^\beta}\frac{\partial h_{\bar{\mathcal{M}}}}{\partial \tilde{p}_\beta}\right) \\
&= \mu_a C_{bd}^a e_i^b e_j^d \frac{\partial h_{\bar{\mathcal{M}}}}{\partial \tilde{p}_j} + \mu_a\left(C_{bd}^a e_i^b A_\beta^d + \frac{\partial e_i^a}{\partial r^\beta}\right)\frac{\partial h_{\bar{\mathcal{M}}}}{\partial \tilde{p}_\beta},
\end{aligned}
$$

which is equation (5.8.40).

Finally, differentiate $h_{\bar{\mathcal{M}}}$ with respect to r^α to obtain

$$
\frac{\partial h_{\bar{\mathcal{M}}}}{\partial r^\alpha} = \tilde{p}_i \frac{\partial \Omega^i}{\partial \tilde{r}^\alpha} + \tilde{p}_\beta \frac{\partial \dot{r}^\beta}{\partial r^\alpha} - \frac{\partial l_c}{\partial r^\alpha} - \frac{\partial \tilde{l}_c}{\partial \dot{r}^\beta}\frac{\partial \dot{r}^\beta}{\partial r^\alpha} - \frac{\partial l_c}{\partial \Omega^i}\frac{\partial \Omega^i}{\partial r^\alpha} = -\frac{\partial l_c}{\partial r_\alpha},
$$

which together with equation (5.8.47) gives

$$
\dot{\tilde{p}}_\alpha = -\frac{\partial h_{\bar{\mathcal{M}}}}{\partial r^\alpha} - \mu_a F_{j\alpha}^a \frac{\partial h_{\bar{\mathcal{M}}}}{\partial \tilde{p}_j} - \mu_a B_{\alpha\beta}^a \frac{\partial h_{\bar{\mathcal{M}}}}{\partial \tilde{p}_\beta},
$$

which is equation (5.8.42).

Remark. Equations (5.8.47) are the same as the reduced Lagrange–d'Alembert equations. The only difference is that here the reduced constrained Lagrangian l_c is a function of $r, \dot{r}, \Omega$, where in Bloch, Krishnaprasad, Marsden, and Murray [1996] and in Koon and Marsden [1997b] it is considered as a function of $r, \dot{r}, p$. Since it is more natural to use the body angular velocity as a variable on the Lagrangian side, the formulation here looks better.

5.8.5 Example (The Snakeboard). The snakeboard is modeled as a rigid body (the board) with two sets of independently actuated wheels, one on each end of the board. The human rider is modeled as a momentum wheel that sits in the middle of the board and is allowed to spin about the vertical axis. Spinning the momentum wheel causes a countertorque to be exerted on the board. The configuration of the board is given by the position and orientation of the board in the plane, the angle of the momentum wheel, and the angles of the back and front wheels. Let (x, y, θ) represent the position and orientation of the center of the board, ψ the angle of the momentum wheel relative to the board, and ϕ_1 and ϕ_2 the angles of the back and front wheels, also relative to the board. Take the distance between the center of the board and the wheels to be r. See Figure 5.8.1.

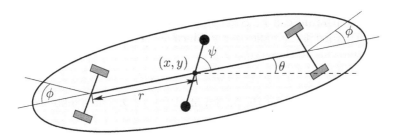

FIGURE 5.8.1. The geometry of the snakeboard.

Following Bloch, Krishnaprasad, Marsden, and Murray [1996], a simplification is made that we shall also assume here, namely that $\phi_1 = -\phi_2$, $J_1 = J_2$. The parameters are also chosen such that $J + J_0 + J_1 + J_2 = mr^2$, where m is the total mass of the board, J is the inertia of the board, J_0 is the inertia of the rotor, and J_1, J_2 are the inertias of the wheels. This simplification eliminates some terms in the derivation but does not affect the essential geometry of the problem. Setting $\phi = \phi_1 = -\phi_2$, then the configuration space becomes $Q = \mathrm{SE}(2) \times S^1 \times S^1$, and the Lagrangian $L : TQ \to \mathbb{R}$ is the total kinetic energy of the system and is given by

$$L(q, \dot{q}) = \frac{1}{2}m(\dot{x}^2 + \dot{y}^2) + \frac{1}{2}mr^2\dot{\theta}^2 + \frac{1}{2}J_0\dot{\psi}^2 + J_0\dot{\psi}\dot{\theta} + J_1\dot{\phi}^2.$$

Clearly, L is independent of the configuration of the board, and hence it is invariant to all possible group actions by the Euclidean group $\mathrm{SE}(2)$.

The Snakeboard Constraint Submanifold. The rolling of the front and rear wheels of the snakeboard is modeled using nonholonomic constraints that allow the wheels to spin about the vertical axis and roll in the direction that they are pointing. The wheels are not allowed to slide in the sideways direction. This gives the constraint one-forms

$$\omega_1(q) = -\sin(\theta + \phi)dx + \cos(\theta + \phi)dy - r\cos\phi\, d\theta,$$
$$\omega_2(q) = -\sin(\theta - \phi)dx + \cos(\theta - \phi)dy + r\cos\phi\, d\theta,$$

which are also invariant under the SE(2) action. The constraints determine the kinematic distribution $\mathcal{D}_q$:

$$\mathcal{D}_q = \text{span}\{\partial_\psi, \partial_\phi, a\partial_x + b\partial_y + c\partial_\theta\},$$

where $a = -2r\cos^2\phi\cos\theta$, $b = -2r\cos^2\phi\sin\theta$, $c = \sin 2\phi$. The tangent space to the orbits of the SE(2) action is given by

$$T_q(\text{Orb}(q)) = \text{span}\{\partial_x, \partial_y, \partial_\theta\}.$$

The intersection between the tangent space to the group orbits and the constraint distribution is thus given by

$$S_q = \mathcal{D}_q \cap T_q(\text{Orb}(q)) = \text{span}\{a\partial_x + b\partial_y + c\partial_\theta\}.$$

The momentum can be constructed by choosing a section of $\mathcal{S} = \mathcal{D} \cap T\,\text{Orb}$ regarded as a bundle over Q. Since $\mathcal{D}_q \cap T_q\,\text{Orb}(q)$ is one-dimensional, the section can be chosen to be

$$\xi_Q^q = a\partial_x + b\partial_y + c\partial_\theta,$$

which is invariant under the action of SE(2) on Q. The nonholonomic momentum is thus given by

$$p = \frac{\partial L}{\partial \dot{q}^i}(\xi_Q^q)^i = ma\dot{x} + mb\dot{y} + mr^2 c\dot{\theta} + J_0 c\dot{\psi}.$$

The kinematic constraints plus the momentum are given by

$$0 = -\sin(\theta + \phi)\dot{x} + \cos(\theta + \phi)\dot{y} - r\cos\phi\dot{\theta},$$
$$0 = -\sin(\theta - \phi)\dot{x} + \cos(\theta - \phi)\dot{y} + r\cos\phi\dot{\theta},$$
$$p = -2mr\cos^2\phi\cos\theta\dot{x} - 2mr\cos^2\phi\sin\theta\dot{y}$$
$$\qquad + mr^2\sin 2\phi\dot{\theta} + J_0\sin 2\phi\dot{\psi}.$$

Adding, subtracting, and scaling these equations, we can write (away from the point $\phi = \pi/2$)

$$\begin{bmatrix} \cos\theta\dot{x} + \sin\theta\dot{y} \\ -\sin\theta\dot{x} + \cos\theta\dot{y} \\ \dot{\theta} \end{bmatrix} + \begin{bmatrix} -\dfrac{J_0}{2mr}\sin 2\phi\dot{\psi} \\ 0 \\ \dfrac{J_0}{mr^2}\sin^2\phi\dot{\psi} \end{bmatrix} = \begin{bmatrix} \dfrac{-1}{2mr}p \\ 0 \\ \dfrac{\tan\phi}{2mr^2}p \end{bmatrix}. \qquad (5.8.49)$$

These equations have the form

$$g^{-1}\dot{g} + A(r)\dot{r} = \Gamma(r)p,$$

where

$$A(r) = -\frac{J_0}{2mr}\sin 2\phi e_x \, d\psi + \frac{J_0}{mr^2}\sin^2 \phi e_\theta \, d\psi,$$

$$\Gamma(r) = \frac{-1}{2mr}e_x + \frac{1}{2mr^2}\tan\phi \, e_\theta.$$

These are precisely the terms that appear in the nonholonomic connection relative to the (global) trivialization (r, g).

Since $\Gamma p = \Omega e$, we can rewrite the constraints using the angular momentum Ω as follows:

$$\begin{bmatrix} \xi^1 \\ \xi^2 \\ \xi^3 \end{bmatrix} = \begin{bmatrix} \dfrac{J_0}{2mr}\sin 2\phi\dot{\psi} \\ 0 \\ -\dfrac{J_0}{mr^2}\sin^2 \phi\dot{\psi} \end{bmatrix} + \begin{bmatrix} -2r\cos^2 \phi\Omega \\ 0 \\ \sin 2\phi\Omega \end{bmatrix}. \qquad (5.8.50)$$

The Snakeboard Reduced Constrained Hamiltonian. From the Lagrangian L we obtain the reduced Lagrangian

$$l(r, \dot{r}, \xi) = \tfrac{1}{2}m((\xi^1)^2 + (\xi^2)^2) + \tfrac{1}{2}mr^2(\xi^3)^2 + \tfrac{1}{2}J_0\dot{\psi}^2 + +J_0\dot{\psi}(\xi^3) + J_1\dot{\phi}^2,$$

where $\xi = g^{-1}\dot{g}$. After plugging in the constraints (5.8.50), we have the reduced constrained Lagrangian

$$l_c(r, \dot{r}, \Omega) = -\frac{J_0^2}{2mr^2}\sin^2 \phi\dot{\psi}^2 + 2mr^2\cos^2 \phi\Omega^2 + \frac{1}{2}J_0\dot{\psi}^2 + +J_1\dot{\phi}^2. \quad (5.8.51)$$

Then the reduced constrained Legendre transform is given by

$$p = \frac{\partial l_c}{\partial \Omega} = 4mr^2\cos^2 \phi\Omega,$$

$$\tilde{p}_\psi = \qquad\qquad\qquad \frac{\partial l_c}{\partial \dot{\psi}} = -\frac{J_0^2}{mr^2}\sin^2 \phi\dot{\psi} + J_0\dot{\psi},$$

$$\tilde{p}_\phi = \qquad\qquad\qquad\qquad \frac{\partial l_c}{\partial \dot{\phi}} = 2J_1\dot{\phi},$$

and its inverse is

$$\Omega = \frac{p}{4mr^2\cos^2 \phi},$$

$$\dot{\psi} = \frac{mr^2\tilde{p}_\psi}{J_0(mr^2 - J_0\sin^2 \phi)},$$

$$\dot{\phi} = \frac{\tilde{p}_\phi}{2J_1}.$$

Therefore, the reduced constrained Hamiltonian $h_{\bar{\mathcal{M}}}$ is

$$h_{\bar{\mathcal{M}}} = p\Omega + \tilde{p}_\psi \dot{\psi} + \tilde{p}_\phi \dot{\phi} - l_c = \frac{\sec^2 \phi}{8mr^2} p^2 + \frac{mr^2}{2J_0(mr^2 - J_0 \sin^2 \phi)} p_\psi^2 + \frac{1}{4J_1} p_\phi^2.$$

The Snakeboard Reduced Poisson Structure Matrix. Recall that in computing the reduced structural matrix we need only to calculate $\{\tilde{p}_\alpha, \tilde{p}_\beta\}$, etc., and then restrict them to $\bar{\mathcal{M}}$. Since

$$p = -2r \cos^2 \phi \cos \theta p_x - 2r \cos^2 \phi \sin \theta p_y + \sin 2\phi p_\theta,$$

$$\tilde{p}_\psi = \frac{J_0}{2mr^2} \sin 2\phi \cos \theta p_x + \frac{J_0}{2mr^2} \sin 2\phi \sin \theta p_y - \frac{J_0}{mr^2} \sin^2 \phi p_\theta + p_\psi,$$

$$\tilde{p}_\phi = p_\phi,$$

we have

$$\{p, \tilde{p}_\phi\} = \{-2r \cos^2 \phi \mu_1, p_\phi\} + \{\sin 2\phi \mu_3, p_\phi\}$$
$$= 2r \sin 2\phi \mu_1 + 2 \cos 2\phi \mu_3. \tag{5.8.52}$$

Similarly, we obtain

$$\{\tilde{p}_\psi, \tilde{p}_\phi\} = \frac{J_0}{mr} \cos 2\phi \mu_1 - \frac{J_0}{mr} \sin 2\phi \mu_3, \tag{5.8.53}$$

$$\{p, \tilde{p}_\psi\} = 0. \tag{5.8.54}$$

As for μ_1, μ_2, μ_3 (restricted to $\bar{\mathcal{M}}$), recall that

$$\mu_1 = \cos \theta p_x + \sin \theta p_y = \cos \theta (m\dot{x}) + \sin \theta (m\dot{y})$$
$$= \frac{J_0}{r} \cos \phi \sin \phi \dot{\psi} - \frac{1}{2r} p = \frac{mr \sin \phi \cos \phi}{mr^2 - J_0 \sin^2 \phi} \tilde{p}_\psi - \frac{1}{2r} p.$$

We can also obtain μ_2, μ_3 in a similar way. Therefore,

$$\begin{bmatrix} \mu_1 \\ \mu_2 \\ \mu_3 \end{bmatrix} = \begin{bmatrix} \dfrac{mr \sin \phi \cos \phi}{(mr^2 - J_0 \sin^2 \phi)} \tilde{p}_\psi \\ 0 \\ \dfrac{mr^2 \cos^2 \phi}{(mr^2 - J_0 \sin^2 \phi)} \tilde{p}_\psi \end{bmatrix} + \begin{bmatrix} \dfrac{-1}{2r} p \\ 0 \\ \dfrac{\tan \phi}{2} p \end{bmatrix}. \tag{5.8.55}$$

So after substituting the constraints (5.8.55) into equations (5.8.52) to (5.8.54), we have

$$\{p, \tilde{p}_\phi\}_{\bar{\mathcal{M}}} = -\tan \phi p + \frac{2mr^2 \cos^2 \phi}{mr^2 - J_0 \sin^2 \phi} \tilde{p}_\psi, \tag{5.8.56}$$

$$\{\tilde{p}_\psi, \tilde{p}_\phi\}_{\bar{\mathcal{M}}} = -\frac{J_0}{2mr^2} p - \frac{J_0 \sin \phi \cos \phi}{mr^2 - J_0 \sin^2 \phi} p_\psi, \tag{5.8.57}$$

$$\{p, \tilde{p}_\psi\}_{\bar{\mathcal{M}}} = 0. \tag{5.8.58}$$

Therefore, the reduced Poisson structure matrix is given by

$$
\begin{pmatrix}
0 & 0 & 0 & -2r\cos^2\phi & 0 & 0 & \frac{J_0}{2mr}\sin 2\phi & 0 \\
0 & 0 & 0 & 0 & 0 & 0 & 0 & 0 \\
0 & 0 & 0 & \sin 2\phi & 0 & 0 & -\frac{J_0}{mr^2}\sin^2\phi & 0 \\
2r\cos^2\phi & 0 & -\sin 2\phi & 0 & 0 & 0 & 0 & \{p,\tilde{p}_\phi\}_{\bar{\mathcal{M}}} \\
0 & 0 & 0 & 0 & 0 & 0 & 1 & 0 \\
0 & 0 & 0 & 0 & 0 & 0 & 0 & 1 \\
-\frac{J_0}{2mr}\sin 2\phi & 0 & \frac{J_0}{mr^2}\sin^2\phi & 0 & -1 & 0 & 0 & \{\tilde{p}_\psi,\tilde{p}_\phi\}_{\bar{\mathcal{M}}} \\
0 & 0 & 0 & -\{p,\tilde{p}_\phi\}_{\bar{\mathcal{M}}} & 0 & -1 & -\{\tilde{p}_\psi,\tilde{p}_\phi\}_{\bar{\mathcal{M}}} & 0
\end{pmatrix},
$$

where $\{p,\tilde{p}_\phi\}_{\bar{\mathcal{M}}}$ and $\{\tilde{p}_\psi,\tilde{p}_\phi\}_{\bar{\mathcal{M}}}$ are given as above by (5.8.56) and (5.8.57).

The Reduced Constrained Hamilton Equations. It is straightforward to find that

$$
\frac{\partial h_{\bar{\mathcal{M}}}}{\partial p} = \frac{\sec^2\phi}{4mr^2}p,
$$

$$
\frac{\partial h_{\bar{\mathcal{M}}}}{\partial \psi} = 0,
$$

$$
\frac{\partial h_{\bar{\mathcal{M}}}}{\partial \phi} = \frac{\sec^2\phi\tan\phi}{4mr^2}p^2 + \frac{mr^2\sin 2\phi}{2(mr^2 - J_0\sin^2\phi)^2}\tilde{p}_\psi^2, \tag{5.8.59}
$$

$$
\frac{\partial h_{\bar{\mathcal{M}}}}{\partial \tilde{p}_\psi} = \frac{mr^2}{J_0(mr^2 - J_0\sin^2\phi)}\tilde{p}_\psi,
$$

$$
\frac{\partial h_{\bar{\mathcal{M}}}}{\partial \tilde{p}_\phi} = \frac{1}{2J_1}\tilde{p}_\phi.
$$

Then by using the formula in (5.8.39) and after some computations, we obtain the momentum equation and the reduced constrained Hamilton equations as follows:

$$
\dot{p} = \left(-\tan\phi\, p + \frac{2mr^2\cos^2\phi}{mr^2 - J_0\sin^2\phi}\tilde{p}_\psi\right)\frac{1}{2J_1}\tilde{p}_\phi,
$$

$$
\dot{\psi} = \frac{mr^2}{J_0(mr^2 - J_0\sin^2\phi)}\tilde{p}_\psi,
$$

$$
\dot{\phi} = \frac{1}{2J_1}\tilde{p}_\phi, \tag{5.8.60}
$$

$$
\dot{\tilde{p}}_\psi = -\left(\frac{J_0}{mr^2}p + \frac{J_0\sin 2\phi}{2(mr^2 - J_0\sin^2\phi)}\tilde{p}_\psi\right)\frac{1}{2J_1}\tilde{p}_\phi,
$$

$$
\dot{\tilde{p}}_\phi = 0.
$$

Also, we can obtain the following reconstruction equations on the Hamiltonian side:

$$\dot{x} = \xi^1 \cos\theta - \xi^2 \sin\theta$$
$$= \left(-\frac{1}{2mr}p + \frac{r\sin 2\phi}{2(mr^2 - J_0 \sin^2\phi)}\tilde{p}_\psi \right)\cos\theta,$$
$$\dot{y} = \xi^1 \sin\theta - \xi^2 \cos\theta$$
$$\qquad\qquad\qquad\qquad\qquad\qquad\qquad\qquad (5.8.61)$$
$$= \left(-\frac{1}{2mr}p + \frac{r\sin 2\phi}{2(mr^2 - J_0 \sin^2\phi)}\tilde{p}_\psi \right)\sin\theta,$$
$$\dot{\theta} = \xi^3 = \frac{\tan\phi}{2mr^2}p - \frac{\sin^2\phi}{mr^2 - J_0 \sin^2\phi}\tilde{p}_\psi.$$

Together, these two sets of equations give us the dynamics of the full constrained systems but in a form that is suitable for control-theoretical purposes. ♦

We remark that for certain nonholonomic systems it is possible to make the system Hamiltonian by a change in time parameterization. For some of the literature in this area see for example Ehlers et al. [2005].

6
Control of Mechanical and Nonholonomic Systems

6.1 Background in Kinematic Nonholonomic Control Systems

Nonholonomic Motion Planning. There is a large literature on nonholonomic motion planning, the construction in the nonholonomic setting of explicit open loop controls as discussed in Chapter 4; see, for example, Li and Canny [1993] and Murray and Sastry [1993] and the works cited therein as well as, for example, Sussmann and Liu [1991], Lafferiere and Sussman [1991], Lewis, Ostrowski, Murray, and Burdick [1994], Ostrowski, Desai and Kumar [1997], Lewis and Murray [1999], Ostrowski [2000], Bullo, Leonard, and Lewis [2000], and Bullo, Lewis, and Lynch [2002]. It is not our intention in this book to give an exhaustive account of the details of this theory. We will instead illustrate some aspects of the theory illustrated by Brockett's generalization of the Heisenberg system as well as systems in so-called chained form.

In nonholonomic motion planning one's goal is to use open-loop control to reach a desired point in phase space. Nonholonomic systems, by virtue of the nonintegrable nature of their constraints, are amenable to rather elegant path planning algorithms. The basic situation considered is usually that of kinematic control systems, where the vector fields defining the system velocities do not span the state space, but nonetheless one *can* move from any point of the space to any other. This is, as we have seen, a fundamental property of nonholonomic systems.

We shall consider the class of completely controllable kinematic systems of the form

$$\dot{x} = \sum_{i}^{m} u_i(t) g_i(x),$$ (6.1.1)

where $x \in \mathbb{R}^n$ for a suitable class of functions g_i on $\mathbb{R}^n$ and a suitable class of functions u_i on $\mathbb{R}^+$.

The *motion planning problem* is to find an efficient algorithm that gives for every pair of points p and q an open loop control $t \mapsto u(t) = (u_1(t), \ldots, u_n(t))$ that steers the system from p to q.

6.1.1 Example (Generalized Heisenberg System). We consider first Brockett's canonical system, a generalization of the Heisenberg system (Brockett [1981]). (We will treat the stabilization problem for this system in detail later in this chapter.) The system is the following:

$$\dot{x} = u,$$ (6.1.2)

$$\dot{Y} = xu^T - ux^T,$$ (6.1.3)

where $x, u \in \mathbb{R}^n$ and $Y \in \mathfrak{so}(n)$, $n \geq 2$. Here $\mathfrak{so}(n)$ is the Lie algebra of $n \times n$ skew-symmetric matrices, and elements of $\mathbb{R}^n$ are viewed as column vectors.

In terms of components, the last equation reads

$$\dot{Y}_{ij} = x_i u_j - x_j u_i.$$ (6.1.4)

The importance of this system is that it is a canonical form for a class of controllable systems of the form $\dot{x} = B(x)u$, $u \in \mathbb{R}^n$, $x \in \mathbb{R}^{n(n+1)/2}$. The class in question is as follows: Let E^0 be the subbundle of the tangent bundle spanned by the control fields, and define the *first derived algebra* to be given by $E^1 = E^0 + [E^0, E^0]$. Then this system is a normal form for the controllable systems of this type, where the first derived algebra of control vector fields spans the tangent space $T\mathbb{R}^{n(n+1)/2}$ at any point. That is, Brockett showed that such a system can be transformed to the form (6.1.2)–(6.1.3) up to a suitable order in a neighborhood of a given point such as the origin.

The key to controlling this system is being able to change Y without changing x. Since x is directly controlled, it is easily changed. We present here a method of changing Y using sinusoids, which is motivated by the optimal control problem of Brockett [1981] (see Section 1.8 for this analysis in the special case of the Heisenberg system). We follow the exposition and some of the ideas of Murray and Sastry [1993], although we use a slightly different form of the equations.

To solve the motion planning problem for this system, the idea is to proceed along loops in x-space, which gradually drives one through Y-space. This is just a reflection of the holonomy in the system as described in Section 3.14. The use of holonomy loops in stabilizing nonholonomic mechanical systems is discussed in Section 6.6.

Motivated by the fact that the optimal solution of the Heisenberg system (Section 1.8) gives a u that consists of sinusoids, we choose the control law

$$u_i = \sum_k a_{ik} \sin kt + \sum_k b_{ik} \cos kt, \qquad k = 1, \ldots, \qquad (6.1.5)$$

where a_{ik} and b_{ik} are real numbers. Since $\dot{x}_i = u_i$, integration gives

$$x_i = -\sum_k \frac{a_{ik}}{k} \cos kt + \sum_k \frac{b_{ik}}{k} \sin kt + C_i, \qquad (6.1.6)$$

where C_i is a constant depending on the initial value of x.

Substituting these equations for $x_i(t)$ and $u_i(t)$ into equation (6.1.4) and integrating yields

$$Y_{ij}(2\pi) = \sum_k \frac{2\pi}{k} \left(b_{ik} a_{jk} - b_{jk} a_{ik} \right) + Y_{ij}(0), \qquad (6.1.7)$$

since all integrals except those of the squares of cosine and sine vanish. Under this input, the x's remain unchanged.

Thus, this gives the following solution to the motion planning problem: First drive the x to the desired final value; then use the control (6.1.5) to drive Y to the desired final value. ◆

Chained Systems. Similar algorithms may also be given for higher-order systems (see Brockett and Dai [1992] and Murray and Sastry [1993]). One such class that may be easily handled is the class of *chained systems*, which are systems of the form

$$\dot{\xi}_1 = v_1,$$
$$\dot{\xi}_2 = v_2,$$
$$\dot{\xi}_3 = \xi_2 v_1, \qquad (6.1.8)$$
$$\vdots$$
$$\dot{\xi}_n = \xi_{n-1} v_1.$$

One can show that a large class of kinematic two-input systems may be put into this form. To make this specific, we state a result proved in Murray and Sastry [1993] and then illustrate the proof of the theorem for the Heisenberg system.

6.1.2 Proposition. *Consider a controllable system*

$$\dot{x} = u_1 g_1(x) + u_2 g_2(x), \qquad (6.1.9)$$

where g_1 and g_2 are linearly independent and smooth. Define the distributions

$$\Delta_0 \equiv \mathrm{span}\left\{g_1, g_2, \mathrm{ad}_{g_1} g_2, \cdots, \mathrm{ad}_{g_1}^{n-2} g_2\right\},$$
$$\Delta_1 \equiv \mathrm{span}\left\{g_2, \mathrm{ad}_{g_1} g_2, \cdots, \mathrm{ad}_{g_1}^{n-2} g_2\right\}, \qquad (6.1.10)$$
$$\Delta_2 \equiv \mathrm{span}\left\{g_2, \mathrm{ad}_{g_1} g_2, \cdots, \mathrm{ad}_{g_1}^{n-3} g_2\right\}.$$

If there exists an open set $U \in \mathbb{R}^n$ such that $\Delta_0(x) = \mathbb{R}^n$ for all $x \in U$, Δ_1, and Δ_2 are involutive on U, and there exists a smooth function $h_1 : U \to \mathbb{R}$ such that $dh_1 \cdot \Delta_1 = 0$ and $\mathcal{L}_{g_1} h_1 = 1$, then there exists a local feedback transformation

$$\xi = \phi(x), \quad u = \beta(x)v \qquad (6.1.11)$$

such that the transformed system is in the chained form (6.1.8).

6.1.3 Example. Consider now the Heisenberg system

$$\begin{aligned}
\dot{x} &= u_1, \\
\dot{y} &= u_2, \qquad (6.1.12) \\
\dot{z} &= xu_2 - yu_1.
\end{aligned}$$

In this case $\Delta_0 = \mathbb{R}^3$, since the system is controllable. Also,

$$\begin{aligned}
\Delta_1 &= \mathrm{span}\left\{\frac{\partial}{\partial x_2} + x_1 \frac{\partial}{\partial x_3}, 2\frac{\partial}{\partial x_3}\right\}, \\
\Delta_2 &= \mathrm{span}\left\{\frac{\partial}{\partial x_2} + x_1 \frac{\partial}{\partial x_3}\right\}.
\end{aligned} \qquad (6.1.13)$$

Now we choose $\xi_1 = h_1 = x_1$. Following the prescription in Murray and Sastry [1993], we construct a function h_2 such that $dh_2 \cdot \Delta_2 = 0$ and $dh_2 \cdot \mathrm{ad}_{g_1}^{n-2} g_2 \neq 0$. This means that

$$\frac{\partial h_2}{\partial x_2} + x_1 \frac{\partial h_2}{\partial x_3} = 0, \qquad \frac{\partial h_2}{\partial x_3} \neq 0, \qquad (6.1.14)$$

which is satisfied by the function

$$h_2 = x_3 - x_1 x_2.$$

Now set

$$\xi_2 = \mathcal{L}_{g_1} h_2 = -2x_2 \qquad (6.1.15)$$

and

$$v_1 = u_1, \quad v_2 = \left(\mathcal{L}_{g_2}^2 h_2\right) u_1 + \left(\mathcal{L}_{g_2} \mathcal{L}_{g_1} h_2\right) u_2 = -2u_2. \qquad (6.1.16)$$

Then

$$\begin{aligned}
\dot{\xi}_1 &= \dot{x}_1 = u_1 = v_1, \\
\dot{\xi}_2 &= -2\dot{x}_2 = v_2, \qquad (6.1.17) \\
\dot{\xi}_3 &= \dot{x}_3 - \dot{x}_1 x_2 - x_1 \dot{x}_2 = -2x_2 u_1 = \xi_2 v_1,
\end{aligned}$$

which puts the system into the chained form desired.   ◆

Extended Systems. Aside from such special classes of systems it is possible to use rather general motion planning algorithms as shown in the work of Sussmann and his collaborators (see, for example, the papers Sussmann and Liu [1991] and Lafferiere and Sussman [1991]. One of the key ideas in this work is to use an **extended** system of the form

$$\dot{x} = v_1 g_1(x) + \cdots + v_m g_m(x) + v_{m+1} g_{m+1}(x) + \cdots + v_r g_r(x), \quad (6.1.18)$$

where $g_{m+1}, \ldots, g_r(x)$ are higher-order Lie brackets of the g_i chosen so that $g_1(x), \ldots, g_r(x)$ span all $\mathbb{R}^n$. The idea is then to compute a motion controller for the extended system (which is easy, since we have an independent control vector field for each independent direction in $\mathbb{R}^n$) and then use that to construct one for the original system.

One approach is to use bracketed functions from the so-called P. Hall basis. Sussmann and Lafferiere then show how to construct motion control algorithms that are exact for nilpotent and nilpotentizable systems and that are approximate, but converge in a suitable sense, in the general case. Sussmann and Liu use a similar approach involving highly oscillatory inputs.

Nonholonomic Stabilization Techniques. We shall show here that it follows from Brockett's necessary conditions (see Section 4.5) that nonholonomic kinematic systems do not satisfy the necessary conditions for the existence of a smooth or indeed continuous feedback stabilization law. (The same holds true for dynamic nonholonomic systems; see Section 6.6). There are thus two basic approaches to achieving feedback to the origin: the use of time-varying feedback and the use of discontinuous feedback. We discuss in Sections 6.2 and 6.3 an approach to the use of the latter in the kinematic setting. Discontinuous stabilization in the dynamic setting is discussed in Section 6.6. In this section we discuss briefly some of the literature on the use of dynamic feedback and combinations of dynamic and discontinuous feedback.

We have (see, for example, Pomet [1992]):

6.1.4 Proposition. *Consider the system*

$$\dot{x} = \sum_{k=1}^{m} u_k g_k, \quad (6.1.19)$$

where $x \in \mathbb{R}^n$, $m < n$, and

$$\text{rank}\{g_1(0), \ldots, g_m(0)\} = m.$$

There exists no continuous feedback that locally asymptotically stabilizes the origin.

Proof. Let v be a nonzero vector linearly independent of $g_1(0), \ldots, g_m(0)$. By continuity there is an $\epsilon > 0$ such that for all $(x, u_1, \ldots, u_m)$ with $\|x\| < \epsilon$

the vector $\sum_{k=1}^{m} g_k(x)$ is not a multiple of v. Hence the map

$$(x, u_1, \ldots, u_m) \to \sum_{k=1}^{m} u_k g_k(x) \qquad (6.1.20)$$

does not map the ϵ-neighborhood of 0 (in (x, u)-space) into a neighborhood of 0 in $\mathbb{R}^n$. Hence the system does not satisfy the necessary condition of Brockett [1983] in Section 4.5. ∎

One way to deal with this problem is to use time-varying feedback. A key result on stabilization by time-varying feedback is that of Coron [1992]. See also the related work of Pomet [1992], for example. Coron's result is as follows:

6.1.5 Theorem (Coron). *Consider the system* (6.1.19) *and suppose that all the vector fields g_i are C^∞ and that the Lie algebra of vector fields generated by the g_i spans $\mathbb{R}^n$ (i.e., the system is controllable).*
Then for any positive T there exists a C^∞ feedback law

$$u(x, t) = (u_1(x, t), \ldots, u_n(x, t))$$

such that

- *$u(0, t) = 0$ for all $t \in \mathbb{R}$,*

- *$u(x, t+T) = u(x, t)$ for all $x \in \mathbb{R}^n$ and $t \in \mathbb{R}$,*

- *the origin is a globally asymptotically stable point of the system* (6.1.19).

Pomet [1992] gives a more easily derived explicit feedback law and constructs an explicit Lyapunov function under a more restrictive controllability condition, namely, that

$$\text{rank}\{g_1(x), g_2(x), \ldots, g_m(x), [g_1, g_2](x), \ldots, [g_1, g_m](x), \ldots,$$
$$\text{ad}_{g_1}^j g_2(x), \ldots, \text{ad}_{g_1}^j g_m(x), \ldots\} = n. \qquad (6.1.21)$$

Here g_1 does not, of course, play a special role and may be replaced with any function g_i or linear combination of g_i.

General controllers constructed in this fashion may have quite slow convergence rates. In M'Closkey and Murray [1993], for example, convergence rates of various control laws are examined, and laws are produced that give exponential convergence. We will not give the details of the general construction here, but indicate the general idea using the Heisenberg system as an example.

Consider again the system

$$\dot{x} = u_1,$$
$$\dot{y} = u_2, \qquad (6.1.22)$$
$$\dot{z} = yu_1.$$

(As discussed above, see (6.1.17). This system is just the Heisenberg system in chained form.)

Then the idea is to use a combination of time dependent and nonsmooth feedback. In the first instance one considers a smooth feedback of the form

$$u_1 = -x + F_1(z)\cos t,$$
$$u_2 = -y + F_2(z)\sin t, \tag{6.1.23}$$

and observes that this gives convergence but not asymptotic convergence. In order to obtain convergence one needs nondifferentiable functions F_1 and F_2. For example, one can choose

$$F_1(z) = \operatorname{sgn}(z)\sqrt{|z|}, \quad F_2(z) = \sqrt{|z|}, \tag{6.1.24}$$

where $\operatorname{sgn}(\cdot)$ is the signum function.

Some related work may be found in Teel, Murray, and Walsh [1995], where again the system (6.1.22) is considered but with controls

$$u_1 = -x + z\sin t,$$
$$u_2 = -y - z^2\cos t.$$

It is shown by realizing the controls by the "exosystem"

$$\dot{w}_1 = w_2, \quad w_1(0) = 0,$$
$$\dot{w}_2 = -w_1, \quad w_2(0) = 1,$$

and using center manifold theory that the system is stabilized and the convergence is exponential in nature. Using the exosystem the control law becomes

$$u_1 = -x + zw_1,$$
$$u_2 = -y + z^2 w_2.$$

Then there is a local center manifold on which the dynamics is given by

$$\dot{z} = -\frac{1}{4}z^3 (w_1 + w_2)^2, \tag{6.1.25}$$

and hence the system is locally asymptotically stable to the origin.

Other interesting work on time-varying controllers for nonholonomic systems may be found, for example, in Canudas De Wit and Sørdalen [1992], Sørdalen and Egeland [1995], Walsh and Bushnell [1995], Morin, Pomet, and Samson [1999], and Morin and Samson [2000].

Exercises

⋄ **6.1-1.** Carry out the motion planning problem for the following variant to the generalized Heisenberg system:

$$\dot{x}_i = u_i, \qquad i = 1, \ldots, n,$$
$$\dot{Y}_{ij} = x_i u_j, \qquad i > j.$$

◇ **6.1-2.** Carry out the motion planning problem for the following system on $\mathbb{R}^5$ (see Murray and Sastry [1993]):

$$\dot{x}_1 = u_1,$$
$$\dot{x}_2 = u_2,$$
$$\dot{x}_{21} = x_2 u_1,$$
$$\dot{x}_{211} = x_{21} u_1,$$
$$\dot{x}_{212} = x_{21} u_2.$$

Hint: For steering x_1, x_2, and x_{21} use sine and cosine controls as in the text. To steer x_{211} independently of the other states use $u_1 = a \sin t$ and $u_2 = b \cos 2t$, and for x_{212} use $u_1 = b \cos 2t$ and $u_2 = a \sin t$.

◇ **6.1-3.** Prove that the chained system (6.1.8) is controllable.

◇ **6.1-4.** Show that the time-dependent controls

$$u_1 = - \left(y + z \cos t \right) y \cos t - \left(yz + x \right),$$
$$u_2 = z \sin t - \left(y + z \cos t \right),$$

stabilize the system (6.1.22). (Hint: Rewrite the systems as a time-independent system on $\mathbb{R}^3 \times S^1$ and use the Lyapunov function

$$V(t, x, y, z)) = \frac{1}{2} \left(y + z \cos t \right)^2 + \frac{1}{2} z^2 + \frac{1}{2} x^2$$

and LaSalle's principle — see Pomet [1992].)

◇ **6.1-5.** Complete the details of the center manifold computation, yielding the dynamics (6.1.25).

6.2   Stabilization of the Heisenberg System

Here we follow Bloch and Drakunov [1994, 1996] in considering a discontinuous approach to the stabilization problem for the nonholonomic integrator or Heisenberg system. This is the prototypical example for which smooth feedback fails. The idea is to use the natural algebraic structure of the system together with ideas from sliding mode theory (see DeCarlo, Zak, and Drakunov [1996], Drakunov and Utkin [1992]). Other work on the discontinuous approach to such systems includes Kolmanovsky, Reyhanoglu, and McClamroch [1994], Khennouf and Canudas de Wit [1995], Astolfi [1996], Brockett [2000], and Agrachev and Liberzon [2001]. Another interesting problem for such systems is the problem of tracking, which was also analyzed in Bloch and Drakunov [1996] and in Bloch and Drakunov [1995]. Related work also includes, for example, Ryan [1990], Sira-Ramirez and

Lischinsky-Arenas [1990], Morse [1997], Zhang and Hirschorn [1997], Sontag [1999], and Reyhanoglu, Cho, and McClamroch [2000]. There are also many other references in the literature. Morse [1997] is a good source of references.

We have the system

$$\dot{x} = u, \tag{6.2.1}$$
$$\dot{y} = v, \tag{6.2.2}$$
$$\dot{z} = xv - yu. \tag{6.2.3}$$

The problem of stabilizing this system, even locally, is not a trivial task, since, as can be easily seen, the linearization in the vicinity of the origin gives the noncontrollable system

$$\dot{x} = u,$$
$$\dot{y} = v,$$
$$\dot{z} = 0.$$

The main difficulty is the fact that stabilization of x and y leads to a zero right-hand side of (6.2.3), and therefore, the variable z cannot be steered to zero. That simple observation implies that to stabilize the system one needs to make z converge "faster" than x and y.

We consider the control law

$$u = -\alpha x + \beta y \, \text{sign}(z), \tag{6.2.4}$$
$$v = -\alpha y - \beta x \, \text{sign}(z), \tag{6.2.5}$$

where α and β are positive constants.

Let us show that there exists a set of initial conditions such that trajectories starting there converge to the origin. To do this, consider a Lyapunov function for the (x, y)-subspace:

$$V = \frac{1}{2}(x^2 + y^2). \tag{6.2.6}$$

The time derivative of V along the trajectories of the system (6.2.3) is negative:

$$\dot{V} = -\alpha x^2 + \beta xy \, \text{sign}(z) - \alpha y^2 - \beta xy \, \text{sign}(z) = -\alpha(x^2 + y^2)$$
$$= -2\alpha V. \tag{6.2.7}$$

Therefore, under the control (6.2.4), (6.2.5) the variables x and y are stabilized.

Now let us consider the variable z. Using (6.2.3), (6.2.4), and (6.2.5), we obtain

$$\dot{z} = xv - yu = -\beta(x^2 + y^2) \, \text{sign}(z) = -2\beta V \, \text{sign}(z). \tag{6.2.8}$$

Since V does not depend on z and is a positive function of time, the absolute value of the variable z will decrease and will reach zero in finite time if the inequality

$$2\beta \int_0^\infty V(\tau)d\tau > |z(0)| \qquad (6.2.9)$$

holds. If $z(0)$ is such that

$$2\beta \int_0^\infty V(\tau)d\tau = |z(0)|, \qquad (6.2.10)$$

then $z(t)$ converges to the origin in infinite time (asymptotically). Otherwise, it converges to some constant nonzero value of the same sign as $z(0)$.

If the above inequality (6.2.9) holds, the system trajectories are directed to the surface $z = 0$, and the variable $z(t)$ is stabilized at the origin in finite time. (The variables x and y, as follows from (6.2.7), always converge to the origin while within that surface.)

This phenomenon is known as **_sliding mode_** (see Utkin [1992]). The manifold $z = 0$ is a stable integral manifold of the closed-loop system (6.2.1)–(6.2.3), (6.2.4), (6.2.5). Its characteristic feature is reachability in finite time. Using a smooth control, even a control satisfying a local Lipschitz condition (in the vicinity of $\{z = 0\}$) such fast convergence cannot be achieved. On the other hand, within the sliding manifold $\{z = 0\}$ the system behavior is described in accordance with the Filippov definition for systems of differential equations with discontinuous right-hand sides (see Filippov [1988]).

The version of this definition that we are using is as follows: We consider the system

$$\dot{x} = f(x), \qquad (6.2.11)$$

with $f(x)$ a discontinuous function composed of a finite number of functions

$$f(x) \equiv f_k(x) \text{ for } x \in \mathcal{M}_k, \qquad (6.2.12)$$

where the open regions $\mathcal{M}_k$ have piecewise smooth boundaries $\partial\mathcal{M}_k$. Then we define the right-hand side of (6.2.11) within $\partial\mathcal{M}_k$ to be

$$\dot{x} = \sum_{k \in I(x)} \mu_k f_k(x). \qquad (6.2.13)$$

The sum is taken over the set $I(x)$ of all k such that $x \in \partial\mathcal{M}_k$ and the variables μ_k satisfy

$$\sum_{k \in I(x)} \mu_k = 1; \qquad (6.2.14)$$

i.e., the right-hand side belongs to the convex closure co$\{f_k(x) : k \in I(x)\}$ of the vector fields $f_k(x)$ for all $k \in I(x)$. Actually, the Filippov definition replaces the differential equation (6.2.11) by a differential inclusion

$$\dot{x} \in \text{co}\{f_k(x) \mid k \in I(x)\} \tag{6.2.15}$$

for the points x belonging to the boundaries $\partial \mathcal{M}_k$. If within the convex closure there exists a vector field tangent to all or some of the boundaries, then there is a solution of the differential inclusion belonging to $\partial \mathcal{M}_k$ that corresponds to the sliding mode.

In the above relatively simple case, the Filippov definition provides a unique solution and implies that the system on the manifold is

$$\dot{x} = -\alpha x,$$
$$\dot{y} = -\alpha y.$$

From (6.2.7) it follows that

$$V(t) = V(0)e^{-2\alpha t} = \frac{1}{2}(x^2(0) + y^2(0))e^{-2\alpha t}. \tag{6.2.16}$$

Substituting this expression in (6.2.9) and integrating, we find that the condition for the system to be stabilized is

$$\frac{\beta}{2\alpha}\left[x^2(0) + y^2(0)\right] \geq |z(0)|. \tag{6.2.17}$$

The inequality

$$\frac{\beta}{2\alpha}(x^2 + y^2) < |z| \tag{6.2.18}$$

defines a parabolic region $\mathcal{P}$ in the state space.

The above derivation can be summarized in the following theorem:

6.2.1 Theorem. *If the initial conditions for the system* (6.2.1)–(6.2.3) *belong to the complement* $\mathcal{P}^c$ *of the region* $\mathcal{P}$ *defined by* (6.2.18)*, then the control* (6.2.4), (6.2.5) *stabilizes the state.*

If the initial data are such that (6.2.18) is true, i.e., the state is inside the paraboloid, we can use any control law that steers it outside. In fact, any nonzero constant control can be applied. Namely, if $u \equiv u_0 = \text{const}$, $v \equiv v_0 = \text{const}$, then

$$x(t) = u_0 t + x_0,$$
$$y(t) = v_0 t + y_0,$$
$$z(t) = t(x_0 v_0 - y_0 u_0) + z_0.$$

With such x, y, and z, the left-hand side of (6.2.18) is quadratic with respect to time t, while the right-hand side is linear. Hence, as the time increases, the state inevitably will leave $\mathcal{P}$.

A global feedback control law in the form of the feedback (although discontinuous) can be described as follows:

$$(u, v)^T = \begin{cases} (u_0, v_0)^T & \text{if } (x, y, z)^T \in \mathcal{P}, \\ 6.2.4), (6.2.5)^T & \text{if } (x, y, z)^T \in \mathcal{P}^c. \end{cases} \quad (6.2.19)$$

6.2.2 Theorem. *The closed system* (6.2.1)–(6.2.3), (6.2.19) *is globally asymptotically stable at the origin.*

Global asymptotic stability means that:

(i) for all initial conditions we have $x(t), y(t), z(t) \to 0$, when $t \to \infty$;

(ii) for all $\varepsilon > 0$ there exists $\delta > 0$ such that $x_0^2 + y_0^2 + z_0^2 < \delta^2$ implies $x^2(t) + y^2(t) + z^2(t) < \varepsilon^2$ for any $t \geq 0$.

We have already shown above that (i) is true, and (ii) follows from the fact that outside $\mathcal{P}$ and on the surface of the paraboloid $\partial \mathcal{P}$ the state monotonically approaches the origin. For initial conditions inside $\mathcal{P}$ we have

$$x^2(t) + y^2(t) + z^2(t) = (u_0 t + x_0)^2 + (v_0 t + y_0)^2 + [(x_0 v_0 - y_0 u_0)t + z_0]^2. \quad (6.2.20)$$

The maximum of the expression (6.2.20) is achieved for $t = 0$ or $t = t_f$, where t_f is the first moment of time when the state reaches $\partial \mathcal{P}$. This moment is defined by an equation

$$\frac{\beta}{2\alpha}(u_0 t_f + x_0)^2 + (v_0 t_f + y_0)^2 = |(x_0 v_0 - y_0 u_0)t_f + z_0|. \quad (6.2.21)$$

As can be easily seen from (6.2.21), for fixed u_0, v_0, the solution of this equation t_f tends to zero if x_0, y_0, z_0 tend simultaneously to zero. That proves (ii).

The parameters α, β define the size of the paraboloid.

Simulations of the algorithm for two types of initial conditions are shown in Figure 6.2.1. The figure shows the trajectories exiting from the set $\mathcal{P}$ under constant control and then being driven to the origin under the feedback (6.2.4), (6.2.5).

When $\frac{\beta}{\alpha} \to \infty$ the parabolic region $\mathcal{P}$ is limited to the z-axis. From that point of view, to stabilize the system (6.2.1–6.2.3), it is reasonable to increase β as the state approaches the origin (if we decrease α, the convergence of x and y will be slower). To realize this idea we can use a control law where α increases when x and y approach the origin:

$$u = -\alpha x + \beta \frac{y}{x^2 + y^2} \text{ sign}(z), \quad (6.2.22)$$

$$v = -\alpha y - \beta \frac{x}{x^2 + y^2} \text{ sign}(z), \quad (6.2.23)$$

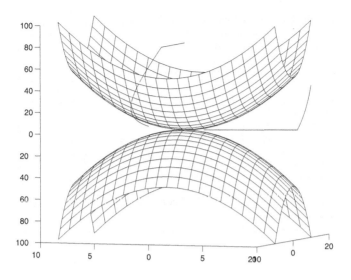

FIGURE 6.2.1. Stabilization of the nonholonomic integrator.

or even

$$u = \alpha x + \beta \frac{y}{x^2 + y^2} z, \tag{6.2.24}$$

$$v = -\alpha y - \beta \frac{x}{x^2 + y^2} z. \tag{6.2.25}$$

(For a detailed analysis in the case (6.2.24), (6.2.25), see Khennouf and Canudas de Wit [1995]).

Then from (6.2.3) we have

$$\dot{z} = -\beta \ \text{sign}(z),$$

for the controls (6.2.22), (6.2.23), or

$$\dot{z} = -\beta z,$$

respectively, for the controls (6.2.24), (6.2.25).

In both cases, the state converges to the origin from any initial conditions, except the ones belonging to the z-axis. But in contrast to (6.2.4), (6.2.5), the control laws (6.2.22), (6.2.23) and (6.2.24), (6.2.25) are unbounded in a neighborhood of the z-axis (on the axis it is not defined). If the initial conditions belong to this set, again we can apply any nonzero constant control for an arbitrarily small period of time and then switch to (6.2.22), (6.2.23) or (6.2.24), (6.2.25). A method of dealing with the boundedness problem is also described by Khennouf and Canudas de Wit [1995].

An ε-stabilizing control (to a neighborhood of the origin) may be obtained by switching α.

Let α be the following function of x and y,

$$\alpha = \alpha_0 \; \text{sign}(x^2 + y^2 - \varepsilon^2), \tag{6.2.26}$$

where $\alpha_0 > 0$, $\beta > 0$ are constants, and let the control be

$$u = -\alpha x + \beta yz, \tag{6.2.27}$$

$$v = -\alpha y - \beta xz. \tag{6.2.28}$$

(One deals with initial data on the z-axis as above.)

Using (6.2.7) we find that from any initial conditions x and y the state reaches an ε-sphere of the (x, y)-space origin:

$$x^2 + y^2 = \text{const} = \varepsilon^2. \tag{6.2.29}$$

After that, the equation for the variable z is

$$\dot{z} = -\beta \varepsilon^2 z. \tag{6.2.30}$$

Therefore, $z \to 0$ as $t \to \infty$, while the variables x and y stay in an ε-vicinity of the origin. Of course, in (6.2.27), (6.2.28) z can be replaced by any function $g(z)$ that guarantees asymptotic stability of the equation

$$\dot{z} = -\beta \varepsilon^2 g(z), \tag{6.2.31}$$

for example, $g(z) = \text{sign}(z)$.

Exercises

⋄ **6.2-1.**   Construct a discontinuous stabilizing controller for the control system (6.1.22).

6.3   Stabilization of a Generalized Heisenberg System

Following Bloch, Drakunov, and Kinyon [2000] we discuss here the stabilization of the canonical generalization of the Heisenberg system (6.1.2), (6.1.3), by discontinuous feedback. We also demonstrate a rather interesting connection with isospectral flows (flows that preserve eigenvalues). Such flows are fundamental to integrable systems theory; see, for example, the discussion of the Toda lattice in Chapter 1. We note that some of this section uses a little more Lie algebra theory than we have used up till now in this book. For the reader unfamiliar with this a useful reference is Sattinger and Weaver [1986]. Most of the section can be read, however, by just ignoring the Lie-algebraic remarks and viewing the Killing form as the trace inner product.

Lie-Algebraic Generalization. We consider here a system that generalizes (6.1.2)–(6.1.3) and can be described as follows. Let $\mathfrak{g}$ be a Lie algebra with a direct sum decomposition $\mathfrak{g} = \mathfrak{m} \oplus \mathfrak{h}$ such that $\mathfrak{h}$ is a Lie subalgebra, $[\mathfrak{h}, \mathfrak{m}] \subseteq \mathfrak{m}$, and $[\mathfrak{m}, \mathfrak{m}] = \mathfrak{h}$. We will consider the following system in $\mathfrak{g}$:

$$\dot{x} = u, \tag{6.3.1}$$

$$\dot{Y} = [u, x], \tag{6.3.2}$$

where $x, u \in \mathfrak{m}$, $Y \in \mathfrak{h}$.

The $\mathfrak{so}(n)$ system (6.1.2)–(6.1.3) is of the type (6.3.1)–(6.3.2), as we now show. Let $\mathfrak{h} = \mathfrak{so}(n)$ and let $\mathfrak{m} = \mathbb{R}^n$. For $x, u \in \mathfrak{m}$, define $[u, x] \equiv xu^T - ux^T \in \mathfrak{h}$. For $Y \in \mathfrak{h}$, $x \in \mathfrak{m}$, define $[Y, x] = [x, Y] \equiv Yx$. It is easy to see that the Lie algebra $\mathfrak{g} \equiv \mathfrak{m} \oplus \mathfrak{h}$ is isomorphic to $\mathfrak{so}(n+1)$: Identify $Y \in \mathfrak{so}(n)$ with the matrix

$$\begin{pmatrix} 0 & 0 \\ 0 & Y \end{pmatrix}$$

and identify $x \in \mathbb{R}^n$ with the matrix

$$\begin{pmatrix} 0 & -x^T \\ x & 0 \end{pmatrix}.$$

The adjoint action of $\mathfrak{h}$ on $\mathfrak{m}$ agrees with the standard action of $\mathfrak{so}(n)$ on $\mathbb{R}^n$, and it is straightforward to check that the desired commutation relations hold.

Our goal is to find a stabilizing control for the system (6.3.1)–(6.3.2). Since this system fails the necessary condition for the existence of a continuous feedback law, our goal here is to find a discontinuous law. This section is based on on Bloch, Drakunov, and Kinyon [1997, 2000].

The General System. Let $\mathfrak{g}$ be a real semisimple Lie algebra with Killing form $B : \mathfrak{g} \times \mathfrak{g} \to \mathbb{R}$. Assume that $\mathfrak{g}$ has a decomposition $\mathfrak{g} = \mathfrak{h} \oplus \mathfrak{m}$, where $\mathfrak{h}$ is a compactly embedded subalgebra that contains no ideals of $\mathfrak{g}$, and $\mathfrak{m}$ is the orthogonal complement of $\mathfrak{h}$ relative to B. Then the commutation relations $[\mathfrak{h}, \mathfrak{m}] \subseteq \mathfrak{m}$ and $[\mathfrak{m}, \mathfrak{m}] = \mathfrak{h}$ hold, the restriction of B to $\mathfrak{h}$ is negative definite, and the representation of $\mathfrak{h}$ on $\mathfrak{m}$ is faithful. Note that if $\mathfrak{g}$ is a simple Lie algebra and $\mathfrak{g} = \mathfrak{h} \oplus \mathfrak{m}$ is a Cartan decomposition, then our hypotheses are satisfied. (See section 2.8 and, e.g., Sattinger and Weaver [1986] for definitions.)

We will consider stabilization of the system (6.3.1)–(6.3.2) in $\mathfrak{g}$, where $x, u \in \mathfrak{m}$, $Y \in \mathfrak{h}$. We may assume without loss of generality that $\mathfrak{g}$ is either of noncompact type or of compact-type. (Indeed, under the given hypotheses, $\mathfrak{g}$ splits into a B-orthogonal direct sum of a compact-type ideal and one of noncompact type. It is straightforward to show that (6.3.1)–(6.3.2) decouples into systems in each ideal, and thus stabilization of (6.3.1)–(6.3.2) follows from stabilization of each of the compact and noncompact cases separately; see Bloch, Drakunov, and Kinyon [2000].) It follows that the restriction of B to $\mathfrak{m}$ is positive definite if $\mathfrak{g}$ is of noncompact type, and negative definite if $\mathfrak{g}$ is of compact type.

Let

$$\epsilon = \begin{cases} 1 & \text{if } \mathfrak{g} \text{ is of noncompact type,} \\ -1 & \text{if } \mathfrak{g} \text{ is of compact type.} \end{cases} \tag{6.3.3}$$

We will use the inner product on $\mathfrak{g}$ defined by the Killing form:

$$\langle x_1 + Y_1, x_2 + Y_2 \rangle \equiv \epsilon B(x_1, x_2) - B(Y_1, Y_2), \tag{6.3.4}$$

for $x_1, x_2 \in \mathfrak{m}$, $Y_1, Y_2 \in \mathfrak{h}$. The corresponding norm will be denoted by $\| \cdot \|$.

(The reader not familiar with the Killing form, compact real forms, etc., can just take $\epsilon = -1$ and assume the inner product and norm to be given by the trace on matrices.)

For $x \in \mathfrak{m}$, let

$$M(x) = \epsilon(\mathrm{ad}_x)^2 \big|_{\mathfrak{h}}. \tag{6.3.5}$$

If $\mathfrak{g}$ is noncompact, then ad_x is B-symmetric, while if $\mathfrak{g}$ is compact, then ad_x is B-skew-symmetric. In either case, $M(x) = \epsilon(\mathrm{ad}_x)^2$ is a nonnegative symmetric operator on $\mathfrak{h}$. Next, for $Y \in \mathfrak{h}$, let

$$N(Y) = -(\mathrm{ad}_Y)^2 \big|_{\mathfrak{m}}. \tag{6.3.6}$$

Since ad_Y is B-skew-symmetric, $N(Y)$ is a nonnegative symmetric operator on $\mathfrak{m}$.

We will make frequent use of two identities relating the operators $M(x)$ and $N(Y)$. First, the Jacobi identity implies

$$[Y, M(x)Y] = \epsilon[x, N(Y)x] \tag{6.3.7}$$

for all $x \in \mathfrak{m}$, $Y \in \mathfrak{h}$. Second, the invariance of the Killing form implies

$$\langle Y, M(x)Y \rangle = \|[Y, x]\|^2 = \langle x, N(Y)x \rangle \tag{6.3.8}$$

for all $x \in \mathfrak{m}$, $Y \in \mathfrak{h}$.

We also require two estimates arising from $M(x)$ and $N(Y)$. For details of their proofs, see Bloch, Drakunov, and Kinyon [2000]. First, we have the inequality

$$\mathrm{tr}(M(x)) \le \|x\|^2 \tag{6.3.9}$$

for all $x \in \mathfrak{m}$. Second, there exists a constant $0 < \eta < 1$ such that

$$\mathrm{tr}(N(Y)) > \eta\|Y\|^2 \tag{6.3.10}$$

for all $Y \in \mathfrak{h}$.

Controls. We consider the following controls for the system (6.3.1)–(6.3.2):

$$u = -\alpha x + \beta[Y, x] + \gamma N(Y)x, \qquad (6.3.11)$$

where $\alpha, \beta, \gamma : \mathfrak{g} \to \mathbb{R}$ are real-valued functions, with $\alpha, \gamma \geq 0$ and $\beta\epsilon \leq 0$. With the control (6.3.11) (and using (6.3.7)), the system (6.3.1)–(6.3.2) becomes

$$\dot{x} = -\alpha x + \beta[Y, x] + \gamma N(Y)x, \qquad (6.3.12)$$

$$\dot{Y} = \beta\epsilon M(x)Y - \gamma\epsilon[Y, M(x)Y]. \qquad (6.3.13)$$

Using (6.3.12) and the skew-symmetry of ad_Y, we easily compute

$$\frac{d}{dt}\|x\|^2 = -2\alpha\|x\|^2 + 2\gamma\langle x, N(Y)x\rangle. \qquad (6.3.14)$$

Let λ_* denote the largest eigenvalue of $N(Y)$. Then $\langle x, N(Y)x\rangle \leq \lambda_*\|x\|^2$ for all $x \in \mathfrak{m}$, and thus the right-hand side of (6.3.14) is nonpositive if $\lambda_*\gamma \geq \alpha$. In this case $\|x\|$ is nonincreasing, and if $\alpha = \gamma = 0$, then $\|x\|$ is constant.

Using (6.3.13), we obtain

$$\frac{d}{dt}\|Y\|^2 = 2\beta\epsilon\langle Y, M(x)Y\rangle. \qquad (6.3.15)$$

Since $\beta\epsilon \leq 0$ and $M(x)$ is a nonnegative operator, the right hand side of (6.3.15) is nonpositive. Thus $\|Y\|$ is nonincreasing in general, and is constant if $\beta = 0$.

Our (necessarily discontinuous) stabilization algorithm will involve switching the control (6.3.11) among the following three cases: (i) $\alpha > 0$, $\beta = \gamma = 0$; (ii) $\alpha = \kappa\lambda_*$, $\gamma = \kappa$, and $\beta = 0$, where, as above, λ_* is the largest eigenvalue of $N(Y)$ and where κ is a positive function; (iii) $\alpha = \gamma = 0$, $\beta\epsilon < 0$. We now discuss the dynamics of the system (6.3.12)–(6.3.13) in each of these cases.

Case I: $\alpha > 0$, $\beta = \gamma = 0$:
In this case, the system (6.3.12)–(6.3.13) is

$$\dot{x} = -\alpha x, \qquad (6.3.16)$$

$$\dot{Y} = 0. \qquad (6.3.17)$$

Here x is driven to 0 radially while Y remains fixed. If Y was not already 0, implementing (6.3.11) with these parameter values will render the system unstabilizable. Hence this case will be used only if $Y \equiv 0$.

Case II: $\alpha = \kappa\lambda_*$, $\gamma = \kappa$, $\beta = 0$:
As noted above, $\kappa > 0$. In this case, the control (6.3.11) has the form

$$u = -\kappa\,(\lambda_* x - N(Y)x), \qquad (6.3.18)$$

while the system (6.3.12)–(6.3.13) is

$$\dot{x} = -\kappa(\lambda_* x - N(Y)x), \tag{6.3.19}$$

$$\dot{Y} = -\kappa\epsilon[Y, M(x)Y]. \tag{6.3.20}$$

In this case, $\|Y\|$ is constant. In addition, (6.3.20) is a Lax equation in Y (see Section 1.12). It follows that the spectrum of ad_Y is constant. Therefore, the spectrum of the operator $N(Y)$ is constant, as are the dimensions of its eigenspaces. In particular, the eigenvalue λ_*, which occurs in (6.3.19), is constant.

Let $0 \leq \lambda_0 < \lambda_1 < \cdots < \lambda_s = \lambda_*$ denote those eigenvalues of $N(Y)$ that are distinct (thus $s \leq \dim \mathfrak{m} - 1$). Let $x = x_0 + x_1 + \cdots + x_s$ be the unique decomposition of x into the eigenspaces of $N(Y)$. Then the differential equation (6.3.19) decouples into the following system of equations in $\mathfrak{m}$:

$$\dot{x}_0 = -\kappa(\lambda_* - \lambda_0)x_0,$$
$$\dot{x}_1 = -\kappa(\lambda_* - \lambda_1)x_1,$$
$$\vdots \tag{6.3.21}$$
$$\dot{x}_{s-1} = -\kappa(\lambda_* - \lambda_{s-1})x_{s-1},$$
$$\dot{x}_s = 0.$$

Since $\kappa(\lambda_* - \lambda_j) > 0$ for $j = 0, 1, \ldots, s - 1$, it follows that $x_j \to 0$ asymptotically. If we let x_* denote the projection of x onto the λ_*-eigenspace of $N(Y)$, that is, $x_* = x_s$, then noting that $x_* \equiv x_* \big|_{t=0}$ is constant, we conclude that

$$x \to x_*$$

asymptotically.

Note that (6.3.19)–(6.3.20) and (6.3.7) imply the following:

$$\dot{Y} = -\kappa[x, N(Y)x] = [x, \dot{x}]. \tag{6.3.22}$$

Since x converges to a λ_*-eigenvector of $N(Y)$, the right-hand side of (6.3.19) converges to 0 and thus $\dot{x}$ converges to 0. Therefore, (6.3.22) implies that $\dot{Y}$ converges to 0.

Summarizing this case, we have that Y evolves isospectrally (with constant spectrum) and with constant norm and asymptotically vanishing velocity, while x is driven to x_*, the (constant) projection of x onto the λ_*-eigenspace of $N(Y)$.

Case III: $\alpha = \gamma = 0$, $\beta\epsilon < 0$:

The system (6.3.12)–(6.3.13) for this case is

$$\dot{x} = \beta[Y, x], \tag{6.3.23}$$

$$\dot{Y} = \beta\epsilon M(x)Y. \tag{6.3.24}$$

In this case, $\|x\|$ is constant. In addition, (6.3.23) is a Lax equation in x, and thus ad_x has constant spectrum. Therefore, the spectrum of the operator $M(x)$ is constant, as are the dimensions of its eigenspaces. Let $0 \le \mu_0 < \mu_1 < \cdots < \mu_r$ denote those eigenvalues of $M(x)$ that are distinct (thus $r \le \dim \mathfrak{h} - 1$). For $Y \in \mathfrak{h}$, let $Y = Y_0 + \cdots + Y_r$ denote the unique decomposition of Y into the eigenspaces of $M(x)$. Then the differential equation (6.3.24) decouples into the following system of equations in $\mathfrak{h}$:

$$
\begin{aligned}
\dot{Y}_0 &= \beta\epsilon\mu_0 Y_0, \\
\dot{Y}_1 &= \beta\epsilon\mu_1 Y_1, \\
&\vdots \\
\dot{Y}_r &= \beta\epsilon\mu_r Y_r.
\end{aligned}
\tag{6.3.25}
$$

Since $\beta\epsilon\mu_j < 0$ for $j = 1, \ldots, r$, we have that $Y_j \to 0$ asymptotically. If $\mu_0 > 0$, then the same applies to Y_0. Otherwise, if $M(x)$ has $\mu_0 = 0$ as an eigenvalue, then Y_0 remains constant. Thus we have either $Y \to 0$ or $Y \to Y_0$ asymptotically, where $Y_0 \equiv Y_0 \big|_{t=0}$ is constant. In either case, if we let $Y_{\#}$ denote the projection of Y onto the nullspace of $M(x)$, then noting that $Y_{\#} \equiv Y_{\#} \big|_{t=0}$ is constant, we conclude that

$$
Y \to Y_{\#}
$$

asymptotically.

Using system (6.3.23)–(6.3.24), we can derive the equation

$$
\frac{d}{dt} M(x)^n Y = \beta\epsilon[Y, M(x)^n Y] + \beta\epsilon M(x)^{n+1} Y
\tag{6.3.26}
$$

for every nonnegative integer n. Indeed, the case $n = 0$ is just (6.3.24). Using the induction hypothesis, we have for $n > 0$,

$$
\begin{aligned}
\frac{d}{dt} M(x)^n Y = {}& \epsilon\Big[\dot{x}, [x, M(x)^{n-1}Y]\Big] + \epsilon\Big[x, [\dot{x}, M(x)^{n-1}Y]\Big] \\
&+ M(x)\Big(\beta\epsilon[Y, M(x)^{n-1}Y] + \beta\epsilon M(x)^n Y\Big).
\end{aligned}
\tag{6.3.27}
$$

Now,

$$
[\dot{x}, [x, M(x)^{n-1}Y]] = \beta[[Y, x], [x, M(x)^{n-1}Y]]
\tag{6.3.28}
$$

and

$$
[x, [\dot{x}, M(x)^{n-1}Y]] = \beta[x, [[Y, x], M(x)^{n-1}Y]],
\tag{6.3.29}
$$

while applying the Jacobi identity repeatedly gives

$$
\begin{aligned}
M(x)[Y, M(x)^{n-1}Y] = {}& [Y, M(x)^n Y] + \epsilon[[x, Y], [x, M(x)^{n-1}Y]] \\
&+ \epsilon[x, [[x, Y], M(x)^{n-1}Y]].
\end{aligned}
\tag{6.3.30}
$$

Substituting (6.3.28), (6.3.29), and (6.3.30) into (6.3.27) and simplifying gives (6.3.26).

Then from (6.3.26),

$$\frac{d}{dt} f(M(x))Y = \beta\epsilon[Y, f(M(x))Y] + \beta\epsilon f(M(x))M(x)Y \qquad (6.3.31)$$

follows immediately for every real analytic function f. As an interesting special case of this, let $p(\mu)$ be the minimal polynomial of $M(x)$ and assume that $\mu_0 = 0$ is an eigenvalue of $M(x)$ (so that Y does not converge to 0). Then $p(\mu) = \mu q(\mu)$ for some polynomial q. Taking $f = q$ in (6.3.31) gives

$$\frac{d}{dt} q(M(x))Y = \beta\epsilon[Y, q(M(x))Y]. \qquad (6.3.32)$$

It follows that the spectrum of $q(M(x))Y$ remains constant; that is, it evolves isospectrally.

Summarizing this case, we have that x evolves isospectrally with constant norm, Y is driven to $Y_\#$, its (constant) projection onto the nullspace of $M(x)$, and $q(M(x))Y$ evolves with constant spectrum.

Remark. It is interesting to compare the system of equations of Case III with the double bracket equations discussed, for example, in Brockett [1988] and Bloch, Brockett, and Ratiu [1992] and briefly in Chapter 1 in connection with the Toda lattice. In these papers the isospectral flow $\dot{L} = [L, [L, N]]$, L, N lying in a compact Lie algebra, N fixed, was considered. This flow is the gradient flow of $\langle L, N \rangle$ on an adjoint orbit of the corresponding Lie group with respect to the so-called normal metric. Equation (6.3.24) is, on the other hand, of the form $\dot{Y} = \beta\epsilon[X, [X, Y]]$, which is *not* isospectral (although it is coupled to the isospectral equation (6.3.23)). Further, as we have seen, we have a different function, $\langle Y, Y \rangle$, decreasing along its flow, which is precisely what is needed in this context.

The Stabilization Algorithm. We now describe our feedback strategy. As before, λ_* denotes the largest eigenvalue of the operator $N(Y)$, x_* denotes the projection of x onto the λ_*-eigenspace of $N(Y)$, and $Y_\#$ denotes the projection of Y onto the nullspace of $M(x)$. Let $\delta > 0$ be a prescribed error tolerance. In informal pseudocode, the algorithm can be described as follows:

---

begin

  while $\|Y\| \geq \delta$

  1. Let $r = \|x\|$. Implement the control (6.3.11) with $\alpha = \lambda_*\kappa$, $\gamma = \kappa$, and $\beta = 0$. Then Y evolves isospectrally with constant norm, while x converges to the constant x_*. If $x_* \neq 0$, then go to Step 3.

2. Let z_* denote a fixed λ_*-eigenvector of $N(Y)$ with

$$\|z_*\| = r\left(1 - 1/\dim \mathfrak{m}\right)^{1/2}.$$

Let $u = -\alpha(x - z_*)$, where $\alpha > 0$. Then x converges to z_* while Y remains constant.

3. Implement the control (6.3.11) with $\alpha = \gamma = 0$, $\beta \epsilon < 0$. Then x evolves isospectrally with constant norm, while Y converges to the constant $Y_\#$.

end while

if $\|x\| \geq \delta$, **then**

4. implement the control (6.3.11) with $\alpha > 0$, $\beta = \gamma = 0$. Then x will converge to 0 radially, while Y remains 0.

end

In Step 1, if α is a constant, then x will converge to x_* in infinite time; if, for example, $\alpha = 1/\|x - x_*\|$, then x will converge in finite time. Similarly, in Step 3, if β is a constant, then Y will converge to $Y_\#$ in infinite time; if, for example, $\beta = 1/\|Y - Y_\#\|$, then Y will converge in finite time. To establish the convergence claim made in Step 2, we simply note that in this case $x(t)$ has the form $x(t) = f(t)z_*$, where $f(t)$ is a scalar-valued function satisfying

$$\dot{f} = -\alpha(f - 1), \quad f(0) = 0.$$

(For instance, if $\alpha > 0$ is constant, we have $f(t) = 1 - e^{-\alpha t}$.) It follows from (6.3.2) that $\dot{Y} = [u, x] = 0$, so that Y is constant, as claimed.

Step 2 is implemented if x converges to 0 in Step 1. One instance where this could happen occurs if the initial value of x is 0, in which case the first implementation of Step 1 is trivial. More generally, the case where the projection of x onto the λ_*-eigenspace of $N(Y)$ is 0 seems to be the natural higher-dimensional analogue of the situation in the Heisenberg system where the initial value starts on the z-axis. As in Steps 1 and 3, Step 2 can also be implemented in finite time.

The λ_*-eigenspace of $N(Y)$ will, in general, have dimension greater than 1 (since the nonzero eigenvalues of the B-skew-symmetric operator ad_Y come in complex conjugate pairs). Thus there is no unique choice of eigenvector z_* in Step 2. Any lexicographic ordering of the eigenvectors relative to a coordinate basis will suffice as a selection scheme. The rationale behind the particular normalization of z_* will be explained below. (Choices of this type occur naturally in stabilizing nonholonomic systems; see Sontag [1998] for comments on this and related robustness issues.)

We will now show that our algorithm successfully stabilizes the system
(6.3.1)–(6.3.2) by showing that each of $\|x\|$ and $\|Y\|$ can be brought to
within the prescribed error tolerance. Note that as soon as the test condi-
tion of the while loop fails, that is, as soon as $\|Y\| < \delta$, then the system
will be stabilized whether Step 4 needs to be executed or not. Thus we may
assume that the initial value of Y satisfies $\|Y\| \geq \delta$ so that the while loop
will be executed at least once. If Y ever converges to 0 in Step 3 because
$Y_{\#} = 0$, then the test condition of the while loop will eventually fail. As
noted, this is enough to guarantee that the system is stabilizable.

Assume that for every iteration of Step 3 we have $Y_{\#} \neq 0$. We will show
that after finitely many iterations of the while loop, the test condition
will fail. In other words, the projection of Y onto the nullspace of $M(x)$ is
eventually arbitrarily small in norm. In fact, we will show a stronger result,
for when this situation occurs, then it turns out that $\|x\|$ is simultaneously
brought to within the error tolerance. Thus as soon as the while loop's test
condition fails, the test condition of the if–then statement (Step 4) will also
fail, and the system will already be stabilized.

Assume first that Step 3 is about to be executed. Since Step 1 and
possibly Step 2 have already been executed, the initial values $x(0) = x_*$ and
$Y(0) = Y_*$ satisfy $N(Y_*)x_* = \lambda_* x_*$. As before, let Y_j denote the projection
of Y onto the μ_j-eigenspace of $M(x)$. Recall that

$$Y_{\#} = Y_0 \equiv Y_0(0)$$

throughout Step 3, and that $Y(t) \to Y_{\#}$ asymptotically. Using the orthogonality of the eigenspaces, we compute

$$\|Y_{\#}\|^2 = \|Y_*\|^2 - \sum_{j=1}^{r} \|Y_j(0)\|^2$$

$$\leq \|Y_*\|^2 - \frac{1}{\sum_{j=0}^{r} \mu_j} \sum_{j=0}^{r} \mu_j \|Y_j(0)\|^2. \qquad (6.3.33)$$

Note that we are using $\mu_0 = 0$. Now using the orthogonality once again,
we compute

$$\sum_{j=0}^{r} \mu_j \|Y_j(0)\|^2 = \left\langle Y_*, \sum_{j=0}^{r} \mu_j Y_j(0) \right\rangle = \langle Y_*, M(x_*)Y_* \rangle$$

$$= \langle x_*, N(Y_*)x_* \rangle \qquad (6.3.34)$$

$$= \lambda_* \|x_*\|^2. \qquad (6.3.35)$$

Here we have used (6.3.8) to obtain (6.3.34). In addition, using (6.3.9), we
have

$$\sum_{j=1}^{r} \mu_j \leq \operatorname{tr}(M(x_*)) \leq \|x_*\|^2. \qquad (6.3.36)$$

Applying (6.3.35) and (6.3.36) to (6.3.33) yields

$$\|Y_\#\|^2 \le \|Y_*\|^2 - \lambda_*. \tag{6.3.37}$$

Now using (6.3.10), we have

$$\lambda_* \ge \frac{1}{\dim \mathfrak{m}} \text{tr}(N(Y_*)) > \frac{\eta}{\dim \mathfrak{m}} \|Y_*\|^2. \tag{6.3.38}$$

Applying (6.3.38) to (6.3.37) gives our final estimate for Step 3:

$$\|Y_\#\|^2 < \left(1 - \frac{\eta}{\dim \mathfrak{m}}\right) \|Y_*\|^2. \tag{6.3.39}$$

Now assume that Step 3 has already been executed and that Step 1 is about to be executed again. Then the initial values $x(0) = x_\#$ and $Y(0) = Y_\#$ in Step 1 satisfy $M(x_\#)Y_\# = 0$. By (6.3.8), this implies $\langle x_\#, N(Y_\#)x_\# \rangle = 0$. As before, let x_j denote the projection of x into the λ_j-eigenspace of $N(Y)$. Recall that $x_* = x_s \equiv x_s(0)$ throughout Step 1, and that $x(t) \to x_*$ asymptotically. Using the orthogonality of the eigenspaces, we compute

$$\|x_*\|^2 = \|x_\#\|^2 - \sum_{j=0}^{s-1} \|x_j(0)\|^2$$

$$\le \|x_\#\|^2 - \frac{1}{\sum_{j=0}^{s}(\lambda_s - \lambda_j)} \sum_{j=0}^{s} (\lambda_s - \lambda_j)\|x_j(0)\|^2. \tag{6.3.40}$$

Using orthogonality again, we compute

$$\sum_{j=0}^{s} (\lambda_s - \lambda_j)\|x_j(0)\|^2 = \lambda_s \|x_\#\|^2 - \left\langle x_\#, \sum_{j=0}^{s} \lambda_s x_j(0) \right\rangle$$

$$= \lambda_s \|x_\#\|^2 - \langle x_\#, N(Y_\#)x_\# \rangle$$

$$= \lambda_s \|x_\#\|^2. \tag{6.3.41}$$

Also,

$$\sum_{j=0}^{s} (\lambda_s - \lambda_j) = s\lambda_s - \sum_{j=0}^{s-1} \lambda_j. \tag{6.3.42}$$

Applying (6.3.41) and (6.3.42) to (6.3.40) gives

$$\|x_*\|^2 \le \left(1 - \frac{\lambda_s}{s\lambda_s - \sum_{j=0}^{s-1} \lambda_j}\right) \|x_\#\|^2. \tag{6.3.43}$$

Finally,

$$\frac{\lambda_s}{s\lambda_s - \sum_{j=0}^{s-1} \lambda_j} \ge \frac{1}{s} \ge \frac{1}{\dim \mathfrak{m}}, \tag{6.3.44}$$

and applying (6.3.44) to (6.3.43) gives our final estimate for Step 1:

$$\|x_*\|^2 \leq \left(1 - \frac{1}{\dim \mathfrak{m}}\right)\|x_\#\|^2. \tag{6.3.45}$$

Now assume that Step 2 is executed because $x = 0$ (that is, $x_* = 0$ in Step 1). Rename $x_* = z_*$, where z_* is the chosen λ_*-eigenvector. Then the normalization of z_* described in Step 2 immediately implies that (6.3.45) holds as an equality.

Define two sequences of real numbers as follows: Let a_j and b_j denote, respectively, the initial values of $\|x\|^2$ and $\|Y\|^2$ prior to the $(j+1)$st iteration of the while loop, where $j = 0, 1, \ldots$. Recall that $\|Y\|$ remains constant during Steps 1 and 2 and $\|x\|$ remains constant during Step 3. Our estimates (6.3.39) and (6.3.45) imply that the sequences $\{a_j\}$ and $\{b_j\}$ satisfy

$$a_{j+1} \leq \left(1 - \frac{1}{\dim \mathfrak{m}}\right)a_j, \tag{6.3.46}$$

$$b_{j+1} < \left(1 - \frac{\eta}{\dim \mathfrak{m}}\right)b_j. \tag{6.3.47}$$

Since

$$0 < 1 - \frac{1}{\dim \mathfrak{m}} < 1 - \frac{\eta}{\dim \mathfrak{m}} < 1, \tag{6.3.48}$$

it follows from (6.3.46)–(6.3.47) that the sequences $\{a_j\}$ and $\{b_j\}$ each converge to 0. In particular, it is immediate that each of $\|x\|$ and $\|Y\|$ can be brought to within the prescribed error tolerance $\delta > 0$ in finitely many iterations of the while loop.

In summary, we have proven the following result.

6.3.1 Theorem. *The algorithm given in Steps 1–4 above globally stabilizes the system* (6.3.1)–(6.3.2).

We remark that while we have used the error tolerance δ above to indicate how the stabilization algorithm works in practice, the formal proof of stability follows from letting δ approach zero.

For an explicit application of this algorithm to the cases $\mathfrak{g} = \mathfrak{so}(n)$ and in particular $\mathfrak{so}(3)$, see the web supplement.

Exercises

◇ **6.3-1.** Consider the system (6.1.2), (6.1.3). Show that stabilization of this system may be reduced to the stabilization of a three-dimensional system that may be stabilized as in the Heisenberg example by the use of the variables

$$V_1 = \frac{1}{2}x^T x, \quad V_2 = \frac{1}{2}x^T Y^T Y x, \quad V_3 = \frac{1}{2}Y^T Y. \tag{6.3.49}$$

(See Bloch and Drakunov [1998].)

⬦ **6.3-2.** Write out the system (6.3.12), (6.3.13) explicitly in the case of $\mathfrak{g} = \mathfrak{so}(3)$ and $\mathfrak{g} = \mathfrak{so}(4)$.

6.4 Controllability, Accessibility, and Stabilizability

In this and subsequent sections we consider a class of nonholonomic dynamic control systems and various control and stabilizability properties, following the work of Bloch, Reyhanoglu, and McClamroch [1992]. There is a huge related literature. We mention here briefly some related work in robotics and control: Li and Montgomery [1988], Li and Canny [1990], Murray and Sastry [1993], Murray, Li and Sastry [1994], Murray [1995], Krishnaprasad and Yang [1991], Reyhanoglu, van der Schaft, McClamroch, and Kolmanovsky [1999], Cortés, Martínez, Ostrowski, and Zhang [2002], Lewis [2000], Cortés and Martínez [2001], Schneider [2000], Schneider [2002]. More references may be found throughout the book.

We consider the class of mechanical (Lagrangian) nonholonomic control systems described by the equations

$$\frac{d}{dt}\frac{\partial L}{\partial \dot{q}^i} - \frac{\partial L}{\partial q^i} = \sum_{j=1}^{m}\lambda_j a_i^j + \sum_{j=1}^{l} b_i^j u_j\,, \tag{6.4.1}$$

$$\sum_{i=1}^{n} a_i^j \dot{q}^i = 0 \qquad j = 1,\dots,m\,. \tag{6.4.2}$$

These equations are a controlled version of the nonholonomic equations in Lagrange multiplier form discussed in Section 5.2. As in that section, we assume here that we have a Lagrangian on the tangent bundle to an arbitrary configuration space Q, given by $L : TQ \to \mathbb{R}$. In coordinates $q^i, i = 1,\dots,n$, on Q with induced coordinates $(q^i, \dot{q}^i)$ for the tangent bundle, we have $L(q^i, \dot{q}^i)$. All computations here will be local, however, and for the moment we will assume $Q = \mathbb{R}^n$. Here L is taken to be the mechanical Lagrangian

$$L = \frac{1}{2} \sum_{i,j=1}^{n} g_{ij}(q)\dot{q}^i \dot{q}^j - V(q). \tag{6.4.3}$$

Hence equation (6.4.1) takes the explicit form

$$g_{ij}\ddot{q}^j + \frac{\partial g_{ij}}{\partial q^k}\dot{q}^k\dot{q}^j - \frac{1}{2}\frac{\partial g_{jk}}{\partial q^i}\dot{q}^j\dot{q}^k + \frac{\partial V}{\partial q^k} = \sum_{j=1}^{m}\lambda_j a_i^j + \sum_{j=1}^{r} b_i^j u_j\,. \tag{6.4.4}$$

For convenience below we shall sometimes rewrite equation (6.4.4) as

$$g_{ij}\ddot{q}^j + f_i(q,\dot{q}) = \sum_{j=1}^{m} \lambda_j a_i^j + \sum_{j=1}^{l} b_i^j u_j . \qquad (6.4.5)$$

All functions are assumed to be smooth. We shall make some assumptions on the controls later on. As in Chapter 5, we shall assume that the constraints may be rewritten as

$$\dot{s}^a + A_\alpha^a(r,s)\dot{r}^\alpha = 0, \qquad a = 1,\ldots,m, \qquad (6.4.6)$$

where $q = (r,s) \in \mathbb{R}^{n-m} \times \mathbb{R}^m$.

As in Chapter 5, denote the distribution defined by the constraints at q by $\mathcal{D}_q$.

6.4.1 Definition (Vershik and Gershkovich [1988]). *Consider the following nondecreasing sequence of locally defined distributions:*

$$N_1 = \mathcal{D},$$
$$N_k = N_{k-1} + \mathrm{span}\{[X,Y] \mid X \in N_1,\ Y \in N_{k-1}\}.$$

Then there is an integer k^* such that

$$N_k = N_{k^*}$$

for all $k > k^*$. If $\dim N_{k^*} = n$ *and $k^* > 1$, then the constraints (6.4.6) are called* **completely nonholonomic**, *and the smallest (finite) number k^* is called the* **degree of nonholonomy**.

We assume (through Section 6.6) that the constraint equations are completely nonholonomic everywhere with nonholonomy degree k^*. Note that for this to hold, $n - m$ must be strictly greater than one. The constraints then define a $(2n-m)$-dimensional smooth submanifold

$$\mathbf{M} = \left\{(q,\dot{q}) \mid a_i^j \dot{q}^i = 0,\ j = 1,\ldots,m\right\} \qquad (6.4.7)$$

of the phase space. This manifold $\mathbf{M}$ plays a critical role in the concept of solutions and the formulation of control and stabilization problems.

As shown in Bloch, Reyhanoglu, and McClamroch [1992], (6.4.1) and (6.4.2) provide a well-posed set of equations.

We subsequently use the notation $(Q(t,q_0,\dot{q}_0), \Lambda(t,q_0,\dot{q}_0))$ to denote the solution of equations (6.4.1), (6.4.2) at time $t \geq 0$ corresponding to the initial conditions $(q_0,\dot{q}_0)$. Thus for each initial condition $(q_0,\dot{q}_0) \in \mathbf{M}$ and each bounded, measurable input function $u : [0,T) \to \mathbb{R}^l$,

$$(Q(t,q_0,\dot{q}_0), \dot{Q}(t,q_0,\dot{q}_0)) \in \mathbf{M}$$

holds for all $t \geq 0$ where the solution is defined.

We say that a solution is an equilibrium solution if it is a constant solution; note that if (q^e, λ^e) is an equilibrium solution, we refer to q^e as an equilibrium configuration.

6.4.2 Theorem. *Suppose that $u(t) = 0$, $t \geq 0$. The set of equilibrium configurations of equations (6.4.1), (6.4.2) is given by*

$$\left\{ q^i \ \middle| \ \frac{\partial V(q, 0)}{\partial q^i} - a_i^j \lambda_j = 0, \ i = 1, \ldots, n, \ \text{for some } \lambda \in \mathbb{R}^m \right\}.$$

We remark that generically we obtain an equilibrium *manifold* with dimension at least m. On the other hand, for certain cases, there may not be even a single equilibrium configuration (e.g., the dynamics of a ball on an inclined plane). However, with our controllability assumptions below we can always introduce an equilibrium manifold of dimension at least m by appropriate choice of input. This is, of course, all in the smooth category.

Exercises

⋄ **6.4-1.** Consider the following kinematic model of a car (see, e.g., Murray and Sastry [1993], Abraham, Marsden, and Ratiu [1988]):

$$\begin{aligned}
\dot{x} &= \cos\theta u_1, \\
\dot{y} &= \sin\theta u_1, \\
\dot{\phi} &= u_2, \\
\dot{\theta} &= \tan\phi u_1.
\end{aligned} \tag{6.4.8}$$

Here (x, y) denotes the position of the rear axle, θ the angle of the car with respect to the horizontal, and ϕ the steering angle with respect to the car body. Show that the system is completely nonholonomic except for $\phi = \pm\pi/2$ and that the degree of nonholonomy is 3.

6.5   Smooth Stabilization to a Manifold

As in the kinematic case discussed in Section 6.1 we can show from the Brockett necessary condition that there is no smooth feedback that will smoothly stabilize a dynamic nonholonomic system. We can state this formally as follows (see Bloch, Reyhanoglu, and McClamroch [1992]):

6.5.1 Theorem. *Let $m \geq 1$ and let $(q^e, 0)$ denote an equilibrium solution in $\mathbf{M}$. The nonholonomic control system, defined by equations (6.4.1), (6.4.2) is not asymptotically stabilizable using C^1 feedback to $(q^e, 0)$.*

Proof. A necessary condition for the existence of a C^1 asymptotically stabilizing feedback law for system (6.5.7), (6.5.8), (6.5.9) is that the image of the mapping

$$(x_1, x_2, x_3, v) \to (x_3, -A(x_1, x_2)x_3, v)$$

contain some neighborhood of zero. No points of the form

$$\begin{pmatrix} 0 \\ \epsilon \\ \alpha \end{pmatrix}, \ \epsilon \neq 0 \text{ and } \alpha \in \mathbb{R}^{n-m} \text{ arbitrary,}$$

are in its image: It follows that Brockett's necessary condition is not satisfied. Hence system (6.5.7), (6.5.8), (6.5.9) cannot be asymptotically stabilized to $(r^e, s^e, 0)$ by a C^1 feedback law. Thus the nonholonomic control system defined by equations (6.4.1), (6.4.2) is not C^1 asymptotically stabilizable to $(q^e, 0)$. ∎

We remark that as in the kinematic case even C^0 (continuous) feedback is ruled out (see, e.g., Zabczyk [1989]).

A corollary of this result is that a single equilibrium solution of (6.4.1), (6.4.2) cannot be asymptotically stabilized using linear feedback, nor can it be asymptotically stabilized using feedback linearization or any other control design approach that uses smooth feedback.

As discussed in Bloch, Reyhanoglu, and McClamroch [1992], it turns out that the best one can achieve in the way of smooth stabilization is stabilization to an equilibrium manifold. We discuss this in this section and turn to the problem of stabilization to a point by nonsmooth feedback in the next section.

We want to consider feedback control of the form $u_i = U_i(q, \dot{q})$, where $U : \mathbf{M} \to \mathbb{R}^l$; the corresponding closed loop is described by

$$g_{ij}\ddot{q}^j + f_i(q, \dot{q}) = \sum_{j=1}^{m} \lambda_j a_i^j + \sum_{j=1}^{l} b_i^j U_j(q, \dot{q}), \tag{6.5.1}$$

$$\sum_{i=1}^{n} a_i^j \dot{q}^i = 0, \qquad j = 1, \dots, m. \tag{6.5.2}$$

The set of equilibrium configurations of equations (6.5.1), (6.5.2) is given by

$$\left\{ q^i \ \middle| \ \frac{\partial V(q, 0)}{\partial q^i} - a_i^j \lambda_j = \sum_{j=1}^{l} b_i^j U_j(q, 0), \ i = 1, \dots, n, \text{ for some } \lambda \in \mathbb{R}^m \right\},$$

which is a smooth submanifold of the configuration space.

We now introduce a suitable stability definition for the closed-loop system.

6.5.2 Definition. *Assume that* $u_i = U_i(q, \dot{q})$. *Let*

$$\mathbf{M}_s = \{(q, \dot{q}) \mid \dot{q} = 0\}$$

be an embedded submanifold of $\mathbf{M}$. *Then* $\mathbf{M}_s$ *is called **locally stable** if for any neighborhood* $\mathbf{U} \supset \mathbf{M}_s$ *there is a neighborhood* $\mathbf{V}$ *of* $\mathbf{M}_s$ *with* $\mathbf{U} \supset \mathbf{V} \supset \mathbf{M}_s$ *such that if* $(q_0, \dot{q}_0) \in \mathbf{V} \cap \mathbf{M}$, *then the solution of equations* (6.5.1), (6.5.2) *satisfies* $(Q(t, q_0, \dot{q}_0), \dot{Q}(t, q_0, \dot{q}_0)) \in \mathbf{U} \cap \mathbf{M}$ *for all* $t \geq 0$. *If in addition,* $(Q(t, q_0, \dot{q}_0), \dot{Q}(t, q_0, \dot{q}_0)) \to (q_s, 0)$ *as* $t \to \infty$ *for some* $(q_s, 0) \in \mathbf{M}_s$, *then we say that* $\mathbf{M}_s$ *is a **locally asymptotically stable equilibrium manifold** of equations* (6.5.1), (6.5.2).

Note that if $(Q(t, q_0, \dot{q}_0), \dot{Q}(t, q_0, \dot{q}_0)) \to (q_s, 0)$ as $t \to \infty$ for some $(q_0, 0) \in \mathbf{M}_s$, it follows that there is $\lambda_s \in \mathbb{R}^m$ such that $\Lambda(t, q_0, \dot{q}_0) \to \lambda_s$ as $t \to \infty$.

The usual definition of local stability corresponds to the case that $\mathbf{M}_s$ is a single equilibrium solution; the more general case is required here.

The existence of a feedback function such that a certain equilibrium manifold is asymptotically stable is of particular interest; hence we introduce the following definition:

6.5.3 Definition. *The system defined by equations* (6.4.1), (6.4.2) *is said to be **locally asymptotically stabilizable to a smooth equilibrium manifold** $\mathbf{M}_s$ in $\mathbf{M}$ if there exists a feedback function $U : \mathbf{M} \to \mathbb{R}^l$ such that for the associated closed loop equations* (6.5.1), (6.5.2), $\mathbf{M}_s$ *is locally asymptotically stable.*

If there exists such a feedback function that is smooth on $\mathbf{M}$, then we say that equations (6.4.1), (6.4.2) are smoothly asymptotically stabilizable to $\mathbf{M}_s$; of course, it is possible (and we subsequently show that it is generic in certain cases) that equations (6.4.1), (6.4.2) might be asymptotically stabilizable to $\mathbf{M}_s$ but not smoothly (even continuously) asymptotically stabilizable to $\mathbf{M}_s$.

Normal Form Equations. We now show that using feedback, one can reduce the controlled nonholonomic equations to a normal form that is easy to analyze.

We recall that the reduced state space is $(2n - m)$-dimensional. The state of the system can be specified by the n-vector of configuration variables and an $(n - m)$-vector of velocity variables. Let $q = (r, s)$ be the partition of the configuration variables corresponding to the constraints introduced previously.

Define the $n \times (n - m)$ matrix C as follows:

$$C_\alpha^i = \begin{cases} \delta_\alpha^i, & i = 1, \ldots, n - m, \\ -A_\alpha^i, & i = n - m + 1, \ldots, n \end{cases} \tag{6.5.3}$$

Then

$$\dot{q}^i = C_\alpha^i \dot{r}^\alpha. \tag{6.5.4}$$

Taking time derivatives yields

$$\ddot{q}^i = C_\alpha^i(q)\ddot{r}^\alpha + \dot{C}_\alpha^i(q)\dot{r}^\alpha.$$

Substituting this into equation (6.4.5) and multiplying both sides of the resulting equation by $C^T(q)$ gives

$$C_\alpha^{iT}(q)g_{ij}(q)C_\beta^j(q)\ddot{r}^\beta = C_\alpha^{iT}(q)[b_i^j(q)u_j - f_i(q,C\dot{r}) - g_{ij}(q)\dot{C}_\alpha^j(q)\dot{r}^\alpha].$$
(6.5.5)

Note that $C_\alpha^{iT}(q)g_{ij}(q)C_\beta^j(q)$ is an $(n-m) \times (n-m)$ symmetric positive definite matrix function.

We also assume that $l = n - m$ (for simplicity) and that the matrix product $C_\alpha^{iT}(q)b_i^j(q)$ is locally invertible. Hence for any $u \in \mathbb{R}^l$ there is a unique $v \in \mathbb{R}^{n-m}$ that satisfies

$$C_\alpha^{iT}(q)g_{ij}(q)C_\beta^j(q)v^\beta = C_\alpha^{iT}(q)[b_i^j(q)u_j - f_i(q,C\dot{r}) - g_{ij}(q)\dot{C}_\alpha^j(q)\dot{r}^\alpha].$$
(6.5.6)

This assumption guarantees that the reduced configuration variables satisfy the linear equations

$$\ddot{r}^\alpha = v^\alpha.$$

Now define the following state variables:

$$x_1^\alpha = r^\alpha, \ x_2^a = s^a, \ x_3^\alpha = \dot{r}^\alpha.$$

These variables thus satisfy the equations

$$\dot{x}_1 = x_3, \tag{6.5.7}$$
$$\dot{x}_2 = -A(x_1,x_2)x_3, \tag{6.5.8}$$
$$\dot{x}_3 = v. \tag{6.5.9}$$

We shall call equations (6.5.7), (6.5.8), (6.5.9) the **normal form** equations for the system (6.4.1), (6.4.2).
Equations (6.5.7), (6.5.8), (6.5.9) define a drift vector field

$$f(x) = (x_3, -A(x_1,x_2)x_3, 0)$$

and control vector fields $g_i(x) = (0,0,e_i)$, where e_i is the ith standard basis vector in $\mathbb{R}^{n-m}$, $i = 1, \ldots, n-m$, according to the standard control system form

$$\dot{x} = f(x) + \sum_{i=1}^{n-m} g_i(x)v_i. \tag{6.5.10}$$

We consider local properties of equations (6.5.7), (6.5.8), (6.5.9), near an equilibrium solution $(x_1^e, x_2^e, 0)$.

Note that the normal form equations (6.5.7), (6.5.8), (6.5.9) are a special case of the normal form equations studied by Byrnes and Isidori [1988].

Stabilization to an Equilibrium Manifold. We now study the problem of stabilization of equations (6.4.1), (6.4.2) to a smooth equilibrium submanifold of **M** defined by

$$\mathbf{N}_e = \{(q, \dot{q}) \mid \dot{q} = 0, \; w(q) = 0\},$$

where $w(q)$ is a smooth $(n - m)$-vector function. We show that, with appropriate assumptions, there exists a smooth feedback such that the closed loop is locally asymptotically stable to $\mathbf{N}_e$.

The smooth stabilization problem is the problem of giving conditions such that there exists a smooth feedback function $U : \mathbf{M} \to \mathbb{R}^l$ such that $\mathbf{N}_e$ is locally asymptotically stable. Of course, we are interested not only in demonstrating that such a smooth feedback exists but also in indicating how such an asymptotically stabilizing smooth feedback can be constructed.

We now assume that we have here nonholonomic control systems whose normal form equations satisfy the property that if $r(t)$ and $\dot{r}(t)$ are exponentially decaying functions, then the solution to

$$\dot{s} = -A(r(t), s)\dot{r}(t)$$

is bounded (all the physical examples of nonholonomic systems, of which we are aware, satisfy this assumption).

Note also that the first and second time derivatives of $w(q)$ are given by

$$\dot{w} = \frac{\partial w(q)}{\partial q} C(q)\dot{r},$$

$$\ddot{w} = \frac{\partial}{\partial q}\left(\frac{\partial w(q)}{\partial q} C(q)\dot{s}\right) C(q)\dot{r} + \frac{\partial w(q)}{\partial q} C(q)v.$$

We have (see Bloch, Reyhanoglu, and McClamroch [1992]) the following theorem:

6.5.4 Theorem. *Assume that the above solution property holds. Then the nonholonomic control system defined by equations (6.4.1) and (6.4.2), is locally asymptotically stabilizable to*

$$\mathbf{N}_e = \{(q, \dot{q}) \mid \dot{q} = 0, \; w(q) = 0\}, \tag{6.5.11}$$

using smooth feedback, if the transversality condition

$$\det\left(\frac{\partial w(q)}{\partial r}\right) \det\left(\frac{\partial w(q)}{\partial q} C(q)\right) \neq 0 \tag{6.5.12}$$

is satisfied.

Proof. It is sufficient to analyze the system in the normal form (6.5.7), (6.5.8), (6.5.9). By the transversality condition, the change of coordinates from $(r, s, \dot{r})$ to $(w, s, \dot{w})$ is a diffeomorphism.

Let

$$v = -\left(\frac{\partial w(q)}{\partial q}C(q)\right)^{-1}\left[\frac{\partial}{\partial q}\left(\frac{\partial w(q)}{\partial q}C(q)\dot{r}\right)C(q)\dot{r}\right.$$
$$\left. + K_1\frac{\partial w(q)}{\partial q}C(q)\dot{r} + K_2 w(q)\right],$$

where K_1 and K_2 are symmetric positive definite $(n-m)\times(n-m)$ constant matrices. Then obviously,

$$\ddot{w} + K_1\dot{w} + K_2 w = 0$$

is asymptotically stable to the origin so that $(w,\dot{w}) \to 0$ as $t \to \infty$. The remaining system variables satisfy equation (6.5.7) of the normal form equations (with $x_2 = s$) and by our assumption on the constraint matrix A, these variables remain bounded for all time. Thus $(q(t),\dot{q}(t)) \to \mathbf{N}_e$ as $t \to \infty$. ∎

6.6   Nonsmooth Stabilization

The results in the previous section demonstrate that smooth feedback can be used to asymptotically stabilize certain smooth manifolds $\mathbf{N}_e$ in $\mathbf{M}$. These results do not guarantee smooth asymptotic stabilization to a single equilibrium solution if $m \geq 1$. In this section we indicate how a single equilibrium can be asymptotically stabilized by use of piecewise analytic feedback. However, this is by no means the only approach to stabilization. As mentioned above, there is a large literature on this subject.

We first demonstrate that the normal form equations (6.5.7), (6.5.8), (6.5.9) and hence the nonholonomic control system defined by equations (6.4.1) and (6.4.2), satisfy certain strong local controllability properties. In particular, we show that the system is strongly accessible and that the system is small-time locally controllable at any equilibrium. These results provide a theoretical basis for the use of inherently nonlinear control strategies and suggest constructive procedures for the desired control strategies. We have (Bloch, Reyhanoglu, and McClamroch [1992]) the following theorems:

6.6.1 Theorem. *Let $m \geq 1$ and let $(q^e,0)$ denote an equilibrium solution in $\mathbf{M}$. The nonholonomic control system defined by equations (6.4.1) and (6.4.2) is strongly accessible at $(q^e,0)$.*

Proof. It suffices to prove that system (6.5.7), (6.5.8), (6.5.9) is strongly accessible at the origin. Let I denote the set $\{1,\ldots,n-m\}$. The drift and control vector fields can be expressed as

$$f = \sum_{j=1}^{n-m} x_{3,j}\tau_j,$$

$$g_i = \frac{\partial}{\partial x_{3,i}}, \ i \in I,$$

where

$$\tau_j = \frac{\partial}{\partial x_{1,j}} - \sum_{i=1}^{n-m} A_i^j(x_1, x_2)\frac{\partial}{\partial x_{2,i}}, \ j \in I,$$

are considered as vector fields on the (x_1, x_2, x_3) state space and the notation $x_{i,j}$ denotes the jth coordinate of the variable x_i. It can be verified that

$$[g_{i_1}, f] = \tau_{i_1}, \ i_1 \in I;$$
$$[g_{i_2}, [f, [g_{i_1}, f]]] = [\tau_{i_2}, \tau_{i_1}], \ i_1, i_2 \in I;$$
$$\vdots$$
$$[g_{i_{k*}}, [f, \ldots, [g_{i_2}, [f, [g_{i_1}, f]]] \ldots]] = [\tau_{i_{k*}}, \ldots, [\tau_{i_2}, \tau_{i_1}] \ldots],$$
$$i_k \in I, 1 \le k \le k^*,$$

hold, where k^* denotes the nonholonomy degree. Let

$$\mathcal{G} = \text{span}\{g_i, i \in I\},$$
$$\mathcal{H} = \text{span}\{[g_{i_1}, f], \ldots, [g_{i_{k*}}, [f, \ldots, [g_{i_2}, [f, [g_{i_1}, f]]] \ldots]]; i_k \in I, 1 \le k \le k^*\}.$$

Note that $\dim \mathcal{G}(0) = n - m$ and $\dim \mathcal{H}(0) = n$, since the distribution defined by the constraints is completely nonholonomic; moreover,

$$\dim\{\mathcal{G}(0) \cap \mathcal{H}(0)\} = 0.$$

It follows that the strong accessibility distribution

$$\mathcal{L}_0 = \text{span}\{X \ : \ X \in \mathcal{G} \cup \mathcal{H}\}$$

has dimension $2n - m$ at the origin. Hence the strong accessibility rank condition (Sussmann and Jurdjevic [1972]) is satisfied at the origin. Thus system (6.5.7), (6.5.8), (6.5.9) is strongly accessible at the origin. Hence the nonholonomic control system (6.4.1), (6.4.2) is strongly accessible at $(q^e, 0)$. ∎

6.6.2 Theorem. *Let $m \ge 1$ and let $(q^e, 0)$ denote an equilibrium solution in* **M**. *The nonholonomic control system defined by equations (6.4.1) and (6.4.2) is small-time locally controllable at $(q^e, 0)$.*

Proof. It suffices to prove that the system (6.5.7), (6.5.8), (6.5.9) is small-time locally controllable at the origin.

The proof involves the notion of the degree of a bracket. To make this notion well-defined we consider, as in Sussmann [1987], a Lie algebra of indeterminates and an associated evaluation map (on vector fields) as follows:

Let $\mathbf{X} = (X_0, \ldots, X_{n-m})$ be a finite sequence of indeterminates. Let $A(\mathbf{X})$ denote the free associative algebra over $\mathbb{R}$ generated by the X_j, let $L(\mathbf{X})$ denote the Lie subalgebra of $A(\mathbf{X})$ generated by $X_0, \ldots, X_{n-m}$, and let $\mathrm{Br}(\mathbf{X})$ be the smallest subset of $L(\mathbf{X})$ that contains $X_0, \ldots, X_{n-m}$ and is closed under bracketing.

Now consider the vector fields $f, g_1, \ldots, g_{n-m}$ on the manifold $\mathbf{M}$. Each $f, g_1, \ldots, g_{n-m}$ is a member of $D(\mathbf{M})$, the algebra of all partial differential operators on $C^\infty(\mathbf{M})$, the space of C^∞ real-valued functions on $\mathbf{M}$. Now let $g_0 = f$, and let $\mathbf{g} = (g_0, \ldots, g_{n-m})$ and define the evaluation map

$$\mathrm{Ev}(\mathbf{g}) : A(\mathbf{X}) \to D(\mathbf{M})$$

obtained by substituting the g_j for the X_j, i.e.,

$$\mathrm{Ev}(\mathbf{g})\left(\sum_I a_I X_I\right) = \sum_I a_I g_I,$$

where $g_I = g_{i_1} g_{i_2} \cdots g_{i_k}$, $I = (i_1, \ldots, i_k)$. Note that the kernel of $\mathrm{Ev}(\mathbf{g})$: $A(\mathbf{X}) \to A(\mathbf{g})$ is the set of all algebraic identities satisfied by the g_i, while the kernel of $\mathrm{Ev}(\mathbf{g})$: $L(\mathbf{X}) \to L(\mathbf{g})$ is the set of Lie-algebraic identities satisfied by g_i.

Now let B be a bracket in $\mathrm{Br}(\mathbf{X})$. We define the ***degree of a bracket*** to be

$$\delta(B) = \sum_{i=0}^{n-m} \delta^i(B),$$

where $\delta^0(B)$, $\delta^1(B), \ldots, \delta^{n-m}(B)$ denote the number of times $X_0, \ldots, X_{n-m}$, respectively, occur in B. The bracket B is called "bad" if $\delta^0(B)$ is odd and $\delta^i(B)$ is even for each $i = 1, \ldots, n - m$. The theorem of Sussmann tells us that the system is small-time locally controllable (STLC) at the origin if it satisfies the accessibility rank condition; and if B is "bad," there exist brackets $C_1, \ldots, C_k$ of lower degree in $\mathrm{Br}(\mathbf{X})$ such that

$$\mathrm{Ev}_0(\mathbf{g})(\beta(B)) = \sum_{i=1}^{k} \xi_i \, \mathrm{Ev}_0(\mathbf{g})(C_i),$$

where Ev_0 denotes the evaluation map at the origin and $(\xi_1, \ldots, \xi_k) \in \mathbb{R}^k$. Here $\beta(B)$ is the symmetrization operator,

$$\beta(B) = \sum_{\pi \in S_{n-m}} \bar{\pi}(B),$$

where $\pi \in S_{n-m}$, the group of permutations of $\{1, \ldots, n-m\}$, and for $\pi \in S_{n-m}$, $\bar{\pi}$ is the automorphism of $L(\mathbf{X})$ that fixes X_0 and sends X_i to $X_{\pi(i)}$.

By (6.6.1), the system is accessible at the origin.

The brackets in $\mathcal{G}$ are obviously "good" (not of the type defined as "bad"), and

$$\delta^0(h) = \sum_{j=1}^{n-m} \delta^j(h) \qquad \forall h \in \mathcal{H};$$

thus $\delta(h)$ is even for all h in $\mathcal{H}$, i.e., $\mathcal{H}$ contains "good" brackets only. It follows that the tangent space $T_0\mathbf{M}$ to $\mathbf{M}$ at the origin is spanned by the brackets that are all "good." Next we show that the brackets that might be "bad" vanish at the origin. First note that f vanishes at the origin. Let B denote a bracket satisfying $\delta(B) > 1$. If B is a "bad" bracket, then necessarily

$$\delta^0(B) \neq \sum_{j=1}^{n-m} \delta^j(B);$$

that is, $\delta(B)$ must be odd. It can be verified that if

$$\delta^0(B) < \sum_{j=1}^{n-m} \delta^j(B),$$

then B is identically zero, and if

$$\delta^0(B) > \sum_{j=1}^{n-m} \delta^j(B),$$

then B is of the form

$$\sum_{i=1}^{n-m} r_i(x_3) Y_i(x_1, x_2),$$

for some vector fields $Y_i(x_1, x_2)$, $i \in I$, where $r_i(x_3)$, $i \in I$, are homogeneous functions of degree

$$\left(\delta^0(B) - \sum_{j=1}^{n-m} \delta^j(B)\right) \text{ in } x_3;$$

thus B vanishes at the origin. Thus the Sussmann condition is satisfied, and the system is small-time locally controllable. ∎

We note that "good" and "bad" brackets and associated symmetric products play a key role in the configuration controllability analysis of Lewis and Murray [1999]. See also Bullo [2001] and Shen [2002].

Piecewise Analytic Stabilizing Controllers. We consider now stabilization for so-called controlled nonholonomic Chaplygin systems.

We first describe the class of controlled nonholonomic Chaplygin systems. If the functions used in defining equations (6.4.1), (6.4.2) do not depend explicitly on the configuration variables s, so that the system is locally described by

$$g_{ij}(r)\ddot{q}^j + f_i(r,\dot{q}) = \sum_{j=1}^{m} \lambda_j a_i^j(r) + \sum_{j=1}^{l} b_i^j(r)u_j, \qquad (6.6.1)$$

$$\dot{s}^a + A_\alpha^a(r)\dot{r}^\alpha = 0, \qquad a = 1,\ldots,m, \qquad (6.6.2)$$

then the uncontrolled system is called a *nonholonomic Chaplygin system* (see Chapter 5 and Neimark and Fufaev [1972]). In terms of the Lagrangian formalism for the problem, this corresponds to the Lagrangian of the free problem being cyclic in (i.e., independent of) the variables s, while the constraints are also independent of s. The cyclic property is an expression of symmetries in the problem, as we have discussed. More generally, if a system can be expressed in the form (6.6.1), (6.6.2) using feedback, then we refer to it as a **controlled nonholonomic Chaplygin system**.

For the nonholonomic Chaplygin system described by equations (6.6.1), (6.6.2), equation (6.5.5) becomes

$$C_\alpha^{iT}(r)g_{ij}(r)C_\beta^j(r)\ddot{r}^\beta = C_\alpha^{iT}(r)\left[b_i^j(r)u_j - f_i(r,C\dot{r}) - g_{ij}(r)\dot{C}_\alpha^j(r)\dot{r}^\alpha\right], \qquad (6.6.3)$$

which is an equation in the variables $(r,\dot{r})$ only. As a consequence, the r variables coordinatize a reduced configuration space for the system (6.6.1), (6.6.2). This reduced configuration space is also referred to as the **base space** (or **shape space**) of the system. The term shape space arises from the theory of coupled mechanical systems, where it refers to the internal degrees of freedom of the system.

As we did earlier, we assume that $r = n - m$ and that the matrix product $C_\alpha^{iT}(q)b_i^j(q)$ is locally invertible; this assumption is not restrictive. Thus it can be shown that the normal form equations for the system (6.6.1), (6.6.2) are given by

$$\dot{x}_1 = x_3, \qquad (6.6.4)$$

$$\dot{x}_2 = -A(x_1)x_3, \qquad (6.6.5)$$

$$\dot{x}_3 = v, \qquad (6.6.6)$$

where $x_1 = r$, $x_2 = \dot{r}$, $x_3 = s$, and v satisfies

$$C_\alpha^{iT}(r)g_{ij}(r)C_\beta^j(r)v^\beta = C_\alpha^{iT}(r)\left[b_i^j(r)u_j - f_i(r,C\dot{r}) - g_{ij}(r)\dot{C}_\alpha^j(r)\dot{r}^\alpha\right]. \qquad (6.6.7)$$

Again we shall make use of these normal form equations to obtain control results.

Relation to Geometric Phases. Clearly, there is no continuous feedback that asymptotically stabilizes a single equilibrium. However, the controllability properties possessed by the system guarantee the existence of a piecewise analytic feedback (see Sussmann [1979]). We now describe the ideas that are employed to construct such a feedback that achieves the desired local asymptotic stabilization of a single equilibrium solution. These ideas are based on the use of holonomy (geometric phase), which has proved useful in a variety of kinematics and dynamics problems (see Section 3.14 and, e.g., Krishnaprasad and Yang [1991], Shapere and Wilczek [1988], and Marsden, Montgomery, and Ratiu [1990]. The key observation here is that the holonomy, the extent to which a closed path in the base space fails to be closed in the configuration space, depends only on the path traversed in the base space and not on the time history of traversal of the path. Related ideas have been used for a class of path planning problems, based on kinematic relations, in Li and Canny [1990], Li and Montgomery [1988], and Krishnaprasad and Yang [1991].

For simplicity, we consider control strategies that transfer any initial configuration and velocity (sufficiently close to the origin) to the zero configuration with zero velocity. The proposed control strategy initially transfers the given initial configuration and velocity to the origin of the $(q_1, \dot{q}_1)$ base phase space. The main point then is to determine a closed path in the q_1 base space that achieves the desired holonomy. We show that the indicated assumptions guarantee that this holonomy construction can be made and that (necessarily piecewise analytic) feedback can be determined that accomplishes the desired control objective.

Let $x^0 = \left(x_1^0, x_2^0, x_3^0\right)$ denote an initial state. We now describe two steps involved in construction of a control strategy that transfers the initial state to the origin.

Step 1: Bring the system to the origin of the (x_1, x_3) base phase space; i.e., find a control that transfers the initial state $\left(x_1^0, x_2^0, x_3^0\right)$ to $\left(0, x_2^T, 0\right)$ in a finite time, for some x_2^T.

Step 2: Traverse a closed path (or a series of closed paths) in the x_1 base space to produce a desired holonomy in the (x_1, x_2) configuration space; i.e., find a control that transfers $\left(0, x_2^T, 0\right)$ to $(0, 0, 0)$.

The desired holonomy condition is given by

$$x_2^T = \oint_\gamma A(x_1)\, dx_1, \tag{6.6.8}$$

where γ denotes a closed path traversed in the base space. The holonomy is reflected in the fact that traversing a closed path in the base space yields a nonclosed path in the full configuration space. Note that here, for notational simplicity in presenting the main idea, we assume that the desired holonomy can be obtained by a single closed path. In general, more than one loop may be required to produce the desired holonomy; for such cases γ can be viewed as a concatenation of a series of closed paths.

Under the weak assumptions mentioned previously, explicit procedures can be given for each of the above two steps. Step 1 is classical; it is step 2, involving the holonomy, that requires special consideration. Explicit characterization of a closed path γ that satisfies the desired holonomy equation (6.6.8) can be given for several specific examples (see below). However, some problems may require a general computational approach. An algorithm based on Lie-algebraic methods as in Lafferiere and Sussman [1991] can be employed to approximately characterize the required closed path. Suppose the closed path γ that satisfies the desired holonomy condition is chosen. Then a feedback algorithm that realizes the closed path in the base space can be constructed, since the base space equations represent $n - m$ decoupled double integrators.

This general construction procedure provides a strategy for transferring an arbitrary initial state of equations (6.6.4), (6.6.5), (6.6.6) to the origin. Implementation of this control strategy in a (necessarily piecewise analytic) feedback form can be accomplished as follows.

Let $a = (a_1, \ldots, a_{n-m})$ and $b = (b_1, \ldots, b_{n-m})$ denote displacement vectors in the x_1 base space and let $\gamma(a, b)$ denote the closed path (in the base space) formed by the line segments from $x_1 = 0$ to $x_1 = a$, from $x_1 = a$ to $x_1 = a + b$, from $x_1 = a + b$ to $x_1 = b$, and from $x_1 = b$ to $x_1 = 0$. Then the holonomy of the parametrized family

$$\{\gamma(a, b) | a, b \in \mathbb{R}^{n-m}\}$$

is determined by the holonomy function $\gamma(a, b) \rightarrow \alpha(a, b)$ given as

$$\alpha(a, b) = -\oint_{\gamma(a,b)} A(x_1) \, dx_1.$$

Now let π_s denote the projection map $\pi_s : (x_1, x_2, x_3) \rightarrow (x_1, x_3)$. In order to construct a feedback control algorithm to accomplish the above two steps, we first define a feedback function $V^{x_1^*}(\pi_s x)$ that satisfies the following condition: For any $\pi_s x(t_0)$ there is $t_1 \geq t_0$ such that the unique solution of

$$\dot{x}_1 = x_3, \qquad \dot{x}_3 = V^{x_1^*}(\pi_s x), \qquad (6.6.9)$$

satisfies $\pi_s x(t_1) = (x_1^*, 0)$. Note that the feedback function is parametrized by the vector x_1^*. Moreover, for each x_1^*, there exists such a feedback function. One such feedback function $V^{x_1^*}(\pi_s x) = (V_1^{x_1^*}(\pi_s x), \ldots, V_{n-m}^{x_1^*}(\pi_s x))$ is given as

$$V_i^{x^*}(\pi_s x) = \begin{cases} -k_i \, \text{sign}(x_{1,i} - x_{1,i}^* + x_{3,i} |x_{3,i}| / 2k_i), & (x_{1,i}, x_{3,i}) \neq (x_{1,i}^*, 0), \\ 0, & (x_{1,i}, x_{3,i}) = (x_{1,i}^*, 0), \end{cases}$$

where k_i, $i = 1, \ldots, n - m$, are positive constants chosen such that the resulting motion, when projected to the base space, constitutes a straight line connecting $x_1(t_0)$ to $x_1(t_1) = x_1^*$.

We specify the control algorithm, with values denoted by v^*, according to the following construction, where x denotes the "current state":

Control algorithm for v^*.

Step 0: Choose (a^*, b^*) to achieve the desired holonomy.

Step 1: Set $v^* = V^{a^*}(\pi_s x)$, until $\pi_s x = (a^*, 0)$; then go to Step 2;

Step 2: Set $v^* = V^{a^*+b^*}(\pi_s x)$, until $\pi_s x = (a^* + b^*, 0)$; then go to Step 3;

Step 3: Set $v^* = V^{b^*}(\pi_s x)$, until $\pi_s x = (b^*, 0)$; then go to Step 4;

Step 4: Set $v^* = V^0(\pi_s x)$, until $\pi_s x = (0, 0)$; then go to Step 0;

We here assumed that the desired holonomy can be obtained by a single closed path. Clearly, the above algorithm can be modified to account for cases for which more than one closed path is required to satisfy the desired holonomy. The algorithm is illustrated in the example below.

Note that the control algorithm is constructed by appropriate switchings between members of the parametrized family of feedback functions. On each cycle of the algorithm the particular functions selected depend on the closed path parameters a^*, b^*, computed in Step 0, to correct for errors in x_2.

The control algorithm can be initialized in different ways. The most natural is to begin with Step 4, since v^* in that step does not depend on the closed path parameters; however, many other initializations of the control algorithm are possible.

Justification that the constructed control algorithm asymptotically stabilizes the origin follows as a consequence of the construction procedure: that switching between feedback functions guarantees that the proper closed path (or a sequence of closed paths) is traversed in the base space so that the origin $(0, 0, 0)$ is necessarily reached in a finite time. This construction of a stabilizing feedback algorithm represents an alternative to the approach by Hermes [1980], which is based on Lie-algebraic properties.

It is important to emphasize that the above construction is based on the a priori selection of simply parametrized closed paths in the base space. The above selection simplifies the tracking problem in the base space, but other path selections could be made, and they would, of course, lead to a different feedback strategy from that proposed above.

We remark that the technique presented in this section can be generalized to some systems that are not Chaplygin. For instance, this generalization is tractable to systems for which equation (20) takes the form

$$\dot{x}_2 = \rho(x_2)A(x_1),$$

where $\rho(x_2)$ denotes certain Lie group representation (see, for example, Marsden, Montgomery, and Ratiu [1990]). The holonomy of a closed path for such systems is given as a path-ordered exponential rather than a path integral.

6.6.3 Example (Control of a Rolling Wheel or Disk). Consider again the control of a vertical wheel rolling without slipping on a plane surface. As before, let x and y denote the coordinates of the point of contact of the wheel on the plane, let φ denote the heading angle of the wheel, measured from the x-axis, and let θ denote the rotation angle of the wheel due to rolling, measured from a fixed reference. Then the equations of motion, with all numerical constants set to unity, are given by

$$
\begin{aligned}
\ddot{x} &= \lambda_1, \\
\ddot{y} &= \lambda_2, \\
\ddot{\theta} &= -\lambda_1 \cos \varphi - \lambda_2 \sin \varphi + u_1, \\
\ddot{\varphi} &= u_2,
\end{aligned}
\tag{6.6.10}
$$

where u_1 denotes the control torque about the rolling axis of the wheel and u_2 denotes the control torque about the vertical axis through the point of contact; the components of the force of constraint arise from the two nonholonomic constraints

$$
\dot{x} = \dot{\theta} \cos \varphi, \quad \dot{y} = \dot{\theta} \sin \varphi,
\tag{6.6.11}
$$

which have nonholonomy degree three at any configuration. The constraint manifold is a six-dimensional manifold and is given by

$$
\mathbf{M} = \left\{ (\theta, \varphi, x, y, \dot{\theta}, \dot{\varphi}, \dot{x}, \dot{y}) \mid \dot{x} = \dot{\theta} \cos \varphi, \ \dot{y} = \dot{\theta} \sin \varphi \right\},
$$

and any configuration is an equilibrium if the controls are zero.

Define the variables

$$
z_1 = \theta, \quad z_2 = \varphi, \quad z_3 = x, \quad z_4 = y, \quad z_5 = \dot{\theta}, \quad z_6 = \dot{\varphi},
$$

so that the reduced differential equations are given by

$$
\begin{aligned}
\dot{z}_1 &= z_5, \\
\dot{z}_2 &= z_6, \\
\dot{z}_3 &= z_5 \cos z_2, \\
\dot{z}_4 &= z_5 \sin z_2, \\
\dot{z}_5 &= \frac{1}{2} u_1, \\
\dot{z}_6 &= u_2.
\end{aligned}
\tag{6.6.12}
$$

Then we have the following result:

Let $z^e = (z_1^e, z_2^e, z_3^e, z_4^e, 0, 0)$ denote an equilibrium solution of the reduced differential equations corresponding to $u = 0$. The rolling wheel dynamics have the following properties:

1. There is a smooth feedback that asymptotically stabilizes the closed loop to any smooth two-dimensional equilibrium manifold in **M** that satisfies the transversality condition.

2. There is no smooth feedback that asymptotically stabilizes z^e.

3. The system is strongly accessible at z^e, since the space spanned by the vectors

$$g_1, \ g_2, \ [g_1, f], \ [g_2, f], \ [g_2, [f, [g_1, f]]], \ [g_2, [f, [g_1, [f, [g_2, f]]]]]$$

   has dimension 6 at z^e.

4. The system is small-time locally controllable at x^e, since the brackets satisfy sufficient conditions for small-time local controllability.

Note here that the base variables are (z_1, z_2). Consider a parametrized rectangular closed path γ in the base space with four corner points of the form

$$(0, 0), \ (z_1, 0), \ (z_1, z_2), \ (0, z_2);$$

i.e., $a = (z_1, 0)$ and $b = (0, z_2)$, following the notation introduced above. By evaluating the holonomy integral in closed form for this case, we find that the holonomy equations are

$$z_3^T = z_1(\cos z_2 - 1),$$

$$z_4^T = z_1 \sin z_2.$$

These equations can be explicitly solved (inverted) to determine a closed path (or a concatenation of closed paths) γ^* that achieves the desired holonomy. One solution can be given as follows: If $z_3^T \neq 0$, then γ^* is the closed path specified by

$$a^* = -\left(\left(\left(z_3^T \right)^2 + \left(x_4^T \right)^2 \right) / 2z_3^T, 0 \right),$$

$$b^* = \left(0, -\sin^{-1} \left(2z_3^T z_4^T / \left(\left(z_3^T \right)^2 + \left(z_4^T \right)^2 \right) \right) \right),$$

and if $x_3^T = 0$, then γ^* is a concatenation of two closed paths specified by

$$a^* = \left(0.5 z_4^T, 0 \right), \qquad b^* = (0, 0.5\pi),$$

$$a^{**} = \left(-0.5 z_4^T, 0 \right), \quad b^{**} = (0, -0.5\pi).$$

Note that the previously described feedback algorithm can be used (with the modification indicated in the general development) to asymptotically stabilize the rolling wheel to the origin.   ♦

In the web supplement a similar stabilization argument is also applied to a knife edge.

Exercises

⋄ **6.6-1.** Verify the accessibility and small-time local controllability results for the vertical penny system.

⋄ **6.6-2.** Verify accesibility and small-time local controllability at the equilibrium $x^e = (x_1^e, x_2^e, x_3^e, 0, 0)$ for the system on $\mathbb{R}^5$ given by

$$\dot{x}_1 = x_2,$$
$$\dot{x}_2 = x_5,$$
$$\dot{x}_3 = x_1 x_5, \qquad\qquad (6.6.13)$$
$$\dot{x}_4 = u_1 + u_2 x_3 - x_1 x_5^2,$$
$$\dot{x}_5 = u_2.$$

⋄ **6.6-3.** Analyze the geometric phase problem for the controlled rigid body with two torques (see Chapter 1 for the equations with one torque). This provides a form of attitude control. For the geometric phase for the free rigid body see Montgomery [1991b] and Marsden and Ratiu [1999]. For the controlled case see Bloch, Leonard, and Marsden [2001] and references therein.

6.7   Nonholonomic Systems on Riemannian Manifolds

We consider here the formulation of controlled classical nonholonomic systems on Riemannian manifolds. Such a formulation is useful for considering systems with nontrivial geometry, such as the rolling ball, and we follow the work of Bloch and Crouch [1995] here. We contrast here kinematic and dynamic control systems in this setting, discuss conservation laws and reduction in the presence of controls, and use the bundle structure to get a nontrivial control result.

Holonomic Control Systems.   First, consider the holonomic or unconstrained case:

Let $(Q, \langle, \rangle)$ be an n-dimensional Riemannian manifold, with metric $g(\ ,\ )$ $= \langle\ ,\ \rangle$. Denote the norm of a tangent vector X at the point p by $\|X_p\| = \langle X_p, X_p \rangle^{\frac{1}{2}}$. The geodesic flow on Q is then given by

$$\frac{D^2 q}{dt^2} = 0, \qquad\qquad (6.7.1)$$

where $\frac{Dq}{dt}$ denotes the covariant derivative. This flow minimizes the integral $\int_0^1 \left\| \frac{Dq}{dt} \right\|^2 dt$ along parametrized paths.

We define a *controlled holonomic system* to be a system of the form

$$\frac{D^2 q}{dt^2} = \sum_{i=1}^{N} u_i X_i, \tag{6.7.2}$$

where $\{X_i\}$ is an arbitrary set of control vector fields, the u_i are functions of time, and $N \le n$. (Note that here we do not consider systems evolving under the influence of a potential, but the analysis is easily extended to include a potential.) Such systems are sometimes now called *affine connection control systems* (see Bullo [2002]).

Equilibrium Controllability.   This Riemannian setup is useful for analyzing controllability concepts for mechanical systems, as shown in the work of Lewis [1995] and Lewis and Murray [1999] (see also Murray [1995] and Shen [2002], Shen, McClamroch, and Bloch [2004]).

6.7.1 Definition.   *The system* (6.7.2) *is said to be **equilibrium controllable** if for any two equilibrium points q_0 and q_1 there exist a time T and set of controls u_i that drive the system from q_0 to q_1 in time T.*

We can also define a configuration notion of a reachable set (Lewis and Murray [1999]) :

6.7.2 Definition.   *Given $q_0 \in Q$ we define $R^Q(q_0, t)$ to be the set of all $q \in Q$ for which there exists an admissible control u such that there is a trajectory of* (6.7.2) *with $q(0) = q_0$, $q(t) = q$ with arbitrary velocity. The **configuration reachable** set from q_0 at time T is defined to be*

$$R_T^Q(q_0) = \bigcup_{0 \le t \le T} R^Q(q_0, t). \tag{6.7.3}$$

6.7.3 Definition.   *The control system* (6.7.2) *is **configuration accessible** if there is exists a time T such that R_T^Q contains an open subset of Q.*

Lewis and Murray [1999] show that this kind of controllability may be analyzed using the *symmetric product* (see also Crouch [1981]) on vector fields X, Y given by

$$\langle X : Y \rangle = \nabla_X Y + \nabla_Y X. \tag{6.7.4}$$

Defining the sequence of vector fields

$$\mathcal{G}^{(1)} = \{X_1, \ldots, X_m\},$$

$$\mathcal{G}^{(i)} = \{\nabla_Y X + \nabla_X Y : X \in \mathcal{G}^{(j)}, Y \in \mathcal{G}^{(k)}, i = j + k\},$$

$$\mathcal{G}^{(\infty)} = \bigcup_{i=1}^{\infty} \mathcal{G}^{(i)},$$

we can establish (see Lewis and Murray [1999]) the following theorem:

6.7.4 Theorem. *If the involutive closure of the vector fields $\mathcal{G}^{(\infty)}$ spans $T_q Q$ for each $q \in Q$, then the system (6.7.2) is locally configuration accessible at each $q \in Q$.*

Local equilibrium controllability can be analyzed using a symmetric version of the "good" and "bad" brackets of Sussmann [1987] that we discussed in Section 6.6. A symmetric product in $\mathcal{G}^{(\infty)}$ is said to be "bad" if it contains an even number of copies of X_i for each $i = 1, \ldots, m$. Otherwise, we say that it is "good." Then Lewis and Murray [1999] prove the following result:

6.7.5 Theorem. *If every bad symmetric product in $\mathcal{G}^{(\infty)}$ evaluated at an equilibrium point can be written as a linear combination of good symmetric products of lower order, then the system (6.7.2) is locally equilibrium controllable.*

See the exercises below for an example.

Nonholonomic Systems.   We now consider the formulation of controlled nonholonomic systems in this Riemannian setting. The global geometric approach here turns out be useful for formulating certain control results and for understanding symmetries in the control setting.

Classical nonholonomic systems are obtained from Lagrange–d'Alembert's principle, as discussed in Section 5.2.

The equations are

$$\frac{D^2 q}{dt^2} = \sum_{k=1}^{m} \lambda_i W_i, \tag{6.7.5}$$

subject to

$$\omega_k \left(\frac{Dq}{dt} \right) = \left\langle W_k, \frac{Dq}{dt} \right\rangle = 0, \qquad 1 \leq k \leq m,$$

where $\omega_k(X) = \langle W_k, X \rangle$ and the λ_i are Lagrange multipliers. The constraints are given by the 1-forms ω_k, $1 \leq k \leq m$, which define a (smooth) distribution H on Q.

We now define a **controlled nonholonomic mechanical system** to be a system of the form

$$\frac{D^2 q}{dt^2} = \sum_{i=1}^{m} \lambda_i W_i + \sum_{i=1}^{N} u_i X_i \tag{6.7.6}$$

subject to

$$\left\langle W_k, \frac{Dq}{dt} \right\rangle = 0 \qquad 1 \leq k \leq m, \tag{6.7.7}$$

where the $u_i(t)$ are controls and the X_i are arbitrary smooth (control) vector fields. In fact, the X_i are not as arbitrary as they appear, as the remark below shows.

Remarks.

1. We now consider which general force fields F are compatible with the nonholonomic constraints, i.e., which $F(q, \dot{q})$ are allowed in the system

$$\frac{D^2 q}{dt^2} = F$$

subject to

$$\left\langle W_k, \frac{Dq}{dt} \right\rangle = 0, \qquad 1 \leq k \leq m. \tag{6.7.8}$$

Differentiating the constraints gives

$$\langle W_k, F \rangle + \left\langle \frac{DW_k}{dt}, \frac{Dq}{dt} \right\rangle = 0, \qquad 1 \leq k \leq m$$

or, if ∇ is the Riemannian connection on $(Q, \langle, \rangle)$,

$$\langle W_k, F \rangle + \langle \nabla_{\dot{q}} W_k, \dot{q} \rangle = 0, \qquad 1 \leq k \leq m. \tag{6.7.9}$$

These are the conditions that the force field must satisfy to be compatible with the constraints. The Lagrange multipliers ensure that the forces satisfy the above relations. This also shows that the number of independent force fields is less than or equal to $n - m$.

2. We define the kinematic system corresponding to the dynamic system (6.7.6), (6.7.7) to be

$$\frac{Dq}{dt} = \sum_{i=1}^{N} u_i \widehat{X}_i, \qquad \widehat{X}_i \in H.$$

For controllability analysis, the assumption is made that H is completely nonholonomic (see Vershik and Gershkovich [1988]). We analyze such systems elsewhere.

3. We note that while the nonholonomic system under discussion is not variational in the Lagrangian sense, it can be seen as the solution of an instantaneous variational problem,

$$\min_{\frac{D^2 q}{dt^2}} \frac{1}{2} \left\| \frac{D^2 q}{dt^2} - \sum_{i=1}^{N} u_i X_i(q) \right\|$$

subject to

$$\left\langle W_k, \frac{Dq}{dt} \right\rangle = 0, \qquad 1 \leq k \leq m. \tag{6.7.10}$$

This yields the nonholonomic equations as well as the following equations for λ_k:

$$\sum_{j=1}^{m}\langle W_k, W_j\rangle\lambda_j = -\left\langle \frac{DW_k}{dt}, \frac{Dq}{dt}\right\rangle - \sum_{i=1}^{N}u_i\langle W_k, X_i\rangle, \quad 1 \le k \le m.$$

(6.7.11)

This approach to obtaining the nonholonomic equations of motion is called **Gauss's principle of least constraint** (see Gauss [1829], Gibbs [1879], and, e.g., Lewis [1996]).   ◆

Symmetries and Conservation Laws. Symmetries in mechanics give rise to constants of the motion, and their role in the nonholonomic setting is particularly interesting, as we have seen in Chapter 5. Here we consider symmetries in the Riemannian setting and in the presence of controls. We obtain a result that is a slight generalization of that in Arnold, Kozlov, and Neishtadt [1988] and that extends to the control setting some of the momentum equation analysis in Chapter 5.

In the Riemannian context isometries (which preserve the metric) are generated by Killing vector fields. (Z is a Killing vector field if $\langle\nabla_Y Z, Y\rangle = 0$ for all vector fields Y; see, for example, Crouch [1981].) Further, a sufficient condition for $\left\langle Z, \frac{Dq}{dt}\right\rangle$ to be a constant of motion for the geodesic flow is that Z be a Killing vector field. For controlled nonholonomic systems we have the following:

6.7.6 Theorem. *Sufficient conditions for* $\left\langle Z, \frac{Dq}{dt}\right\rangle$ *to be a constant of motion for the controlled nonholonomic system* (6.7.6) *are:*

(i) $Z \in H$;

(ii) $Z_q \in \mathrm{Span}\{X_1, \ldots, X_N\}^{\perp}$;

(iii) Z *is a Killing vector field.*

Proof.

$$\frac{d}{dt}\left\langle Z, \frac{Dq}{dt}\right\rangle = \left\langle \frac{DZ}{dt}, \frac{Dq}{dt}\right\rangle + \left\langle Z, \frac{D^2q}{dt^2}\right\rangle.$$

The first term is zero by (iii), and the second is zero by (6.7.6), (i), and (ii).

Note that when $Q = \mathbb{R}^n$ and the metric g is independent of q_i, then $\partial/\partial q_i$ is a Killing vector field.   ∎

Reduction for Nonholonomic Systems on Riemannian Manifolds. We discuss here an approach to reduction for nonholonomic control systems; this just extends to this control setting the analysis in Chapter 5.

We introduce a bundle structure in Q,

$$
\begin{array}{c}
Q \\
\downarrow \pi \\
B
\end{array}
$$

with fiber F, $\dim B = r$, and $\dim F = n - r$. This structure must be compatible with the constraints in the sense that $\pi_* H_q = T_{\pi(q)} B$, $\forall q \in Q$ (thus $\dim H = n - m \geq \dim B = r$). Our aim is to reduce the dynamical system (6.7.6), (6.7.7) so that evolution on the fiber is given by a first-order equation. To do this we introduce a further assumption, namely, that one of the followig holds:

(1) $\dim H = \dim B$, i.e., $n = m + r$. In this case $\widehat{H} \equiv H$ clearly defines a horizontal distribution on the bundle.

(2) $\dim H - \dim B = n - m - r = s > 0$, and there exist s linearly independent vector fields $Z_1, \ldots, Z_s$ that satisfy conditions (i)–(iii) of Theorem 6.7.6. In particular,

$$
\left\langle Z_i, \frac{Dq}{dt} \right\rangle = c_i \tag{6.7.12}
$$

are constants of the motion for (6.7.6), (6.7.7).

We define a distribution $\widehat{H}_0$ on Q by setting

$$
\begin{aligned}
X \in \widehat{H}_0 \quad \text{if} \quad \langle W_k, X \rangle &= 0, & 1 \leq k \leq m, \\
\langle Z_k, X \rangle &= 0, & 1 \leq k \leq s.
\end{aligned} \tag{6.7.13}
$$

Condition (i) of Lemma 1 ensures that $\widehat{H}_0$ is r-dimensional. We also require that $\widehat{H}_0$ define a horizontal distribution of the bundle, i.e.,

$$
T_q F \cap \widehat{H}_0 = \{0\} \quad \forall q \in Q. \tag{6.7.14}
$$

We further define the r-dimensional affine connection $\widehat{H}$ on Q by setting

$$
\begin{aligned}
q \in \widehat{H} \quad \text{if} \quad \langle W_k, X \rangle &= 0, & 1 \leq k \leq m \\
\langle Z_k, X \rangle &= c_k & 1 \leq k \leq s.
\end{aligned} \tag{6.7.15}
$$

In either case (1) or (2) we have, as a direct sum of affine subspaces, $\widehat{H}_q \oplus T_q F = T_q Q$, and so any vector field on Q can be decomposed uniquely into components $Y_q = Y_q^H + Y_q^F$ with $Y_q^H \in \widehat{H}_q$ and $Y_q^F \in F_q$. With this structure we can now decompose the velocity:

$$
\frac{Dq}{dt} = \dot{q}^H + \dot{q}^F, \quad \dot{q}^H \in \widehat{H}_q, \quad \dot{q}^F \in T_q F. \tag{6.7.16}
$$

Using (6.7.7) we obtain

$$\langle W_k, \dot{q}^F \rangle = -\langle W_k, \dot{q}^H \rangle, \qquad 1 \le k \le m,$$
$$\langle Z_k, \dot{q}^F \rangle = \langle Z_k, \dot{q}^H \rangle + c_k, \quad 1 \le k \le s. \qquad (6.7.17)$$

Our rank conditions on Z_k and W_k allow us to solve these equations, giving

$$\dot{q}^F = f_F \left(q, \dot{q}^H \right). \qquad (6.7.18)$$

Equations (6.7.6) now define the second-order set of equations

$$\frac{D\dot{q}^H}{dt} = f_H(q, \dot{q}, u). \qquad (6.7.19)$$

We have now provided a reduction from the $2n$ first-order equations (6.7.6) to $n + r$ first-order equations.

Now locally we can write $\dot{q}^H = \frac{D}{dt} q^B$ for some trajectory $q^B(t) \in B$. In some cases we may be able to rewrite the equations globally in terms of a trajectory $q(t) = \left(q^B(t), q^F(t) \right)$, $q^B \in B$, $q^F \in F$.

The class of Chaplygin control systems introduced in Bloch, Reyhanoglu, and McClamroch [1992] and discussed above correspond to the case where $H = \widehat{H}$ and there exists a trivial bundle structure $Q = \mathbb{R}^n$ with the Euclidean structure. In this case, since $\partial/(\partial q_i^F)$ is Killing and symmetries correspond to invariance with respect to q_i^F, a global prescription can be given, in which the equations become

$$\dot{q}^F = f_F \left(q^B, \dot{q}^B \right),$$
$$\frac{D\dot{q}^B}{dt} = f_H \left(q^B, \dot{q}^B, u \right), \qquad (6.7.20)$$

and in fact, f_F is affine in $\dot{q}^B$.

In certain related cases, such as when Q is a principal bundle with its structure group G being a group of isometries and $H = \widehat{H}$ is G-invariant, we can also obtain a global reduction.

Another such case is the class of **controlled** nonabelian Chaplygin systems. The class of **nonabelian Chaplygin control systems** is defined as follows. We are again given a system of the form above, but with the following additional properties: The manifold Q is acted on by a Lie group G, and the bundle structure above is a principal bundle with fiber G. The group G acts by isometries on the metric on Q, and H is invariant under G in the sense that $H_{g \cdot q} = g_* H_q$. Finally, the control vector fields should also be invariant under G. (The uncontrolled class of these "nonabelian" Chaplygin systems was discussed in Chapter 5; see also Koiller [1992].)

Now, such control systems fall into a class of control systems on principal bundles whose controllability may be assessed by considering the projection of the system to the base. An analysis of this class of systems is due to San

Martin and Crouch [1984]. Precisely, consider an analytic nonlinear control system $\dot{x} = f(x, u)$ defined on a connected principal fiber bundle $Q(G, B)$ with base B and structure group G. Denote the set of vector fields $f(\cdot, u)$ on Q by D and suppose that D is projectable in the sense that for each $X \in D$ there exists a unique vector field X' on B satisfying $\pi_* X = X' \circ \pi$. Denote the projected set of vector fields by D'. Note that this is automatically satisfied for systems invariant under G and thus for the nonholonomic Chaplygin systems defined above. We then have the following:

6.7.7 Theorem (San Martin–Crouch). *Let $Q(B, G)$ be a connected principal fiber bundle with G a compact Lie group and let D be a G-invariant projectable family of vector fields on Q defining a control system Σ_D that is accessible. Denote by Σ'_D the system on B defined by $D' = \pi(D)$. Then Σ_D is controllable if and only if Σ'_D is controllable.*

6.7.8 Theorem. *An accessible nonabelian Chaplygin control system with compact structure group that is controllable on the base of its principal bundle is controllable.*

6.7.9 Example (**The Rolling Ball**). We consider again the controlled rolling ball on the plane discussed in Chapter 1. We now illustrate the ideas above in this specific physical setting.

Recall that we use the coordinates x, y for the linear horizontal displacement and $P \in \mathrm{SO}(3)$ for the angular displacement of the ball. Thus P gives the orientation of the ball with respect to inertial axes $\mathbf{e}_1$, $\mathbf{e}_2$, $\mathbf{e}_3$ fixed in the plane, where the $\mathbf{e}_i$ are the standard basis vectors aligned with the x-,y-, and z-axes respectively.

We let $\boldsymbol{\omega} \in \mathbb{R}^3$ denote the angular velocity of the ball with respect to inertial axes. In particular, the ball may spin freely about the z-axis, and the z-component of angular momentum is conserved. If J denotes the inertia tensor of the ball with respect to the body axes, then $\mathbb{J} = P^T J P$ denotes the inertia tensor of the ball with respect to the inertial axes, and $\mathbb{J}\boldsymbol{\omega}$ is the angular momentum of the ball with respect to the inertial axes. The conservation law alluded to above is expressed as

$$\mathbf{e}_3^T \mathbb{J} \boldsymbol{\omega} = c. \tag{6.7.21}$$

The nonholonomic constraints are expressed as

$$\begin{aligned} \mathbf{e}_2^T \boldsymbol{\omega} + \dot{x} &= 0, \\ \mathbf{e}_1^T \boldsymbol{\omega} - \dot{y} &= 0. \end{aligned} \tag{6.7.22}$$

The kinematics for the rotating ball we express as $\dot{P} = \hat{\Omega} P$, where $\Omega = P\boldsymbol{\omega}$ is the angular velocity in the body frame.

We now wish to write down equation (6.7.6) for this example using the global coordinates $q = (x, y, P)$. The metric on $Q = \mathrm{SO}(3) \times \mathbb{R}^2$ is defined by

$$\left\langle \left(\hat{\Omega}_1 P, \dot{x}_1, \dot{y}\right), \left(\hat{\Omega}_2 P, \dot{x}_2, \dot{y}_2\right)\right\rangle = \Omega_1^T J \Omega_2 + \dot{x}_1^T \dot{x}_2 + \dot{y}_1^T \dot{y}_2. \qquad (6.7.23)$$

An explicit expression for the Riemannian connection on SO(3) may be found by consulting Arnold [1989], p. 327. Equation (6.7.6) becomes

$$\hat{\Omega} P - (\widehat{J^{-1}\hat{\Omega}J\Omega})P = \lambda_1 (\widehat{J^{-1}P\mathbf{e}_1})P + \lambda_2(\widehat{J^{-1}P\mathbf{e}_2})P,$$
$$m\ddot{x} = \lambda_2 + u_1, \qquad (6.7.24)$$
$$m\ddot{y} = -\lambda_1 + u_2.$$

In inertial coordinates $\omega = P^T \Omega$, the system is

$$\dot{\omega} = \mathbf{J}^{-1}\hat{\omega}\mathbf{J}\omega + \lambda_1 \mathbf{J}^{-1}\mathbf{e}_1 + \lambda_2 \mathbf{J}^{-1}\mathbf{e}_2,$$
$$m\ddot{x} = \lambda_2 + u_1,$$
$$m\ddot{y} = -\lambda_1 + u_2, \qquad (6.7.25)$$
$$\dot{P} = P\hat{\omega}.$$

From the equations it is easy to see that indeed (6.7.21) is a constant of the motion. In fact Z, in formula (6.7.12) is given by $Z = (\hat{P\mathbf{e}_3})P$. It is easy to check conditions (i) and (ii) of Theorem 6.7.6, and with more effort, one can verify condition (iii).

In this example there are two admissible reductions, as described above. The first takes $B = $ SO(3) and the fiber $F = \mathbb{R}^2$, and the reduction of system (6.7.25) is obtained simply by substituting the two constraints for the second-order equations. This corresponds to the case $H = \hat{H}$ discussed above.

For the second reduction we take $B = \mathbb{R}^2$, $F = $ SO(3), and we now employ the constraints and the constant of motion to obtain the following expression for ω:

$$\omega = \dot{x}(\alpha_1 \mathbf{e}_3 - \mathbf{e}_2) + \dot{y}(\mathbf{e}_1 - \alpha_2 \mathbf{e}_3) + \alpha_3 \mathbf{e}_3,$$

where

$$\alpha_1 = \frac{\mathbf{e}_3^T J \mathbf{e}_1}{\mathbf{e}_3^T J \mathbf{e}_3}, \quad \alpha_2 = \frac{\mathbf{e}_3^T J \mathbf{e}_2}{\mathbf{e}_3^T J \mathbf{e}_3}, \quad \alpha_3 = \frac{c}{\mathbf{e}_3^T J \mathbf{e}_3}. \qquad (6.7.26)$$

The reduced equations become, after substituting for the multipliers,

$$m\ddot{x} = \lambda_2 + u_1,$$
$$m\ddot{y} = -\lambda_1 + u_2, \qquad (6.7.27)$$
$$\dot{P} = P(\dot{x}(\alpha_2 \mathbf{e}_3 - \mathbf{e}_2) + \widehat{\dot{y}(\mathbf{e}_1 - \alpha_1 \mathbf{e}_3)} + \alpha_3 \mathbf{e}_3).$$

In this case $\hat{H}$ is obtained through Assumption (2) above with $s = 1$. Note also that here we obtain seven first-order ODEs rather than the eight obtained from the previous reduction. ♦

Other aspects of control of nonholonomic systems including optimal control and stabilization by energy methods are described in the following chapters.

Exercises

◇ **6.7-1.** (See, e.g., Murray [1995].) Show that the system (a model of a controlled planar rigid body)

$$\dot{x} = \cos\theta u_1 - \sin\theta u_2,$$
$$\dot{y} = \sin\theta u_1 + \cos\theta u_2,$$
$$\dot{\theta} = u_1$$

is locally configuration accessible and equilibrium controllable.

◇ **6.7-2.** Verify condition (iii) in Theorem 6.7.6 for the rolling ball.

7
Optimal Control

Given a set of nonholonomic constraints, there are two interesting associated problems. One of these is nonvariational (namely, the Lagrange–d'Alembert principle) appropriate for the dynamics of constrained mechanical systems, which we studied extensively in Chapter 5, while the other is variational, which is appropriate for optimal control problems. In this chapter we concentrate on these optimal control problems.

This chapter gives a selection of techniques and results in optimal control theory that are optimization problems for mechanical systems, including nonholonomic systems. The topics treated here are a small and admittedly biased selection from a vast literature.[1] For related work of the authors on optimal control on Lie algebras, see the topic "Optimal Control on Lie Algebras and Adjoint Orbits" in the Internet Supplement section of this book's website.

7.1 Variational Nonholonomic Problems

Suppose a submanifold of the tangent bundle is given as the zero set of a set of constraints on the bundle. Suppose also that we are given a Lagrangian or, more generally, an objective function that we wish to minimize or maximize. Then we can proceed in the following two ways:

---

[1]See, for example, Pontryagin [1959], Sussmann [1998a,b], and references therein.

(1) We can consider the conditional variational problem of minimizing a functional subject to the trajectories lying in the given submanifold and obtain the Euler–Lagrange equations via the Lagrange method of appending the constraints to the Lagrangian via Lagrange multipliers.

(2) We can project, via a suitable projection, the vector field of the unconditional problem on the whole tangent bundle at every point to the tangent space of a given submanifold.

The vector fields arising from these two approaches will not, of course, coincide in general, even though both are tangent to the constraint submanifold. The two approaches were compared earlier in Sections 1.3 and 1.4, where in the latter section we compared the two types of dynamics for the vertical rolling disk. The first method gives us variational nonholonomic problems, while (real) nonholonomic mechanics are obtained by a procedure of the second type. In fact, this is implemented by the Lagrange–d'Alembert principle, as we have seen in Chapters 1 and 5. As we saw, nonholonomic mechanics is not variational, since while we allow all possible variations in taking the variations of the Lagrangian, the variations have to lie in the nonintegrable constraint distribution and are thus not independent of one another or reducible to constraints on the configuration variables.

The Lagrange Problem. Variational nonholonomic problems, on the other hand, are equivalent to the classical Lagrange problem of minimizing a functional over a class of curves with fixed extreme points and satisfying a given set of equalities.

More precisely, we have the following (see, e.g., Bloch and Crouch [1994]): Let Q be a smooth manifold and TQ its tangent bundle with coordinates $(q^i, \dot{q}^i)$. Let $L : TQ \to \mathbb{R}$ be a given smooth Lagrangian and let $\Phi : TQ \to \mathbb{R}^{n-m}$ be a given smooth function.

7.1.1 Definition. *The **Lagrange problem** is given by*

$$\min_{q(\cdot)} \int_0^T L(q, \dot{q})dt \qquad (7.1.1)$$

subject to the fixed endpoint conditions $q(0) = 0$, $q(T) = q_T$, and subject to the constraints

$$\Phi(q, \dot{q}) = 0.$$

7.1.2 Example (The Falling Cat Problem). The falling cat problem is an abstraction of the problem of how a falling cat should optimally (in some sense) move its body parts so that it achieves a 180° reorientation during its fall.

In this case we begin with a Riemannian manifold Q (the configuration space of the problem) with a free and proper isometric action of a Lie group G on Q (the group SO(3) for the falling cat). Let $\mathcal{A}$ denote the mechanical connection; that is, it is the principal connection whose horizontal space is the metric orthogonal to the group orbits (see Section 3.10). The quotient space $Q/G = X$, the shape space, inherits a Riemannian metric from that on Q. Given a curve $c(t)$ in Q, we shall denote the corresponding curve in the shape space X by $r(t)$.

The problem under consideration is as follows:

Isoholonomic Problem (Falling Cat problem). *Fixing two points $q_1, q_2 \in Q$, among all curves $q(t) \in Q$, $0 \le t \le 1$, such that $q(0) = q_0, q(1) = q_1$, and $\dot{q}(t) \in \mathrm{hor}_{q(t)}$ (horizontal with respect to the mechanical connection $\mathcal{A}$), find the curve or curves $q(t)$ such that the energy of the shape space curve, namely,*

$$\frac{1}{2} \int_0^1 \|\dot{r}\|^2 dt,$$

is minimized.

We shall examine the solution of this problem in Section 7.5.   ◆

As we shall see in Section 7.3, many optimal control problems can be cast in this form. Note that these problems are certainly variational, over a restricted class of curves satisfying $\Phi(q, \dot{q}) = 0$.

Local Solution. We can proceed to solve the Lagrange problem locally by forming the modified Lagrangian

$$\Lambda(q, \dot{q}, \lambda) = L(q, \dot{q}) + \lambda \cdot \Phi(q, \dot{q}), \tag{7.1.2}$$

with $\lambda \in \mathbb{R}^{n-m}$. The Euler–Lagrange equations then take the form

$$\frac{d}{dt} \frac{\partial}{\partial \dot{q}} \Lambda(q, \dot{q}, \lambda) - \frac{\partial}{\partial q} \Lambda(q, \dot{q}, \lambda) = 0, \tag{7.1.3}$$

$$\Phi(q, \dot{q}) = 0. \tag{7.1.4}$$

The case we are particularly interested in is the case of classical (linear in the velocity) nonholonomic constraints:

$$\omega_i(q, \dot{q}) = \sum_{k=1}^{n} a_{ik}(q) \dot{q}^k = 0, \qquad i = 1, \ldots, n - m. \tag{7.1.5}$$

In the case that these constraints are integrable (equivalent to functions of q only) and L is physical, i.e., it is a holonomic mechanical system, this system will represent physical dynamics. In the nonholonomic case, these equations will not be physical; one needs the Lagrange–d'Alembert principle, as we have seen in Chapters 1, 3, and 5. The following theorem gives the differential equations for the Lagrange problem.

7.1.3 Theorem. *A solution of the Lagrange problem* Definition 7.1.1 *with constraints of the form* (7.1.5) *satisfies the following equations:*

$$\frac{d}{dt}\frac{\partial}{\partial \dot{q}_i}L - \frac{\partial}{\partial q_i}L + \sum_{j=1}^{n-m}\left(\frac{d}{dt}\lambda_j\right)a_{ji} + \sum_{j=1}^{n-m}\lambda_j\left(\dot{a}_{ji} - \sum_{k=1}^{n}\frac{\partial a_{jk}}{\partial q_i}\dot{q}_k\right) = 0$$

(7.1.6)

with the constraints

$$\sum_{k=1}^{n}a_{ik}\dot{q}^k = 0.$$

(7.1.7)

Contrast these equations of motion with the nonholonomic equations of motion with Lagrange multipliers obtained in Chapters 1 and 5 from the Lagrange–d'Alembert principle:

$$\frac{d}{dt}\frac{\partial}{\partial \dot{q}_i}L - \frac{\partial}{\partial q_i}L = \sum_{j=1}^{n-m}\lambda_j a_{ji}.$$

(7.1.8)

Observe that if we (formally) set $\lambda_j = 0$ and $\dot{\lambda}_j = \lambda_j$ in the variational nonholonomic equations, we recover the nonholonomic equations of motion. It is precisely the omission of the λ_j term that destroys the variational nature of the nonholonomic equations.

7.1.4 Examples. Here we recall from Chapter 1 two examples that will be used to illustrate the theory above: the vertical rolling penny (or unicycle) and the rolling (homogeneous) ball (see Bloch and Crouch [1993]).

A. (Rolling Disk or Unicycle.)
We consider again the vertical disk discussed in Section 1.4, this time without controls. The variational problems yielded the augmented Lagrangian

$$L = \frac{1}{2}m\left(\dot{x}^2 + \dot{y}^2\right) + \frac{1}{2}I\dot{\theta}^2 + \frac{1}{2}J\dot{\varphi}^2 + \mu_1\left(\dot{x} - R\dot{\theta}\cos\varphi\right) + \mu_2\left(\dot{y} - R\dot{\theta}\sin\varphi\right),$$

giving the Lagrange equations

$$m\ddot{x} + \dot{\mu}_1 = 0,$$
$$m\ddot{y} + \dot{\mu}_2 = 0,$$
$$J\ddot{\varphi} + R\mu_1\dot{\theta}\sin\varphi - R\mu_2\dot{\theta}\cos\varphi = 0,$$
$$I\ddot{\theta} - R\frac{d}{dt}(\mu_1\cos\varphi + \mu_2\sin\varphi) = 0.$$

(7.1.9)

As we saw in Section 1.4, we obtain

$$\mu_1 = -mR\dot{\theta}\cos\varphi + A,$$
$$\mu_2 = -mR\dot{\theta}\sin\varphi + B,$$

where A and B are integration constants giving the equations

$$J\ddot{\varphi} = R\dot{\theta}(A\sin\varphi - B\cos\varphi),$$
$$(I + mR^2)\ddot{\theta} = R\dot{\varphi}(-A\sin\varphi + B\cos\varphi).$$

Note that we may obtain the nonholonomic equations of motion (1.4.3) by setting the constants of integration for the multipliers A and B equal to zero. However, there is not always so simple a relationship between the variational and nonholonomic equations. For more details see Cardin and Favretti [1996], Favretti [1998], Martinez, Cortes, and De León [2000], and Martínez, Cortés, and De León [2001].

Moreover, setting $\mu_j = 0$ and $\mu_j = \mu_j$ in equations (7.1.9) gives the equations

$$m\ddot{x} = 0,$$
$$m\ddot{y} = 0,$$
$$J\ddot{\varphi} = 0,$$
$$I\ddot{\theta} = R(\mu_1\cos\varphi + \mu_2\sin\varphi),$$

which are precisely the nonholonomic mechanical equations for the vertical rolling disk (1.4.3), as the theory above indicated.

B.   (The Rolling Ball)

Here we treat the example of a controlled rolling ball on the plane as a variational nonholonomic problem. The nonholonomic mechanical system was treated in Section 1.9.

We will use the coordinates x, y for the linear horizontal displacement and $P \in SO(3)$ for the angular displacement of the ball. Thus P gives the orientation of the ball with respect to inertial axes $\mathbf{e}_1, \mathbf{e}_2, \mathbf{e}_3$, where the $\mathbf{e}_i$ are the standard basis vectors aligned with the x-, y-, and z-axes, respectively. In particular, P maps the position of a fixed point in the ball measured in the inertial axes to a fixed reference position. This is not the definition given in many standard texts (e.g., Abraham and Marsden [1978]), where the inertial and body frames are interchanged. It does give a right-invariant description of the kinematics expressed in the body frame, which is useful from some points of view.

Let $\boldsymbol{\omega} \in \mathbb{R}^3$ denote the angular velocity of the ball with respect to inertial axes. In particular, the ball may spin freely about the z-axis and the z-component of angular momentum is conserved. If J denotes the inertia tensor of the ball with respect to the body axes, then $\mathbf{J} = P^T J P$ denotes the inertia tensor of the ball with respect to the inertial axes, and $\mathbf{J}\boldsymbol{\omega}$ is the angular momentum of the ball with respect to the inertial axes. The conservation law alluded to above is expressed as

$$\mathbf{e}_3^T \mathbf{J}\boldsymbol{\omega} = c. \tag{7.1.10}$$

The nonholonomic constraints are expressed as

$$\mathbf{e}_2^T \boldsymbol{\omega} + \dot{x} = 0,$$
$$\mathbf{e}_1^T \boldsymbol{\omega} - \dot{y} = 0. \tag{7.1.11}$$

Note that these do not include constraints on the spin about the z-axis, which can be additionally imposed through applied torques (see, e.g., Brockett and Dai [1992]).

The kinematics of the rotating ball may be expressed as $\dot{P} = S(\nu)P$, where $\nu = P\boldsymbol{\omega}$ is the angular velocity in the body frame and $S(\nu)$ is the skew-symmetric matrix satisfying $a \times b = S(b)a$ for all $a, b \in \mathbb{R}^3$ (see, for example, Crouch [1984]). Here we will explicitly derive the Euler–Lagrange equations for the variational nonholonomic problem, from which we may write down the mechanical nonholonomic system.

To obtain the variational control system we first write down the Lagrangian in the following form, where m denotes the mass of the ball:

$$L = \frac{1}{2}\nu^T J\nu + \mu_1 \left(\nu^T Pe_1 - \dot{y}\right) + \mu_2 \left(\nu^T Pe_2 + \dot{x}\right)$$
$$+ \frac{1}{2}m\left(\dot{x}^2 + \dot{y}^2\right) + \operatorname{trace} Q^T(\dot{P} - S(\nu)P). \tag{7.1.12}$$

Note that we have expressed the constraints (7.1.11) in terms of ν, and we have treated the kinematic equations themselves as constraints, and have therefore introduced an extra Lagrange multiplier in the form of a matrix Q. (The inner product on the space of 3×3 matrices is just the trace form: $\langle Q, P \rangle = \operatorname{trace} Q^T P$.) In order to manipulate the Lagrangian (7.1.12) it is convenient to use the identity

$$a^T A b = \operatorname{trace}\left(ba^T A\right) = \operatorname{trace}\left(Aba^T\right).$$

The forced Euler–Lagrange equations corresponding to this Lagrangian can now be written as

$$\dot{q}^T + Q^T S(\nu) - \mu_1 e_1 \nu^T - \mu_2 e_2 \nu^T = 0, \tag{7.1.13}$$
$$w^T (J\nu + \mu_1 Pe_1 + \mu_2 Pe_2) - \operatorname{trace} Q^T S(w)P = 0, \quad \forall w \in \mathbb{R}^3, \tag{7.1.14}$$
$$m\ddot{x} + \mu_2 = u_1,$$
$$m\ddot{y} - \mu_1 = u_2. \tag{7.1.15}$$

Differentiating equation (7.1.14) yields

$$w^T (J\dot{\nu} + \dot{\mu}_1 Pe_1 + \dot{\mu}_2 Pe_2 + \mu_1 S(\nu)Pe_1 + \mu_2 S(\nu)Pe_2)$$

$$-\operatorname{trace} \dot{q}^T S(w)P - \operatorname{trace} Q^T S(w)S(\nu)P = 0,$$

and substituting from (7.1.13) gives

$$w^T(J\dot{\nu} + \dot{\mu}_1 Pe_1 + \dot{\mu}_2 Pe_2 + \mu_1 S(\nu)Pe_1 + \mu_2 S(\nu)Pe_2)$$
$$+ \text{ trace } Q^T(S(\nu)S(w) - S(w)S(\nu))P$$
$$- \mu_1 \nu^T S(w)Pe_1 - \mu_2 \nu^T S(w)Pe_2 = 0. \qquad (7.1.16)$$

But the Jacobi identity for the cross product yields

$$S(\nu)S(w) - S(w)S(\nu) = S(S(\nu)w), \qquad (7.1.17)$$

and from (49) we obtain

$$\text{trace } Q^T S(S(\nu)w)P = -w^T S(\nu)(J\nu + \mu_1 Pe_1 + \mu_2 Pe_2),$$

so (7.1.16) implies the following system of equations describing the variational controlled rolling ball:

$$J\dot{\nu} = S(\nu)J\nu - \dot{\mu}_1 Pe_1 - \dot{\mu}_2 Pe_2 - \mu_1 S(\nu)Pe_1 - \mu_2 S(\nu)Pe_2,$$
$$\dot{P} = S(\nu)P,$$
$$m\ddot{x} = -\dot{\mu}_2 + u_1, \qquad e_2^T P^T \nu + \dot{x} = 0, \qquad (7.1.18)$$
$$m\ddot{y} = \dot{\mu}_1 + u_2, \qquad e_1^T P^T \nu - \dot{y} = 0.$$

Following the prescription described above, we can write down the equations describing the nonholonomic controlled rolling ball in the form

$$J\dot{\nu} = S(\nu)J\nu + \lambda_1 Pe_1 + \lambda_2 Pe_2,$$
$$\dot{P} = S(\nu)P,$$
$$m\ddot{x} = \lambda_2 + u_1, \qquad e_2^T P^T \nu + \dot{x} = 0, \qquad (7.1.19)$$
$$m\ddot{y} = -\lambda_1 + u_2, \qquad e_1^T P^T \nu - \dot{y} = 0.$$

Note that equations (7.1.18) and (7.1.19) can be rewritten in terms of the angular velocity $\boldsymbol{\omega}$; the variational equations become

$$\mathbf{J}\dot{\boldsymbol{\omega}} = S(\boldsymbol{\omega})\mathbf{J}\boldsymbol{\omega} - \dot{\mu}_1 e_1 - \dot{\mu}_2 e_2 - \mu_1 S(\boldsymbol{\omega})e_1 - \mu_2 S(\boldsymbol{\omega})e_2,$$
$$\dot{P} = PS(\boldsymbol{\omega}),$$
$$m\ddot{x} = -\dot{\mu}_2 + u_1, \qquad e_2^T \boldsymbol{\omega} + \dot{x} = 0,$$
$$m\ddot{y} = \dot{\mu}_1 + u_2, \qquad e_1^T \boldsymbol{\omega} - \dot{y} = 0,$$

while the nonholonomic equations are simply obtained using the usual prescription.   ◆

Exercises

⋄ **7.1-1.** Show how the variational constrained rolling ball equations become the nonholonomic mechanical rolling ball equations discussed in Section 1.9 if we make the Lagrange multiplier substitution discussed after equation (7.1.8).

⋄ **7.1-2.** Compute the variational equations of motion for the ball on a rotating plate. (See Section 1.9 for the nonholonomic mechanical case and also Lewis and Murray [1995].)

Note that more exercises comparing variational dynamics and nonholonomic mechanics may be found in Chapter 1.

7.2 Optimal Control and the Maximum Principle

This section discusses the maximum principle, which gives necessary conditions for optimal control problems. In the literature, optimal control problems including such problems as the Bernoulli minimum time problem are typically cast in a different setting from the classical variational problems, more closely associated with mechanics.

The basic difference lies in the way in which the trajectories are formulated; in the optimal control setting the trajectories are "parametrized" by the controlled vector field, while in the traditional variational setting trajectories are simply "constrained." We discuss this in detail in this section. The other basic difference is that the necessary conditions for extremals in the optimal control setting are typically expressed using a Hamiltonian formulation using the Pontryagin maximum principle, rather than the Lagrangian setting (see Pontryagin [1959], Pontryagin, Boltyanskii, Gamkrelidze and Mischenko [1962], Lee and Markus [1976], and Sussmann [1998a, 1998b].

A General Formulation of Optimal Control Problems. We state a typical optimal control problem,

$$\min_{u(\cdot)} \int_0^T g(x, u)dt, \qquad (7.2.1)$$

subject to the following conditions:

(i) a differential equation constraint $\dot{x} = f(x, u)$, and a state space constraint $x \in M$, and a constraint on the controls $u \in \Omega \subset \mathbb{R}^k$;

(ii) the endpoint conditions: $x(0) = x_0$ and $x(T) = x_T$, where f and $g \geq 0$ are smooth, Ω is a closed subset of $\mathbb{R}^k$, and M is a smooth manifold of dimension n that is the state space of the system. The integrand g is sometimes referred to as the *cost function*.

7.2.1 Example (Optimal Control of the Heisenberg System).

We recall this optimal control problem from Section 1.8:

Optimal Steering Problem. Given a number $a > 0$, find time-dependent controls u_1, u_2 that steer the trajectory starting at $(0,0,0)$ at time $t = 0$ to the point $(0,0,a)$ after a given time $T > 0$ and that among all such controls minimizes

$$\frac{1}{2} \int_0^T \left(u_1^2 + u_2^2 \right) dt, \tag{7.2.2}$$

subject to the dynamics

$$\dot{x} = u_1,$$
$$\dot{y} = u_2,$$
$$\dot{z} = xu_2 - yu_1. \tag{7.2.3}$$

A complete solution to this problem was given in Section 1.8.   ◆

The Pontryagin Maximum Principle. To state necessary conditions dictated by the Pontryagin maximum principle, we introduce a parametrized Hamiltonian function on T^*M:

$$\hat{H}(x, p, u) = \langle p, f(x, u) \rangle - p_0 g(x, u), \tag{7.2.4}$$

where $p_0 \geq 0$ is a fixed positive constant, and $p \in T^*M$. Note that p_0 is the multiplier of the cost function and that $\hat{H}$ is linear in p.

We denote by $t \mapsto u^*(t)$ a curve that satisfies the following relationship along a trajectory $t \mapsto (x(t), p(t))$ in T^*M:

$$H(x(t), p(t), u^*(t)) = \max_{u \in \Omega} \hat{H}(x(t), p(t), u). \tag{7.2.5}$$

Then if u^* is defined implicitly as a function of x and p by equation (7.2.5), we can define H^* by

$$H^*(x(t), p(t), t) = H(x(t), p(t), u^*(t)). \tag{7.2.6}$$

The time-varying Hamiltonian function H^* defines a time-varying Hamiltonian vector field X_{H^*} on T^*M with respect to the canonical symplectic structure on T^*M.

One statement of Pontryagin's maximum principle gives necessary conditions for extremals of the problem (7.2.1) as follows: An extremal trajectory $t \mapsto x(t)$ of the problem (7.2.1) is the projection onto M of a trajectory of the flow of the vector field X_{H^*} that satisfies the boundary condition (7.2.1) (ii), and for which $t \mapsto (p(t), p_0)$ is not identically zero on $[0, T]$.

The extremal is called *normal* when $p_0 \neq 0$ (in which case we may set $p_0 = 1$ by normalizing the Hamiltonian function). When $p_0 = 0$ we call the extremal *abnormal*, corresponding to the case where the extremal is determined by constraints alone. Such a case is discussed in Section 7.5.

If the extremal control function u^* is not determined by the system (7.2.5) along the extremal trajectory, then the extremal is said to be *singular*, in which case further (higher-order) necessary conditions are needed to determine u^*. See, for example, the work of Krener [1977].

In most of this chapter (see Section 7.5, however) we are interested only in nonsingular situations in which $\Omega = \mathbb{R}^k$ (or indeed, more general vector bundles), since it is these cases that occur in the mechanical situations we are interested in. There has, however, been much interesting work on abnormal extremals recently; see, for example, the work of Montgomery [1994] and Sussmann [1996] and Section 7.5.

We also suppose that the data are sufficiently regular that u^* is determined uniquely from the condition

$$0 \equiv \frac{\partial \hat{H}}{\partial u}(x(t), p(t), u^*(t)), \quad t \in [0, T]. \tag{7.2.7}$$

(Since u^* maximizes the function $\hat{H}$, its partial derivative in u evaluated at u^* must vanish.)

It follows from the implicit function theorem that there exists a function k such that $u^*(t) = k(x(t), p(t))$. We then set

$$H(x, p) \stackrel{\Delta}{=} \hat{H}(x, p, k(x, p)). \tag{7.2.8}$$

Thus along an extremal,

$$H(x(t), p(t)) = H^*(x(t), p(t), t). \tag{7.2.9}$$

We briefly motivate our statement of the Pontryagin maximum principle in the presence of regularity conditions alluded to above: In particular, we assume that $\Omega = \mathbb{R}^m$ and that $u^*(t)$ is uniquely determined by the condition (7.2.7). Treating the optimal control problem (7.2.1) as a variational problem with constraints, we augment the cost function and constraints (in the form of the constraining state differential equation) by multipliers $p_0 \in \mathbb{R}^+$ and $p \in T_M^*$. We obtain necessary conditions in the form

$$\delta \int_0^T \left(p \left(f(x, u) - \dot{x} \right) - p_0 g(x, u) \right) dt = 0, \tag{7.2.10}$$

where the variations are taken over pairs (x, u) satisfying the constraints $\dot{x} = f(x, u)$ and the boundary conditions $x(0) = x_0$, $x(T) = x_T$.

We may restate the condition (7.2.10) as

$$\delta \int_0^T \left(\hat{H}(x, p, u) - p\dot{x} \right) dt = 0. \tag{7.2.11}$$

Under the assumed regularity we may eliminate the variation with respect to u, and from (7.2.9) the necessary condition becomes

$$\delta \int_0^T \left(\hat{H}(x, p) - p\dot{x} \right) dt = 0. \tag{7.2.12}$$

This is, of course, just Hamilton's principle (see Definition 3.3.1) for the Hamiltonian H, which yields necessary conditions in terms of the usual Hamiltonian equations. Now from (7.2.7) and (7.2.9) the Hamiltonian equations for H may be replaced by the Hamiltonian equations for H^*, resulting in the statement of the maximum principle above. Note that whereas $\hat{H}$ and H^* are affine in p, H is in general *not* affine in p.

The main point of the Pontryagin maximum principle is that the result stated above is true under far less severe regularity conditions and in particular where Ω is a proper subset of $\mathbb{R}^n$. Minimum time problems such as the Bernoulli problem make sense typically only in cases where Ω is a proper subset of $\mathbb{R}^n$. For a full treatment of the maximum principle when the constraint equation $\dot{x} = f(x, u)$ is a linear equation see Sontag [1990]. For the general nonlinear case see Sussmann [1998a].

Exercises

$\diamond$ **7.2-1.** Consider the system

$$\dot{x} = -2x + u. \tag{7.2.13}$$

Show that the optimal control that transfers the system from $x(0) = 1$ to $x(1) = 0$ so as to minimize $\int_0^1 u^2 dt$ is given by $u^8(t) = -4e^{2t} / \left(e^4 - 1 \right)$. (See Barnett [1978].)

$\diamond$ **7.2-2.** The system

$$\dot{x}_1 = x_2,$$
$$\dot{x}_2 = -x_2 + u$$

is to be transferred from $x(0) = 0$ to the line $ax_1 + bx_2 = c$ at time T so as to minimize $\int_0^T u^2 dt$ for a, b, c, T given. Use the maximum principle to compute u^*. (See Barnett [1978].)

7.3 Variational Nonholonomic Systems and Optimal Control

Variational nonholonomic problems (i.e., constrained variational problems) are equivalent to optimal control problems under certain regularity conditions. This issue was investigated in depth in Bloch and Crouch [1994], employing the classical results (Rund [1966], Bliss [1930]), which relate classical constrained variational problems to Hamiltonian flows, although not optimal control problems. We outline the simplest relationship as discussed in Bloch and Crouch [1994].

Consider a modified Lagrangian

$$\Lambda(q, \dot{q}, \lambda) = L(q, \dot{q}) + \lambda \cdot \Phi(q, \dot{q}) \qquad (7.3.1)$$

with Euler–Lagrange equations

$$\frac{d}{dt} \frac{\partial}{\partial \dot{q}} \Lambda(q, \dot{q}, \lambda) - \frac{\partial}{\partial q} \Lambda(q, \dot{q}, \lambda) = 0,$$
$$\Phi(q, \dot{q}) = 0. \qquad (7.3.2)$$

We will rewrite this equation in Hamiltonian form and show that the resulting equations are equivalent to the equations of motion given by the maximum principle for a suitable optimal control problem.

Set

$$p = \frac{\partial}{\partial \dot{q}} \Lambda(q, .\dot{q}, \lambda) \qquad (7.3.3)$$

and consider this equation together with the constraints

$$\Phi(q, \dot{q}) = 0. \qquad (7.3.4)$$

We wish to solve (7.3.3) and (7.3.4) for $(\dot{q}, \lambda)$.

Now assume that on an open set U the matrix

$$\begin{bmatrix} \dfrac{\partial^2}{\partial \dot{q}^2} \Lambda(q, \dot{q}, \lambda) & \dfrac{\partial}{\partial \dot{q}} \Phi(q, \dot{q})^T \\[3mm] \dfrac{\partial}{\partial \dot{q}} \Phi(q, \dot{q}) & 0 \end{bmatrix} \qquad (7.3.5)$$

has full rank. (This generalizes the usual Legendre condition that $\frac{\partial^2}{\partial \dot{q}^2} L(q, \dot{q})$ has full rank.) By the implicit function theorem, we can solve for $\dot{q}$ and λ:

$$\dot{q} = \phi(q, p),$$
$$\lambda = \psi(q, p). \qquad (7.3.6)$$

We now have the following theorem:

7.3.1 Theorem ((Carathéodory [1967], Rund [1966], Arnold, Kozlov, and Neishtadt [1988], Bloch and Crouch [1994])). *Under the transformation (7.3.6), the Euler–Lagrange system (7.3.2) is transformed to the Hamiltonian system*

$$\dot{q} = \frac{\partial}{\partial p}H(q,p),$$

$$\dot{p} = -\frac{\partial}{\partial q}H(q,p), \qquad (7.3.7)$$

where

$$H(q,p) = p \cdot \phi(q,p) - L(q,\phi(q,p)). \qquad (7.3.8)$$

Proof. $\Phi(q,\phi(q,p)) = 0$ implies

$$\frac{\partial \Phi}{\partial q} + \frac{\partial \Phi}{\partial \dot{q}}\frac{\partial \phi}{\partial q} = 0,$$

$$\frac{\partial \Phi}{\partial \dot{q}}\frac{\partial \phi}{\partial p} = 0.$$

Hence, using (7.3.3), we have

$$\frac{\partial H}{\partial p} = \phi + \left(p - \frac{\partial L}{\partial \dot{q}}\right) \cdot \frac{\partial \phi}{\partial p} = \dot{q} + \lambda \cdot \left(\frac{\partial \Phi}{\partial \dot{q}}\frac{\partial \phi}{\partial p}\right) = \dot{q}.$$

Similarly,

$$\frac{\partial H}{\partial q} = -\frac{\partial L}{\partial q} + \left(p - \frac{\partial L}{\partial \dot{q}}\right) \cdot \frac{\partial \phi}{\partial q} = -\frac{\partial L}{\partial q} + \lambda \cdot \left(\frac{\partial \Phi}{\partial \dot{q}}\frac{\partial \phi}{\partial q}\right)$$

$$= -\left(\frac{\partial L}{\partial q} + \lambda \cdot \frac{\partial \Phi}{\partial q}\right) = -\frac{\partial \Lambda}{\partial q} = -\dot{p}. \qquad ∎$$

We now compare this to the optimal control setup.

7.3.2 Definition. *Let the optimal control problem be given by*

$$\min_{u(\cdot)} \int_0^T g(q,u)dt \qquad (7.3.9)$$

subject to $q(0) = 0$, $q(T) = q_T$,

$$\dot{q} = f(q,u),$$

where $q \in \mathbb{R}^n$, $u \in \mathbb{R}^m$.

Then we have the following:

7.3.3 Theorem. *The Lagrange problem and optimal control problem generate the same (regular) extremal trajectories, provided that:*

(i) $\Phi(q, \dot{q}) = 0$ *if and only if there exists a u such that* $\dot{q} = f(q, u)$.

(ii) $L(q, f(q, u)) = g(q, u)$.

(iii) *The optimal control u^* is uniquely determined by the condition*

$$\frac{\partial \hat{H}}{\partial u}(q, p, u^*) = 0, \qquad (7.3.10)$$

where

$$\frac{\partial^2 \hat{H}}{\partial u^2}(q, p, u^*)$$

is of full rank and

$$\hat{H}(q, p, u) = \langle p, f(q, u) \rangle - g(q, u) \qquad (7.3.11)$$

is the Hamiltonian function given by the maximum principle.

Proof. By (iii) we may use the equation

$$p \cdot \frac{\partial f}{\partial u}(q, u^*) - \frac{\partial g}{\partial u}(q, u^*) = 0$$

to deduce that there exists a function r such that $u^* = r(q, p)$.

The extremal trajectories are now generated by the Hamiltonian

$$\overline{H}(q, p) = \hat{H}(q, p, r(x, p)) = p \cdot f(q, r(q, p)) - g(q, r(q, p)). \qquad (7.3.12)$$

Then the result follows, and we have

$$\overline{H}(q, p) = H(q, p),$$
$$f(q, r(q, p)) = \phi(q, p),$$
$$g(q, r(q, p)) = L(q, \phi(q, p)). \qquad \blacksquare$$

7.4 Kinematic Sub-Riemannian Optimal Control Problems

The Kinematic Sub-Riemannian Optimal Control Problem. Here we consider the optimal control of underactuated kinematic control problems of the type discussed in Chapter 6.

The problem is referred to as sub-Riemannian in that it gives rise to a geodesic flow with respect to a singular metric (see the work of Strichartz

[1983, 1987] and Montgomery [2002]) and references therein. This problem has an interesting history in control theory (see Brockett [1973a, 1981], Baillieul [1975]). See also Bloch, Crouch, and Ratiu [1994] and Sussmann [1996] and further references in the text below.

We consider control systems of the form

$$\dot{x} = \sum_{i=1}^{m} X_i u_i, \quad x \in M, \quad u \in \Omega \subset \mathbb{R}^m, \qquad (7.4.1)$$

where Ω contains an open subset that contains the origin, M is a smooth manifold of dimension n, and each of the vector fields in the collection $F := \{X_1, \ldots, X_k\}$ is complete.

We assume that the system satisfies the accessibility rank condition and is thus controllable, since there is no drift term. Then we can pose the optimal control problem

$$\min_{u(\cdot)} \int_0^T \frac{1}{2} \sum_{i=1}^{m} u_i^2(t) dt \qquad (7.4.2)$$

subject to the dynamics (7.4.1) and the endpoint conditions $x(0) = x_0$ and $x(T) = x_T$.

To view this as a constrained variational problem we make some additional regularity assumptions. These are not necessary, but even when they hold, they produce a very rich class of problems.

Assumption.

(i) The system defined by (7.4.1) satisfies the accessibility rank condition.

(ii) The dimension of the distribution D_F defined by the span of $X_1, \ldots, X_k$ is constant on M and equal to k. (Thus the vector fields $X_1, \ldots, X_k$ are everywhere independent.)

(iii) There exist exactly $n - k = m$ one-forms on M $\omega_1, \ldots, \omega_m$ such that the codistribution

$$D_F^\perp(x) = \{\bar{\omega} \in T_x^* M; \quad \bar{\omega} D_F(x) = 0\}$$

is spanned by $\omega_1, \ldots, \omega_m$ everywhere. (This condition implies that M is parallelizable.)

Since D_F has constant dimension on M, we may define a norm on each subspace $D_F(x)$; if $X \in D_F(x)$ and $X = \sum_{i=1}^{k} \alpha_i X_i(x)$, then we define

$$|X| := \sum_{i=1}^{k} \alpha_i^2.$$

This norm defines an inner product on $D_F(x)$, denoted by $\langle \cdot, \cdot \rangle_x$, which can be extended to a metric on M. The optimal control problem (7.4.2) is now equivalent to the following constrained variational problem when the assumptions (i), (ii), (iii) hold:

$$\min_{x(\cdot)} \frac{1}{2} \int_0^T \langle \dot{x}, \dot{x} \rangle_x dt \qquad (7.4.3)$$

subject to the condition that $x(\cdot)$ is a piecewise C^1 curve in M such that $x(0) = x_0$, $x(T) = x_T$, and $\omega_i(x)(\dot{x}) = 0$, $1 \le i \le m$. This problem is often referred to as the **sub-Riemannian geodesic problem**, to distinguish it from the Riemannian geodesic problem, in which the constraints are absent. These problems were studied by Griffiths [1983] from the constrained variational viewpoint, and from the optimal control viewpoint by Brockett[1981, 1983]. In the sub-Riemannian geodesic problem, abnormal extremals play an important role. See work by Hermann [1962], Strichartz [1983], Montgomery[1994 1995], Sussmann [1996], Agrachev and Sarychev [1996], Agrachev and Sarychev [1998].

The singular nature of the sub-Riemannian geodesic problem is manifested in many ways, such as the existence of distinct abnormal extremals and the singular nature of the sub-Riemannian geodesic ball, as first investigated by Brockett [1981]. If we define a metric on M by setting

$$d(x_0, x_T) = \min_{x(\cdot)} \int_0^T |\dot{x}| dt, \dot{x} \in D_F(x), x(0) = x_0, x(T) = x_T,$$
$$B_\epsilon^F(x_0) = \{\bar{x} \in M; d(\bar{x}, x_0) \le \epsilon\},$$

then the sub-Riemannian geodesic ball $S_\epsilon^F(x_0)$ is simply the boundary of $B_\epsilon^F(x_0)$.

7.4.1 Example. For the Heisenberg sub-Riemannian geodesic problem that we saw expressed as an optimal control problem in equations (7.2.2), (7.2.3) (see also Chapter 1) we have

$$F = \{(\partial/\partial x) - y(\partial/\partial z), (\partial/\partial y) + x(\partial/\partial z)\},$$

while $D_F^\perp$ is spanned by $\omega = x \, dy - y \, dx - dz$. Here $S_\epsilon^F(0)$ has a singularity along the z-axis. This class of problem continues to evoke a great deal of interest, especially in the areas of establishing when abnormal extremals are optimal and obtaining a precise description of $S_\epsilon^F(x_0)$. We shall return to the notion of abnormal extremals later.   ◆

Formulation on Riemannian Manifolds. We now set up the problem on a Riemannian manifold as follows, although the cases of most interest usually have more structure, such as a group structure. In the following subsections we will look at such special cases. We follow here the approach of Bloch, Crouch, and Ratiu [1994].

Let M be a Riemannian manifold of dimension n with metric denoted by $\langle \cdot, \cdot \rangle$. The corresponding Riemannian connection and covariant derivative will be denoted by ∇ and $D/\partial t$, respectively. Now assume that M is such that there exist smooth vector fields $X^1(q), \ldots, X^n(q)$ satisfying $\langle X^i(q), X^j(q) \rangle = \delta_{ij}$, an orthonormal frame for $T_q M$ for all $q \in M$. This, of course, limits the class of manifolds we consider (to parallelizable manifolds), but is satisfied for the main case of interest to us, namely, when M is a Lie group G. As we have seen in Chapters 1, 4, and 6, this theory applies to a large class of practical kinematic control problems of interest, essentially all underactuated systems where the velocities are directly controlled. This includes the Heisenberg system, the kinematic rolling penny, and the controlled knife edge.

We now define the *kinematic control system* on M by

$$\frac{dq}{dt} = \sum_{i=1}^{m} u_i X^i(q), \quad m < n. \tag{7.4.4}$$

The optimal control problem for (7.4.4) is defined by

$$\min_u \int_0^T \frac{1}{2} \sum_{i=1}^{m} u_i^2(t)dt; \quad q(0) = q_0, \quad q(T) = q_T, \tag{7.4.5}$$

subject to (7.4.4).

This may be posed as a variational problem on M as follows: Define the constraints

$$\omega_k \left(\frac{dq}{dt} \right) = \left\langle X^k, \frac{dq}{dt} \right\rangle = 0, \quad m < k \leq n, \tag{7.4.6}$$

and let

$$Z_t = \sum_{k=m+1}^{n} \lambda_k(t) X^k, \tag{7.4.7}$$

where the λ_k are Lagrange multipliers. By the orthonormality of the X^i the optimal control problem then becomes

$$\min_q J(q) = \min_q \int_0^T \left(\frac{1}{2} \left\langle \frac{dq}{dt}, \frac{dq}{dt} \right\rangle + \left\langle Z_t, \frac{dq}{dt} \right\rangle \right) dt, \tag{7.4.8}$$

$$\left\langle Z_t, \frac{dq}{dt} \right\rangle = 0. \tag{7.4.9}$$

We now briefly derive necessary conditions for the regular extremals of this variational problem following Milnor [1963] and Crouch and Silva-Leite [1991]. (For interesting recent work on abnormal extremals see, for example, Bryant and Hsu [1993], Montgomery [1994], and Sussmann [1996].)

Firstly, we have to define the variations we are going to use: The tangent space to the space Ω of C^2 curves satisfying the boundary conditions of

(7.4.5) is denoted by $T_q\Omega$. It is the space of C^1 vector fields $t \to W_t$ along $q(t)$ satisfying $W_0 = 0 = W_T$. The curve $t \to \frac{DW_t}{\partial t}$ in TM is continuous. Exponentiating a vector field in $T_q\Omega$ we obtain a one-parameter variation of q:

$$\alpha : [0, T] \times (-\epsilon, \epsilon) \to M, \tag{7.4.10}$$

$$\alpha_u(t) = \alpha(t, u) = \exp_{q(t)}(uW_t), \tag{7.4.11}$$

where exp is the exponential mapping (integral curve) on M. Note that

$$\alpha_u(0) = q(0) = q_0, \quad \alpha_u(T) = q(T) = q_T, \quad \alpha_0(t) = q(t),$$

$$\frac{\partial \alpha_0(t)}{\partial u} = W_t, \quad 0 \le t \le T.$$

The necessary conditions for regular extremals are obtained from

$$\frac{d}{du} J(\alpha_u)|_{u=0} = 0, \tag{7.4.12}$$

where

$$J(\alpha_u) = \int_0^T \left(\frac{1}{2} \left\langle \frac{\partial \alpha_u}{\partial t}, \frac{\partial \alpha_u}{\partial t} \right\rangle + \left\langle Z_t(\alpha_u), \frac{\partial \alpha_u}{\partial t} \right\rangle \right) dt. \tag{7.4.13}$$

Now

$$\frac{DJ(\alpha_u)}{du}\bigg|_{u=0} = \int_0^T \left(\left\langle \frac{dq}{dt}, \frac{DW_t}{\partial t} \right\rangle + \left\langle \nabla_{W_t} Z_t, \frac{dq}{dt} \right\rangle + \left\langle Z_t, \frac{D}{\partial t} W_t \right\rangle \right) dt$$

$$= \int_0^T \left(-\left\langle \frac{D}{dt} V_t, W_t \right\rangle - \left\langle \frac{D}{\partial t} Z_t, W_t \right\rangle \right.$$

$$\left. - \langle \nabla_{Z_t} V_t, W_t \rangle + \langle [W_t, Z_t], V_t \rangle \right) dt, \tag{7.4.14}$$

where

$$V_t = \frac{dq}{dt} = \sum_{i=1}^m v_i(t) X^i(q).$$

Note that in this computation we use $\nabla_W Z = \nabla_Z W + [W, Z]$ and $Z[\langle V, W \rangle] = \langle \nabla_Z V, W \rangle + \langle V, \nabla_Z W \rangle$.

Thus these are the necessary conditions on a general Riemannian manifold. We now specialize further:

Necessary Conditions on a Compact Semisimple Lie Group. In this paragraph we show that if the underlying manifold is a compact semisimple Lie group, then the sub-Riemannian optimal control problem discussed in the previous section may be reduced to a computation in the Lie algebra.

Now let $M = G$, G a compact semisimple Lie group, with Lie algebra $\mathfrak{g}$, and let $\langle\langle \cdot, \cdot \rangle\rangle = -\frac{1}{2}\kappa(\cdot, \cdot)$, where κ is the Killing form on $\mathfrak{g}$. Let J be a positive definite linear mapping $J : \mathfrak{g} \to \mathfrak{g}$ satisfying

$$\langle\langle JX, Y \rangle\rangle = \langle\langle X, JY \rangle\rangle, \tag{7.4.15}$$

$$\langle\langle JX, X \rangle\rangle \geq 0 \quad (= 0 \text{ if and only if } X = 0). \tag{7.4.16}$$

Now we can define a right-invariant metric on G as follows: If $X, Y \in \mathfrak{g}$ and R_g is right translation on G by $g \in G$, then

$$X_g^r = X^r(g) = R_{g*}X \quad \text{and} \quad Y_g^r = Y^r(g) = R_{g*}Y$$

are corresponding right-invariant vector fields. Now

$$\langle X^r(g), Y^r(g) \rangle = \langle\langle X, JY \rangle\rangle \tag{7.4.17}$$

defines a right-invariant metric on G. Corresponding to the right-invariant metric $\langle \cdot, \cdot \rangle$ there is a unique Riemannian connection ∇ (see, e.g., Nomizu [1954] and earlier sections), and ∇ defines a bilinear form on $\mathfrak{g}$:

$$(X, Y) \to \nabla_X Y = \frac{1}{2}\{[X, Y] + J^{-1}[X, JY] + J^{-1}[Y, JX]\}, \quad X, Y \in \mathfrak{g}. \tag{7.4.18}$$

The expression for ∇ on right-invariant vector fields on G is

$$(\nabla_{X^r} Y^r)(g) = (\nabla_X Y)_g^r. \tag{7.4.19}$$

We now show how to reduce the variational problem to one in the Lie algebra: Choose an orthonormal basis e_i on $\mathfrak{g}$, $\langle\langle e_i, Je_j \rangle\rangle = \delta_{ij}$, and extend it to a right-invariant orthonormal frame on $T_g G$, $X^i(g) = R_{g*}e_i \equiv X^{ir}(g)$. We consider again the computation of $\frac{d}{du}J(\alpha_u)\,|_{u=0}$. Suppose in $\mathfrak{g}$,

$$V_t = \sum_{i=1}^m v_i(t)e_i, \quad \dot{V}_t = \sum_{i=1}^m \dot{v}_i(t)e_i, \tag{7.4.20}$$

and similarly for $W_t = \sum_{i=1}^n w_i(t)e_i$ and Z_t. Then at the group level we have for $\frac{dg}{dt} = V_t^r$,

$$\frac{DV_t^r}{dt} = \frac{D}{dt}\left(\sum_{i=1}^m v_i(t)X^i\right) = (\dot{V}_t + \nabla_{V_t} V_t)_g^r = (\dot{V}_t + J^{-1}[V_t, JV_t])_g^r \tag{7.4.21}$$

by (7.4.18) and (7.4.19), and

$$\frac{DZ_t^r}{dt} = (\dot{Z}_t + \nabla_{V_t} Z_t)_g^r. \tag{7.4.22}$$

Now, by (7.4.17),

$$\frac{d}{du} J(\alpha_u) \mid_{u=0} = \int_0^T (-\langle\!\langle \dot{V}_t + J^{-1}[V_t, JV_t] + \dot{Z}_t + \nabla_{V_t} Z_t + \nabla_{Z_t} V_t, \; JW_t\rangle\!\rangle$$

$$+ \langle\!\langle [W_t, Z_t], JV_t\rangle\!\rangle)dt. \tag{7.4.23}$$

Using

$$\langle\!\langle [X, Y], Z\rangle\!\rangle + \langle\!\langle Y, [X, Z]\rangle\!\rangle = 0 \tag{7.4.24}$$

the necessary conditions are thus

$$\dot{V}_t + J^{-1}[V_t, JV_t] + \dot{Z}_t + \nabla_{V_t} Z_t + \nabla_{Z_t} V_t + J^{-1}[JV_t, Z_t] = 0. \tag{7.4.25}$$

By (7.4.18)

$$\nabla_{V_t} Z_t + \nabla_{Z_t} V_t = J^{-1}[V_t, JZ_t] + J^{-1}[Z_t, JV_t].$$

Hence the necessary conditions on $\mathfrak{g}$ are

$$\dot{V}_t + J^{-1}[V_t, JZ_t] + \dot{Z}_t + J^{-1}[V_t, JV_t] = 0 \tag{7.4.26}$$

with the constraint

$$\left\langle \frac{dg}{dt}, Z_t \right\rangle = \langle\!\langle V_t, JZ_t\rangle\!\rangle = 0. \tag{7.4.27}$$

Equations (7.4.26) are identical to equations derived in Brockett [1973a] in the case $J = I$. We can see this as follows:
Write the system (7.4.4) as

$$\frac{dg}{dt} = \sum_{i=1}^m u_i B^i g, \quad m < n, \tag{7.4.28}$$

where $B_i \in \mathfrak{g}$ and the $X^i(g) = B^i g$ are thus right invariant vector fields on G. Then $V_t = \sum_{i=1}^m u_i(t) B^i$. Now set $L_t = V_t + Z_t$. Then equation (7.4.26) becomes

$$\dot{L}_t = J^{-1}[JL_t, V_t], \tag{7.4.29}$$

or

$$\dot{L}_t = \sum_{i=1}^m u_i J^{-1}[JL_t, B^i]. \tag{7.4.30}$$

Setting $J = I$ we recover precisely Brockett's equations (in the case of zero drift). Note also that in the case $J = I$ our equations (7.4.26) assume the symmetric form

$$\dot{V}_t + [V_t, Z_t] + \dot{Z}_t = 0. \tag{7.4.31}$$

Brockett obtained his equations by applying the maximum principle to Lie groups, while we have taken a direct variational approach.

The Case of Symmetric Space Structure. Suppose now that G/K is a Riemannian symmetric space (see e.g., Helgason [2001]), G as above, K a closed subgroup of G with Lie algebra $\mathfrak{k}$. Then $\mathfrak{g} = \mathfrak{p} \oplus \mathfrak{k}$ with $[\mathfrak{p}, \mathfrak{p}] \subset \mathfrak{k}$, $[\mathfrak{p}, \mathfrak{k}] \subset \mathfrak{p}$, $[\mathfrak{k}, \mathfrak{k}] \subset \mathfrak{k}$, and $\langle\!\langle \mathfrak{k}, \mathfrak{p} \rangle\!\rangle = 0$. We now want to consider the necessary conditions (7.4.26) in this case. We shall see that they simplify in an intriguing fashion, giving us a singular case of the so-called generalized rigid body equations.

The generalized rigid body equations are a natural generalization of the classical 3-dimensional rigid body equations discussed in Chapter 1. We recall from Manakov [1976] and Ratiu [1980] that the left-invariant generalized rigid body equations on $\mathrm{SO}(n)$ may be written as

$$\dot{Q} = Q\Omega,$$
$$\dot{M} = [M, \Omega], \tag{7.4.32}$$

where $Q \in \mathrm{SO}(n)$ denotes the configuration space variable (the attitude of the body), $\Omega = Q^{-1}\dot{Q} \in \mathfrak{so}(n)$ is the body angular velocity, and

$$M := J(\Omega) = \Lambda\Omega + \Omega\Lambda \in \mathfrak{so}(n)$$

is the body angular momentum. Here $J : \mathfrak{so}(n) \to \mathfrak{so}(n)$ is the symmetric (with respect to the inner product defined by the Killing form), positive definite, and hence invertible operator defined by

$$J(\Omega) = \Lambda\Omega + \Omega\Lambda,$$

where Λ is a diagonal matrix satisfying $\Lambda_i + \Lambda_j > 0$ for all $i \neq j$. For $n = 3$ the elements of Λ_i are related to the standard diagonal moment of inertia tensor I by $I_1 = \Lambda_2 + \Lambda_3$, $I_2 = \Lambda_3 + \Lambda_1$, $I_3 = \Lambda_1 + \Lambda_2$.

For further details see the Internet Supplement and Bloch, Crouch, Marsden, and Ratiu [2002].

Assume that $e_1, \ldots, e_m$ is a basis for $\mathfrak{p}$ and $e_{m+1}, \ldots, e_n$ is a basis for $\mathfrak{k}$. Suppose that $J : \mathfrak{p} \to \mathfrak{p}$ and $J : \mathfrak{k} \to \mathfrak{k}$. Then $\langle\!\langle V_t, JZ_t \rangle\!\rangle = 0$ for

$$Z_t = \sum_{i=m+1}^{n} \lambda_i(t)e_i \in \mathfrak{k} \quad \text{and} \quad V_t = \sum_{k=1}^{m} v_i(t)e_i \in \mathfrak{p}.$$

Since $\dot{V}_t + J^{-1}[V_t, JZ_t] \in \mathfrak{p}$ and $\dot{Z}_t + J^{-1}[V_t, JV_t] \in \mathfrak{k}$, the necessary conditions (7.4.26) become

$$\dot{V}_t = J^{-1}[JZ_t, V_t],$$
$$\dot{Z}_t = J^{-1}[JV_t, V_t], \tag{7.4.33}$$

or, if we define $P_t = JV_t$ and $Q_t = JZ_t$,

$$\dot{P}_t = [Q_t, J^{-1}P_t],$$
$$\dot{Q}_t = [P_t, J^{-1}P_t]. \tag{7.4.34}$$

We will now show that equations (7.4.34) are Hamiltonian with respect to the Lie–Poisson structure on $\mathfrak{g}$.

Recall that for F, H functions on $\mathfrak{g}$, their $(-)$ Lie–Poisson bracket is given by

$$\{F, H\}(X) = -\langle\!\langle X, [\nabla F(X), \nabla H(X)]\rangle\!\rangle, \quad X \in \mathfrak{g}, \qquad (7.4.35)$$

where $dF(X) \cdot Y = \langle\!\langle \nabla F(X), Y \rangle\!\rangle$.

For $H(X)$ a given Hamiltonian, we thus have the Lie–Poisson equations $\dot{F}(X) = \{F, H\}(X)$. Letting $F(X) = \langle\!\langle A, X \rangle\!\rangle$, $A \in \mathfrak{g}$, we obtain

$$\langle\!\langle A, \dot{X}\rangle\!\rangle = -\langle\!\langle X, [A, \nabla H(X)]\rangle\!\rangle = \langle\!\langle A, [X, \nabla H(X)]\rangle\!\rangle \qquad (7.4.36)$$

and hence

$$\dot{X} = [X, \nabla H(X)]. \qquad (7.4.37)$$

For $H(M) = \frac{1}{2}\langle\!\langle M, J^{-1}M \rangle\!\rangle$, $M \in \mathfrak{g}$, and J as in the previous subsection, we obtain the generalized rigid body equations

$$\dot{M} = [M, J^{-1}M]. \qquad (7.4.38)$$

Now for $X = P + Q \in \mathfrak{p} \oplus \mathfrak{k}$, let $H(X) = H(P) = \frac{1}{2}\langle\!\langle P, J^{-1}P \rangle\!\rangle$, $P \in \mathfrak{p}$. Then $\nabla H(X) = J^{-1}P \in \mathfrak{p}$, and equations (7.4.37) become

$$(Q_t + P_t)^{\cdot} = [Q_t + P_t, J^{-1}P_t], \qquad (7.4.39)$$

or

$$\dot{P}_t = [Q_t, J^{-1}P_t],$$
$$\dot{Q}_t = [P_t, J^{-1}P_t], \qquad (7.4.40)$$

which are precisely equations (7.4.34).

Thus equations (7.4.34) are Lie–Poisson with respect to the "singular" Hamiltonian $H(P)$. Summarizing then, we have the following result:

7.4.2 Theorem. *The optimal trajectories for the singular optimal control problem (7.4.4), (7.4.5) on a Riemannian symmetric space are given by equations (7.4.34). These equations are Lie–Poisson with respect to a singular rigid body Hamiltonian on $\mathfrak{g}$.*

We see, therefore, that we can obtain the singular optimal trajectories by letting $J \mid_{\mathfrak{k}} \to \infty$ in the full rigid body Hamiltonian $H(X) = \frac{1}{2}\langle\!\langle X, J^{-1}X \rangle\!\rangle$, thus obtaining the singular Hamiltonian

$$H(P) = \frac{1}{2}\langle\!\langle P, J^{-1}P \rangle\!\rangle.$$

This observation also enables us to obtain the singular rigid body equations directly by a limiting process from the full rigid body equations. The key is the correct choice of angular velocity and momentum variables corresponding to the Lie algebra decomposition $\mathfrak{g} = \mathfrak{p} \oplus \mathfrak{k}$.

In the notation of equation (7.4.34) we write an arbitrary element of $\mathfrak{g}$ as $M = JV + Q$, $JV \in \mathfrak{p}$, $Q \in \mathfrak{k}$. Then the generalized rigid body equations (7.4.38) become

$$J\dot{V}_t = [Q_t, V_t] + [JV_t, J^{-1}Q_t],$$
$$\dot{Q}_t = [Q_t, J^{-1}Q_t] + [JV_t, V_t]. \qquad (7.4.41)$$

Letting $J \mid_{\mathfrak{k}} \to \infty$ we obtain

$$J\dot{V}_t = [Q_t, V_t],$$
$$\dot{Q}_t = [JV_t, V_t]. \qquad (7.4.42)$$

We note that this is a mixture between the Lagrangian and Hamiltonian pictures. While the variables in $\mathfrak{k}$ are momenta (and should really be viewed as lying in $\mathfrak{k}^*$), the variables in $\mathfrak{p}$ are velocities. These variables not only are the natural ones in which to take the limit in the full rigid body equations, but are natural from the point of view of the maximum principle, for the variables Q correspond to the constraints and therefore are naturally viewed as costates.

As mentioned in the previous section, the necessary conditions above may also be derived directly from the maximum principle developed for Lie groups, yielding an invariant maximum principle (see, e.g., Brockett [1973a], Jurdjevic [1991]).

The Hamiltonian in the maximum principle of the system (7.4.34) is precisely $\frac{1}{2}\langle P_t, J^{-1}P_t \rangle$. This is just the sum of the Hamiltonians corresponding to each of the vector fields X_i.

We would now like to consider the optimal control problem as $J \mid_{\mathfrak{k}} \to \infty$. Write $M \in \mathfrak{g}$ as $M = JZ + P$, $Z \in \mathfrak{k}$, $P \in \mathfrak{p}$, which we can do, since $J : \mathfrak{p} \to \mathfrak{p}$ and $J : \mathfrak{k} \to \mathfrak{k}$. Then the Hamiltonian (in the maximum principle) becomes

$$H(M) = \langle M, J^{-1}M \rangle = \langle JZ + P, J^{-1}(JZ + P) \rangle$$
$$= \langle JZ, Z \rangle + \langle P, J^{-1}P \rangle. \qquad (7.4.43)$$

Letting $J \mid_{\mathfrak{k}} \to \infty$ we see that the cost becomes infinite unless $Z = 0$, i.e., unless the constraints are satisfied.

7.4.3 Example.   We consider a simple but nontrivial example: the symmetric space $SO(3)/SO(2)$. In this case $\mathfrak{g} = \mathfrak{k} \oplus \mathfrak{p}$ becomes $\mathfrak{so}(3) = \mathfrak{so}(2) \oplus \mathbb{R}^2$ relative to a given choice of z-axis used to embed $SO(2)$ into $SO(3)$. We may thus represent matrices in $\mathfrak{so}(3)$ as

$$\begin{bmatrix} 0 & -\omega_3 & \omega_2 \\ \omega_3 & 0 & -\omega_1 \\ -\omega_2 & \omega_1 & 0 \end{bmatrix} \qquad (7.4.44)$$

with the lower 2×2 block in $\mathfrak{so}(2)$.

This example illustrates the importance of writing the optimal equations in the natural variables $M = JV + Q$ in order to understand the limiting process in equations (7.4.41) and (7.4.42).

We write here

$$
M = \begin{bmatrix} 0 & -J_3\omega_3 & J_2\omega_2 \\ J_3\omega_3 & 0 & -m_1 \\ -J_2\omega_2 & m_1 & 0 \end{bmatrix}. \tag{7.4.45}
$$

Here $Q_t \in \mathfrak{so}(2)$ has "momentum" variable m_1. Then the equations

$$
(JV_t + Q_t)^{\cdot} = [JV_t + Q_t, V_t] \tag{7.4.46}
$$

become, for $\mathfrak{g} = \mathfrak{so}(3)$,

$$
\begin{aligned}
\dot{m}_1 &= (J_2 - J_3)\omega_2\omega_3, \\
J_2\dot{\omega}_2 &= -m_1\omega_3, \\
J_3\dot{\omega}_3 &= m_1\omega_2.
\end{aligned} \tag{7.4.47}
$$

The full rigid body equations in these variables are

$$
(JV_t + Q_t)^{\cdot} = [JV_t + Q_t, V_t + J^{-1}Q_t], \tag{7.4.48}
$$

which for $\mathfrak{g} = \mathfrak{so}(3)$ are

$$
\begin{aligned}
\dot{m}_1 &= (J_2 - J_3)\omega_2\omega_3, \\
J_2\dot{\omega}_2 &= \left(\frac{J_3}{J_1} - 1\right)\omega_3 m_1, \\
J_3\dot{\omega}_3 &= \left(1 - \frac{J_2}{J_1}\right)m_1\omega_2,
\end{aligned} \tag{7.4.49}
$$

which clearly approaches to (7.4.47) as $J_1 \to \infty$.

Note that if we write the rigid body equations in the usual form,

$$
J(V_t + Z_t)^{\cdot} = [J(V_t + Z_t), V_t + Z_t], \tag{7.4.50}
$$

we obtain, for $\mathfrak{g} = \mathfrak{so}(3)$,

$$
\begin{aligned}
J_1\dot{\omega}_1 &= (J_2 - J_3)\omega_2\omega_3, \\
J_2\dot{\omega}_2 &= (J_3 - J_1)\omega_1\omega_3, \\
J_3\dot{\omega}_3 &= (J_1 - J_2)\omega_2\omega_1.
\end{aligned} \tag{7.4.51}
$$

In this formulation, where we do not distinguish between $\mathfrak{p}$ and $\mathfrak{k}$, the limiting process described above is not obvious. The same is true for the rigid body in the momentum representation.

We remark that this set of equations, despite its singular nature, is still integrable, for we still have two conserved quantities, the Hamiltonian $H(\omega) = J_2\omega_2^2 + J_3\omega_3^2 (= \frac{1}{2}\langle P, J^{-1}P\rangle)$ and the Casimir

$$C(\omega) = m_1^2 + J_2^2\omega_2^2 + J_3^2\omega_2^2.$$

(Recall that a Casimir function for a Poisson structure is a function that commutes with every other function under the Poisson bracket; Section 3.5.)

It is interesting to consider the case $J = I$. Equations (7.4.34) then become

$$\dot{P}_t = [\mathcal{Q}_t, P_t],$$
$$\dot{\mathcal{Q}}_t = 0. \qquad (7.4.52)$$

Hence $\mathcal{Q}_t = Q$ is constant.

Similarly, considering (7.4.33), we obtain

$$\dot{V}_t = [Z_t, V_t], \quad Z_t = Z, \qquad (7.4.53)$$

Z a constant. This is, of course, solvable: $V_t = \mathrm{Ad}_{e^{Zt}} V_0$ and $u_i(t) = \langle\!\langle e_i, \mathrm{Ad}_{e^{Zt}} V_0\rangle\!\rangle$.

Consider again the case $SO(3)/SO(2)$. Since $V_t \in \mathbb{R}^2$ and $Z_t \in \mathfrak{so}(2)$, we may set

$$Z = \begin{bmatrix} 0 & 0 & 0 \\ 0 & 0 & -\phi \\ 0 & \phi & 0 \end{bmatrix}, \qquad (7.4.54)$$

where ϕ is fixed. Then

$$e^{Zt} = \begin{bmatrix} 1 & 0 & 0 \\ 0 & \cos\phi t & -\sin\phi t \\ 0 & \sin\phi t & \cos\phi t \end{bmatrix}. \qquad (7.4.55)$$

Hence the optimal evolution of V_t (or equivalently the optimal controls) is given by rotation. This recovers precisely the result of Baillieul, who indeed analyzed the case $J = I$ in dimension 3. (See Baillieul [1975], page III-5.) ◆

7.5 Optimal Control and a Particle in a Magnetic Field

We begin by considering the connection between optimal control of the Heisenberg system and the motion of a particle in a magnetic field. This will be seen to be a special case of the more general motion of a particle in a magnetic or Yang–Mills field. The control analysis of the Heisenberg model goes back to Brockett [1981] and Baillieul [1975], while a modern treatment of the relationship with a particle in a magnetic field may be found in Montgomery [1993], for example. A nice treatment of the pure mechanical aspects of a particle in a magnetic field may be found in Marsden and Ratiu [1999].

That the Heisenberg equations are a particular case of planar charged particle motion in a magnetic field may be seen by considering the slightly more general problem below. (Compare this analysis with that of original Heisenberg system in Section 1.8.)

We now consider the optimal control problem

$$\min \int (u^2 + v^2)dt \tag{7.5.1}$$

subject to the equations

$$\dot{x} = u,$$
$$\dot{y} = v,$$
$$\dot{z} = A_1 u + A_2 v, \tag{7.5.2}$$

where $A_1(x, y)$ and $A_2(x, y)$ are smooth functions of x and y. (Thus $A_1 = y$ and $A_2 = -x$ recovers the Heisenberg equations.)

Form the augmented Lagrangian

$$L\left(x, \dot{x}, y, \dot{y}, z, \dot{z}, \lambda, \dot{\lambda}\right) = \tfrac{1}{2}\left(\dot{x}^2 + \dot{y}^2\right) + \frac{\lambda}{2}\left(\dot{z} - A_1\dot{x} - A_2\dot{y}\right).$$

Then the Euler–Lagrange equation for z yields $\lambda = \text{const}$, while the equations for x and y are

$$\ddot{x} = \frac{\lambda}{2}\Big(\frac{\partial A_2}{\partial x} - \frac{\partial A_1}{\partial y}\Big)\dot{y},$$
$$\ddot{y} = \frac{\lambda}{2}\Big(\frac{\partial A_1}{\partial y} - \frac{\partial A_2}{\partial x}\Big)\dot{x}. \tag{7.5.3}$$

Now let $\mathbf{A}$ be the vector $(A_1, A_2, 0)$. Then

$$(\nabla \times \mathbf{A})_z = \frac{\partial A_2}{\partial x} - \frac{\partial A_1}{\partial y} \equiv B_z.$$

Hence the Euler–Lagrange equations may be rewritten

$$\ddot{x} = \frac{\lambda}{2}B_z\dot{y},$$
$$\ddot{y} = -\frac{\lambda}{2}B_z\dot{x}, \tag{7.5.4}$$

or, if $\mathbf{v} = (\dot{x}, \dot{y}, 0)$ and $\mathbf{B} = (0, 0, B_z)$, as

$$\frac{d\mathbf{v}}{dt} + \mathbf{v} \times \frac{\lambda}{2}\mathbf{B} = 0. \tag{7.5.5}$$

This is indeed the motion of a planar charged particle in a magnetic field (see, e.g., Marsden and Ratiu [1999]), where the Lagrange multiplier is identified with a multiple of the charge.

Solution by the Maximum Principle. This problem may also be solved via the maximum principle. To do this, form the Hamiltonian

$$H = p_1 u + p_2 v + p_3(A_1 u + A_2 v) - \frac{u^2}{2} - \frac{v^2}{2}. \tag{7.5.6}$$

The optimality conditions

$$\frac{\partial H}{\partial u} = 0 = \frac{\partial H}{\partial v}$$

yield

$$u = p_1 + p_3 A_1,$$
$$v = p_2 + p_3 A_2. \tag{7.5.7}$$

Hence the optimal Hamiltonian becomes

$$H = \frac{1}{2}\left\{(p_1 + A_1 p_3)^2 + (p_2 + A_2 p_3)^2\right\}. \tag{7.5.8}$$

This is the Hamiltonian for a particle in a magnetic field. Note that to obtain the sign in, say, Marsden and Ratiu [1999] we replace A_i by $-A_i$ in the $\dot{z}$ equation. Note also that p_3 is a cyclic variable and hence is constant in time. This may be chosen to have the value e/c, the charge over the speed of light, as in equations for the particle.

Rigid Extremals. A particularly interesting phenomenon that occurs in sub-Riemannian optimal control problems is the existence of rigid extremals, i.e., isolated extremals that admit no allowable variations. In such a situation, we can obtain an optimal solution to the optimal control problem that does not satisfy the optimal control equations. We can rephrase this, as does Montgomery [1994], by saying that one has a (sub-Riemannian) geodesic that does not satisfy the geodesic equations.

We will just consider here the example of Montgomery without attempting to give a general analysis. The problem has a rather interesting history; for some details on this see the paper of Montgomery and, for example, Strichartz [1983].

We follow here the treatment in Montgomery, casting things in the language of optimal control theory. Since we will be considering an optimal trajectory with cylindrical geometry, it is best to phrase the problem in cylindrical coordinates (r, θ, z) in $\mathbb{R}^3$. We thus consider the optimal control problem

$$\min_{u,v} \int (u^2 + r^2 v^2) dt$$

subject to
$$\dot{r} = u,$$
$$\dot{\theta} = v, \qquad\qquad (7.5.9)$$
$$\dot{z} = -A(r)v,$$

where $A(r)$ is a smooth function of r with a single nondegenerate maximum at $r = 1$. Thus we require $\frac{dA}{dr}\big|_{r=1} = 0$ and $\frac{d^2 A}{dr^2}\big|_{r=1} < 0$. We can take, for example,

$$A = \frac{1}{2} r^2 - \frac{1}{4} r^4.$$

The control vector fields are

$$X_1 = \frac{\partial}{\partial r},$$
$$X_2 = \frac{\partial}{\partial \theta} - A(r) \frac{\partial}{\partial z}. \qquad (7.5.10)$$

The system is controllable, since

$$[X_1, X_2] = \frac{-dA}{dr} \frac{\partial}{\partial z}, \qquad\qquad (7.5.11)$$

so X_1, X_2, and $[X_1, X_2]$ span $\mathbb{R}^3$ everywhere except at $r = 1$. But for $r = 1$,

$$[X_1, [X_1, X_2]] = \frac{-d^2 A}{dr^2} \frac{\partial}{\partial z} \neq 0,$$

so the control vector fields still span $\mathbb{R}^3$.

Now consider the helices with pitch $A(1)$ given by $(r, \theta, z) = (1, \theta, -A(1)\theta)$. These curves clearly lie in the constraint distribution. In what follows we will sketch some arguments due to Montgomery that show that this helix is minimizing but does not satisfy the optimal control (or sub-Riemannian geodesic) equations.

Consider firstly the geodesic equations. They are given in Cartesian form by equations (7.5.5), where

$$B_z = \frac{\partial A_2}{\partial x} - \frac{\partial A_1}{\partial y}.$$

To compute B_z here observe that we need to replace $A(r)d\theta$ by $A_1 dx + A_2 dy$. Now

$$A_1 dx + A_2 dy = A(r)d\theta \qquad\qquad (7.5.12)$$

implies

$$\left(\frac{-\partial A_1}{\partial y} + \frac{\partial A_2}{\partial x} \right) dx \wedge dy = \frac{dA}{dr}\, dr \wedge d\theta. \tag{7.5.13}$$

But $dx \wedge dy = r\, dr \wedge d\theta$. Hence

$$\frac{1}{r}\frac{dA}{dr} = \frac{\partial A_2}{\partial x} - \frac{\partial A_1}{\partial y} = B_z.$$

For our given curve $r = 1$, and hence B_z is zero. But $\ddot{x}$ and $\ddot{y}$ are clearly not zero. Hence this curve cannot satisfy the geodesic equations.

We now show that the helix is an isolated point in the space of all piecewise C^1 curves in the constraint distribution with fixed endpoints. Since it is isolated, it is automatically a local minimum. However, it is not isolated in the C^0 or H^1 topologies. Montgomery, however, shows that it is still a local minimum, at least for short enough arcs; see Montgomery [1990].

Now consider a curve γ in the constraint distribution connecting the points $(x_0, y_0, 0)$ and (x_0, y_0, z_1). The projected curve onto the (x, y)-plane is thus closed.

Since $dz = -A\, d\theta$ along the curve, we have, using Stokes's theorem,

$$z_1 = -\int A\, d\theta = -\iint_\Delta \frac{dA}{dr} dr\, d\theta = -\iint_\Delta B(r) r\, dr\, d\theta,$$

where Δ is the region in the plane enclosed by the projected curve. Following our earlier magnetic analogy, this quantity can be viewed as the flux through the region by the plane enclosed by the curve.

Now recall that in our case

$$B = \frac{1}{r}\frac{dA}{dr} = 1 - r^2, \tag{7.5.14}$$

and thus B is positive in the interior of the projected unit disk and negative on the exterior. Hence, if we perturb the helix so as to push part of the projected curve into the interior of the disk, we subtract flux since B is positive in the interior. On the other hand if we push the projected curve to the exterior of the disk, we add negative flux. In either case, z_1 decreases, violating the fixed endpoint conditions. Hence *there are no allowable piecewise smooth variations of the helix*; that is, it is indeed rigid.

The Falling Cat Theorem.   The solution to this related problem discussed earlier in Section 7.1 is given as follows:

7.5.1 Theorem.   (Montgomery [1984, 1990, 1991a]). *If $q(t)$ is a (regular) optimal trajectory for the isoholonomic problem, then there exists a curve $\lambda(t) \in \mathfrak{g}^*$ such that the reduced curve $r(t)$ in $X = Q/G$ together with $\lambda(t)$ satisfies **Wong's equations***

$$\dot{p}_\alpha = -\lambda_a \mathcal{B}^a_{\alpha\beta}\dot{r}^\beta - \frac{1}{2}\frac{\partial g^{\beta\gamma}}{\partial r^\alpha} p_\beta p_\gamma,$$

$$\dot{\lambda}_b = -\lambda_a C^a_{db} \mathcal{A}^d_\alpha \dot{r}^\alpha,$$

where $g_{\alpha\beta}$ is the local representation of the metric on the base space X; that is,

$$\frac{1}{2}\|\dot{r}\|^2 = \frac{1}{2}g_{\alpha\beta}\dot{r}^\alpha\dot{r}^\beta,$$

$g^{\beta\gamma}$ *is the inverse of the matrix $g_{\alpha\beta}$, p_α is defined by*

$$p_\alpha = \frac{\partial l}{\partial \dot{r}^\alpha} = g_{\alpha\beta}\dot{r}^\beta,$$

and where we write the components of $\mathcal{A}$ as $\mathcal{A}^b_\alpha$ and similarly for its curvature $\mathcal{B}$.

Proof. As with the Heisenberg system, by general principles in the calculus of variations, given an optimal solution $q(t)$, there is a Lagrange multiplier $\lambda(t)$ such that the new action function defined on the space of curves with fixed endpoints by

$$\mathfrak{S}[q(\,\cdot\,)] = \int_0^1 \left[\frac{1}{2}\|\dot{r}(t)\|^2 + \langle\lambda(t), \mathcal{A}\dot{q}(t)\rangle\right] dt$$

has a critical point at this curve. Using the integrand as a Lagrangian, identifying $\Omega = \mathcal{A}\dot{q}$, and applying the reduced Euler–Lagrange equations discussed in Section 3.11 to the reduced Lagrangian

$$l(r, \dot{r}, \Omega) = \frac{1}{2}\|\dot{r}\|^2 + \langle\lambda, \Omega\rangle$$

then gives Wong's equations by the following simple calculations:

$$\frac{\partial l}{\partial \dot{r}^\alpha} = g_{\alpha\beta}\dot{r}^\beta; \quad \frac{\partial l}{\partial r^\alpha} = \frac{1}{2}\frac{\partial g^{\beta\gamma}}{\partial r^\alpha}\dot{r}^\beta\dot{r}^\gamma; \quad \frac{\partial l}{\partial \Omega^a} = \lambda_a.$$

The constraints are $\Omega = 0$, and so the reduced Euler–Lagrange equations become

$$\frac{d}{dt}\frac{\partial l}{\partial \dot{r}^\alpha} - \frac{\partial l}{\partial r^\alpha} = -\lambda_a(\mathcal{B}^a_{\alpha\beta}\dot{r}^\beta),$$

$$\frac{d}{dt}\lambda_b = -\lambda_a(\mathcal{E}^a_{\alpha b}\dot{r}^\alpha) = -\lambda_a C^a_{db}\mathcal{A}^d_\alpha\dot{r}^\alpha.$$

But

$$\frac{d}{dt}\frac{\partial l}{\partial \dot{r}^\alpha} - \frac{\partial l}{\partial r^\alpha} = \dot{p}_\alpha - \frac{1}{2}\frac{\partial g_{\beta\gamma}}{\partial r^\alpha}\dot{r}^\beta\dot{r}^\gamma = \dot{p}_\alpha + \frac{1}{2}\frac{\partial g^{\kappa\sigma}}{\partial r^\alpha}g_{\kappa\beta}g_{\sigma\gamma}\dot{r}^\beta\dot{r}^\gamma$$

$$= \dot{p}_\alpha + \frac{1}{2}\frac{\partial g^{\beta\gamma}}{\partial r^\alpha}p_\beta p_\gamma,$$

and so we have the desired equations. ∎

Remark. There is a rich literature on Wong's equations, and it was an important ingredient in the development of reduction theory. Some references are Sternberg [1977], Guillemin and Sternberg [1978], Weinstein [1978b], Montgomery, Marsden, and Ratiu [1984], Montgomery [1984], Koon and Marsden [1997a], and Cendra, Holm, Marsden, and Ratiu [1998].

7.6   Optimal Control of Mechanical Systems

We discuss here various optimal control problems of mechanical systems (as opposed to the kinematic case).

We recall from Section 6.7 the following formalism for a general holonomic mechanical system with inputs on a Riemannian manifold M, a smooth (infinitely differentiable), n-dimensional manifold with a Riemannian metric denoted by $g(\cdot, \cdot)$.

The norm of a vector $X_p \in T_pM$ will be denoted by

$$\|X_p\|_g = \sqrt{g(X_p, X_p)}.$$

The notion of an inertia tensor is modeled by a bundle mapping

$$J : TM \to TM$$

such that J is the identity on M. Thus for each $p \in M$ we have a linear mapping

$$J_p : T_pM \to T_pM.$$

We assume that for each p, J_p is an isomorphism satisfying for each $X_p, Y_p \in T_pM$:

(i) $g(J_pX_p, Y_p) = g(X_p, J_pY_p)$;
(ii) $g(J_pX_p, X_p) \geq 0$ ($= 0$ if and only if $X_p = 0$).

From J we may define another Riemannian metric on M by setting

$$\langle X, Y \rangle = g(JX, Y)$$

for all vector fields X, Y on M. We refer to g as the ambient metric, and $\langle \cdot, \cdot \rangle$ as the mechanical metric. The norm of a vector $X_p \in T_pM$ with respect to the mechanical metric will be denoted by

$$\|X_p\| = \sqrt{\langle X_p, X_p \rangle}.$$

The mechanical metric determines a unique Riemannian connection on M, denoted by ∇, and thereby determines a covariant derivative, denoted by $D/\partial t$.

Now let $\hat{X}_i$, $1 \le i \le N$, be $N \le n$ independent vector fields on M, and let $u_i(\cdot)$, $1 \le i \le N$, be input or control functions (real-valued functions of time). We then model a force field F by setting

$$F_{q(t)} = \sum_{i=1}^{N} u_i(t) \langle \hat{X}_i(q(t)), \cdot \rangle.$$

Let $J^* : TM \to T^*M$ denote the bundle isomorphism determined on fibers by

$$J_p^* X_p = \langle X_p, \cdot \rangle = g(J_p X_p, \cdot).$$

Thus we obtain a control system of the form

$$\frac{D^2 q}{\partial t^2} = \sum_{i=1}^{N} u_i \hat{X}_i(q). \tag{7.6.1}$$

Equation (7.6.1) represents a general holonomic mechanical system with inputs. We ignore for the purposes of exposition here a possible potential term in this equation, but it may be added without any extra difficulty.

When $F \equiv 0$ the equations (7.6.1) reduce to the geodesic equations on the Riemannian manifold $(M, \langle \cdot, \cdot \rangle)$:

$$\frac{D^2 q}{\partial t^2} = 0.$$

This flow is known to be an extremal of the variational problem

$$\min_{q} \int_0^T \left\| \frac{dq}{dt} \right\|^2 dt \tag{7.6.2}$$

over trajectories $q(t)$ with $q(0) = q_0, q(T) = q_T$.

Optimal Control on Riemannian Manifolds. We now introduce a natural optimal control problem for the system (7.6.1). To do this, we first define a norm on fibers of T^*M in the usual way:

$$\|F_q\| = \sup_{\|W_q\|_g \ne 0} \frac{|F(W_q)|}{\|W_q\|_g}. \tag{7.6.3}$$

Note that we use the ambient metric in this definition. We introduce the minimum force control problem as

$$\min_{q} \int_0^T \frac{1}{2} \|F_{q(t)}\|^2 dt \tag{7.6.4}$$

subject to the dynamics (7.6.1) and the boundary conditions

$$q(0) = q_0, \quad \frac{dq}{dt}(0) = \dot{q}_0, \quad q(T) = q_T, \quad \frac{dq}{dt}(T) = \dot{q}_T. \tag{7.6.5}$$

From (7.6.3) we obtain

$$\|F_q\| = \sup_{\|W_q\|_g \neq 0} \frac{g(J_q \frac{D^2 q}{\partial t^2}, W_q)}{\sqrt{g(W_q, W_q)}} = \left\| J_q \frac{D^2 q}{\partial t^2} \right\|_g.$$

Thus the cost functional (7.6.4) may be reformulated as

$$\min_q \int_0^T \frac{1}{2} \left\langle \frac{D^2 q}{\partial t^2}, J_q \frac{D^2 q}{\partial t^2} \right\rangle dt. \tag{7.6.6}$$

It is now natural to consider the formulation (7.6.1) of the holonomic control system. It is convenient to assume a little more structure for the force field F; we modify the definition as follows:

$$F_{q(t)} = \sum_{i=1}^N u_i(t) \left\langle J_{q(t)}^{-1} X_i(q(t)), \cdot \right\rangle, \tag{7.6.7}$$

where X_i, $1 \leq i \leq n$, is an orthonormal basis of vector fields with respect to the ambient metric,

$$g(X_i, X_j) = \delta_{ij}, \qquad 1 \leq i, j \leq n. \tag{7.6.8}$$

Clearly, in general, such a choice of basis cannot be made globally. However, if M is a Lie group, for example, then such a choice is indeed possible. With the force field (7.6.7) the system (7.6.1) may be rewritten as

$$\frac{D^2 q}{\partial t^2} = \sum_{i=1}^N J_q^{-1} X_i(q) u_i(t). \tag{7.6.9}$$

The orthonormality assumption (7.6.8) now implies that

$$\left\langle \frac{D^2 q}{\partial t^2}, J_q \frac{D^2 q}{\partial t^2} \right\rangle = \sum_{i=1}^N u_i^2(t).$$

It follows that for system (7.6.9) the minimum force control problem is defined by the cost functional

$$\min_u \int_0^T \frac{1}{2} \sum_{i=1}^N u_i^2(t) \, dt \tag{7.6.10}$$

subject to the boundary conditions (7.6.5).

In the case $N = n$ this optimal control problem corresponds to the higher-order variational problem posed by the functional (7.6.6), with boundary conditions (7.6.5). Similar higher-order variational problems have been treated in various contexts, most notably as the minimum curvature problem (Griffiths [1983], Jurdjevic [1991]) and in the context of interpolation problems in Gabriel and Kajiya [1988], Noakes, Heinzinger, and Paden [1989], and Crouch and Silva-Leite [1991]. In the latter three works a simpler functional is considered, namely,

$$\min_q \int_0^T \frac{1}{2} \left\| \frac{D^2 q}{\partial t^2} \right\|^2 dt. \tag{7.6.11}$$

The extremals of such functionals satisfy an equation of the form

$$\frac{D^4 q}{\partial t^2} + R\left(\frac{D^2 q}{\partial t^2}, \frac{Dq}{\partial t}\right) \frac{Dq}{\partial t} \equiv 0, \tag{7.6.12}$$

where R is the curvature tensor associated with the connection ∇. It follows that for $N = n$, the minimum force control problem introduced above is a natural higher-order version of the classical variational problem (7.6.2), which is often interpreted as a minimum energy problem.

Thus we wish to consider the variational problem (7.6.6)

$$\min_q J_2(q) = \min_q \int_0^T \frac{1}{2} \left\langle \frac{D^2 q}{\partial t^2}, J_q \frac{D^2 q}{\partial t^2} \right\rangle dt \tag{7.6.13}$$

subject to the boundary conditions (7.6.5)

$$q(0) = q_0, \quad \frac{dq}{dt}(0) = \dot{q}_0, \quad q(T) = q_T, \quad \frac{dq}{dt}(T) = \dot{q}_T.$$

We examine the conditions

$$\frac{d}{du} J_2(\alpha_u) \bigg|_{u=0} = 0,$$

where α is a variation of a trajectory q meeting the boundary conditions. Assuming that $\frac{\partial \alpha_0}{\partial u}(t) = Z_t$, we obtain

$$\frac{dJ_2}{du}(\alpha_u) \bigg|_{u=0} = \int_0^T \left[\left\langle \frac{D}{\partial u} \frac{D^2}{\partial t^2} \alpha_u, J_q \frac{D^2 q}{\partial t^2} \right\rangle \bigg|_{u=0} \right. $$
$$\left. + \frac{1}{2} \left\langle \frac{D^2 q}{\partial t^2}, (\nabla_{Z_t} J)_q \left(\frac{D^2 q}{\partial t^2}\right) \right\rangle \right] dt.$$

The first term in the integrand is rewritten, using the techniques in Milnor [1963], in the form

$$\left\langle \frac{D^2}{\partial t^2} Z_t + R\left(Z_t, \frac{Dq}{\partial t}\right) \frac{Dq}{\partial t}, J_q \frac{D^2 q}{\partial t^2} \right\rangle.$$

Using an identity for the curvature tensor R (p. 53 of Milnor [1963]), integration by parts, and the boundary conditions

$$Z_0 = Z_T = \frac{DZ_0}{\partial t} = \frac{DZ_T}{\partial t} = 0,$$

we obtain

$$\frac{d}{du} J_2(\alpha_u)\bigg|_{u=0} = \int_0^T \left[\left\langle Z_t, \frac{D^2}{\partial t^2} J_q \frac{D^2 q}{\partial t^2} + R\left(J_q \frac{D^2 q}{\partial 2}, \frac{Dq}{\partial t} \right) \frac{Dq}{\partial t} \right\rangle \right.$$
$$\left. + \frac{1}{2} \left\langle \frac{D^2 q}{\partial t^2}, (\nabla_{Z_t} J)_q \left(\frac{D^2 q}{\partial t^2} \right) \right\rangle \right] dt.$$

Define the mapping $J' : TM \times TM \to TM$ by setting

$$\langle J'(X,Y), Z \rangle = \frac{1}{2} \langle (\nabla_Z J)(X), Y \rangle$$

for all vector fields X, Y, and Z on M. We obtain the following result.

7.6.1 Lemma. *Necessary conditions for the variational problem (7.6.6) subject to the boundary conditions (7.6.5) are given by*

$$\frac{D^2}{\partial t^2} J_q \frac{D^2 q}{\partial t^2} + R\left(J_q \frac{D^2 q}{\partial t^2}, \frac{Dq}{\partial t} \right) \frac{Dq}{\partial t} + J'_q \left(\frac{D^2 q}{\partial t^2}, \frac{D^2 q}{\partial t^2} \right) \equiv 0. \qquad (7.6.14)$$

This equation generalizes equation (7.6.12), corresponding to the class where J is the identity.

Optimal Control on Lie Groups   We now specialize to the case where M is a compact semisimple Lie group G, with right-invariant mechanical metric defined by a positive definite mapping J on the Lie algebra $\mathfrak{g}$. In this case we have the following result.

7.6.2 Lemma. *In the situation above, for $X, Y, Z \in \mathfrak{g}$,*

$$(\nabla_{Z_g^r} J_g)(X_g^r) = [(\nabla_Z J)(X)]_g^r,$$
$$J'_g(X_g^r, Y_g^r) = [J'(X,Y)]_g^r,$$

where

$$(\nabla_Z J)(X) = \tfrac{1}{2}\{[JZ, X] + J[X, Z] + J^{-1}[Z, J^2 X] + J^{-1}[JX, JZ]\} \qquad (7.6.15)$$

and

$$J'(X,Y) = \tfrac{1}{4}\{[X, JY] + [Y, JX] + J^{-1}[J^2 Y, X] + J^{-1}[J^2 X, Y]\}. \qquad (7.6.16)$$

Proof.

$$0 = Z^r \left\langle J_g(X_g^r), Y_g^r \right\rangle$$

$$= \left\langle \nabla_{Z_g^r} J_g(X_g^r), Y_g^r \right\rangle + \left\langle J_g(\nabla_{Z_g^r} X_g^r), Y_g^r \right\rangle + \left\langle J_g(X_g^r), \nabla_{Z_g^r} X_g^r \right\rangle$$

$$= \langle\!\langle JY, (\nabla_Z J)(X) \rangle\!\rangle + \langle\!\langle JY, J(\nabla_Z X) \rangle\!\rangle + \langle\!\langle JX, J(\nabla_Z Y) \rangle\!\rangle,$$

where $\langle\!\langle \cdot, \cdot \rangle\!\rangle = -\frac{1}{2}\kappa(\cdot, \cdot)$, where κ is the Killing form on $\mathfrak{g}$.

Corresponding to the right-invariant metric $\langle \cdot, \cdot \rangle$ there exists a unique Riemannian connection ∇. Explicit formulas for ∇ are given in Arnold [1989] and Nomizu [1954].

Specifically, ∇ defines a bilinear form on $\mathfrak{g}$,

$$(X, Y) \mapsto \nabla_X Y = \tfrac{1}{2}\left\{ [X, Y] + J^{-1}[X, JY] + J^{-1}[Y, JX] \right\}, \qquad (7.6.17)$$

where ∇ is extended to right-invariant vector fields on G by setting

$$(\nabla_{X^r} Y^r)(g) = (\nabla_X Y)_g^r, \quad g \in G, \quad X, Y \in \mathfrak{g},$$

Thus we obtain

$$\langle\!\langle JY, (\nabla_Z J)(X) \rangle\!\rangle = \left\langle\!\!\left\langle JY, -\tfrac{1}{2}\{J[Z, X] + [Z, JX] + [X, JZ]\} \right\rangle\!\!\right\rangle$$
$$+ \left\langle\!\!\left\langle JX, -\tfrac{1}{2}\{J[Z, Y] + [Z, JY] + [Y, JZ]\} \right\rangle\!\!\right\rangle.$$

Since the operator ad satisfies

$$\langle\!\langle \mathrm{ad}_Y(X), Z \rangle\!\rangle + \langle\!\langle X, \mathrm{ad}_Y(Z) \rangle\!\rangle = 0, \ X, Y, Z \in \mathfrak{g}, \qquad (7.6.18)$$

we may be rewrite this as

$$-\tfrac{1}{2}\langle\!\langle JY, \{J[Z, X] + [Z, JX] + [X, JZ]$$
$$- J^{-1}[Z, J^2 X] - [Z, JX] + J^{-1}[JZ, JX]\} \rangle\!\rangle,$$

from which we obtain

$$\left\langle J_g'(X_g^r, Y_g^r), Z_g^r \right\rangle = \langle\!\langle J'(X, Y), JZ \rangle\!\rangle = \tfrac{1}{2} \langle\!\langle (\nabla_Z J)(X), JY \rangle\!\rangle.$$

Thus

$$\langle\!\langle J'(X, Y), JZ \rangle\!\rangle$$
$$= \tfrac{1}{4} \langle\!\langle JY, \{[JZ, X] + J[X, Z] + J^{-1}[Z, J^2 X] + J^{-1}[JX, JZ]\} \rangle\!\rangle.$$

Using the identity (7.6.18) we obtain (7.6.16). ∎

In this special case we can reformulate the necessary conditions (7.6.14) on the Lie algebra $\mathfrak{g}$, using a right-invariant frame on G to express vector fields. In particular, if

$$W_t = \sum_{i=1}^{L} w_i(t) X_i \in \mathfrak{g},$$

then we set

$$\frac{DW_t}{\partial t} = \frac{\partial W_t}{\partial t} + \nabla_{V_t} W_t,$$

where

$$\frac{\partial W_t}{\partial t} = \sum_{i=1}^{L} \dot{w}_i(t) X_i,$$

and ∇ is given in equation (7.6.17). Thus

$$\frac{Dg}{\partial t} = [V_t]^r_g, \qquad \frac{D^2 g}{\partial t^2} = \left(\frac{\partial V_t}{\partial t} + \nabla_{V_t} W_t\right)^r_g = \left(\frac{DV_t}{\partial t}\right)^r_g.$$

From Lemma 7.6.2, equation (7.6.14) can be written in this special case as the following system of equations on $\mathfrak{g}$:

$$\frac{D^2}{\partial t^2} J \frac{DV_t}{\partial t} + R\left(J\frac{DV_t}{\partial t}, V_t\right) V_t + \frac{1}{2}\left[\frac{DV_t}{\partial t}, J\frac{DV_t}{\partial t}\right]$$

$$+ \frac{1}{2} J^{-1}\left[J^2 \frac{DV_t}{\partial t}, \frac{DV_t}{\partial t}\right] = 0. \tag{7.6.19}$$

Nonholonomic Optimal Control and the Rolling Ball. Using a synthesis of the techniques used above for the Heisenberg system and the falling cat problem, Koon and Marsden [1997a] generalized these problems to the nonholonomic case. In addition, these methods allow one to treat the falling cat problem even in the case that the angular momentum is not zero.

In this process the momentum equation plays the role of the constraint. It is inserted as a first-order differential constraint on the nonholonomic momentum. See also the work Koon and Marsden [1997a] as well as the Internet Supplement. Some related work on the optimal control of nonholonomic systems with symmetry may be found in Cortés and Martínez [2000].

Here we consider optimal control of the rolling ball (see the introduction and Section 7.3 for notation). We suppose here that $J = \alpha I_3$, I_3 the 3×3 identity matrix, in which case the equations become, after suitable state feedback, the following control system on $\mathbb{R}^4 \times SO(3)$:

$$\ddot{x} = \widetilde{u}_1,$$

$$\ddot{y} = \widetilde{u}_2, \tag{7.6.20}$$

$$\dot{P} = PS(-\dot{x}\mathbf{e}_2 + \dot{y}\mathbf{e}_1 + \bar{c}\mathbf{e}_3); \quad \bar{c} = c/\alpha.$$

This is evidently controllable (it is accessible, and the uncontrolled trajectories are periodic).

The obvious minimum energy control problem is

$$\min_u \int_0^T \frac{1}{2}(\widetilde{u}_1^2 + \widetilde{u}_2^2)dt \tag{7.6.21}$$

subject to (7.6.20).

This may be viewed as the following constrained variational problem on $SO(3)$. Set $q = P$ and

$$\frac{D\dot{q}}{dt} = \dot{y}X_1 - \dot{x}X_1 + \bar{c}X_3,$$

$$X_i(q) = PS(e_i). \qquad (7.6.22)$$

Then the variational problem may be posed as

$$\min_q \int_0^T \left\langle \frac{D^2q}{dt^2}, \frac{D^2q}{dt^2} \right\rangle dt$$

subject to (7.6.22).

In contrast with minimum energy holonomic control problems, this nonholonomic problem introduces constraints into the variational problem. However, this now falls into the class of problems analyzed by Crouch and Leite [1991]. Defining the vector of Lagrange multipliers by $\Lambda = \sum_i \lambda_i X_i$, the resulting extremals satisfy

$$x^{(4)} = \bar{c}y^{(3)} + \lambda_3\dot{y} - \bar{c}\lambda_1,$$

$$y^{(4)} = -\bar{c}x^{(3)} - \lambda_3\dot{x} - \bar{c}\lambda_2,$$

$$\dot{\lambda}_3 = -\dot{y}x^{(3)} + \dot{x}y^{(3)} - \lambda_2\dot{y} - \lambda_1\dot{x}, \qquad (7.6.23)$$

$$\dot{\lambda}_1 = 0,$$

$$\dot{\lambda}_2 = 0.$$

We remark that these equations are variational and are *not* equivalent to a nonholonomic problem where one takes the accelerations to lie in some distribution on the second tangent bundle.

Exercises

⋄ **7.6-1.** Compute the optimal control equations for the minimum energy control problem for the dynamic rolling penny.

8
Stability of Nonholonomic Systems

In Chapter 3 we briefly discussed the energy-momentum method for analyzing stability of relative equilibria of mechanical systems. In the nonholonomic case, while energy is conserved, momentum generally is not. In some cases, however, the momentum equation is integrable, leading to invariant surfaces that make possible an energy-momentum analysis similar to that in the holonomic case. When the momentum equation is not integrable, one can get asymptotic stability in certain directions, and the stability analysis is rather different from that in the holonomic case. Nonetheless, to show stability we will make use of the conserved energy and the dynamic momentum equation.

8.1 The Nonholonomic Energy-Momentum Method

Here we analyze the stability of relative equilibria for nonholonomic mechanical systems with symmetry using an energy-momentum analysis for nonholonomic systems that is analogous to that for holonomic systems given in Simo, Lewis, and Marsden [1991] and discussed briefly in Chapter 3. This section is based on the paper Zenkov, Bloch, and Marsden [1998] and follows the the spirit of the paper by Bloch, Krishnaprasad, Marsden, and Murray [1996], hereinafter referred to as [BKMM] as in Chapter 5. We will illustrate our energy-momentum stability analysis with a low-dimensional model example, and then with several mechanical examples of interest including the falling disk, the roller racer, and the rattleback top.

As discussed in Chapter 5, symmetries do not always lead to conservation laws as in the classical Noether theorem, but rather to an interesting *momentum equation*. The momentum equation has the structure of a parallel transport equation for the momentum corrected by additional terms. This parallel transport occurs in a certain vector bundle over shape space; this geometry is explained in detail in Cendra, Marsden, and Ratiu [2001b]. In some instances such as the Routh problem of a sphere rolling inside a surface of revolution (see Zenkov [1995]) this equation is *pure transport*, and in fact is integrable (the curvature of the transporting connection is zero). This leads to nonexplicit conservation laws. In other important instances, the momentum equation is *partially integrable* in a sense that we shall make precise. Our goal is to make use of, as far as possible, the energy-momentum approach to stability for Hamiltonian systems.

Some History of the Energy-Momentum Method. This method which was discussed in some detail in Section 3.12 goes back to fundamental work of Routh (and many others in this era), and in more modern works, that of Arnold [1966a] and Smale [1970], and Simo, Lewis, and Marsden [1991] (see, for example, Marsden [1992] for an exposition and additional references). This method has also been important in control, as detailed and referenced elsewhere in this book (but see for example, Bloch, Krishnaprasad, Marsden, and Alvarez [1992], Bloch, Leonard, and Marsden [2000], Ortega, van der Schaft, Mashcke and Escobar [1999], van der Schaft [1986]). Other useful references on stability that contain many references of interest are Mikhailov and Parton [1990] and Merkin [1997].

Because of the nature of the momentum equation, the analysis we present is rather different in several important respects. In particular, our energy-momentum analysis varies according to the structure of the momentum equation, and correspondingly, we divide our analysis into several parts.

8.1.1 Example (A Mathematical Example). We now consider an instructive, but (so far as we know) nonphysical example. Unlike the rolling disk, for example (see below), it has asymptotically stable relative equilibria, and is a simple example that exhibits the richness of stability in nonholonomic systems. The general theorems presented later are well illustrated by this example, and the reader may find it helpful to return to it again later.

Consider a Lagrangian on $T\mathbb{R}^3$ of the form

$$L\left(r^1, r^2, s, \dot{r}^1, \dot{r}^2, \dot{s}\right) = \frac{1}{2}\left\{\left(1 - \left[a\left(r^1\right)\right]^2\right)\left(\dot{r}^1\right)^2 - 2a\left(r^1\right)b\left(r^1\right)\dot{r}^1\dot{r}^2\right.$$

$$\left. + \left(1 - \left[b\left(r^1\right)\right]^2\right)\left(\dot{r}^2\right)^2 + \dot{s}^2\right\} - V\left(r^1\right),$$

$$(8.1.1)$$

where a, b, and V are given real-valued functions of a single variable. We consider the nonholonomic constraint

$$\dot{s} = a\left(r^1\right)\dot{r}^1 + b\left(r^1\right)\dot{r}^2. \tag{8.1.2}$$

Using the definitions (see Section 5.2 and equation (5.2.8)), straightforward computations show that

$$B_{12} = \frac{\partial b}{\partial r^1} = -B_{12},$$

where we suppress the index for the s-variable, since there is only one such variable.

The constrained Lagrangian is

$$L_c = \frac{1}{2}\left\{\left(\dot{r}^1\right)^2 + \left(\dot{r}^2\right)^2\right\} - V(r^1),$$

and the equations of motion, namely,

$$\frac{d}{dt}\frac{\partial L_c}{\partial \dot{r}^\alpha} - \frac{\partial L_c}{\partial r^\alpha} = -\dot{s}B_{\alpha\beta}\dot{r}^\beta,$$

become

$$\frac{d}{dt}\frac{\partial L_c}{\partial \dot{r}^1} - \frac{\partial L_c}{\partial r^1} = -\dot{s}B_{12}\dot{r}^2, \qquad \frac{d}{dt}\frac{\partial L_c}{\partial \dot{r}^2} = \dot{s}B_{12}\dot{r}^1.$$

The Lagrangian is independent of r^2, and correspondingly, we introduce the nonholonomic momentum defined by

$$p = \frac{\partial L_c}{\partial \dot{r}^2}.$$

Taking into account the constraint equation and the equations of motion above, we can rewrite the equations of motion in the form

$$\ddot{r}^1 = -\frac{\partial V}{\partial r^1} - \frac{\partial b}{\partial r^1}\left(a\left(r^1\right)\dot{r}^1 + b\left(r^1\right)p\right)p, \tag{8.1.3}$$

$$\dot{p} = \frac{\partial b}{\partial r^1}\left(a\left(r^1\right)\dot{r}^1 + b\left(r^1\right)p\right)\dot{r}^1. \tag{8.1.4}$$

Observe that the momentum equation does not, in any obvious way, imply a conservation law.

A *relative equilibrium* is a point (r_0, p_0) that is an equilibrium modulo the variable r^2; thus, from equations (8.1.3) and (8.1.4), we require $\dot{r}^1 = 0$ and

$$\frac{\partial V}{\partial r^1}\left(r_0^1\right) + \frac{\partial b}{\partial r^1}b\left(r_0^1\right)p_0^2 = 0.$$

We shall see in Section 8.4 that the relative equilibria defined by these conditions are Lyapunov stable and in addition asymptotically stable in certain directions if the following two stability conditions are satisfied:

(i) The energy function $E = \frac{1}{2}\left(\dot{r}^1\right)^2 + \frac{1}{2}p^2 + V$, which has a relative critical point at (r_0, p_0) (i.e., when E is restricted to leaves of the integrable transport distribution described in Section 8.4), has a positive definite second derivative at this point.

(ii) The derivative of E along the flow of the auxiliary system

$$\ddot{r}^1 = -\frac{\partial V}{\partial r^1}\frac{\partial b}{\partial r^1}\left(a\left(r^1\right)\dot{r}^1 + b\left(r^1\right)p\right)p, \quad \dot{p} = \frac{\partial b}{\partial r^1}\left(r^1\right)p\dot{r}^1$$

is strictly negative.   ◆

While the above example is mathematical in nature, there are two physical examples that are key for this chapter; these are the roller racer and the rattleback, which were introduced in Sections 1.10 and 1.11. We shall return to these shortly.

8.2   Overview

In this part of the book we make the following assumption:

Skew Symmetry Assumption.   *We assume that the tensor $C_{ab}^c I^{ad}$ is skew-symmetric in c, d.* (See Theorem 5.7.3 for the definition of these quantities.)

This holds for many physical examples and certainly for the systems discussed here. (Exceptions include systems with no shape space such as the homogeneous sphere on the plane and certain cases of the "Suslov" problem of an inhomogeneous rigid body subject to a linear constraint in the angular velocities.) The preceding assumption is an intrinsic (coordinate-independent) condition, since $C_{ab}^c I^{ad}$ represents an intrinsic bilinear map of $\left(\mathfrak{g}^{\mathcal{D}}\right)^* \times \left(\mathfrak{g}^{\mathcal{D}}\right)^*$ to $\left(\mathfrak{g}^{\mathcal{D}}\right)^*$.

Under this assumption, the terms quadratic in p in the momentum equation vanish. We may write the equations of motion in terms of the constrained Routhian R defined in Section 5.7, equation(5.7.3). We obtain (see equations (5.7.4) and (5.7.5))

$$\frac{d}{dt}\frac{\partial R}{\partial \dot{r}^\alpha} - \frac{\partial R}{\partial r^\alpha} = -\mathcal{D}_{b\alpha}^c I^{bd}p_c p_d - \mathcal{B}_{\alpha\beta}^c p_c \dot{r}^\beta$$

$$- \mathcal{D}_{\beta ab}I^{bc}p_c \dot{r}^\beta - \mathcal{K}_{\alpha\beta\gamma}\dot{r}^\beta \dot{r}^\gamma, \qquad (8.2.1)$$

$$\frac{d}{dt}p_b = \mathcal{D}_{b\alpha}^c p_c \dot{r}^\alpha + \mathcal{D}_{\alpha\beta b}\dot{r}^\alpha \dot{r}^\beta. \qquad (8.2.2)$$

As a result, *the dimension of the family of the relative equilibria equals the number of components of the (nonholonomic) momentum map.*

In the case where $C^c_{ab} = 0$ (such as when we have cyclic variables, that is, internal abelian symmetries) the matrix $C^c_{ab} I^{ad}$ vanishes, and the preceding equations of motion are the same as those obtained by Karapetyan [1983].

Below, three principal cases will be considered:

1. **Pure Transport Case.** In this case, terms quadratic in $\dot{r}$ are not present in the momentum equation, so it is in the form of a transport equation; i.e., the momentum equation is an equation of parallel transport, and the equation itself defines the relevant connection.

   Under certain integrability conditions given below, the transport equation defines invariant surfaces, which allow us to use a type of energy-momentum method for stability analysis in a similar fashion to the manner in which the holonomic case uses the level surfaces defined by the momentum map. The key difference is that in our case, the additional invariant surfaces do not arise from conservation of momentum. In this case, one gets stable, but not asymptotically stable, relative equilibria. Examples include the rolling disk, a body of revolution rolling on a horizontal plane, and the Routh problem.

2. **Integrable Transport Case.** In this case, terms quadratic in $\dot{r}$ are present in the momentum equation, and thus it is not a pure transport equation. However, in this case, we assume that the transport part is integrable. As we shall also see, in this case relative equilibria may be asymptotically stable. We are able to find a generalization of the energy-momentum method that gives conditions for asymptotic stability. An example is the roller racer.

3. **Nonintegrable Transport Case.** Again, the terms quadratic in $\dot{r}$ are present in the momentum equation, and thus it is not a pure transport equation. However, the transport part is not integrable. Again, we are able to demonstrate asymptotic stability using the Lyapunov–Malkin theorem and to relate it to an energy-momentum-type analysis under certain eigenvalue hypotheses, as we will see in Section 8.5. An example is the rattleback top. Another example is an inhomogeneous sphere with a center of mass lying off the planes spanned by the principal axis body frame as discussed in Markeev [1992].

   These eigenvalue hypotheses do not hold in some examples, such as the inhomogeneous (unbalanced) Kovalevskaya sphere rolling on the plane.

In the sections below where these different cases are discussed we will make clear at the beginning of each section what the underlying hypotheses on the systems are by listing the key hypotheses and labeling them by H1, H2, and H3.

8.3   The Pure Transport Case.

Here we assume that:

H1 $\mathcal{D}_{\alpha\beta b}$ are skew-symmetric in α, β. Under this assumption, the momentum equation can be written as the vanishing of the connection one-form defined by $dp_b - \mathcal{D}_{b\alpha}^c p_c dr^\alpha$.

H2 The curvature of the preceding connection form is zero.

An interesting example of this case is that of Routh's problem of a sphere rolling without slipping in a surface of revolution. See Zenkov [1995].

Under the above two assumptions, the distribution defined by the momentum equation is integrable, and so we get invariant surfaces, which makes further reduction possible. These hypotheses enable us to use the energy-momentum method in a way that is similar to the holonomic case.

If the number of shape variables is one, the above connection is integrable, because it may be treated as a system of linear ordinary differential equations with coefficients depending on the shape variable r:

$$\frac{dp_b}{dr} = \mathcal{D}_b^c p_c.$$

As a result, we obtain an integrable nonholonomic system, because after solving the momentum equation for p_b and substituting the result in the equation for the shape variable, the latter equation may be viewed as a Lagrangian system with one degree of freedom, which is integrable.

Energy-Momentum for The Pure Transport Case. We now develop the energy-momentum method for the case in which the momentum equation is pure transport. Under the assumptions H1 and H2 made so far, the equations of motion become

$$\frac{d}{dt}\frac{\partial R}{\partial \dot{r}^\alpha} - \frac{\partial R}{\partial r^\alpha} = -\mathcal{D}_{b\alpha}^c I^{bd} p_c p_d - \mathcal{B}_{\alpha\beta}^c p_c \dot{r}^\beta - \mathcal{K}_{\alpha\beta\gamma}\dot{r}^\beta\dot{r}^\gamma, \qquad (8.3.1)$$

$$\frac{d}{dt}p_b = \mathcal{D}_{b\alpha}^c p_c \dot{r}^\alpha. \qquad (8.3.2)$$

A *relative equilibrium* is a point $(r, \dot{r}, p) = (r_0, 0, p_0)$ that is a fixed point for the dynamics determined by equations (8.3.1) and (8.3.2). Under assumption H1 the point (r_0, p_0) is seen to be a critical point of the amended potential (see equation (5.7.8)).

Because of our zero curvature assumption H2, the solutions of the momentum equation lie on surfaces of the form $p_a = P_a(r^\alpha, k_b)$, $a, b = 1, \ldots, m$, where k_b are constants labeling these surfaces.

Using the functions $p_a = P_a(r^\alpha, k_b)$ we introduce the **reduced amended potential** $U_k(r^\alpha) = U(r^\alpha, P_a(r^\alpha, k_b))$. We think of the function $U_k(r^\alpha)$ as being the restriction of the function U to the invariant manifold

$$Q_k = \{(r^\alpha, p_a) \mid p_a = P_a(r^\alpha, k_b)\}.$$

8.3.1 Theorem. *Let assumptions H1 and H2 hold and let (r_0, p_0), where $p_0 = P(r_0, k^0)$, be a relative equilibrium. If the reduced amended potential $U_{k^0}(r)$ has a nondegenerate minimum at r_0, then this equilibrium is Lyapunov stable.*

Proof. First, we show that the relative equilibrium

$$r^\alpha = r_0^\alpha, \quad p_a^0 = P_a(r_0^\alpha, k_b^0) \tag{8.3.3}$$

of the system (8.3.1), (8.3.2) is stable modulo perturbations consistent with Q_{k^0}. Consider the phase flow restricted to the invariant manifold Q_{k^0}, where k^0 corresponds to the relative equilibrium. Since $U_{k^0}(r^\alpha)$ has a nondegenerate minimum at r_0^α, the function $E|_{Q_{k^0}}$ is positive definite. By Theorem 5.7.4 its derivative along the flow vanishes. Using $E|_{k^0}$ as a Lyapunov function, we conclude that equations (8.3.1), (8.3.2), restricted to the manifold Q_{k^0}, have a stable equilibrium point r_0^α on Q_{k^0}.

To finish the proof, we need to show that equations (8.3.1), (8.3.2), restricted to nearby invariant manifolds Q_k, have stable equilibria on these manifolds.

If k is sufficiently close to k^0, then by the properties of families of Morse functions (see Milnor [1963]), the function $U_k : Q_k \to \mathbb{R}$ has a nondegenerate minimum at the point r^α that is close to r_0^α. This means that for all k sufficiently close to k^0 the system (8.3.1), (8.3.2) restricted to Q_k has a stable equilibrium r^α. Therefore, the equilibrium (8.3.3) of equations (8.3.1), (8.3.2) is stable.

The stability here cannot be asymptotic, since the dynamical systems on Q_k have a positive definite conserved quantity: the reduced energy function. ∎

Remark. Even though in general $P_a(r^\alpha, k_b)$ cannot be found explicitly, the types of critical points of U_k may be explicitly determined as follows. First of all, note that

$$\frac{\partial p_a}{\partial r^\alpha} = \mathcal{D}_{b\alpha}^c p_c$$

as long as $(r^\alpha, p_a) \in Q_k$. Therefore,

$$\frac{\partial U_k}{\partial r^\alpha} = \nabla_\alpha U,$$

where

$$\nabla_\alpha = \frac{\partial}{\partial r^\alpha} + \mathcal{D}_{b\alpha}^c p_c \frac{\partial}{\partial p_b}. \tag{8.3.4}$$

Then the relative equilibria satisfy the condition

$$\nabla_\alpha U = 0,$$

while the condition for stability

$$\frac{\partial^2 U_k}{\partial r^2} \gg 0$$

(i.e., positive definiteness) becomes the condition

$$\nabla_\alpha \nabla_\beta U \gg 0.$$

In the commutative case this was shown by Karapetyan [1983].

Now we give the stability condition in a form similar to that of the energy-momentum method for holonomic systems given in Simo, Lewis, and Marsden [1991].

8.3.2 Theorem (The nonholonomic energy-momentum method). *Under assumptions H1 and H2, the point $q_e = (r_0^\alpha, p_a^0)$ is a relative equilibrium if and only if there is a $\xi \in \mathfrak{g}^{q_e}$ such that q_e is a critical point of the* **augmented energy** $E_\xi : \mathcal{D}/G \to \mathbb{R}$ *(i.e., E_ξ is a function of $(r, \dot{r}, p)$), defined by*

$$E_\xi = E - \langle p - P(r, k), \xi \rangle.$$

This equilibrium is stable if $\delta^2 E_\xi$ restricted to $T_{q_e} Q_k$ is positive definite (here δ denotes differentiation with respect to all variables except ξ).

Proof.   A point $q_e \in Q_k$ is a relative equilibrium if $\partial_{r^\alpha} U_k = 0$. This condition is equivalent to $d\left(E|_{Q_k}\right) = 0$. The last equation may be represented as $d(E - \langle p - P(r, k), \xi \rangle) = 0$ for some $\xi \in \mathfrak{g}^{q_e}$. Similarly, the condition for stability $d^2 U_k \gg 0$ is equivalent to $d^2\left(E|_{Q_k}\right) \gg 0$, which may be represented as $\left(\delta^2 E_\xi\right)|_{T_{q_e} Q_k} \gg 0$.   ∎

Note that if the momentum map is preserved by the dynamics, then the formula for E_ξ becomes

$$E_\xi = E - \langle p - k, \xi \rangle,$$

which is the same as the formula for the augmented energy E_ξ for holonomic systems.

8.3.3 Example (The Falling Rolling Disk).   There are several examples that illustrate the ideas above. For instance, the falling disk, Routh's problem, and a body of revolution rolling on a horizontal plane are systems where the momentum equation defines an integrable distribution and we are left with only one shape variable. Since the stability properties of all these systems are similar, we consider here only the rolling disk. For the body of revolution on the plane see Chaplygin [1897a] and Karapetyan [1983]. For the Routh problem, see Zenkov [1995].

Consider again the disk rolling without sliding on the xy-plane. Recall that we have the following: Denote the coordinates of contact of the disk in the xy-plane by x, y. Let θ, φ, and ψ denote the angle between the plane of the disk and the vertical axis, the "heading angle" of the disk, and "self-rotation" angle of the disk respectively, as introduced earlier.

The Lagrangian and the constraints in these coordinates are given by

$$L = \frac{m}{2}\left[(\xi - R(\dot{\varphi}\sin\theta + \dot{\psi}))^2 + \eta^2\sin^2\theta + (\eta\cos\theta + R\dot{\theta})^2\right]$$
$$+ \frac{1}{2}\left[A(\dot{\theta}^2 + \dot{\varphi}^2\cos^2\theta) + B(\dot{\varphi}\sin\theta + \dot{\psi})^2\right] - mgR\cos\theta,$$
$$\dot{x} = -\dot{\psi}R\cos\varphi,$$
$$\dot{y} = -\dot{\psi}R\sin\varphi,$$

where $\xi = \dot{x}\cos\varphi + \dot{y}\sin\varphi + R\dot{\psi}$, $\eta = -\dot{x}\sin\varphi + \dot{y}\cos\varphi$. Note that the constraints may be written as $\xi = 0$, $\eta = 0$.

This system is invariant under the action of the group $G = \mathrm{SE}(2)\times\mathrm{SO}(2)$; the action by the group element (a,b,α,β) is given by

$$(\theta,\varphi,\psi,x,y) \mapsto (\theta,\varphi+\alpha,\psi+\beta, x\cos\alpha - y\sin\alpha + a, x\sin\alpha + y\cos\alpha + b).$$

Obviously,

$$T_q\,\mathrm{Orb}(q) = \mathrm{span}\left(\frac{\partial}{\partial\varphi}, \frac{\partial}{\partial\psi}, \frac{\partial}{\partial x}, \frac{\partial}{\partial y}\right),$$

and

$$\mathcal{D}_q = \mathrm{span}\left(\frac{\partial}{\partial\theta}, \frac{\partial}{\partial\varphi}, R\cos\varphi\frac{\partial}{\partial x} + R\sin\varphi\frac{\partial}{\partial y} - \frac{\partial}{\partial\psi}\right),$$

which imply

$$\mathcal{S}_q = \mathcal{D}_q \cap T_q\,\mathrm{Orb}(q) = \mathrm{span}\left(\frac{\partial}{\partial\varphi}, -R\cos\varphi\frac{\partial}{\partial x} - R\sin\varphi\frac{\partial}{\partial y} + \frac{\partial}{\partial\psi}\right).$$

Choose vectors $(1,0,0,0)$, $(0,1,0,0)$, $(0,0,1,0)$, $(0,0,0,1)$ as a basis of the Lie algebra $\mathfrak{g}$ of the group G. The corresponding generators are

$$\partial_x, \quad \partial_y, \quad -y\partial_x + x\partial_y + \partial_\varphi, \quad \partial_\psi.$$

Taking into account that the generators ∂_φ, $-R\cos\varphi\,\partial_x - R\sin\varphi\,\partial_y + \partial_\psi$ correspond to the elements $(y,-x,1,0)$, $(-R\cos\varphi, -R\sin\varphi, 0, 1)$ of the Lie algebra $\mathfrak{g}$, we obtain the following momentum equations:

$$\dot{p}_1 = mR^2\cos\theta\,\dot{\theta}\dot{\psi},$$
$$\dot{p}_2 = -mR^2\cos\theta\,\dot{\theta}\dot{\varphi}, \tag{8.3.5}$$

where

$$p_1 = A\dot{\varphi}\cos^2\theta + (mR^2 + B)(\dot{\varphi}\sin\theta + \dot{\psi})\sin\theta,$$
$$p_2 = (mR^2 + B)(\dot{\varphi}\sin\theta + \dot{\psi}), \tag{8.3.6}$$

into which the constraints have been substituted. One may notice that

$$p_1 = \frac{\partial l_c}{\partial \dot{\varphi}}, \quad p_2 = \frac{\partial l_c}{\partial \dot{\psi}}.$$

Solving (8.3.6) for $\dot{\varphi}$, $\dot{\psi}$ and substituting the solutions in the equations (8.3.5) we obtain another representation of the momentum equations:

$$\frac{dp_1}{dt} = mR^2 \cos \theta \left(-\frac{\sin \theta}{A \cos^2 \theta} p_1 + \left(\frac{1}{mR^2 + B} + \frac{\sin^2 \theta}{A \cos^2 \theta} \right) p_2 \right) \dot{\theta},$$

$$\frac{dp_2}{dt} = mR^2 \cos \theta \left(-\frac{1}{A \cos^2 \theta} p_1 + \frac{\sin \theta}{A \cos^2 \theta} p_2 \right) \dot{\theta}. \tag{8.3.7}$$

The right-hand sides of (8.3.7) do not have terms quadratic in the shape variable θ. The distribution, defined by (8.3.7), is integrable and defines two integrals of the form $p_1 = P_1(\theta, k_1, k_2)$, $p_2 = P_2(\theta, k_1, k_2)$. It is known that these integrals may be written down explicitly in terms of the hypergeometric function. See Appel [1900], Chaplygin [1897a], and Korteweg [1899] for details.

To carry out stability analysis, we use the remark following Theorem 8.3.1. Using formulae (8.3.6), we obtain the amended potential

$$U(\theta, p) = \frac{1}{2} \left[\frac{(p_1 - p_2 \sin \theta)^2}{A \cos^2 \theta} + \frac{p_2^2}{B + mR^2} \right] + mgR \cos \theta.$$

Straightforward computation shows that the positive definiteness condition for stability, $\nabla^2 U \gg 0$, of a relative equilibrium $\theta = \theta_0$, $p_1 = p_1^0$, $p_2 = p_2^0$ becomes

$$\frac{B}{A(mR^2 + B)} (p_2^0)^2 + \frac{mR^2 \cos^2 \theta_0 + 2A \sin^2 \theta_0 + A}{A^2} (p_1^0 - p_2^0 \sin \theta_0)^2$$
$$- \frac{(mR^2 + 3B) \sin \theta_0}{A(mR^2 + B) \cos^2 \theta_0} (p_1^0 - p_2^0 \sin \theta_0) p_2^0 - mgR \cos \theta_0 > 0.$$

Note that this condition guarantees stability here relative to θ, $\dot{\theta}$, p_1, p_2; in other words, we have stability modulo the action of SE(2) × SO(2). ◆

The falling disk may be considered as a limiting case of the body of revolution that also has an integrable pure transport momentum equation (this example is treated in Chaplygin [1897a] and Karapetyan [1983]). The rolling disk has also been analyzed by O'Reilly [1996] and Cushman, Hermans, and Kemppainen [1996]. O'Reilly considered bifurcation of relative equilibria, the stability of vertical stationary motions, as well as the possibility of sliding.

8.4   The Nonpure Transport Case

In this section we consider the case in which the coefficients $\mathcal{D}_{\alpha\beta b}$ are not skew-symmetric in α, β and the two subcases where the transport part of the momentum equation is integrable or is not integrable, respectively. In either case one may obtain asymptotic stability.

8.4.1 Example (The Mathematical Example Continued). We first consider stability for the mathematical example discussed earlier and discuss the role of the Lyapunov–Malkin theorem. Recall from Section 8.1.1 that the equations of motion are

$$\ddot{r} = -\frac{\partial V}{\partial r} - \frac{\partial b}{\partial r}\left(a(r)\dot{r} + b(r)p\right)p,$$
$$\dot{p} = \frac{\partial b}{\partial r}\left(a(r)\dot{r} + b(r)p\right)\dot{r}; \tag{8.4.1}$$

here and below we write r instead of r^1.

Recall also that a point $r = r_0$, $p = p_0$ is a relative equilibrium if r_0 and p_0 satisfy the condition

$$\frac{\partial V}{\partial r}(r_0) + \frac{\partial b}{\partial r}b(r_0)\,p_0^2 = 0.$$

Introduce coordinates u_1, u_2, v in a neighborhood of this equilibrium by

$$r = r_0 + u_1, \quad \dot{r} = u_2, \quad p = p_0 + v.$$

The linearized equations of motion are

$$\dot{u}_1 = u_2,$$
$$\dot{u}_2 = \mathcal{A}u_2 + \mathcal{B}u_1 + \mathcal{C}v,$$
$$\dot{v} = \mathcal{D}u_2,$$

where

$$\mathcal{A} = -\frac{\partial b}{\partial r}ap_0,$$

$$\mathcal{B} = -\frac{\partial^2 V}{\partial r^2} - \left[\frac{\partial^2 b}{\partial r^2}b + \left(\frac{\partial b}{\partial r}\right)^2\right]p_0^2,$$

$$\mathcal{C} = -2\frac{\partial b}{\partial r}bp_0,$$

$$\mathcal{D} = \frac{\partial b}{\partial r}bp_0,$$

and where V, a, b, and their derivatives are evaluated at r_0. The characteristic polynomial of these linearized equations is calculated to be

$$\lambda[\lambda^2 - \mathcal{A}\lambda - (\mathcal{B} + \mathcal{C}\mathcal{D})].$$

It obviously has one zero root. The two others have negative real parts if

$$\mathcal{B} + \mathcal{C}\mathcal{D} < 0, \quad \mathcal{A} < 0. \tag{8.4.2}$$

These conditions imply linear stability. We discuss the meaning of these conditions later.

Next, we make the substitution $v = y + \mathcal{D}u_1$, which defines the new variable y. The (nonlinear) equations of motion become

$$\dot{u}_1 = u_2,$$
$$\dot{u}_2 = \mathcal{A}u_2 + (\mathcal{B} + \mathcal{C}\mathcal{D})u_1 + \mathcal{C}y + \mathcal{U}(u, y),$$
$$\dot{y} = \mathcal{Y}(u, y),$$

where $\mathcal{U}(u, y)$, $\mathcal{Y}(u, y)$ stand for nonlinear terms, and $\mathcal{Y}(u, y)$ vanishes when $u = 0$. By Lemma 2.4.6 there exists a further substitution $u = x + \varphi(y)$ such that the equations of motion in coordinates x, y become

$$\dot{x} = Px + X(x, y),$$
$$\dot{y} = Y(x, y),$$

where $X(x, y)$ and $Y(0, y)$ satisfy the conditions $X(0, y) = 0$, $Y(0, y) = 0$. Here,

$$P = \begin{pmatrix} 0 & 1 \\ \mathcal{B} + \mathcal{C}\mathcal{D} & \mathcal{A} \end{pmatrix}.$$

This form enables us to use the Lyapunov–Malkin theorem and conclude that the linear stability implies nonlinear stability and in addition that we have asymptotic stability with respect to the variables x_1, x_2.

To find a Lyapunov-function-based approach for analyzing the stability of the mathematical example, we introduce a modified dynamical system and use its energy function and momentum to construct a Lyapunov function for the original system. This modified system is introduced for the purpose of finding the Lyapunov function and is not used in the stability proof. We will generalize this approach below, and this example may be viewed as motivation for the general approach.

Consider then the new system obtained from the Lagrangian (8.1.1) and the constraint (8.1.2) by setting $a(r) = 0$. Notice that L_c stays the same, and therefore, the equation of motion may be obtained from (8.4.1):

$$\ddot{r} = -\frac{\partial V}{\partial r} - \frac{\partial b}{\partial r}b(r)p^2, \quad \dot{p} = \frac{\partial b}{\partial r}b(r)p\dot{r}.$$

The condition for existence of the relative equilibria also stays the same. However, a crucial observation is that for the new system, the momentum equation is now integrable, in fact explicitly, so that in this example,

$$p = k\exp(b^2(r)/2).$$

Thus, we may proceed and use this invariant surface to perform reduction. The amended potential, defined by $U(r, p) = V(r) + \frac{1}{2}p^2$, becomes

$$U_k(r) = V(r) + \frac{1}{2}\left(k \exp(b^2(r)/2)\right)^2.$$

Consider the function

$$W_k = \frac{1}{2}(\dot{r})^2 + U_k(r) + \epsilon(r - r_0)\dot{r}.$$

If ϵ is small enough and U_k has a nondegenerate minimum, then so does W_k. Suppose that the matrix P has no eigenvalues with zero real parts. Then by Theorem 2.4.4 equations (8.4.1) have a local integral $p - \mathcal{P}(r, \dot{r}, c)$. Differentiate W_k along the vector field determined by (8.4.1). We obtain

$$\dot{W}_k = -\epsilon\left(\frac{\partial^2 V}{\partial r^2}(r_0) + \left(\frac{\partial^2 b}{\partial r^2}b(r_0) + 2\left(\frac{\partial b}{\partial r}b(r_0)\right)^2 + \left(\frac{\partial b}{\partial r}(r_0)\right)^2\right)p_0^2\right)$$
$$- \frac{\partial b}{\partial r}a(r_0)p_0\dot{r}^2 + \epsilon\dot{r}^2 + \{\text{higher-order terms}\}.$$

Therefore, W_k is a Lyapunov function for the flow restricted to the local invariant manifold $p = \mathcal{P}(r, \dot{r}, c)$ if

$$\frac{\partial^2 V}{\partial r^2}(r_0) + \left(\frac{\partial^2 b}{\partial r^2}b(r_0) + 2\left(\frac{\partial b}{\partial r}b(r_0)\right)^2 + \left(\frac{\partial b}{\partial r}(r_0)\right)^2\right)p_0^2 > 0 \qquad (8.4.3)$$

and

$$\frac{\partial b}{\partial r}a(r_0)p_0\dot{r}^2 > 0. \qquad (8.4.4)$$

Notice that the Lyapunov conditions (8.4.3) and (8.4.4) are the same as conditions (8.4.2).

Introduce the operator

$$\nabla_r = \frac{\partial}{\partial r} + \frac{\partial b}{\partial r}b(r)p\frac{\partial}{\partial p}$$

(cf. Karapetyan [1983]).

Then condition (8.4.3) may be rewritten as

$$\nabla_r^2 U > 0,$$

which is the same as the condition for stability of stationary motions of a nonholonomic system with an integrable momentum equation (recall that this means that there are no terms quadratic in $\dot{r}$, only transport terms defining an integrable distribution). The left-hand side of formula (8.4.4) may be viewed as a derivative of the energy function

$$E = \frac{1}{2}\dot{r}^2 + \frac{1}{2}p^2 + V.$$

along the flow

$$\ddot{r} = -\frac{\partial V}{\partial r} - \frac{\partial b}{\partial r}\left(a(r)\dot{r} + b(r)p\right)p, \quad \dot{p} = \frac{\partial b}{\partial r}b(r)p\dot{r},$$

or as a derivative of the amended potential U along the vector field defined by the nontransport terms of the momentum equations

$$\dot{p} = \frac{\partial b}{\partial r}a(r)\dot{r}^2. \qquad \blacklozenge$$

The Nonholonomic Energy-Momentum Method. We now generalize the energy-momentum method discussed above for the mathematical example to the general case in which the transport part of the momentum equation is integrable.

Here we assume hypothesis H2 in the present context, namely:

H2 The curvature of the connection form associated with the transport part of the momentum equation, namely, $dp_b - \mathcal{D}^c_{b\alpha}p_c dr^\alpha$, is zero.

The momentum equation in this situation is

$$\frac{d}{dt}p_b = \mathcal{D}^c_{b\alpha}p_c\dot{r}^\alpha + \mathcal{D}_{\alpha\beta b}\dot{r}^\alpha\dot{r}^\beta.$$

Hypothesis H2 implies that the form due to the transport part of the momentum equation defines an integrable distribution. Associated with this distribution there is a family of integral manifolds

$$p_a = P_a(r^\alpha, k_b)$$

with P_a satisfying the equation $dP_b = \mathcal{D}^c_{b\alpha}P_c dr^\alpha$. Note that these manifolds *are not invariant manifolds of the full system* under consideration because the momentum equation has nontransport terms. Substituting the functions $P_a(r^\alpha, k_b)$, $k_b = \text{const}$, into $E(r, \dot{r}, p)$, we obtain a function

$$V_k(r^\alpha, \dot{r}^\alpha) = E(r^\alpha, \dot{r}^\alpha, P_a(r^\alpha, k_b))$$

that depends only on r^α, $\dot{r}^\alpha$ and parametrically on k. This function will not be our final Lyapunov function but will be used to construct one in the proof to follow.

Pick a relative equilibrium $r^\alpha = r_0^\alpha$, $p_a = p_a^0$. In this context we introduce the following definiteness assumptions:

H3 At the equilibrium $r^\alpha = r_0^\alpha$, $p_a = p_a^0$ the two symmetric matrices $\nabla_\alpha\nabla_\beta U$ and $(\mathcal{D}_{\alpha\beta b} + \mathcal{D}_{\beta\alpha b})I^{bc}p_c$ are positive definite.

8.4.2 Theorem. *Under assumptions H2 and H3, the equilibrium $r^\alpha = r_0^\alpha$, $p_a = p_a^0$ is Lyapunov stable. Moreover, the system has local invariant* *manifolds that are tangent to the family of manifolds defined by the integrable transport part of the momentum equation at the relative equilibria. The relative equilibria that are close enough to r_0, p_0 are asymptotically stable in the directions defined by these invariant manifolds. In addition, for initial conditions close enough to the equilibrium $r^\alpha = r_0^\alpha$, $p_a = p_a^0$, the perturbed solution approaches a nearby equilibrium.*

Proof. The substitution $p_a = p_a^0 + y_a + \mathcal{D}_{aa}^b(r_0)p_b^0 u^\alpha$, where $u^\alpha = r^\alpha - r_0^\alpha$, eliminates the linear terms in the momentum equation. In fact, with this substitution, the equations of motion (8.2.1), (8.2.2) become

$$\frac{d}{dt}\frac{\partial R}{\partial \dot r^\alpha} - \frac{\partial R}{\partial r^\alpha} = -\mathcal{D}_{ba}^c I^{bd}p_c p_d - \mathcal{B}_{\alpha\beta}^c p_c \dot r^\beta$$
$$- \mathcal{D}_{\beta ab}I^{bc}p_c \dot r^\beta - \mathcal{K}_{\alpha\beta\gamma}\dot r^\beta \dot r^\gamma,$$
$$\frac{d}{dt}y_b = \mathcal{D}_{ba}^c y_c \dot r^\alpha + (\mathcal{D}_{ba}^c - \mathcal{D}_{ba}^c(r_0))p_c^0 \dot r^\alpha + \mathcal{D}_{\alpha\beta b}\dot r^\alpha \dot r^\beta.$$

One can check that H3 implies the hypotheses of Theorem 2.4.4 (see Zenkov, Bloch, and Marsden [1998] for details of this computation). Thus, the above equations have local integrals $y_a = f_a(r, \dot r, c)$, where the functions f_a are such that $\partial_r f_a = \partial_{\dot r} f_a = 0$ at the equilibria. Therefore, the original equations (8.2.1), (8.2.2) have n local integrals

$$p_a = \mathcal{P}_a(r^\alpha, \dot r^\alpha, c_b), \quad c_b = \text{const}, \tag{8.4.5}$$

where $\mathcal{P}_a$ are such that

$$\frac{\partial \mathcal{P}}{\partial r^\alpha} = \frac{\partial P}{\partial r^\alpha}, \quad \frac{\partial \mathcal{P}}{\partial \dot r^\alpha} = 0$$

at the relative equilibria.

We now use the $V_k(r^\alpha, \dot r^\alpha)$ to construct a Lyapunov function to determine the conditions for asymptotic stability of the relative equilibrium $r^\alpha = r_0^\alpha$, $p_a = p_a^0$. We will do this in a fashion similar to that used by Chetaev [1959] and Bloch, Krishnaprasad, Marsden, and Ratiu [1994].

Without loss of generality, suppose that $g_{\alpha\beta}(r_0) = \delta_{\alpha\beta}$. Introduce the function

$$W_k = V_k + \epsilon \sum_{\alpha=1}^{\sigma} u^\alpha \dot r^\alpha.$$

Consider the following two manifolds at the equilibrium (r_0^α, p_a^0): the integral manifold of the transport equation

$$Q_{k^0} = \left\{ p_a = P_a(r^\alpha, k^0) \right\},$$

and the local invariant manifold

$$\mathcal{Q}_{c^0} = \left\{ p_a = \mathcal{P}_a(r^\alpha, \dot r^\alpha, c^0) \right\}.$$

Restrict the flow to the manifold $\mathcal{Q}_{c^0}$. Choose $(r^\alpha, \dot{r}^\alpha)$ as local coordinates on $\mathcal{Q}_{c^0}$; then V_{k^0} and W_{k^0} are functions defined on $\mathcal{Q}_{c^0}$. Since

$$\frac{\partial U_{k^0}}{\partial r^\alpha}(r_0) = \nabla_\alpha U(r_0, p_0) = 0$$

and

$$\frac{\partial^2 U_{k^0}}{\partial r^\alpha \partial r^\beta}(r_0) = \nabla_\alpha \nabla_\beta U(r_0, p_0) \gg 0,$$

the function V_{k^0} is positive definite in some neighborhood of the relative equilibrium $(r_0^\alpha, 0) \in \mathcal{Q}_{c^0}$. The same is valid for the function W_{k^0} if ϵ is small enough.

Now we show that $\dot{W}_{k^0}$ (as a function on $\mathcal{Q}_{c^0}$) is negative definite. Calculate the derivative of W_{k^0} along the flow:

$$\dot{W}_{k^0} = g_{\alpha\beta}\dot{r}^\alpha \ddot{r}^\beta + \frac{1}{2}\dot{g}_{\alpha\beta}\dot{r}^\alpha \dot{r}^\beta + I^{ab}P_a \dot{P}_b$$

$$+ \frac{1}{2}\dot{I}^{ab}P_a P_b + \dot{V} + \epsilon \sum_{\alpha=1}^{\sigma}\left((\dot{r}^\alpha)^2 + u^\alpha \ddot{r}^\alpha\right). \tag{8.4.6}$$

Using the explicit representation of equation (8.2.1), we obtain

$$g_{\alpha\beta}\ddot{r}^\beta + \dot{g}_{\alpha\beta}\dot{r}^\beta = \frac{1}{2}\frac{\partial g_{\beta\gamma}}{\partial r^\alpha}\dot{r}^\beta \dot{r}^\gamma - \frac{\partial V}{\partial r^\alpha} - \frac{1}{2}\frac{\partial I^{ab}}{\partial r^\alpha}P_a P_b - \mathcal{D}_{ba}^c I^{bd}P_c P_d$$

$$- \mathcal{D}_{\beta ab}I^{bc}P_c \dot{r}^\beta - \mathcal{B}_{\alpha\beta}^c P_c \dot{r}^\beta - \mathcal{K}_{\alpha\beta\gamma}\dot{r}^\beta \dot{r}^\gamma. \tag{8.4.7}$$

Therefore,

$$g_{\alpha\beta}\dot{r}^\alpha \ddot{r}^\beta + \frac{1}{2}\dot{g}_{\alpha\beta}\dot{r}^\alpha \dot{r}^\beta + I^{ab}P_a \dot{P}_b + \frac{1}{2}\dot{I}^{ab}P_a P_b + \dot{V}$$

$$= -\frac{1}{2}\dot{g}_{\alpha\beta}\dot{r}^\alpha \dot{r}^\beta + \frac{1}{2}\frac{\partial g_{\beta\gamma}}{\partial r^\alpha}\dot{r}^\alpha \dot{r}^\beta \dot{r}^\gamma - \frac{\partial V}{\partial r^\alpha}\dot{r}^\alpha - \frac{1}{2}\frac{\partial I^{ab}}{\partial r^\alpha}P_a P_b \dot{r}^\alpha$$

$$- \mathcal{D}_{ba}^c I^{bd}P_c P_d \dot{r}^\alpha - \mathcal{D}_{\beta ab}I^{bc}P_c \dot{r}^\alpha \dot{r}^\beta - \mathcal{B}_{\alpha\beta}^c P_c \dot{r}^\alpha \dot{r}^\beta$$

$$- \mathcal{K}_{\alpha\beta\gamma}\dot{r}^\alpha \dot{r}^\beta \dot{r}^\gamma + I^{ab}\mathcal{D}_{ba}^c P_a P_c \dot{r}^\alpha + \frac{1}{2}\frac{\partial I^{ab}}{\partial r^\alpha}P_a P_b \dot{r}^\alpha + \dot{V}.$$

Using the skew symmetry of $\mathcal{B}_{\alpha\beta}^c$ and $\mathcal{K}_{\alpha\beta\gamma}$ with respect to α, β and canceling the terms

$$-\frac{1}{2}\dot{g}_{\alpha\beta}\dot{r}^\alpha \dot{r}^\beta + \frac{1}{2}\frac{\partial g_{\beta\gamma}}{\partial r^\alpha}\dot{r}^\alpha \dot{r}^\beta \dot{r}^\gamma - \frac{\partial V}{\partial r^\alpha}\dot{r}^\alpha + \dot{V},$$

we obtain

$$g_{\alpha\beta}\dot{r}^\alpha \ddot{r}^\beta \frac{1}{2}\dot{g}_{\alpha\beta}\dot{r}^\alpha \dot{r}^\beta + I^{ab}P_a \dot{P}_b + \frac{1}{2}\dot{I}^{ab}P_a P_b + \dot{V}$$

$$= -\mathcal{D}_{\beta ab}I^{bc}P_c \dot{r}^\alpha \dot{r}^\beta + \left(\frac{1}{2}\frac{\partial I^{ab}}{\partial r^\alpha} + I^{ac}\mathcal{D}_{ca}^b\right)(P_a P_b - P_a P_b)\dot{r}^\alpha. \tag{8.4.8}$$

Substituting (8.4.8) in (8.4.6) and determining $\ddot{r}^\alpha$ from (8.4.7),

$$\dot{W}_{k^0} = -\mathcal{D}_{\beta\alpha b}I^{bc}\mathcal{P}_c\dot{r}^\alpha\dot{r}^\beta + \epsilon\sum_{\alpha=1}^{\sigma}(\dot{r}^\alpha)^2$$

$$-\epsilon\sum_{\gamma=1}^{\sigma}g^{\alpha\beta}u^\gamma\left(\frac{\partial V}{\partial r^\alpha} + \frac{1}{2}\frac{\partial I^{ab}}{\partial r^\alpha}\mathcal{P}_a\mathcal{P}_b + \mathcal{D}_{b\alpha}^c I^{bd}\mathcal{P}_c\mathcal{P}_d\right)$$

$$+\epsilon\sum_{\gamma=1}^{\sigma}g^{\alpha\gamma}u^\gamma\left(-\dot{g}_{\alpha\beta}\dot{r}^\beta + \frac{1}{2}\frac{\partial g_{\beta\gamma}}{\partial r^\alpha}\dot{r}^\beta\dot{r}^\gamma - \mathcal{B}_{\alpha\beta}^c\mathcal{P}_c\dot{r}^\beta\right.$$

$$\left. - \mathcal{D}_{\beta\alpha h}I^{bc}\mathcal{P}_c\dot{r}^\beta - \mathcal{K}_{\alpha\beta\gamma}\dot{r}^\beta\dot{r}^\gamma\right)$$

$$+\frac{1}{2}\frac{\partial I^{ab}}{\partial r^\alpha}\left(P_aP_b - \mathcal{P}_a\mathcal{P}_b\right)\dot{r}^\alpha + I^{ac}\mathcal{D}_{c\alpha}^b\left(P_aP_b - \mathcal{P}_a\mathcal{P}_b\right)\dot{r}^\alpha. \quad (8.4.9)$$

Since

$$\frac{\partial V}{\partial r^\alpha} + \frac{1}{2}\frac{\partial I^{ab}}{\partial r^\alpha}\mathcal{P}_a\mathcal{P}_b + \mathcal{D}_{b\alpha}^c I^{bd}\mathcal{P}_c\mathcal{P}_d = 0$$

at the equilibrium and the linear terms in the Taylor expansions of $\mathcal{P}$ and P are the same,

$$\frac{\partial V}{\partial r^\alpha} + \frac{1}{2}\frac{\partial I^{ab}}{\partial r^\alpha}\mathcal{P}_a\mathcal{P}_b + \mathcal{D}_{b\alpha}^c I^{bd}\mathcal{P}_c\mathcal{P}_d = F_{\alpha\beta}u^\beta + \{\text{nonlinear terms}\}, \quad (8.4.10)$$

where

$$F_{\alpha\beta} = \frac{\partial}{\partial r^\beta}\left(\frac{\partial V}{\partial r^\alpha} + \frac{1}{2}\frac{\partial I^{ab}}{\partial r^\alpha}P_aP_b + \mathcal{D}_{b\alpha}^c I^{bd}P_cP_d\right)$$

$$= \frac{\partial^2 V}{\partial r^\alpha\partial r^\beta} + \frac{1}{2}\frac{\partial^2 I^{ab}}{\partial r^\alpha\partial r^\beta}P_aP_b + \frac{\partial I^{ab}}{\partial r^\alpha}P_a\frac{\partial P_b}{\partial r^\beta}$$

$$+ \frac{\partial}{\partial r^\beta}\left(\mathcal{D}_{b\alpha}^c I^{bd}\right)P_cP_d + \mathcal{D}_{b\alpha}^c I^{bd}\left(\frac{\partial P_c}{\partial r^\beta}P_d + P_c\frac{\partial P_d}{\partial r^\beta}\right)$$

$$= \frac{\partial^2 V}{\partial r^\alpha\partial r^\beta} + \frac{1}{2}\frac{\partial^2 I^{ab}}{\partial r^\alpha\partial r^\beta}P_aP_b + \frac{\partial I^{ab}}{\partial r^\alpha}P_a\mathcal{D}_{b\beta}^c P_c$$

$$+ \frac{\partial}{\partial r^\beta}\left(\mathcal{D}_{b\alpha}^c I^{bd}\right)P_cP_d + \mathcal{D}_{b\alpha}^c I^{bd}\left(\mathcal{D}_{c\beta}^a P_d + P_c\mathcal{D}_{d\beta}^a P_a\right)$$

$$= \nabla_\alpha\nabla_\beta U.$$

In the last formula all the terms are evaluated at the equilibrium.

Taking into account that $g_{\alpha\beta} = \delta_{\alpha\beta} + O(u)$, that the Taylor expansion of $P_aP_b - \mathcal{P}_a\mathcal{P}_b$ starts from the terms of the second order, and using (8.4.10), we obtain from (8.4.9)

$$\dot{W}_{k^0} = -\mathcal{D}_{\beta\alpha b}I^{bc}(r_0)p_c^0\dot{r}^\alpha\dot{r}^\beta - \epsilon F_{\alpha\beta}u^\alpha u^\beta + \epsilon\sum_{\alpha=1}^{\sigma}(\dot{r}^\alpha)^2$$

$$- \epsilon\left(\mathcal{D}_{\beta\alpha b}I^{bc}(r_0)p_c^0 + \mathcal{B}_{\alpha\beta}^c(r_0)p_c^0\right)u^\alpha\dot{r}^\beta + \{\text{cubic terms}\}.$$

Therefore, the condition $(\mathcal{D}_{\alpha\beta b} + \mathcal{D}_{\beta\alpha b})I^{bc}p_c^0 \gg 0$ implies that $\dot{W}_{k^0}$ is negative definite if ϵ is small enough and positive. Thus, W_{k^0} is a Lyapunov function for the flow on $\mathcal{Q}_{c^0}$, and therefore the equilibrium $(r_0^\alpha, 0)$ for the flow on $\mathcal{Q}_{c^0}$ is asymptotically stable.

Using the same arguments we used in the proof of Theorem 8.3.1, we conclude that the equilibria on the nearby invariant manifolds $\mathcal{Q}_k$ are asymptotically stable as well.   ∎

There is an alternative way to state the above theorem, which uses the basic intuition we used to find the Lyapunov function.

8.4.3 Theorem (The nonholonomic energy-momentum method). *Under the assumption that H2 holds, the point $q_e = (r_0^\alpha, p_a^0)$ is a relative equilibrium if and only if there is a $\xi \in \mathfrak{g}^{q_e}$ such that q_e is a critical point of the augmented energy $E_\xi = E - \langle p - P(r, k), \xi \rangle$. Assume that*

(i) *$\delta^2 E_\xi$ restricted to $T_{q_e}\mathcal{Q}_k$ is positive definite (here δ denotes differentiation by all variables except ξ);*

(ii) *the quadratic form defined by the flow derivative of the augmented energy is negative definite at q_e.*

Then H3 holds, and this equilibrium is Lyapunov stable and asymptotically stable in the directions of the invariant manifolds (8.4.5).

Proof. We have already shown in Theorem 8.3.2 that positive definiteness of $\delta^2 E_\xi|_{T_{q_e}\mathcal{Q}_k}$ is equivalent to the condition $\nabla_\alpha \nabla_\beta U \gg 0$. To complete the proof, we need to show that the requirement (ii) of the theorem is equivalent to the condition $(\mathcal{D}_{\alpha\beta b} + \mathcal{D}_{\beta\alpha b})I^{bc}(r_0)p_c^0 \gg 0$. Compute the flow derivative of E_ξ:

$$\dot{E}_\xi = \dot{E} - \langle \dot{p} - \dot{P}, \xi \rangle = -\langle \dot{p} - \dot{P}, \xi \rangle$$
$$= -(\mathcal{D}_{a\alpha}^b p_b \dot{r}^\alpha + \mathcal{D}_{\alpha\beta a} \dot{r}^\alpha \dot{r}^\beta - \mathcal{D}_{a\alpha}^b P_b \dot{r}^\alpha)\xi^a + \text{higher-order terms.}$$

Since at the equilibrium $p = P$, $\xi^a = I^{ab}p_b$, and $\dot{E} = 0$ (Theorem 5.7.4), we obtain

$$\dot{E}_\xi = -\mathcal{D}_{\alpha\beta a}I^{ab}(r_0)p_b^0 \dot{r}^\alpha \dot{r}^\beta + \text{higher-order terms.}$$

The condition of negative definiteness of the quadratic form determined by $\dot{E}_\xi$ is thus equivalent to $(\mathcal{D}_{\alpha\beta b} + \mathcal{D}_{\beta\alpha b})I^{bc}(r_0)p_c^0 \gg 0$.   ∎

For some examples, such as the roller racer, we need to consider a degenerate case of the above analysis. Namely, we consider a nongeneric case, when $U = \frac{1}{2}I^{ab}(r)p_a p_b$ (the original system has no potential energy), and the components of the locked inertia tensor I^{ab} satisfy the condition

$$\frac{1}{2}\frac{\partial I^{ab}}{\partial r^\alpha} + I^{ac}\mathcal{D}_{c\alpha}^b = 0. \tag{8.4.11}$$

Consequently, the covariant derivatives of the amended potential are equal to zero, and the equations of motion (8.2.1), (8.2.2) become

$$\frac{d}{dt}\left(g_{\alpha\beta}\dot{r}^{\beta}\right) - \frac{1}{2}\frac{\partial g_{\beta\gamma}}{\partial r^{\alpha}}\dot{r}^{\beta}\dot{r}^{\gamma} = -\mathcal{D}_{\beta\alpha b}I^{bc}p_{c}\dot{r}^{\beta} - \mathcal{B}_{\alpha\beta}^{c}p_{c}\dot{r}^{\beta} - \mathcal{K}_{\alpha\beta\gamma}\dot{r}^{\beta}\dot{r}^{\gamma},$$

$$(8.4.12)$$

$$\frac{d}{dt}p_{b} = \mathcal{D}_{b\alpha}^{c}p_{c}\dot{r}^{\alpha} + \mathcal{D}_{\alpha\beta b}\dot{r}^{\alpha}\dot{r}^{\beta}.$$

$$(8.4.13)$$

Thus, we obtain an $(m + \sigma)$-dimensional *manifold* of equilibria $r = r_0$, $p = p_0$ of these equations. Further, we cannot apply Theorem 8.4.2 because the condition $\nabla^2 U \gg 0$ fails. However, we can do a similar type of stability analysis as follows.

As before, set

$$V_k = E(r, \dot{r}, P(r, k)) = \frac{1}{2}g_{\alpha\beta}\dot{r}^{\alpha}\dot{r}^{\beta} + \frac{1}{2}I^{ab}(r)P_a(r, k)P_b(r, k).$$

Note that P satisfies the equation

$$\frac{\partial P_b}{\partial r^{\alpha}} = \mathcal{D}_{b\alpha}^{c}P_c,$$

which implies that

$$\frac{\partial}{\partial r^{\alpha}}\left(\frac{1}{2}I^{ab}(r)P_aP_b\right) = \frac{1}{2}\frac{\partial I^{ab}}{\partial r^{\alpha}}P_aP_b + I^{ab}P_a\frac{\partial P_b}{\partial r^{\alpha}}$$

$$= \left(\frac{1}{2}\frac{\partial I^{ab}}{\partial r^{\alpha}} + I^{ab}\mathcal{D}_{b\alpha}^{c}\right)P_aP_b = 0.$$

Therefore

$$\frac{1}{2}I^{ab}P_aP_b = \text{const}$$

and

$$V_k = \frac{1}{2}g_{\alpha\beta}\dot{r}^{\alpha}\dot{r}^{\beta}$$

(up to an additive constant). Thus, V_k is a positive definite function with respect to $\dot{r}$. Compute $\dot{V}_k$:

$$\dot{V}_k = g_{\alpha\beta}\dot{r}^{\alpha}\ddot{r}^{\beta} + \dot{g}_{\alpha\beta}\dot{r}^{\alpha}\dot{r}^{\beta} = -\mathcal{D}_{\beta\alpha b}I^{bc}p_c\dot{r}^{\alpha}\dot{r}^{\beta} + O(\dot{r}^3).$$

Suppose that $(\mathcal{D}_{\alpha\beta b} + \mathcal{D}_{\beta\alpha b})(r_0)I^{bc}(r_0)p_c^0 \gg 0$. Now the linearization of equations (8.4.12) and (8.4.13) about the relative equilibria given by setting $\dot{r} = 0$ has $m + \sigma$ zero eigenvalues corresponding to the r and p directions. Since the matrix corresponding to the $\dot{r}$-directions of the linearized system is of the form $D + G$, where D is positive definite and symmetric (in fact, $D = \frac{1}{2}(\mathcal{D}_{\alpha\beta b} + \mathcal{D}_{\beta\alpha b})(r_0)I^{bc}(r_0)p_c^0)$ and G is skew-symmetric, the determinant of $D + G$ is not equal to zero. This follows from the observation that

$x^t(D + G)x = x^t Dx > 0$ for D positive definite and G skew-symmetric. Thus using Theorem 2.4.4, we find that the equations of motion have local integrals

$$r = \mathcal{R}(\dot{r}, k), \quad p = \mathcal{P}(\dot{r}, k).$$

Therefore, V_k restricted to a common level set of these integrals is a Lyapunov function for the restricted system. Thus, an equilibrium $r = r_0$, $p = p_0$ is stable with respect to r, $\dot{r}$, p and asymptotically stable with respect to $\dot{r}$ if

$$(\mathcal{D}_{\alpha\beta b} + \mathcal{D}_{\beta\alpha b})I^{bc}(r_0)p_c^0 \gg 0. \tag{8.4.14}$$

Summarizing, we have the following theorem:

8.4.4 Theorem. *Under assumption H2 if $U = 0$ and assuming that the conditions (8.4.11) and (8.4.14) hold, the nonholonomic equations of motion have an $(m+\sigma)$-dimensional manifold of equilibria parametrized by r and p. An equilibrium $r = r_0$, $p = p_0$ is stable with respect to r, $\dot{r}$, p and asymptotically stable with respect to $\dot{r}$.*

8.4.5 Example (The **Roller Racer**). The roller racer provides an example and illustration of Theorem 8.4.4. Recall that the Lagrangian and the constraints are

$$L = \frac{1}{2}m(\dot{x}^2 + \dot{y}^2) + \frac{1}{2}I_1\dot{\theta}^2 + \frac{1}{2}I_2(\dot{\theta} + \dot{\phi})^2$$

and

$$\dot{x} = \cos\theta\left(\frac{d_1\cos\phi + d_2}{\sin\phi}\dot{\theta} + \frac{d_2}{\sin\phi}\dot{\phi}\right),$$
$$\dot{y} = \sin\theta\left(\frac{d_1\cos\phi + d_2}{\sin\phi}\dot{\theta} + \frac{d_2}{\sin\phi}\dot{\phi}\right).$$

The configuration space is $SE(2) \times SO(2)$ with the angles measured counterclockwise, and as observed earlier, the Lagrangian and the constraints are invariant under the left action of $SE(2)$ on the first factor of the configuration space.

The nonholonomic momentum is

$$p = m(d_1\cos\phi + d_2)(\dot{x}\cos\theta + \dot{y}\sin\theta) + [(I_1 + I_2)\dot{\theta} + I_2\dot{\phi}]\sin\phi.$$

See Tsakiris [1995] for details of this calculation. The momentum equation is

$$\dot{p} = \frac{((I_1 + I_2)\cos\phi - md_1(d_1\cos\phi + d_2))}{m(d_1\cos\phi + d_2)^2 + (I_1 + I_2)\sin^2\phi}p\dot{\phi}$$
$$+ \frac{(d_1 + d_2\cos\phi)(I_2 d_1\cos\phi - I_1 d_2)}{m(d_1\cos\phi + d_2)^2 + (I_1 + I_2)\sin^2\phi}\dot{\phi}^2.$$

Rewriting the Lagrangian using p instead of $\dot\theta$, we obtain the energy function for the roller racer:

$$E = \frac{1}{2}g(\phi)\dot\phi^2 + \frac{1}{2}I(\phi)p^2,$$

where

$$g(\phi) = I_2 + \frac{md_2^2}{\sin^2\phi} - \frac{\left[m(d_1\cos\phi + d_2)d_2 + I_2\sin^2\phi\right]^2}{\sin^2\phi\left[m(d_1\cos\phi + d_2)^2 + (I_1 + I_2)\sin^2\phi\right]}$$

and

$$I(\phi) = \frac{1}{(d_1\cos\phi + d_2)^2 + (I_1 + I_2)\sin^2\phi}. \tag{8.4.15}$$

The amended potential is given by

$$U = \frac{p^2}{2[(d_1\cos\phi + d_2)^2 + (I_1 + I_2)\sin^2\phi]},$$

which follows directly from (5.7.8) and (8.4.15).

Straightforward computations show that the locked inertia tensor $I(\phi)$ satisfies the condition (8.4.11). Thus the roller racer has a two-dimensional manifold of relative equilibria parametrized by ϕ and p. These relative equilibria are motions of the roller racer in circles about the point of intersection of lines through the axles. For such motions, p is the system momentum about this point, and ϕ is the relative angle between the two bodies.

Thus, we may apply the energy-momentum stability conditions (8.4.14) obtained above for the degenerate case. Multiplying the coefficient of the nontransport term of the momentum equation, evaluated at ϕ_0, by $I(\phi_0)p_0$ and omitting a positive denominator, we obtain the condition for stability of a relative equilibrium $\phi = \phi_0$, $p = p_0$ of the roller racer:

$$(d_1 + d_2\cos\phi_0)(I_2d_1\cos\phi_0 - I_1d_2)p_0 > 0.$$

Note that this equilibrium is stable modulo SE(2) and in addition asymptotically stable with respect to $\dot\phi$.   ♦

8.5   General Case—the Lyapunov–Malkin Method

In this section we study stability in a fairly general setting: We do not a priori assume the hypotheses H1 (skewness of $\mathcal{D}_{\alpha\beta b}$ in α, β), H2 (a curvature is zero), and H3 (definiteness of second variations). Thus, we consider the most general case, when the connection due to the transport part of the momentum equation is not necessarily flat and when the nontransport terms of the momentum equation are not equal to zero. In the case where $\mathfrak{g}^{q_e}$ is commutative, this analysis was done by Karapetyan [1980]. The extension of this to the noncommutative case was done in Zenkov, Bloch, and Marsden [1998], and we refer the reader to this paper for more details. In this section we shall be content to give the general form of the linearization of the equations and discuss the rattleback example.

Linearization Computation. We start by computing the linearization of equations (8.2.1) and (8.2.2) about a given relative equilibrium (r_0, p_0). Introduce coordinates u^α, v^α, and w_a in a neighborhood of the equilibrium $r = r_0$, $p = p_0$ by the formulas

$$r^\alpha = r_0^\alpha + u^\alpha, \quad \dot{r}^\alpha = v^\alpha, \quad p_a = p_a^0 + w_a.$$

The linearized momentum equation is

$$\dot{w}_b = \mathcal{D}_{ba}^c(r_0)p_c^0 v^\alpha.$$

To find the linearization of (8.2.1), we start by rewriting its right-hand side explicitly. Since $R = \frac{1}{2}g_{\alpha\beta}\dot{r}^\alpha\dot{r}^\beta - \frac{1}{2}I^{ab}p_ap_b - V$, equation (8.2.1) becomes

$$g_{\alpha\beta}\ddot{r}^\beta + \dot{g}_{\alpha\beta}\dot{r}^\alpha\dot{r}^\beta - \frac{1}{2}\frac{\partial g_{\beta\gamma}}{\partial r^\alpha}\dot{r}^\beta\dot{r}^\gamma + \frac{1}{2}\frac{\partial I^{ab}}{\partial r^\alpha}p_ap_b + \frac{\partial V}{\partial r^\alpha}$$
$$= -\mathcal{D}_{ca}^a I^{cd}p_ap_d - \mathcal{D}_{\beta ac}I^{ca}p_a\dot{r}^\beta - \mathcal{B}_{\alpha\beta}^a p_a\dot{r}^\beta - \mathcal{K}_{\alpha\beta\gamma}\dot{r}^\beta\dot{r}^\gamma.$$

Keeping only the linear terms, we obtain

$$g_{\alpha\beta}(r_0)\ddot{r}^\beta + \frac{\partial^2 V}{\partial r^\alpha\partial r^\beta}(r_0)\,u^\beta + \frac{1}{2}\frac{\partial^2 I^{ab}}{\partial r^\alpha\partial r^\beta}(r_0)p_a^0p_b^0\,u^\beta + \frac{\partial I^{ab}}{\partial r^\alpha}(r_0)p_a^0\,w_b$$
$$= -\mathcal{D}_{ca}^a I^{cd}(r_0)p_a^0\,w_d - \mathcal{D}_{ca}^a I^{cd}(r_0)p_d^0\,w_a - \frac{\partial \mathcal{D}_{ca}^a I^{cd}}{\partial r^\beta}(r_0)p_a^0p_d^0\,u^\beta$$
$$- \mathcal{D}_{\beta ac}I^{ca}(r_0)p_a^0\,v^\beta - \mathcal{B}_{\alpha\beta}^a(r_0)p_a^0\,v^\beta.$$

Next, introduce matrices $\mathcal{A}$, $\mathcal{B}$, $\mathcal{C}$, and $\mathcal{D}$ by

$$\mathcal{A}_{\alpha\beta} = -\left(\mathcal{D}_{\beta ac}I^{ca}(r_0)p_a^0 + \mathcal{B}_{\alpha\beta}^a(r_0)p_a^0\right),$$
$$\mathcal{B}_{\alpha\beta} = -\left(\frac{\partial^2 V}{\partial r^\alpha\partial r^\beta}(r_0) + \frac{1}{2}\frac{\partial^2 I^{ab}}{\partial r^\alpha\partial r^\beta}(r_0)p_a^0p_b^0 + \frac{\partial \mathcal{D}_{ca}^a I^{cb}}{\partial r^\beta}(r_0)p_a^0p_b^0\right),$$

$$\tag{8.5.1}$$

$$\mathcal{C}_\alpha^a = -\left(\frac{\partial I^{ab}}{\partial r^\alpha}(r_0)p_b^0 + \mathcal{D}_{ca}^b I^{ca}(r_0)p_b^0 + \mathcal{D}_{ca}^a I^{cb}(r_0)p_b^0\right), \tag{8.5.2}$$

$$\mathcal{D}_{a\alpha} = \mathcal{D}_{a\alpha}^c(r_0)p_c^0. \tag{8.5.3}$$

Using this notation and making a choice of r^α such that $g_{\alpha\beta}(r_0) = \delta_{\alpha\beta}$, we can represent the equation of motion in the form

$$\dot{u}^\alpha = v^\alpha, \tag{8.5.4}$$

$$\dot{v}^\alpha = \mathcal{A}^\alpha_\beta v^\beta + \mathcal{B}^\alpha_\beta u^\beta + \mathcal{C}^{\alpha a} w_a + \mathcal{V}^\alpha(u, v, w), \tag{8.5.5}$$

$$\dot{w}_a = \mathcal{D}_{a\alpha} v^\alpha + \mathcal{W}_a(u, v, w), \tag{8.5.6}$$

where $\mathcal{V}$ and $\mathcal{W}$ stand for nonlinear terms, and where

$$\mathcal{A}^\alpha_\beta = \delta^{\alpha\gamma} \mathcal{A}_{\gamma\beta},$$
$$\mathcal{B}^\alpha_\beta = \delta^{\alpha\gamma} \mathcal{B}_{\gamma\beta},$$
$$\mathcal{C}^{\prime\alpha a} = \delta^{\alpha\gamma} \mathcal{C}^a_\gamma.$$

(If $g_{\alpha\beta}(r_0) \neq \delta_{\alpha\beta}$, then here $\mathcal{A}^\alpha_\beta = g^{\alpha\gamma} \mathcal{A}_{\gamma\beta}$, $\mathcal{B}^\alpha_\beta = g^{\alpha\gamma} \mathcal{B}_{\gamma\beta}$, $\mathcal{C}^{\alpha a} = g^{\alpha\gamma} \mathcal{C}^a_\gamma$.) Note that

$$\mathcal{W}_a = \left(\mathcal{D}^c_{a\alpha}(p^0_c + w_c) - \mathcal{D}_{a\alpha} \right) v^\alpha + \mathcal{D}_{\alpha\beta a} v^\alpha v^\beta. \tag{8.5.7}$$

The next step is to eliminate the linear terms from (8.5.6). Putting

$$w_a = \mathcal{D}_{a\alpha} u^\alpha + z_a,$$

(8.5.6) becomes

$$\dot{z}_a = \mathcal{Z}_a(u, v, z),$$

where $\mathcal{Z}_a(u, v, z)$ represents nonlinear terms. Formula (8.5.7) leads to

$$\mathcal{Z}_a(u, v, z) = \mathcal{Z}_{a\alpha}(u, v, z) v^\alpha.$$

In particular, $\mathcal{Z}_a(u, 0, z) = 0$. Equations (8.5.4), (8.5.5), (8.5.6) in the variables u, v, z become

$$\dot{u}^\alpha = v^\alpha,$$
$$\dot{v}^\alpha = \mathcal{A}^\alpha_\beta v^\beta + (\mathcal{B}^\alpha_\beta + \mathcal{C}^{\alpha a} \mathcal{D}_{a\beta}) u^\beta + \mathcal{C}^{\alpha a} z_a + \mathcal{V}^\alpha(u, v, z_a + \mathcal{D}_{a\alpha} u^\alpha),$$
$$\dot{z}_a = \mathcal{Z}_a(u, v, w).$$

Application of the Lyapunov–Malkin Theorem. Using Lemma 2.4.6, we find a substitution $x^\alpha = u^\alpha + \phi^\alpha(z)$, $y^\alpha = v^\alpha$ such that in the variables x, y, z we obtain

$$\dot{x}^\alpha = y^\alpha + X^\alpha(x, y, z),$$
$$\dot{y}^\alpha = \mathcal{A}^\alpha_\beta y^\beta + (\mathcal{B}^\alpha_\beta + \mathcal{C}^{\alpha a} \mathcal{D}_{a\beta}) x^\beta + Y^\alpha(x, y, z), \tag{8.5.8}$$
$$\dot{z}_a = Z_a(x, y, z),$$

where the nonlinear terms $X(x, y, z)$, $Y(x, y, z,)$, $Z(x, y, z)$ vanish if $x = 0$ and $y = 0$. Therefore, we can apply the Lyapunov–Malkin theorem and obtain the following result:

8.5.1 Theorem.   *The equilibrium $x = 0$, $y = 0$, $z = 0$ of the system (8.5.8) is stable with respect to x, y, z and asymptotically stable with respect to x, y if all eigenvalues of the matrix*

$$\begin{pmatrix} 0 & I \\ \mathcal{B} + \mathcal{CD} & \mathcal{A} \end{pmatrix} \tag{8.5.9}$$

have negative real parts.

8.5.2 Example (The Rattleback).   Here we outline the stability theory of the rattleback to illustrate the results discussed above. The details may be found in Karapetyan [1980, 1981], Markeev [1992], and Zenkov, Bloch, and Marsden [1998].

Recall from Section 1.11 that the Lagrangian and the constraints are

$$
\begin{aligned}
L = \frac{1}{2} & \left[A\cos^2\psi + B\sin^2\psi + m(\gamma_1\cos\theta - \zeta\sin\theta)^2 \right] \dot{\theta}^2 \\
+ \frac{1}{2} & \left[(A\sin^2\psi + B\cos^2\psi)\sin^2\theta + C\cos^2\theta \right] \dot{\phi}^2 \\
+ \frac{1}{2} & \left(C + m\gamma_2^2\sin^2\theta \right) \dot{\psi}^2 + \frac{1}{2}m\left(\dot{x}^2 + \dot{y}^2 \right) \\
+ & m(\gamma_1\cos\theta - \zeta\sin\theta)\gamma_2\sin\theta\,\dot{\theta}\dot{\psi} + (A - B)\sin\theta\sin\psi\cos\psi\,\dot{\theta}\dot{\phi} \\
+ & C\cos\theta\,\dot{\phi}\dot{\psi} + mg(\gamma_1\sin\theta + \zeta\cos\theta)
\end{aligned}
$$

and

$$\dot{x} = \alpha_1\dot{\theta} + \alpha_2\dot{\psi} + \alpha_3\dot{\phi}, \quad \dot{y} = \beta_1\dot{\theta} + \beta_2\dot{\psi} + \beta_3\dot{\phi},$$

where the terms were defined in Section 1.11.

Using the Lie algebra element corresponding to the generator

$$\xi_Q = \alpha_3\partial_x + \beta_3\partial_y + \partial_\phi$$

we find the nonholonomic momentum to be

$$
\begin{aligned}
p = I(\theta, \psi)\dot{\phi} & + \left[(A - B)\sin\theta\sin\psi\cos\psi - m(\gamma_1\sin\theta + \zeta\cos\theta)\gamma_2 \right]\dot{\theta} \\
& + \left[C\cos\theta + m(\gamma_2^2\cos\theta + \gamma_1(\gamma_1\cos\theta - \zeta\sin\theta)) \right]\dot{\psi},
\end{aligned}
$$

where

$$
\begin{aligned}
I(\theta, \psi) = & (A\sin^2\psi + B\cos^2\psi)\sin^2\theta + C\cos^2\theta \\
& + m(\gamma_2^2 + (\gamma_1\cos\theta - \zeta\sin\theta)^2).
\end{aligned}
$$

The amended potential (see (5.7.8)) becomes

$$U = \frac{p^2}{2I(\theta, \psi)} - mg(\gamma_1\sin\theta + \zeta\cos\theta).$$

The relative equilibria of the rattleback are: $\theta = \theta_0$, $\psi = \psi_0$, $p = p_0$, where θ_0, ψ_0, p_0 satisfy the conditions

$$mg(\gamma_1 \cos\theta_0 - \zeta \sin\theta_0)I^2(\theta_0, \psi_0)$$
$$+ \left[(A\sin^2\psi_0 + B\cos^2\psi_0 - C)\sin\theta_0 \cos\theta_0\right.$$
$$\left. - m(\gamma_1 \cos\theta_0 - \zeta\sin\theta_0)(\gamma_1\sin\theta_0 + \zeta\cos\theta_0)\right]p_0^2 = 0,$$
$$mg\gamma_2 I^2(\theta_0, \psi_0) + [(A - B)\sin\theta_0 \sin\psi_0 \cos\psi_0$$
$$- m\gamma_2(\gamma_1\sin\theta_0 + \zeta\cos\theta_0)]p_0^2 = 0,$$

which follow from the equations $\nabla_\theta U = 0$, $\nabla_\psi U = 0$.

In particular, consider the relative equilibria: $\theta = c\pi/2$, $\psi = 0$, $p = p_0$, which represent the rotations of the rattleback about the vertical axis of inertia. For such relative equilibria $\xi = \zeta = 0$, and therefore the conditions for existence of relative equilibria are trivially satisfied with an arbitrary value of p_0. Omitting the computations of the linearized equations for the rattleback (see Karapetyan [1980] and Zenkov, Bloch, and Marsden [1998] for details), and the corresponding characteristic polynomial, we just state here the Routh–Hurwitz conditions (see, e.g., Gantmacher [1959]) for all eigenvalues to have negative real parts:

$$\left(R - P\frac{p_0^2}{B^2}\right)\frac{p_0^2}{B^2} - S > 0, \qquad S > 0, \tag{8.5.10}$$

$$(A - C)(r_2 - r_1)p_0 \sin\alpha \cos\alpha > 0. \tag{8.5.11}$$

If these conditions are satisfied, then the relative equilibrium is stable, and it is asymptotically stable with respect to θ, $\dot\theta$, ψ, $\dot\psi$.

In the above formulas r_1, r_2 stand for the radii of curvature of the body at the contact point, α is the angle between the horizontal inertia axis ξ and the r_1-curvature direction, and

$$P = (A + ma^2)(C + ma^2),$$
$$R = \left[(A + C - B + 2ma^2)^2\right.$$

$$- (A + C - B + 2ma^2)ma(r_1 + r_2) + m^2 a^2 r_1 r_2\right]\frac{p_0^2}{B^2}$$

$$- \left[(A - B)\frac{p_0^2}{B^2} + m(a - r_1\sin^2\alpha - r_2\cos^2\alpha)\left(g + a\frac{p_0^2}{B^2}\right)\right](A + ma^2)$$

$$- \left[(C - B)\frac{p_0^2}{B^2} + m(a - r_2\sin^2\alpha - r_1\cos^2\alpha)\left(g + a\frac{p_0^2}{B^2}\right)\right](C + ma^2),$$

$$S = (A - B)(C - B)\frac{p_0^4}{B^4} + m^2(a - r_1)(a - r_2)\left(g + m\frac{p_0^2}{B^2}\right)^2$$

$$+ m\frac{p_0^2}{B^2}\left(g + a\frac{p^2}{B^2}\right)\left[A(a - r_1\cos^2\alpha - r_2\sin^2\alpha)\right.$$

$$+ C\left(a - r_1\sin^2\alpha - r^2\cos^2\alpha\right) - B(2a - r_1 - r_2)\right].$$

Conditions (8.5.10) impose restrictions on the mass distribution, the magnitude of the angular velocity, and the shape of the rattleback only. Condition (8.5.11) distinguishes the direction of rotation corresponding to the stable relative equilibrium.

The rattleback is also capable of performing stationary rotations with its center of mass moving at a constant rate along a circle. A similar argument gives the stability conditions in this case. The details may be found in Karapetyan [1981] and Markeev [1992].   ◆

8.6   Euler–Poincaré–Suslov Equations

An important special case of the reduced nonholonomic equations is the case where there is no shape space at all and the configuration space is $Q = G$, a Lie group. Again, the qualitative behavior of these systems is very interesting, and in particular, the systems may or may not be measure-preserving and can exhibit asymptotic stability with respect to some of the variables, as we show in this section.

Formulation of the Equations.   In this case the basic equations are the Euler–Poincaré equations

$$\frac{d}{dt}p_b = C_{ab}^c I^{ad} p_c p_d = C_{ab}^c p_c \omega^a, \qquad (8.6.1)$$

where $p_a = I_{ab}\omega^b$, $\omega \in \mathfrak{g}$, $p \in \mathfrak{g}^*$ to which we append the left-invariant constraint

$$\langle \mathbf{a}, \omega \rangle = a_i \omega^i = 0. \qquad (8.6.2)$$

Here $\mathbf{a} = a_i e^i \in \mathfrak{g}^*$ and $\omega = \omega^i e_i$ where e_i, $i = 1, \ldots, n$ is a basis for $\mathfrak{g}$ and e^i its dual basis.

As mentioned above we will examine here the question of when such equations exhibit asymptotic behavior. In this context it is key to consider whether or not the equations of motion has an invariant measure. If there is an invariant measure, there can be no contraction of the phase space and hence no asymptotic stability. Unlike Hamiltonian equations of motion, nonholonomic systems need not preserve volume in the phase space, since nonholonomic systems are not Poisson; See section 5.8. The existence of an invariant measure is also key in understanding integrability; we do not really consider this fascinating subject in this book, but see, for example, Arnold, Kozlov, and Neishtadt [1988] and references therein, Koiller [1992], Federov and Kozlov [1995], Hermans [1995], Zenkov [1995], Jovanović [1998], Bloch [2000], Zenkov and Bloch [2003], Schneider [2000].

Unconstrained Case.   Following Kozlov [1988b] it is convenient to consider the unconstrained case first. One considers the question of whether the equations (8.6.1) have an absolutely continuous invariant measure $f d^n \omega$ with summable density f. If f is a positive function of class C^1, one calls the invariant measure an *integral invariant*. A measure on a group G is said to be *left-invariant* if

$$\int_G f(gh) d\omega = \int_G f(h) d\omega$$

for every integrable function on the group and elements $g, h \in G$, and similarly for the right-invariant case.

A group G whose Lie algebra has structure coefficients C_{bc}^a is said to be *unimodular* if it has a bilaterally invariant measure. A criterion for unimodularity is $C_{ac}^c = 0$ (using the Einstein summation convention). One can show that any compact connected Lie group has a bi-invariant measure, unique up to a constant factor. However, compactness in not a necessary condition for the existence of a bi-invariant measure; for example, the Euclidean group in the plane, SE(2), has such a measure. See, e.g., Sattinger and Weaver [1986] for more details on this topic.

Kozlov proves the following theorem:

8.6.1 Theorem. *The Euler–Poincaré equations have an integral invariant if and only if the group G is unimodular.*

Neither direction is hard to prove, but we content ourselves with proving sufficiency here.

Proof of Sufficiency. We know (Liouville's theorem) that a flow of a vector differential equation $\dot{x} = f(x)$ is volume-preserving if $\operatorname{div} f = 0$. In this case the divergence of the right-hand side of equation (8.6.1) is $C_{ac}^c I^{ad} p_d$ (by the skew symmetry of C_{ab}^c in a, b). ∎

Necessity follows from Kozlov's observation (Kozlov [1988b]) that a system of differential equations with homogeneous right-hand side has an integral invariant if and only if its phase flow preserves the standard measure.

Constrained Case. Turning to the case where we have the constraint (8.6.2) using the Lagrange multiplier approach discussed in Section 5.2 we have the equations (Euler–Poincaré–Suslov)

$$\frac{d}{dt} p_b = C_{ab}^c I^{ad} p_c p_d + \lambda a_b = C_{ab}^c p_c \omega^a + \lambda a_b \qquad (8.6.3)$$

together with the constraint (8.6.2). This defines a system on the hyperplane defined by the constraints. One can then formulate a condition for the existence of an invariant measure in the case where the group is compact and we identify $\mathfrak{g}^*$ with $\mathfrak{g}$ by the Killing form (see Section 2.8).

8.6.2 Theorem (Kozlov [1988b]).   *The equations (8.6.3) have an invariant measure if and only if the vector* **a** *is an eigenvector for the operator* $\mathrm{ad}_{I^{-1}\mathbf{a}}$, *i.e., if*

$$[I^{-1}\mathbf{a}, \mathbf{a}] = \mu\mathbf{a} \tag{8.6.4}$$

for $\mu \in \mathbb{R}$.

The proof follows from solving for the constraint and computing the divergence of the right-hand side of the flow. A similar result can be obtained in the noncompact case; see Jovanović [1998] and Zenkov and Bloch [2003]. One can also, of course, formulate the equations of motion in the presence of several constraints.

In Zenkov and Bloch [2003] measure preservation for general nonholonomic systems where one has internal degrees of freedom as well as group variables is discussed.

8.6.3 Example (Euler–Poincaré–Suslov Problem on $SO(3)$**).**   In this case the problem can be formulated as the standard Euler equations

$$I\dot{\omega} = I\omega \times \omega, \tag{8.6.5}$$

where $\omega = (\omega_1, \omega_2, \omega_3)$ is the system angular velocity in a frame where the inertia matrix is of the form $I = \mathrm{diag}(I_1, I_2, I_3)$ and the system is subject to the constraint

$$a \cdot \omega = 0, \tag{8.6.6}$$

where $a = (a_1, a_2, a_3)$.

The nonholonomic equations of motion are then given by

$$I\dot{\omega} = I\omega \times \omega + \lambda a \tag{8.6.7}$$

subject to the constraint (8.6.6). We can easily solve for λ:

$$\lambda = -\frac{I^{-1}a \cdot (I\omega \times \omega)}{I^{-1}a \cdot a}. \tag{8.6.8}$$

If $a_2 = a_3 = 0$ (a constraint that is an eigenstate of the moment of inertia operator), one gets evolution with constant angular velocity. This is the only situation when the classical Suslov problem is measure preserving.
◆

The Generalized Suslov problem.   The $SO(n)$ generalization is also interesting and is known as the Suslov problem (see Federov and Kozlov [1995].) This system may also exhibit asymptotic stability. We only give somewhat brief comments here. (A different, nonasymptotic form is analyzed in Zenkov and Bloch [2000].

The equations of motion are those of an n-dimensional rigid body (see equations (7.4.32)) with skew-symmetric angular velocity matrix Ω with entries Ω_{ij} and symmetric moment of inertia matrix $I = I_{ij}$. One then introduces the constraints $\Omega_{ij} = 0, i, j \geq 2$. The resulting nonholonomic equations of motion are

$$(I_{11} + I_{22})\dot{\Omega}_{12} = I_{12}\left(\Omega_{13}^2 + \Omega_{14}^2 + \cdots + \Omega_{1n}^2\right)$$
$$- (I_{13}\Omega_{13} + I_{14}\Omega_{14} + \cdots + I_{1n}\Omega_{1n})\,\Omega_{12},$$
$$(I_{11} + I_{33})\dot{\Omega}_{13} = I_{13}\left(\Omega_{12}^2 + \Omega_{14}^2 + \cdots + \Omega_{1n}^2\right)$$
$$- (I_{12}\Omega_{12} + I_{14}\Omega_{14} + \cdots + I_{1n}\Omega_{1n})\,\Omega_{13}, \qquad (8.6.9)$$
$$\cdots$$
$$(I_{11} + I_{nn})\dot{\Omega}_{1n} = I_{1n}\left(\Omega_{12}^2 + \Omega_{13}^2 + \cdots + \Omega_{1n-1}^2\right)$$
$$- (I_{12}\Omega_{12} + I_{13}\Omega_{13} + \cdots + I_{1n-1}\Omega_{1n-1})\,\Omega_{1n}.$$

This system has the energy integral

$$H = \frac{1}{2}\left((I_{11} + I_{22})\Omega_{12}^2 + (I_{11} + I_{33})\Omega_{13}^2 + \cdots + (I_{11} + I_{nn})\Omega_{nn}^2\right).$$
$$(8.6.10)$$

Defining the momenta $M_{1j} = (I_{11} + I_{jj})\Omega_{1j}$ by the Legendre transform, we can write the system as one of almost Poisson form $\dot{M} = J(M)\nabla H(M)$, where the almost Poisson matrix (in the angular velocity variables) is

$$J(\Omega) = \begin{bmatrix} 0 & I_{13}\Omega_{12} - I_{12}\Omega_{13} & \cdots & I_{1n}\Omega_{12} - I_{12}\Omega_{1n} \\ I_{12}\Omega_{13} - I_{13}\Omega_{12} & 0 & \cdots & I_{1n}\Omega_{13} - I_{13}\Omega_{1n} \\ \cdots & \cdots & \cdots & \cdots \\ I_{12}\Omega_{1n} - I_{1n}\Omega_{12} & \cdots & \cdots & 0 \end{bmatrix}.$$
$$(8.6.11)$$

Further, the system exhibits asymptotic behavior as indicated by the fact that the function

$$F = (I_{11} + I_{22})I_{12}\Omega_{12} + (I_{11} + I_{33})I_{13}\Omega_{13} + \cdots + (I_{11} + I_{nn})I_{1n}\Omega_{1n} \quad (8.6.12)$$

satisfies

$$\dot{F} = \sum_{i<j}^{n}(I_{1i}\Omega_{1j} - I_{1j}\Omega_{1})^2 \qquad (8.6.13)$$

along the flow and is positive everywhere except at points of the line $\{\Omega_{12} = I_{12}\mu, \ldots, \Omega_{1n} = I_{1n}\mu\}$, $\mu \in \mathbb{R}$.

Thus motion occurs on the energy ellipsoid and is asymptotic to a point on the line above intersecting the ellipsoid.

Asymptotic Dynamics, the Chaplygin Sleigh, and the Toda Lattice. Here we discuss the Chaplygin sleigh and its relationship with the Toda lattice flow (see Bloch [2000]).

Recall firstly the (reduced) Chaplygin sleigh equations from Section 1.7:

$$\dot{v} = a\omega^2,$$
$$\dot{\omega} = -\frac{ma}{I + ma^2}v\omega. \qquad (8.6.14)$$

Normalizing, we have the equations

$$\dot{v} = \omega^2,$$
$$\dot{\omega} = -v\omega. \tag{8.6.15}$$

We now recall the Toda lattice equations defined in Section 1.12 In the two-dimensional case the matrices in the Lax pair are simply

$$L = \begin{pmatrix} b_1 & a_1 \\ a_1 & -b_1 \end{pmatrix}, \quad B = \begin{pmatrix} 0 & a_1 \\ -a_1 & 0 \end{pmatrix},$$

and the equations of motion are given by

$$\dot{b}_1 = 2a_1^2,$$
$$\dot{a}_1 = -2a_1 b_1. \tag{8.6.16}$$

As discussed in Section 1.12 these equations may be solved explicitly by factorization. For the initial data $b_1 = 0$, $a_1 = c$, explicitly carrying out the factorization yields the solution

$$b_1(t) = -c\frac{\sinh 2ct}{\cosh 2ct}, \qquad a_1(t) = \frac{c}{\cosh 2ct}. \tag{8.6.17}$$

Scaling time by a factor of two, we have the following immediate observation:

8.6.4 Proposition.   *The Chaplygin sleigh equations are precisely equivalent to the two-dimensional Toda lattice equations except for the fact that there is no sign restriction on the variable ω. Hence the system can be written in Lax pair form and solved explicitly.*

As described in Neimark and Fufaev [1972] the Chaplygin sleigh equations may be explicity solved also as follows: We have

$$\frac{d}{dt}\left(\frac{\dot{\omega}}{\omega}\right) = -\dot{v} = -\omega^2.$$

Multiplying the preceding equation by $\dot{\omega}/\omega$ and integrating, we obtain

$$\left(\frac{\dot{\omega}}{\omega}\right)^2 = c^2 - \omega^2 \tag{8.6.18}$$

for a suitable constant c. Define the angle ψ by setting $\omega = c\cos\psi$. For suitable choice of c, integration of equation (8.6.18) gives

$$ct = \int_0^\psi \frac{d\theta}{\cos\theta} = \frac{1}{2}\ln\frac{1+\sin\psi}{1-\sin\psi}, \tag{8.6.19}$$

which can be inverted and solved to give precisely the same solution as obtained by the factorization method of Toda.

We now make some remarks on the asymptotic behavior of the Toda lattice and the Chaplygin sleigh. For more details, see Bloch [2000]. As mentioned earlier, the equations of the Toda lattice and the Chaplygin sleigh are identical except for the sign restriction on the variable a_1 (in Toda). But this difference is crucial. Observe firstly that the Toda lattice system has a set of equilibria defined by setting a_i equal to zero in the closure of its phase space. (In the two-dimensional case this set is a line, as it is for the Chaplygin sleigh.) Further, the Toda matrix L tends to a diagonal matrix with the eigenvalues of the matrix defining the initial data ordered by magnitude on the diagonal.

How is it possible then that one can obtain asymptotic behavior in this Hamiltonian system? Vital here is the fact that the system is integrable and that one is evolving on a level set of the integrals: a Lagrange submanifold of the phase space. By the Arnold–Liouville theorem for integrability (see, e.g., Arnold [1989] and Abraham and Marsden [1978]), this submanifold is diffeomorphic to a product of circles and lines. For aperiodic Toda it is in fact diffeomorphic to a set of lines, and the flow on this set of lines is a gradient flow. For the two-dimensional Toda the level set is $b_1^2 + a_1^2 =$ const; $a_1 > 0$. Key also is the fact that the set defined by $a_i = 0$ is reached only asymptotically but is not actually in the phase space. Thus the system is asymptotically stable only in a generalized sense.

On the other hand, for the Chaplygin sleigh one is allowing a_1 (or ω in the sleigh notation) to take any value in $\mathbb{R}$. Hence the flow in the phase plane can asymptotically approach the v-axis from either half-plane, but cannot, of course, cross this axis; see the figure in Section 1.7. The flow is in fact the union of two flows: the standard Toda lattice flow and the flow with $a_1 < 0$. The phase space of the Chaplygin sleigh is the compactification by a line of the union of the phase spaces of two Toda lattice systems, one with $a_1 > 0$ and another with $a_1 < 0$. Such "signed" Toda flows are of interest in general; see Tomei [1984], Davis [1987], Bloch [2000], and references given therein.

We remark that the Suslov problem is an example of a class of systems called LL systems – left-invariant systems with left-invariant constraints. This is a useful class of systems to study – see for example Fedorov and Zenkov [2005]. This is in contrast to LR systems – left-invariant systems with right-invariant constraints, see for example Fedorov and Jovanović [2004]. This latter class includes systems such as the Chaplygin sphere, a nonhomogeneous ball rolling on the plane; see, for example, Schneider [2002]).

Exercises

⋄ **8.6-1.** Show that SO(3) has a bi-invariant measure by testing the relevant condition on the structure coefficients.

◇ **8.6-2.**   Show that SE(2) has a bi-invariant measure.

◇ **8.6-3.**   Compute the the linearized flow equations about the origin for the Euler–Poincaré–Suslov equations on SO(3) and analyze the linear stability.

9
Energy-Based Methods for Stabilization of Controlled Lagrangian Systems

In this final chapter we briefly discuss two recent energy-based methods for stabilizing second-order nonlinear systems and their application to nonholonomic systems. The first is the method of controlled Lagrangians (or Hamiltonians) and "matching." This is developing into a rather large subject, which we just touch on here in order to explain something of the role of connections in the subject and its potential applications to nonholonomic systems. The second is a geometric approach to averaging second-order systems that arise as models of controlled *superarticulated* (or *underactuated*) mechanical (Lagrangian) systems. While not yet constituting a complete theory, the results of this chapter may be thought of as intrinsically second-order versions of the results on kinematically nonholonomic systems presented in Chapters 4 and 6. Needless to say, this chapter just touches on the vast subject of energy-based stabilization.

9.1 Controlled Lagrangian Methods

The idea of designing control laws that minimize energy-like functions is an old one, dating at least as far back as the work of Lyapunov (Lyapunov [1992], Lyapunov [1907]). For mechanical systems, the approach is especially natural, with the designs being related to actual physical energies (kinetic and potential). The main idea in the approach to control design that we shall treat in the present chapter is to consider classes of control designs for Lagrangian (or Hamiltonian) systems with respect to which the controlled system dynamics remains governed by a set of Euler–Lagrange (respectively canonical) equations. Within this class of controls, energy methods can be used to prescribe and analyze stable controlled motions.

While our treatment of *controlled Lagrangian methods* is intended to be self-contained, a complete discussion is beyond the scope of the book, and we refer to primary sources for many details. In particular, Bloch, Leonard, and Marsden [1997, 2000] and Chang, Bloch, Leonard, Marsden, and Woolsey [2002] (and references therein) treat feedback control designs, while other recent work—on oscillation mediated control of Lagrangian and Hamiltonian systems (Baillieul [1993], Baillieul [1995], Weibel [1997], Weibel and Baillieul [1998b], Baillieul [1998], Bullo [2001])—has introduced a class of robust open-loop designs that can be discussed within essentially the same framework. The aim of the present chapter is to provide a brief but systematic treatment of both approaches to the control of Lagrangian and Hamiltonian systems. Further references to work of Ortega, van der Schaft, and others in this area are given below.

The Setup for Control of Underactuated Lagrangian Systems. In general we use the following setting for the analysis of controlled Lagrangian systems (see Chang, Bloch, Leonard, Marsden, and Woolsey [2002]):

9.1.1 Definition. *A* **controlled Lagrangian** **(CL)** *system is a triple* (L, F, W) *where the function* $L : TQ \to \mathbb{R}$ *is the Lagrangian, the fiber-preserving map* $F : TQ \to T^*Q$ *is an external force, and* $W \subset T^*Q$ *is a subbundle of* T^*Q, *called the* **control bundle**, *representing the* **actuation directions**.

Sometimes, we will identify the subbundle W with the set of bundle maps from TQ to W. The fact that W may be smaller than the whole space corresponds to the system being *underactuated*. The equations of motion of the system (L, F, W) may be written as

$$\mathcal{EL}(L)(q, \dot{q}, \ddot{q}) = F(q, \dot{q}) + u, \qquad (9.1.1)$$

with a control u selected from W. Here $\mathcal{EL}$ is the Euler–Lagrange operator with coordinate form as in equations (9.1.2) and (9.1.3). When we choose a specific control map $u : TQ \to W$ (so that u is a function of $(q^i, \dot{q}^i)$), then we call the triple (L, F, u) a **closed-loop Lagrangian system**. We will typically be interested in such feedback controls in this discussion. In the special case where W is integrable (that is, its annihilator $W^\circ \subset TQ$ is integrable in the usual Frobenius sense) and we choose coordinates appropriately, then a CL system, that is, equations (9.1.1), can be locally written in coordinates as

$$\frac{d}{dt}\frac{\partial L}{\partial \dot{q}^i} - \frac{\partial L}{\partial q^i} = F_i + u_i, \quad i = 1, \ldots, k, \qquad (9.1.2)$$

$$\frac{d}{dt}\frac{\partial L}{\partial \dot{q}^i} - \frac{\partial L}{\partial q^i} = F_i, \qquad i = k+1, \ldots, n. \qquad (9.1.3)$$

Here the coordinates $q^1, \ldots, q^k$ are chosen so that $dq^1, \ldots, dq^k$ span W, so W is k-dimensional in this case. The external forces can include gyroscopic forces, friction forces, etc.

The principal aim of much of the current research on nonlinear control methods is to understand (and prescribe control laws for) the dynamic response of the q-variables in (9.1.3). While various approaches—each having certain advantages—have been proposed, the present chapter is confined to a class of powerful techniques developed within the framework of analytical mechanics and closely related to the methods discussed in earlier chapters. More specifically, we shall provide an exposition of recent methods that might all be referred to as *controlled Lagrangian methods*. The unifying idea is that control inputs are used to alter (and prescribe) the form of the Lagrangian governing the behavior of the q-variables. Hence both the design and analysis of control laws may be carried out using the powerful machinery of analytical mechanics. Equations (9.1.2)–(9.1.3) provide mathematical descriptions of a wide and interesting class of controlled mechanical systems, examples of which will be studied in detail.

9.1.2 Example (Cart with Pendulum on an Inclined Plane). Consider the free pendulum on a cart on an inclined plane. This is depicted in Figure 9.1.1.

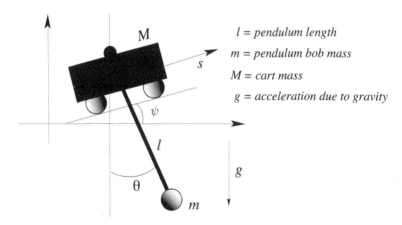

FIGURE 9.1.1. The pendulum on a cart on an inclined plane.

Let s denote the position of the cart along the incline, and let θ denote the angle of the pendulum with respect to the downward pointing vertical. The configuration space for this system is $Q = \mathbb{R}^1 \times S^1$, with the first factor being the cart position s and the second factor being the pendulum angle θ. We assume that s is directly controlled, while θ is controlled only indirectly through dynamic interactions within the mechanism. Clearly, this can be cast in the form (9.1.2), (9.1.3), and this is done by first writing the *Lagrangian*

$$L(s, \theta; \dot{s}, \dot{\theta}) = \frac{1}{2}(\alpha\dot{\theta}^2 + 2\beta\cos(\theta - \psi)\dot{\theta}\dot{s} + \gamma\dot{s}^2) + D\cos\theta - \gamma g s \sin\psi.$$

The constants are defined in terms of the physical parameters by

$$\alpha = m\ell^2, \quad \beta = m\ell, \quad \gamma = m + M, \quad D = mg\ell.$$

The (controlled) equations of motion are the corresponding Euler–Lagrange equations:

$$\begin{pmatrix} \gamma & \beta\cos(\theta - \psi) \\ \beta\cos(\theta - \psi) & \alpha \end{pmatrix} \begin{pmatrix} \ddot{s} \\ \ddot{\theta} \end{pmatrix}$$
$$+ \begin{pmatrix} -\beta\sin(\theta - \psi)\dot{\theta}^2 + \gamma g \sin\psi \\ D\sin\theta \end{pmatrix} = \begin{pmatrix} u \\ 0 \end{pmatrix}. \qquad (9.1.4)$$

$\blacklozenge$

In the remainder of the chapter the cart–pendulum example will serve to illustrate a number of design methods. In the next two sections we shall discuss feedback designs based on ***matching***. The power of these methods lies in the preservation of the Lagrangian structure of the closed-loop dynamics. In Section 9.4 we shall discuss control designs in which high-frequency oscillations of the inputs u and/or the configuration variable r produce prescribed stable responses of q. The latter of these methods may be viewed as a second-order analogue of the approach discussed in Sections 4.3 and 4.4.

9.2   Feedback Design and Matching

An introduction to the use of structured feedback that preserves the Lagrangian or Hamiltonian form of physical models is provided by using the energy–Casimir method to stabilize a rigid body with rotors (see Bloch, Krishnaprasad, Marsden, and Alvarez [1992]). There has been much work on rigid body stabilization using various techniques such as center manifold theory (see, e.g., Aeyels [1985]). Here we simply consider the system as an illustration of the energy–Casimir method.

9.2.1 Example (Rigid Body with a Rotor).   Recall (see Section 1.9) that this system consists of a rigid body with a rotor aligned along the minor principal axis of the body. The Lagrangian for this system is

$$L = \frac{1}{2}\left(\lambda_1\Omega_1^2 + \lambda_2\Omega_2^2 + I_3\Omega_3^2 + J_3(\Omega_3 + \dot{\alpha})^2\right), \qquad (9.2.1)$$

where $I_1 > I_2 > I_3$ are the rigid body moments of inertia, $J_1 = J_2$ and J_3 are the rotor moments of inertia,

$$\lambda_i = I_i + J_i, \quad \Omega = (\Omega_1, \Omega_2, \Omega_3)$$

is the body angular velocity vector of the carrier, and α is the relative angle of the rotor.

The body angular momenta are obtained by differentiating the Lagrangian with respect to the corresponding angular velocities:

$$\Pi_1 = \lambda_1 \Omega_1,$$
$$\Pi_2 = \lambda_2 \Omega_2,$$
$$\Pi_3 = \lambda_3 \Omega_3 + J_3 \dot{\alpha},$$
$$l_3 = J_3(\Omega_3 + \dot{\alpha}).$$

Since the energy is entirely kinetic, the corresponding Hamiltonian is given by substituting the preceding expressions for the momenta into (9.2.1):

$$H = \frac{1}{2}\left(\frac{\Pi_1^2}{\lambda_1} + \frac{\Pi_2^2}{\lambda_2} + \frac{(\Pi_3 - l_3)^2}{I_3}\right) + \frac{l_3^2}{2J_3}. \qquad (9.2.2)$$

One can approach the stabilization problem from either the Lagrangian or Hamiltonian viewpoint. For the Lagrangian point of view and its generalization to Euler–Poincaré systems see Bloch, Leonard, and Marsden [2000]. Here we recall briefly the Hamiltonian picture from Bloch, Krishnaprasad, Marsden, and Alvarez [1992].

If $u = 0$, then the system has an S^1 symmetry corresponding to rotations of the rotor, and correspondingly, l_3 is a constant of motion. The reduced system obtained by taking the quotient by this S^1 symmetry is Hamiltonian, where the reduced Hamiltonian is obtained from (9.2.2) by simply setting l_3 equal to a constant:

$$H_{l_3} = \frac{1}{2}\left(\frac{\Pi_1^2}{\lambda_1} + \frac{\Pi_2^2}{\lambda_2} + \frac{(\Pi_3 - l_3)^2}{I_3}\right) + \frac{l_3^2}{2J_3}.$$

Choose the feedback control law

$$u = ka_3\Pi_1\Pi_2,$$

where k is a gain parameter; then the system retains the S^1 symmetry and $P_k = l_3 - k\Pi_3$ is a new conserved quantity. The closed-loop equations are

$$\dot{\Pi}_1 = \Pi_2\left(\frac{(1-k)\Pi_3 - P_k}{I_3}\right) - \frac{\Pi_3\Pi_2}{\lambda_2},$$
$$\dot{\Pi}_2 = -\Pi_1\left(\frac{(1-k)\Pi_3 - P_k}{I_3}\right) + \frac{\Pi_1\Pi_3}{\lambda_1},$$
$$\dot{\Pi}_3 = a_3\Pi_1\Pi_2.$$

These equations are Hamiltonian with

$$H = \frac{1}{2}\left(\frac{\Pi_1^2}{\lambda_1} + \frac{\Pi_2^2}{\lambda_2} + \frac{((1-k)\Pi_3 - P_k)^2}{(1-k)I_3}\right) + \frac{1}{2}\frac{P_k^2}{J_3(1-k)},$$

using the Lie–Poisson (rigid body) Poisson structure on $\mathfrak{so}(3)^*$.

Two special cases are of interest: $k = 0$, the uncontrolled case, and $k = J_3/\lambda_3$, the *driven case*, where $\dot{\alpha} = \text{constant}$.

Now consider the case $P = 0$ and the special equilibrium $(0, M, 0)$. For $k > 1 - J_3/\lambda_2$, the equilibrium $(0, M, 0)$ is stable. This is proved by the energy–Casimir method. We look at $H + C$, where $C = \varphi(\|\Pi\|^2)$. Pick φ such that

$$\delta(H + C)|_{(0,M,0)} = 0.$$

Then one computes that $\delta^2(H + C)$ is negative definite if $k > 1 - J_3/\lambda_2$ and $\varphi''(M^2) < 0$.   ◆

Controlled Lagrangians and Matching.   There has been a great deal of work on recently on so-called *matching* methods for nonlinear stabilization; see Bloch, Leonard, and Marsden [1997], Bloch, Leonard, and Marsden [1998], Bloch, Leonard, and Marsden [1999b], Bloch, Leonard, and Marsden [2000], Bloch, Chang, Leonard, and Marsden [2001], Hamberg [1999], Auckly, Kapitanski, and White [2000], Ortega, Loria, Nicklasson and Sira-Ramirez [1998], Ortega, van der Schaft, Mashcke and Escobar [1999], and Chang, Bloch, Leonard, Marsden, and Woolsey [2002], to list a few of the references. This generalizes the work done on the satellite with momentum wheels discussed at the beginning of this section. Essentially, the idea is to introduce a feedback into the control system such that the controlled system is still variational with respect to a modified Lagrangian or Hamiltonian. The work of Bloch, Leonard, and Marsden had concentrated on the Lagrangian side, and several of the other papers have focused on the Hamiltonian side. Finally, Chang, Bloch, Leonard, Marsden, and Woolsey [2002] showed that the two approaches are in fact equivalent.

The modified Lagrangian has several adjustable parameters, and modifying them to correspond to the forced (controlled) system is called *matching*. Having the systems in modified Lagrangian form enables one to use energy methods for the analysis of stabilization. In addition, one sometimes requires that symmetries be respected and hence momenta conserved, as in the satellite with momentum wheels. Finally, one adds dissipation to obtain asymptotic stability. We will not detail all this in this book but refer to the reader to the website, the papers, and a forthcoming monograph. Here we will just give a brief summary and an example of how this works for the simple pendulum on a cart when one wants to respect symmetries. We also describe a simple application to nonholonomic systems, which are most relevant to the theme of this book. This application to nonholonomic systems (a unicycle with rotor) is discussed in detail in Zenkov, Bloch, and Marsden [2002b].

Brief Summary of the Method

Here we describe the essence of the simplest case of the method as in Bloch, Leonard, and Marsden [2000]. This illustrates an interesting use of the theory of connections.

The Setting. Suppose our system has configuration space Q and that a Lie group G acts freely and properly on Q. It is useful to keep in mind the case in which $Q = S \times G$ with G acting only on the second factor by acting on the left by group multiplication.

For example, for the inverted planar pendulum on a cart (which we consider in detail below), we have $Q = S^1 \times \mathbb{R}$ with $G = \mathbb{R}$, the group of reals under addition (corresponding to translations of the cart), while for a rigid spacecraft with a rotor we have $Q = \mathrm{SO}(3) \times S^1$, where now the group is $G = S^1$, corresponding to rotations of the rotor.

Our goal will be to control the variables lying in the *shape space* Q/G (in the case in which $Q = S \times G$, then $Q/G = S$) using controls that act directly on the variables lying in G. We assume that the Lagrangian is invariant under the action of G on Q, where the action is on the factor G alone. In many specific examples, such as those given below, the invariance is equivalent to the Lagrangian being cyclic in the G-variables. Accordingly, this produces a conservation law for the free system. Our construction will preserve the invariance of the Lagrangian, thus providing us with a *controlled* conservation law.

The essence of the modification of the Lagrangian involves changing the metric tensor $g(\cdot, \cdot)$ that defines the kinetic energy of the system $\frac{1}{2}g(\dot{q}, \dot{q})$.

Our method relies on a special decomposition of the tangent spaces to the configuration manifold and a subsequent "controlled" modification of this split. We can describe this as follows:

Horizontal and Vertical Spaces. The tangent space to Q can be split into a sum of horizontal and vertical parts defined as follows: For each tangent vector v_q to Q at a point $q \in Q$, we can write a unique decomposition

$$v_q = \mathrm{Hor}\, v_q + \mathrm{Ver}\, v_q \qquad (9.2.3)$$

such that the vertical part is tangent to the orbits of the G-action and where the horizontal part is the metric orthogonal to the vertical space; that is, it is uniquely defined by requiring the identity

$$g(v_q, w_q) = g(\mathrm{Hor}\, v_q, \mathrm{Hor}\, w_q) + g(\mathrm{Ver}\, v_q, \mathrm{Ver}\, w_q) \qquad (9.2.4)$$

where v_q and w_q are arbitrary tangent vectors to Q at the point $q \in Q$. This choice of horizontal space coincides with that given by the *mechanical connection* that we introduced in Section 3.10. We refer to Marsden [1992] for further details regarding the mechanical connection. One can think intuitively of this decomposition of vectors as a decomposition into a piece in the symmetry, or group direction (the vertical piece), and one in the shape, or internal direction (the horizontal piece). In terms of the coordinate description (9.1.2), (9.1.3), tangent vectors in the direction of the controlled coordinates are *vertical*, while vectors in the direction of the uncontrolled (*shape*) variables are *horizontal*.

The Controlled Lagrangian.   For the kinetic energy of our controlled Lagrangian, we use a modified version of the right-hand side of equation (9.2.4). The potential energy remains unchanged. The modification consists of three ingredients:

1. a different choice of horizontal space, denoted by Hor_τ;

2. a change $g \to g_\sigma$ of the metric acting on horizontal vectors; and

3. a change $g \to g_\rho$ of the metric acting on vertical vectors.

We now make the following definition:

9.2.2 Definition.   *Let τ be a Lie-algebra-valued horizontal one-form on Q, that is, a one-form with values in the Lie algebra $\mathfrak{g}$ of G that annihilates vertical vectors. This means that for all vertical vectors v, the infinitesimal generator $[\tau(v)]_Q$ corresponding to $\tau(v) \in \mathfrak{g}$ is the zero vector field on Q. The τ-horizontal space at $q \in Q$ consists of tangent vectors to Q at q of the form*

$$\mathrm{Hor}_\tau v_q = \mathrm{Hor}\, v_q - [\tau(v)]_Q(q),$$

which also defines $v_q \mapsto \mathrm{Hor}_\tau(v_q)$, the τ-horizontal projection. The τ-vertical projection operator is defined by

$$\mathrm{Ver}_\tau(v_q) := \mathrm{Ver}(v_q) + [\tau(v)]_Q(q).$$

Notice that from these definitions and (9.2.3), we have

$$v_q = \mathrm{Hor}_\tau(v_q) + \mathrm{Ver}_\tau(v_q), \tag{9.2.5}$$

just as we did with τ absent. In fact, this new horizontal subspace can be regarded as defining a new connection, the τ-*connection*. The horizontal space itself, which by abuse of notation we also write as just Hor or Hor_τ, of course depends on τ also, but the vertical space does not: It is the tangent to the group orbit. On the other hand, the *projection* map $v_q \mapsto \mathrm{Ver}_\tau(v_q)$ does depend on τ.

9.2.3 Definition.   *Given g_σ, g_ρ and τ, we define the **controlled Lagrangian** to be the following Lagrangian, which has the form of a modified kinetic energy minus the potential energy:*

$$L_{\tau,\sigma,\rho}(v) = \frac{1}{2}\left[g_\sigma(\mathrm{Hor}_\tau v_q, \mathrm{Hor}_\tau v_q) + g_\rho(\mathrm{Ver}_\tau v_q, \mathrm{Ver}_\tau v_q)\right] - V(q), \tag{9.2.6}$$

where V is the potential energy.

The equations corresponding to this Lagrangian will be our closed-loop equations. The new terms appearing in those equations corresponding to the directly controlled variables are interpreted as control inputs. The modifications to the Lagrangian are chosen so that no new terms appear in the equations corresponding to the variables that are not directly controlled. We refer to this process as "matching."

Another way of expressing what we are doing here is the following. A principal connection on a bundle $Q \rightarrow Q/G$ may be thought of as a Lie-algebra-valued one-form, and one can obtain a new connection by adding to it a horizontal one-form τ. The new horizontal space described in the preceding definition is exactly of this sort.

Special Controlled Lagrangians. Here we consider a special class of controlled Lagrangians in which we take $g_\rho = g$. Further, in certain examples of interest, including the inverted pendulum on a cart, we not only can choose $g_\rho = g$ (i.e., there is no g_ρ modification needed), but we can also choose the metric g_σ to modify the original metric g only in the group directions by a scalar factor σ. In this case, the controlled Lagrangian takes the form

$$L_{\tau,\sigma}(v) = L(v + [\tau(v)]_Q(q)) + \frac{\sigma}{2} g([\tau(v)]_Q, [\tau(v)]_Q). \qquad (9.2.7)$$

The Inverted Pendulum on a Cart

The system we consider is the inverted pendulum on a cart that we described in Chapter 1. We remind the reader of the setup here:

The Lagrangian. Recall that we let s denote the position of the cart on the s-axis and let θ denote the angle of the pendulum with the vertical, as in Figure 9.1.1. In the notation depicted in Figure 9.1.1, $\psi = 0$.

The configuration space for this system is $Q = S \times G = S^1 \times \mathbb{R}$. For notational convenience we rewrite the Lagrangian as

$$L(\theta, s, \dot{\theta}, \dot{s}) = \frac{1}{2}(\alpha\dot{\theta}^2 + 2\beta\cos\theta\dot{s}\dot{\theta} + \gamma\dot{s}^2) + D\cos\theta, \qquad (9.2.8)$$

where $\alpha = ml^2$, $\beta = ml$, $\gamma = M + m$, and $D = mgl$ are constants. The momentum conjugate to θ is

$$p_\theta = \frac{\partial L}{\partial \dot{\theta}} = \alpha\dot{\theta} + \beta(\cos\theta)\dot{s}$$

, and the momentum conjugate to s is

$$p_s = \frac{\partial L}{\partial \dot{s}} = \gamma\dot{s} + \beta(\cos\theta)\dot{\theta}.$$

The relative equilibrium defined by $\theta = \pi$, $\dot{\theta} = 0$, and $\dot{s} = 0$ is unstable, since $D > 0$.

The equations of motion for the pendulum–cart system with a control force u acting on the cart (and no direct forces acting on the pendulum) are, as we saw earlier,

$$\frac{d}{dt}\frac{\partial L}{\partial \dot{\theta}} - \frac{\partial L}{\partial \theta} = 0,$$

$$\frac{d}{dt}\frac{\partial L}{\partial \dot{s}} = u,$$

that is,

$$\frac{d}{dt}p_\theta + \beta \sin \theta \dot{s}\dot{\theta} + D \sin \theta = 0;$$

that is,

$$\frac{d}{dt}(\alpha\dot{\theta} + \beta \cos \theta \dot{s}) + \beta \sin \theta \dot{s}\dot{\theta} + D \sin \theta = 0 \qquad (9.2.9)$$

and

$$\frac{d}{dt}p_s = \frac{d}{dt}(\gamma\dot{s} + \beta \cos \theta \dot{\theta}) = u.$$

The Controlled Lagrangian. Next, we form the controlled Lagrangian by modifying only the kinetic energy of the free pendulum–cart system according to the procedure given in the preceding section. This involves a nontrivial choice of τ and g_σ, but in this case, as we have remarked, it is sufficient to let $g_\rho = g$.

The most general s-invariant horizontal one-form τ is given by $\tau = k(\theta)d\theta$, and we choose g_σ to modify g in the group direction by a constant scalar factor σ (in general, σ need not be a constant, but it is for the present class of examples). Using (9.2.7), we let

$$L_{\tau,\sigma} := \frac{1}{2}\left(\alpha\dot{\theta}^2 + 2\beta \cos \theta(\dot{s} + k\dot{\theta})\dot{\theta} + \gamma(\dot{s} + k\dot{\theta})^2\right) + \frac{\sigma}{2}\gamma k^2\dot{\theta}^2 + D\cos\theta.$$
$$(9.2.10)$$

Notice that the variable s is still cyclic. Following the guidelines of the theory, we look for the feedback control by looking at the change in the conservation law. Associated with the new Lagrangian $L_{\tau,\sigma}$, we have the conservation law

$$\frac{d}{dt}\left(\frac{\partial L_{\tau,\sigma}}{\partial \dot{s}}\right) = \frac{d}{dt}(\beta \cos \theta \dot{\theta} + \gamma(\dot{s} + k\dot{\theta})) = 0, \qquad (9.2.11)$$

which we can rewrite in terms of the conjugate momentum p_s for the uncontrolled Lagrangian as

$$\frac{d}{dt}p_s = u := -\frac{d}{dt}(\gamma k(\theta)\dot{\theta}). \qquad (9.2.12)$$

Thus, we identify the term on the right-hand side with the *control force* exerted on the cart.

Using the controlled Lagrangian and equation (9.2.11), the θ equation is computed to be

$$\left(\alpha - \frac{\beta^2}{\gamma}\cos^2\theta + \sigma\gamma k^2(\theta)\right)\ddot{\theta} + \left(\frac{\beta^2}{\gamma}\cos\theta\sin\theta + \sigma\gamma k(\theta)k'(\theta)\right)\dot{\theta}^2$$

$$+ D\sin\theta = 0. \qquad (9.2.13)$$

Matching. The next step is to make choices of k and σ so that equation (9.2.13) using the controlled Lagrangian agrees with the θ equation for the controlled cart (9.2.9) with the control law given by equation (9.2.12). The θ equation for the controlled cart is

$$\left(\alpha - \frac{\beta^2}{\gamma}\cos^2\theta - \beta k(\theta)\cos\theta\right)\ddot{\theta} + \left(\frac{\beta^2}{\gamma}\cos\theta\sin\theta - \beta\cos\theta k'(\theta)\right)\dot{\theta}^2$$

$$+ D\sin\theta = 0. \qquad (9.2.14)$$

Comparing equations (9.2.13) and (9.2.14) we see that we require (twice)

$$\sigma\gamma[k(\theta)]^2 = -\beta k(\theta)\cos\theta. \qquad (9.2.15)$$

Since σ was assumed to be a constant, we set

$$k(\theta) = \kappa\frac{\beta}{\gamma}\cos\theta, \qquad (9.2.16)$$

where κ is a dimensionless constant (so $\sigma = -1/\kappa$).

The Control Law. Substituting for $\ddot{\theta}$ and k in (9.2.12) we obtain the desired nonlinear control law:

$$u = \frac{\kappa\beta\sin\theta\left(\alpha\dot{\theta}^2 + \cos\theta D\right)}{\alpha - \frac{\beta^2}{\gamma}(1 + \kappa)\cos^2\theta}. \qquad (9.2.17)$$

Stabilization. By examining either the energy or the linearization of the closed-loop system, one can see that the equilibrium $\theta = \dot{\theta} = \dot{s} = 0$ is stable if

$$\kappa > \frac{\alpha\gamma - \beta^2}{\beta^2} = \frac{M}{m} > 0. \qquad (9.2.18)$$

In summary, *we get a stabilizing feedback control law for the inverted pendulum, provided that κ satisfies the inequality* (9.2.18).

A simple calculation shows that the denominator of u is nonzero for θ satisfying $\sin^2\theta < E/F$, where $E = \kappa - (\alpha\gamma - \beta^2)/\beta^2$ (which is positive if the stability condition holds) and $F = \kappa + 1$. This range of θ tends to the range $-\pi/2 < \theta < \pi/2$ for large κ.

The above remark suggests that the region of stability (or attraction when damping control is added) is the whole range of non-downward-pointing states. In fact, we assert that this method produces large computable domains of attraction for stabilization (see Chang, Bloch, Leonard, Marsden, and Woolsey [2002]).

This approach has advantages because *it is done within the context of mechanics*; one can understand the stabilization in terms of the effective creation of an *inverted energy well* by the feedback control. (Our feedback in general creates a maximum for balance systems, since for these systems the equilibrium is a maximum of the potential energy, which we do not modify.)

There are numerous powerful extensions and generalizations of these ideas discussed in the references mentioned. We have introduced just the main ideas so that we can apply them to a class of nonholonomic systems. We do this in the next section.

9.3   Stabilization of a Class of Nonholonomic Systems

To illustrate matching in the nonholonomic context we consider a simple setting where we can use the integrability of the momentum equation. This section is based on Zenkov, Bloch, and Marsden [2002b], where more details may be found.

We consider a class of systems that satisfy the following assumptions:

(A1) The controls act on a subset of the shape variables of the system.

(A2) The momentum equation is in the form of a parallel transport equation.

(A3) The connection in the parallel transport equation is flat.

(A4) The actuated variables in the Lagrangian are cyclic.

The (reduced constrained) Routhian (see Chapter 5) of the system is taken to be

$$\mathcal{R}(r, \dot{r}, p) = \frac{1}{2} g_{\alpha\beta}(r) \dot{r}^\alpha \dot{r}^\beta - U(r, p),$$

where the amended potential $U(r, p)$ (see (5.7.8)) is defined by

$$U(r, p) = \frac{1}{2} I^{ab}(r) p_a p_b + V(r),$$

where $I^{ab}(r)$ are the components of the inverse locked inertia tensor and $V(r)$ is the potential energy.

Now we divide the shape variables into unactuated and actuated variables $r^{\alpha'}$ and $r^{\alpha''}$, respectively, and let $u_{\alpha''}$ be the control inputs. Defining the operators ∇_α by

$$\nabla_\alpha = \frac{\partial}{\partial r^\alpha} + \mathcal{D}^b_{a\alpha} p_b \frac{\partial}{\partial p_a},$$

as in Chapter 8 and using Assumption A4 that the actuated variables $r^{\alpha''}$ are cyclic, we obtain reduced equations of the form

$$\frac{d}{dt}\frac{\partial \mathcal{R}}{\partial \dot{r}^{\alpha'}} = \nabla_{\alpha'}\mathcal{R}, \tag{9.3.1}$$

$$\frac{d}{dt}\frac{\partial \mathcal{R}}{\partial \dot{r}^{\alpha''}} = u_{\alpha''}, \tag{9.3.2}$$

$$\dot{p}_a = \mathcal{D}_{a\alpha'}^b p_b \dot{r}^{\alpha'}. \tag{9.3.3}$$

Note that our definition of cyclic variables allows only noncyclic shape velocities to be present in the momentum equation (9.3.3).

Elimination of the Momentum Variables. Since the momentum equation is in the form of a parallel transport equation, it defines a distribution

$$dp_a = \mathcal{D}_{a\alpha'}^c p_c dr^{\alpha'}. \tag{9.3.4}$$

We assume now (Assumption A3) that the curvature of this distribution vanishes (the distribution is flat). This defines the global invariant manifolds Q_c of (9.3.1)–(9.3.3),

$$p_a = \mathcal{P}_a(r^{\alpha'}, c_b), \qquad c_b = \text{const.} \tag{9.3.5}$$

Each of these invariant manifolds is diffeomorphic to the tangent bundle $T(Q/G)$ of the original system's shape space. The dynamics on these invariant manifolds are governed by the equations

$$\frac{d}{dt}\frac{\partial \mathcal{L}_c}{\partial \dot{r}^{\alpha'}} = \frac{\partial \mathcal{L}_c}{\partial r^{\alpha'}}, \qquad \frac{d}{dt}\frac{\partial \mathcal{L}_c}{\partial \dot{r}^{\alpha''}} = u_{\alpha''}, \tag{9.3.6}$$

where

$$\mathcal{L}_c(r^{\alpha'}, \dot{r}^{\alpha}) = \mathcal{R}(r^{\alpha'}, \dot{r}^{\alpha}, \mathcal{P}_a(r^{\alpha'}, c_b)).$$

We thus obtain a family of underactuated controlled Lagrangian systems on Q/G depending on the vector parameter c. The Lagrangians $\mathcal{L}_c$ of these systems are represented by the formula

$$\mathcal{L}_c = \frac{1}{2}g_{\alpha\beta}\dot{r}^{\alpha}\dot{r}^{\beta} - U(r^{\alpha'}, \mathcal{P}(r^{\alpha'}, c)). \tag{9.3.7}$$

This is in a form where we can apply directly the kinetic energy shaping discussed above. Note the independence of the kinetic energy from the vector parameter c. Hence kinetic energy shaping can be accomplished for the whole family of Lagrangians $\mathcal{L}_c$ at once. In particular, if kinetic shaping is sufficient for stabilization, the control law obtained this way is represented by the same formula for all of the systems in (9.3.6).

9.3.1 Example (Stabilization of the Unicycle with Rotor). We now apply the theory developed above to the problem of stabilization of the slow vertical steady-state motions of the unicycle with rotor.

The dynamical model of a homogeneous disk on a horizontal plane with a rotor was discussed briefly in Chapter 1. The rotor is free to rotate in the plane orthogonal to the disk. The rod connecting the centers of the disk and rotor keeps the direction of the radius of the disk through the contact point with the plane.

The configuration space for this system is

$$Q = S^1 \times S^1 \times S^1 \times \mathrm{SE}(2),$$

which we parametrize with coordinates $(\theta, \chi, \psi, \phi, x, y)$. As in Figure 9.3.1, θ is the tilt of the unicycle itself, and ψ and χ are the angular positions of the wheel of the unicycle and the rotor, respectively. The variables (ϕ, x, y), regarded as a point in $\mathrm{SE}(2)$, represent the angular orientation and position of the point of contact of the wheel with the ground.

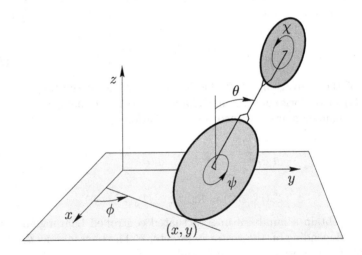

FIGURE 9.3.1. The configuration variables for the unicycle with rotor.

This mechanical system is $\mathrm{SO}(2) \times \mathrm{SE}(2)$-invariant; the group $\mathrm{SO}(2)$ represent the symmetry of the wheel, that is, the symmetry in the ψ variable, while the group $\mathrm{SE}(2)$ represents the Euclidean symmetry of the overall system. The action by the group element (α, β, a, b) on the configuration space is given by

$$(\theta, \chi, \psi, \phi, x, y) \mapsto$$
$$(\theta, \chi, \psi + \alpha, \phi + \beta, x\cos\beta - y\sin\beta + a, x\sin\beta + y\cos\beta + b).$$

We will use the following notation:

M = the mass of the disk,

R = the radius of the disk,

A, B = the principal moments of inertia of the disk,

$\mathcal{A}, \mathcal{B}$ = the principal moments of inertia of the rotor,

r = the rod length,

μ = the rotor mass.

Here A and $\mathcal{A}$ are moments of inertia about axes lying in the disks and passing though the center, and B and $\mathcal{B}$ are moments of inertia about axes through the center and perpendicular to the disks.

The Lagrangian is

$$L = K_{\mathrm{d}} + K_{\mathrm{r}} + \frac{M}{2}v_M^2 + \frac{\mu}{2}v_\mu^2 - V,$$

where

$$K_{\mathrm{d}} = \frac{1}{2}\left[A(\dot{\theta}^2 + \dot{\phi}^2\cos^2\theta) + B(\dot{\phi}\sin\theta + \dot{\psi})^2\right],$$

$$K_{\mathrm{r}} = \frac{1}{2}\left[\mathcal{A}(\dot{\phi}^2\sin^2\theta) + \mathcal{B}(\dot{\chi} + \dot{\theta})^2\right],$$

$$v_M^2 = (\dot{x} - R\dot{\phi}\sin\theta\cos\phi)^2 + (\dot{y} - R\dot{\phi}\sin\theta\sin\phi)^2$$
$$+ 2R\dot{\phi}\cos\theta(\dot{y}\cos\phi - \dot{x}\sin\phi) + R^2\dot{\theta}^2,$$

$$v_\mu^2 = (\dot{x} - (R+r)\dot{\phi}\sin\theta\cos\phi)^2$$
$$+ (\dot{y} - (R+r)\dot{\phi}\sin\theta\sin\phi)^2$$
$$+ 2(R+r)\dot{\phi}\cos\theta(\dot{y}\cos\phi - \dot{x}\sin\phi) + (R+r)^2\dot{\theta}^2,$$

$$V = MgR\cos\theta + \mu g(R+r)\cos\theta.$$

The constraints are given by

$$\dot{x} = -\dot{\psi}R\cos\phi, \qquad \dot{y} = -\dot{\psi}R\sin\phi.$$

Hence the reduced Lagrangian is

$$L_c = \frac{1}{2}\left(\alpha\dot{\theta}^2 + 2\beta\dot{\theta}\dot{\chi} + \beta\dot{\chi}^2 + I_{11}(\theta)\dot{\phi}^2 + 2I_{12}\dot{\phi}\dot{\psi} + I_{22}\dot{\psi}^2\right) - V(\theta),$$

where

$$\alpha = A + MR^2 + \mu(R+r)^2 + \mathcal{B}, \qquad \beta = \mathcal{B},$$

are the components of the *shape metric*, and

$$I_{11} = A\cos^2\theta + \mathcal{A} + (B + MR^2 + \mu R(R+r))\sin^2\theta,$$

$$I_{12} = (B + MR^2 + \mu R(R+r))\sin\theta,$$

$$I_{22} = B + MR^2 + \mu R^2,$$

are the components of the locked inertia tensor. The components of the nonholonomic momentum are

$$p_1 = \frac{\partial L_c}{\partial \dot{\phi}} = I_{11}\dot{\phi} + I_{12}\dot{\psi},$$

$$p_2 = \frac{\partial L_c}{\partial \dot{\psi}} = I_{12}\dot{\phi} + I_{22}\dot{\psi}.$$

For the unicycle with rotor, p_1 is the vertical (i.e., orthogonal to the xy-plane) component of the angular momentum of the system, while p_2 is the component of the disk's angular momentum along the normal direction to the disk.

The reduced dynamics of the unicycle is governed by equations (9.3.1)–(9.3.3) with $r^1 = \theta$, $r^2 = \chi$, and the Routhian

$$\mathcal{R} = \frac{1}{2}\left(\alpha\dot{\theta}^2 + 2\beta\dot{\theta}\dot{\chi} + \beta\dot{\chi}^2 - I^{ab}(\theta)p_a p_b\right) - V(\theta).$$

Here I^{ab} are the components of the inverse inertia tensor.

The shape equations for (θ, χ) describe the motion of the rod and rotor system, while the momentum equations for (p_1, p_2) model the (coupled) wheel dynamics. The coefficients $\mathcal{D}^b_{a\alpha}$ in (9.3.3) for the unicycle with rotor are computed to be

$$\mathcal{D}^a_{11} = I^{2a}(MR + \mu(R + r))R\cos\theta,$$
$$\mathcal{D}^a_{21} = -I^{1a}(MR + \mu(R + r))R\cos\theta.$$

The slow vertical steady-state motions of this system are represented by the relative equilibria

$$\theta = 0, \qquad \dot{\chi} = 0, \qquad p_1 = 0, \qquad p_2 = p_2^0.$$

This system satisfies all Assumption 1–4. In particular, momentum equations define an integrable distribution. The dynamics on the invariant manifolds Q_c are governed by the equations

$$\frac{d}{dt}\frac{\partial L_c}{\partial \dot{\theta}} = \frac{\partial L_c}{\partial \theta}, \qquad \frac{d}{dt}\frac{\partial L_c}{\partial \dot{\chi}} = u_c, \qquad (9.3.8)$$

where

$$\mathcal{L}_c = \frac{1}{2}(\alpha\dot{\theta}^2 + 2\beta\dot{\theta}\dot{\chi} + \beta\dot{\chi}^2) - U_c(\theta),$$

and

$$U_c(\theta) = \frac{1}{2}I^{ab}(\theta)\mathcal{P}_a(\theta, c), \mathcal{P}_b(\theta, c) + V(\theta)$$

is the amended potential for the unicycle with rotor restricted to the invariant manifolds (9.3.5). Observe that this Lagrangian is now **identical** in form to that for the inverted pendulum on a cart discussed in Section 9.2. We can now apply the method there directly, constructing controlled Lagrangians of the form

$$\widetilde{\mathcal{L}}_c = \frac{1}{2}(\alpha\dot{\theta}^2 + 2\beta\dot{\theta}(\dot{\chi} + k\dot{\theta}) + \beta(\dot{\chi} + k\dot{\theta})^2) + \frac{\sigma}{2}(k\dot{\theta})^2 - U_c(\theta),$$

where k and σ are constants.

Again consulting the pendulum on a cart analysis, see equation (9.2.18), we can conclude stability of the relative equilibria $\theta = 0$, $p_1 = 0$, $p_2 = p_2^0$ using the nonholonomic energy–momentum method applied to the controlled Routhian

$$\tilde{R} = \frac{1}{2}\left(\alpha\dot{\theta}^2 + 2\beta\dot{\theta}(\dot{x} + k\dot{\theta}) + \beta(\dot{x} + k\dot{\theta})^2\right)$$
$$+ \frac{\sigma}{2}(k\dot{\theta})^2 - \frac{1}{2}I^{ab}P_aP_b - V(\theta),$$

if we choose a control parameter k satisfying

$$k > \frac{\alpha - \beta}{\beta^2}. \qquad \blacklozenge$$

9.4 Averaging for Controlled Lagrangian Systems

In the previous sections we examined the use of *structured feedback* to alter the form of the Lagrangian in order to achieve certain control design objectives. We now turn our attention, by contrast, to *open-loop designs*, which may also be analyzed within the context of Lagrangian and Hamiltonian systems.

Recall that in Chapter 4 we discussed an interesting and important feature of nonlinear control systems wherein an oscillatory input can be used to generate a net velocity in a direction determined by taking Lie brackets of the controlled vector fields. For controlled Lagrangian systems of the form (9.1.2)–(9.1.3), we shall study the second-order analogue of this phenomenon, where oscillatory inputs are used to create "synthetic force fields" that organize the system dynamics in prescribed ways. While the theory is currently less complete than in the first-order case, it has the advantage that for a broad class of systems, the response to high-frequency forcing may be understood in terms of a bifurcation and critical point analysis of an energy-like function called the **averaged potential**. To emphasize the distinction between controlled and uncontrolled configuration variable, we adopt the notation r for the first k components $q_1, \ldots, q_k$ in (9.1.2)–(9.1.3) and q for the uncontrolled components $q_{k+1}, \ldots, q_n$." (Properly referenced.)

9.4.1 Example (A Simple Example).   To illustrate the second-order theory, we consider the Mathieu equation

$$\ddot{q}(t) + (\delta + \epsilon u(t))q(t) = 0, \qquad (9.4.1)$$

where the function $u(\cdot)$ is assumed to be periodically varying.

There are several ways to analyze this system as a controlled Lagrangian system. For instance, if we write

$$L(q, \dot{q}, \dot{r}) = \frac{1}{2}\dot{q}^2 + \epsilon q \dot{r} \dot{q} + \frac{1}{2}\dot{r}^2 - \frac{\delta^2}{2} q^2,$$

the Euler–Lagrange equations specialize precisely to (9.4.1), provided we replace the variable $u(t)$ with $\ddot{r}(t)$. For a physical interpretation, we refer to the cart–pendulum system described by equation (9.1.4) in which we set the coefficients $\alpha = \gamma = 1$, $\beta = \epsilon/2$, $D = \delta$; $\psi = \pi/2$, $s = r$, $\theta = q$, and we linearize with respect to q about the equilibrium $q = 0$. The example illustrates a general observation that it is frequently useful both in theory and in working with actual physical systems to view $r(\cdot)$ rather than $u(\cdot)$ as the control input. This will be discussed in greater detail below.

The qualitative behavior and stability of solutions to equation (9.4.1) have been widely studied; see, for instance, Stoker [1950]. While the stability of (9.4.1) has been most deeply examined in the specific case of periodic forcing $u(t) = \cos t$, there is also interest in various other types of input wave forms such as pulse-trains, square waves, sawtooth waves, etc., which arise naturally in the context of modern small-scale electromechanical actuation technologies. For the square-wave input

$$u(t) = \begin{cases} -1, & \text{if } 0 \le t < \frac{1}{2}, \\ 1, & \text{if } \frac{1}{2} \le t < 1, \end{cases} \qquad (9.4.2)$$

the regions in the (δ, ϵ) parameter space for which the origin $(q, \dot{q}) = (0, 0)$ is a stable rest point for (9.4.1) are depicted in Figure 9.4.1. Qualitatively, this picture is quite similar to the classical rendering of the stability regions corresponding to the forcing $u(t) = \cos t$. The boundary of the leftmost stability region is approximately parabolic near the origin and satisfies $\delta = -\epsilon^2/48 + o(\epsilon^2)$ (cf. Stoker [1950], pp. 208–213).   ♦

A General Strategy.   This example illustrates an important point that is discussed in detail below. In the general setting of (9.1.2)–(9.1.3), the input $u(\cdot)$ influences the q-coordinate motions only through the dynamic interactions with $r(\cdot)$, $\dot{r}(\cdot)$, and $\ddot{r}(\cdot)$, and it is thus possible to view the entries in the triple $(r, \dot{r}, \ddot{r})$ as the control inputs (to (9.1.3)). This has proven to be fruitful in recent work on oscillation-mediated control of Lagrangian systems. It is also a viewpoint that emerges naturally from the intrinsic definition of second-order control systems discussed in the web supplement.

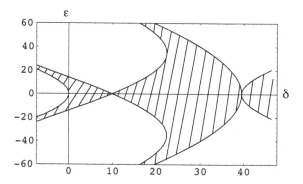

FIGURE 9.4.1. Stable and unstable parameter space regions for Mathieu's equation with square-wave forcing.

In order to treat $(r, \dot{r}, \ddot{r})$ as the control input, it is assumed that there is enough control authority so that an input $u(\cdot)$ can be designed to realize any desired C^2 trajectory $r(\cdot)$ via the dynamics (9.1.2). While this assumption must be verified on a case-by-case basis, the theory discussed below predicts that for a wide class of systems, high-frequency periodic motion in the controlled variables $r(\cdot)$ induces stable localized motions in the uncontrolled ("shape") variables $q(\cdot)$.

To enlarge the class of trajectories $r(\cdot)$, say, to include those for which $\ddot{r}(\cdot)$ is piecewise constant (as in the above example), one has to construct an appropriate approximation. The technical details are relatively unimportant, and it is straightforward to faithfully produce such trajectories for actual physical systems in the laboratory.

Another important (but perhaps subtle) point about the Mathieu equation example is that it is the Lagrangian system itself (equations (9.4.1)) rather than the control input (9.4.2) to which we have applied a parametric stability analysis. The primary aim of the present approach is to outline design procedures that apply to families of (oscillatory) control signals that are parametrized by physically important variables such as amplitude, frequency, and rms covariance.

The stability analysis of such parametrized families of control proceeds by making appropriate coordinate and time-scale changes that transform the model (9.1.3) to a form that allows averaging and stability analysis to be carried out. The design of control inputs is completed via a parametric analysis of the averaged Lagrangian system, and this is completely analogous to the feedback design procedure described above in Section 9.2.

In the following subsections we describe the basic theory of oscillation-mediated control of Lagrangian systems using two essentially different control design approaches, which we label respectively *force-controlled* and *acceleration-controlled* designs. In force-controlled systems, an oscillatory input $u(\cdot)$ drives the combined dynamics (9.1.2), (9.1.3). In acceleration-controlled systems, the triple $(r, \dot{r}, \ddot{r})$ is regarded as the input to a reduced-order Lagrangian in terms of which the equations of motion are given by an Euler–Lagrange equation that is essentially the same as (9.1.3).

For high-frequency inputs of this form, the terms involving $\ddot{r}$ have the greatest effect on the dynamics (9.1.3), and it is for this reason that these are called *acceleration-controlled* systems. As one would expect, the case in which a prescribed oscillatory input $u(\cdot)$ is "filtered" by the r-dynamics (9.1.2) produces q-dynamics that differ markedly from what one obtains by applying a similar oscillatory wave form $r(\cdot)$ (together with $\dot{r}$, $\ddot{r}$) directly to (9.1.3). A remarkable feature of the case in which $(r, \dot{r}, \ddot{r})$ is viewed as the input is that the response of (9.1.3) to high-frequency oscillatory forcing may typically be characterized in terms of the critical point structure of an energy-like function called the *averaged potential* associated with (9.1.3). This simple and elegant characterization is not always possible in the case in which $u(\cdot)$ is the oscillatory input.

Controlled Lagrangian Systems—with and without Dissipation.
We begin by examining the effect of a high-frequency input $u(\cdot)$ in (9.1.2)–(9.1.3), deferring the acceleration-controlled case until later. The starting point for this analysis is classical averaging theory interpreted within the framework of the geometric mechanics that has been developed in the earlier chapters of this book. We continue to distinguish between the m controlled configuration variables r and the n uncontrolled configuration variables q and write $y^T = (r^T, q^T)$. Then, as in (4.6.18), the controlled Lagrangian equations are

$$M(y)\ddot{y} + \hat{\Gamma}(y, \dot{y}) + \frac{\partial V}{\partial y} = \begin{pmatrix} u \\ 0 \end{pmatrix}, \qquad (9.4.3)$$

where

$$\hat{\Gamma}(y, \dot{y}) = \left(\hat{\Gamma}_1(y, \dot{y}), \ldots, \hat{\Gamma}_n(y, \dot{y}) \right)^T,$$

with

$$\hat{\Gamma}_k(y, \dot{y}) = \sum_{i,j} \Gamma_{ijk} \dot{y}_i \dot{y}_j,$$

and

$$\Gamma_{ijk} = \frac{1}{2} \left(\frac{\partial m_{ki}}{\partial y_j} + \frac{\partial m_{kj}}{\partial y_i} - \frac{\partial m_{ij}}{\partial y_k} \right).$$

Averaging Theory.   We now summarize a few facts from averaging theory. See, for example, Sanders and Verhulst [1985]. In the standard treatment, classical averaging methods are used to characterize the dynamics of first-order differential equations of the form

$$\dot{x} = \epsilon f(x, t, \epsilon); \quad x(0) = x_0. \qquad (9.4.4)$$

Here f is a function which is periodic in t (the almost periodic case can also be treated by the methods described here but we do not consider this generalization in the text). The theory describes the degree to which trajectories of this equation may be approximated by trajectories of the associated *autonomous averaged system*

$$\dot{\xi} = \epsilon \bar{f}(\xi), \qquad (9.4.5)$$

where

$$\bar{f}(\xi) = \frac{1}{T} \int_0^T f(\xi, t, 0)\, dt.$$

Under mild regularity assumptions, trajectories of the x and ξ equations that start at the same initial value $x(0) = \xi(0)$ remain "close" over the time interval $0 \le t < \mathcal{O}(1/\epsilon)$.

In neighborhoods of (9.4.5), one can, in many cases, conclude considerably more about the degree to which solutions of (9.4.5) approximate solutions of (9.4.4). If ξ_0 is a hyperbolic equilibrium of (9.4.5), then there exists an $\epsilon_0 > 0$ such that for all $0 < \epsilon < \epsilon_0$, (9.4.4) has a corresponding hyperbolic periodic solution $x(t) = \xi_0 + \mathcal{O}(\epsilon)$ with the same (hyperbolic) stability characteristics as ξ_0. In particular, if ξ_0 is an equilibrium point of (9.4.5) and $\frac{\partial \bar{f}}{\partial y}(\xi_0)$ has all its eigenvalues in the left half-plane, then there is an asymptotically stable periodic solution $x(t) = \xi_0 + \mathcal{O}(\epsilon)$ of (9.4.4) such that $|x(t) - \xi_0| \sim \mathcal{O}(\epsilon)$ for $0 \le t < \infty$. Because one can appeal to this strong approximation result, it is convenient in some cases to introduce dissipation into the models (9.1.2)–(9.1.3), which are rendered as (9.4.3) in our present notation.

Rayleigh Dissipation.   To include rate-dependent (Rayleigh) dissipation (see equation (3.3.9)) we rewrite (9.4.3) as

$$M(y)\ddot{y} + \hat{\Gamma}(y, \dot{y}) + \frac{\partial \hat{D}}{\partial \dot{y}} + \frac{\partial V}{\partial y} = \begin{pmatrix} u \\ 0 \end{pmatrix}, \qquad (9.4.6)$$

where we assume

$$\hat{D}(y, \dot{y}) = \frac{1}{2} \dot{y}^T D(y) \dot{y}$$

is a dissipation function that is quadratic in the velocities with $D(y)$ assumed to be positive definite.

Oscillation-Mediated Control: The Force-Controlled Case   To rewrite (9.4.6) in a form to which it is convenient to apply the above averaging results, we write it as a first-order system by letting $y_1 = y$, $y_2 = \dot{y}$. Then

$$\begin{pmatrix} \dot{y}_1 \\ \dot{y}_2 \end{pmatrix} = \begin{pmatrix} y_2 \\ -M^{-1}(y_1)\left(\hat{\Gamma}(y_1, y_2) + D(y_1)y_2 + \frac{\partial V}{\partial y_1}\right) \end{pmatrix}$$
$$+ \begin{pmatrix} 0 \\ M^{-1}(y_1)\begin{pmatrix} u \\ 0 \end{pmatrix} \end{pmatrix}. \tag{9.4.7}$$

Let $y_1 = \begin{pmatrix} r \\ q \end{pmatrix}$, where r are the controlled and q the uncontrolled variables, and assume that $V(\cdot)$ depends only on the last n q-components of y_1. Hence

$$\frac{\partial V}{\partial y_1} = \begin{pmatrix} 0 \\ \frac{\partial V}{\partial q} \end{pmatrix}.$$

Let $w(\cdot)$ be any bounded, piecewise continuous function, and let $\varphi(t, z)$ be the flow associated with the differential equation

$$\frac{d}{dt}\begin{pmatrix} z_1 \\ z_2 \end{pmatrix} = \begin{pmatrix} 0 \\ M^{-1}(z_1)\begin{pmatrix} w \\ 0 \end{pmatrix} \end{pmatrix}.$$

In this case, noting that z_1 is constant, φ may be explicitly written

$$\varphi(t, z) = \begin{pmatrix} z_1 \\ z_2 + M^{-1}(z_1)\begin{pmatrix} v(t) \\ 0 \end{pmatrix} \end{pmatrix},$$

where $v(t) = \int^t w(s)\,ds$ and $z = (z_1, z_2)^T$. To apply classical averaging techniques, suppose $w(t)$ is any piecewise continuous periodic function, and consider inputs to (9.4.3) of the form $u(t) = (1/\epsilon)\,w(t/\epsilon)$. Next, define the variable $\tilde{x}(t)$ in terms of this flow by writing

$$y(t) = \varphi(t/\epsilon, \tilde{x}(t)).$$

Then $\tilde{x}(\cdot)$ satisfies a differential equation

$$\frac{\partial \varphi}{\partial \tilde{x}}\dot{\tilde{x}}(t) = f(\varphi(t/\epsilon, \tilde{x}(t)),$$

where $f(y)$ is the vector field defined on $\mathbb{R}^{2(m+n)}$ by

$$f(y_1, y_2) = \begin{pmatrix} y_2 \\ -M^{-1}(y_1)\left(\hat{\Gamma}(y_1, y_2) + D(y_1)y_2 + \frac{\partial V}{\partial y_1}\right) \end{pmatrix}.$$

Let $\tau = t/\epsilon$, and $x(\tau) = \tilde{x}(t)$. Then we may rewrite our equation for x in terms of the "slow" time variable τ, and apply classical averaging theory to the resulting equation

$$\frac{dx}{d\tau} = \epsilon \frac{\partial\varphi}{\partial x}^{-1} \Big|_{(\tau, x(\tau))} f(\varphi(\tau, x(\tau))). \tag{9.4.8}$$

The main result on averaging force-controlled systems will be an immediate consequence of the following two lemmas (see, e.g., Sanders and Verhulst [1985]).

9.4.2 Lemma. *Assume that the integrated input $v(\tau) = \int^{\tau} w(s)\,ds$ is a zero-mean periodic function of period $T > 0$. Then the* autonomous *averaged system* associated with *(9.4.8) may be explicitly written*

$$\frac{d}{d\tau}\begin{pmatrix} x_1 \\ x_2 \end{pmatrix} = \epsilon \left\{ \begin{pmatrix} x_2 \\ -M^{-1}(x_1)\left(\hat{\Gamma}(x_1, x_2) + D(x_1)x_2 + \dfrac{\partial V}{\partial q}\right) \end{pmatrix} \right.$$
$$+ \begin{pmatrix} 0 \\ \overline{-\left\{\dfrac{\partial}{\partial x_1}\left[M^{-1}(x_1)\begin{pmatrix} v \\ 0 \end{pmatrix}\right]\right\} \cdot M^{-1}(x_1)\begin{pmatrix} v \\ 0 \end{pmatrix}} \end{pmatrix}$$
$$\left. + \begin{pmatrix} 0 \\ \overline{-M^{-1}(x_1)\,\hat{\Gamma}\left(x_1, M^{-1}(x_1)\begin{pmatrix} v \\ 0 \end{pmatrix}\right)} \end{pmatrix} \right\},$$

where the overbar indicates the result of simple averaging over one period.

A remark on this lemma may be helpful. Namely, if $g(x, \tau)$ is periodic in τ with fundamental period $T > 0$, we write

$$\overline{g(x)} = \overline{g(x, \tau)} = (1/T)\int_0^T g(x, \tau)\,d\tau.$$

The terms

$$\overline{\left\{\dfrac{\partial}{\partial x_1}\left[M^{-1}(x_1)\begin{pmatrix} v \\ 0 \end{pmatrix}\right]\right\} \cdot M^{-1}(x_1)\begin{pmatrix} v \\ 0 \end{pmatrix}}$$
$$+ \overline{M^{-1}(x_1)\,\hat{\Gamma}\left(x_1, M^{-1}(x_1)\begin{pmatrix} v \\ 0 \end{pmatrix}\right)}, \tag{9.4.9}$$

which are quadratic in v, account for the net averaged effect of the oscillatory input u on the dynamics (9.4.3).

Additional insight is obtained by writing the averaged system as a (Lagrangian) second-order differential equation. This may be done using the following lemma.

9.4.3 Lemma. *Write the first m columns of $M^{-1}(x_1)$ as vector fields $Y_1(x_1), \ldots, Y_m(x_1)$, so that*

$$M^{-1}(x_1) \begin{pmatrix} v \\ 0 \end{pmatrix} = \sum_{a=1}^{m} v_a \, Y_a(x_1).$$

The terms in (9.4.8) which are quadratic in $v(\tau)$ and which give rise to (9.4.9) may be rewritten

$$\left\{ \frac{\partial}{\partial x_1} \left[M^{-1}(x_1) \begin{pmatrix} v \\ 0 \end{pmatrix} \right] \right\} \cdot M^{-1}(x_1) \begin{pmatrix} v \\ 0 \end{pmatrix}$$

$$+ M^{-1}(x_1) \, \hat{\Gamma} \left(x_1, M^{-1}(x_1) \begin{pmatrix} v \\ 0 \end{pmatrix} \right) = \sum_{a,b=1}^{m} \langle Y_a : Y_b \rangle \, v_a v_b,$$

where $\langle Y_a : Y_b \rangle$ is a vector product defined componentwise by

$$2 \langle Y_a : Y_b \rangle^i = \sum_{j=1}^{n} \left(\frac{\partial Y_a^i}{\partial x_1^j} Y_b^j + \frac{\partial Y_b^i}{\partial x_1^j} Y_a^J \right) + \sum_{j,k=1}^{n} \Gamma_{jk}^i \left(Y_a^j Y_b^k + Y_a^k Y_b^j \right),$$

and where we denote $\Gamma_{k\ell}^i = \sum_j m^{ij} \Gamma_{k\ell j}$ with m^{ij} being the ij-th component of M^{-1}.

The proof of this lemma involves a straightforward but tedious computation. Together, the lemmas lead immediately to the main approximation result for force-controlled systems.

9.4.4 Theorem. *Consider the controlled Lagrangian dynamics (9.4.6) with (quadratic, positive definite) dissipation function $\hat{D}(y, \dot{y})$ and with input $u(t) = (1/\epsilon) w(t/\epsilon)$. Suppose that the integrated input $v(\tau) = \int^\tau w(s)\, ds$ is a zero-mean periodic function of period $T > 0$. Then the trajectories $y(t)$ of (9.4.6) can be approximated by trajectories $x(t)$ of the averaged Lagrangian system*

$$M(x)\ddot{x} + \hat{\Gamma}(x, \dot{x}) + D(x)\dot{x} + \begin{pmatrix} 0 \\ \frac{\partial V}{\partial q} \end{pmatrix} + M(x) \sum_{a,b=1}^{m} \langle Y_a : Y_b \rangle \, \overline{v_a v_b} = 0, \quad (9.4.10)$$

in the sense that if both (9.4.6) and (9.4.10) have the same initial conditions

$$(y(0), \dot{y}(0)) = (x(0), \dot{x}(0)) = (y_0, \dot{y}_0),$$

then

$$|y(t) - x(t)| \sim \mathcal{O}(\epsilon) \qquad \text{for } 0 \le t < \infty.$$

The bracket terms $\langle \cdot : \cdot \rangle$ of the preceding lemma and theorem are precisely the *symmetric products* defined in equation (6.7.4) (see e.g., Lewis and Murray [1999]). It is easy to see that $\langle Y_a : Y_b \rangle = \langle Y_b : Y_a \rangle$, and this product may also be written in a coordinate-free form as

$$\langle Y_a : Y_b \rangle = \frac{1}{2}(\nabla_{Y_a} Y_b + \nabla_{Y_b} Y_a).$$

The theorem on averaging force-controlled systems is essentially due to Bullo [2001], which we refer to for further details on both the theorem and the role of the symmetric product in the theory of averaged Lagrangian systems.

If we set the dissipation function $D(q) = 0$, we have noted above that general principles in averaging theory (Sanders and Verhulst [1985]) imply that trajectories of (9.4.10) approximate trajectories of the nonautonomous system (9.4.3) on the time interval $0 \le t < \mathcal{O}(1)$. According to the theorem, this approximation may be immediately extended to the semi-infinite time interval $0 \le t \le \infty$ if we assume, on the other hand, that $D(q)$ is positive definite. In the paragraph that follows, we shall be primarily interested in stable motion confined to neighborhoods of equilibrium solutions (rest points) of the averaged dynamics (9.4.10).

In terms of the unreduced dynamics (9.1.2)–(9.1.3) with $v(t) = \int^t u(s)\,ds$, the mapping $\rho : T_r^* R \to T_y^* Y$ defined by

$$u \longmapsto M(y) \sum_{a,b=1}^{m} \langle Y_a : Y_b \rangle \overline{v_a v_b}$$

provides a succinct summary description of the way in which high-frequency inputs $u(\cdot)$ influence the dynamics of the q-variables in our system.

Oscillation-Mediated Control: The Acceleration-Controlled Case.
A widely studied alternative to the above averaging analysis for (9.1.2)–(9.1.3) involves a reduced-order Lagrangian model that is equivalent to (9.1.3) alone with $r(\cdot)$ (together with $\dot{r}(\cdot)$ and $\ddot{r}(\cdot)$) playing the role of oscillatory input. See, for example, Baillieul[1990, 1993, 1995, 1998], Weibel and Baillieul [1998b], Weibel, Kaper and Baillieul [1997].

We note that if $r(\cdot)$ is a periodic function of fundamental frequency ω, the amplitude of $\dot{r}$ scales as $\omega \cdot \mathcal{O}(r)$, and $\ddot{r}$ scales as $\omega^2 \cdot \mathcal{O}(r)$. Acceleration terms clearly have the dominant influence on the system's response, and thus when the triple $(r, \dot{r}, \ddot{r})$ is viewed as the input to (9.1.3) we refer to it as an *acceleration-controlled* Lagrangian system. Assume that the Lagrangian appearing in (9.1.2)–(9.1.3) has the form

$$L(r,q,\dot{r},\dot{q}) = \tfrac{1}{2}(\dot{r}^T, \dot{q}^T) \begin{pmatrix} \mathcal{N}(q,r) & \mathcal{A}(q,r) \\ \mathcal{A}(q,r)^T & \mathcal{M}(q,r) \end{pmatrix} \begin{pmatrix} c\dot{r} \\ \dot{q} \end{pmatrix} - V(q,r). \quad (9.4.11)$$

Equation (9.1.3) relates the state variables q and $\dot{q}$ to the inputs $(r, \dot{r}, \ddot{r})$, and this dynamical relationship may also be obtained by applying the Euler–

Lagrange operator

$$\frac{d}{dt}\frac{\partial}{\partial\dot{q}} - \frac{\partial}{\partial q}$$

to the *reduced Lagrangian*

$$\mathcal{L}(r,\dot{r};q,\dot{q}) = \tfrac{1}{2}\dot{q}^T\mathcal{M}(q,r)\dot{q} + \dot{r}^T\mathcal{A}(q,r)\dot{q} - \mathcal{V}_a(q;r,\dot{r}), \qquad (9.4.12)$$

where $\mathcal{V}_a(q;r,\dot{r}) = V(q) - \left(\dot{r}^T\mathcal{N}(r,q)\dot{r}\right)/2$ is a time-varying potential similar to the *augmented potential*. The q-dynamics prescribed by (9.1.3) is *formally* equivalent to the Euler–Lagrange equations of the reduced Lagrangian:

$$\frac{d}{dt}\frac{\partial\mathcal{L}}{\partial\dot{q}} - \frac{\partial\mathcal{L}}{\partial q} = 0.$$

Writing this out in detail in terms of coordinates, we have

$$\sum_{j=1}^{n} m_{kj}\ddot{q}_j + \sum_{\ell=1}^{m} a_{\ell k}\dot{v}_\ell + \sum_{i,j=1}^{n} \Gamma_{ijk}\dot{q}_i\dot{q}_j + \sum_{j=1}^{n}\sum_{\ell=1}^{m} \hat{\Gamma}_{\ell jk}v_\ell\dot{q}_j = F(t), \qquad (9.4.13)$$

where

$$\Gamma_{ijk} = \frac{1}{2}\left(\frac{\partial m_{ki}}{\partial q_j} + \frac{\partial m_{kj}}{\partial q_i} - \frac{\partial m_{ij}}{\partial q_k}\right),$$

$$\hat{\Gamma}_{\ell jk} = \frac{\partial m_{kj}}{\partial r_\ell} + \frac{\partial a_{\ell k}}{\partial q_j} - \frac{\partial a_{\ell j}}{\partial q_k},$$

and a_{ij} and m_{ij} are the ijth entries in the $m \times n$ and $n \times n$ matrices $\mathcal{A}(q,x)$ and $\mathcal{M}(q,x)$, respectively. $F(t)$ is a vector of *generalized forces*

$$F_i(t) = -\frac{\partial\mathcal{V}_a}{\partial q_i} - \sum_{k,\ell=1}^{m}\frac{\partial a_{\ell i}}{\partial r_k}v_\ell v_k.$$

We have referred to this representation of the system dynamics as being *formal*, because the generalized forces may be thought of as coming from a time-varying potential only if

$$\frac{\partial^2 a_{\ell i}}{\partial q_j \partial r_k} = \frac{\partial^2 a_{\ell j}}{\partial q_i \partial r_k}$$

for all $k,\ell = 1,\dots,m$ and $i,j = 1,\dots,n$.

Hamiltonian Form. To find the analogue of the variational equation (9.4.8), to which averaging theory applies, we write the equations of motion in Hamiltonian form, in terms of the conjugate momentum

$$p = \frac{\partial\mathcal{L}}{\partial\dot{q}} = \mathcal{M}(q)\dot{q} + \mathcal{A}^T(q)\dot{r}.$$

This gives rise to the noncanonical Hamiltonian

$$\mathcal{H}(q,p;r,\dot{r}) = \tfrac{1}{2}(p - \mathcal{A}^T\dot{r})^T\mathcal{M}^{-1}(p - \mathcal{A}^T\dot{r}) + \mathcal{V}_a. \tag{9.4.14}$$

Letting $v = \dot{r}$, we expand equation (9.4.14) and apply simple averaging to yield the *averaged Hamiltonian*

$$\overline{\mathcal{H}}(q,p) = \tfrac{1}{2}p^T\overline{\mathcal{M}^{-1}}p - \overline{v^T\mathcal{A}\mathcal{M}^{-1}}p + \tfrac{1}{2}\overline{v^T\mathcal{A}\mathcal{M}^{-1}\mathcal{A}^T v} + \overline{\mathcal{V}}_a. \tag{9.4.15}$$

As above, the overbars indicate that simple averages over one period (T) have been taken with q and p regarded as constants for the purpose of averaging each term in the expression.

For the averaged Hamiltonian (9.4.15), there is an obvious decomposition into kinetic and potential energy terms in the case that $\overline{v^T\mathcal{A}\mathcal{M}^{-1}} = 0$:

$$\overline{\mathcal{H}}(q,p) = \underbrace{\frac{1}{2}p^T\overline{\mathcal{M}^{-1}}p}_{\substack{\text{avg. kin.}\\\text{energy}}} + \underbrace{\frac{1}{2}\overline{v^T\mathcal{A}\mathcal{M}^{-1}\mathcal{A}v} + \overline{\mathcal{V}}_a}_{\text{averaged potential}}.$$

In the case that $\overline{v^T\mathcal{A}\mathcal{M}^{-1}} \neq 0$, it remains possible to formally decompose the averaged Hamiltonian (9.4.15) into the sum of averaged kinetic and potential energies. If $\bar{v} \neq 0$, then the corresponding input variable $r(t)$ will not be periodic, and there will be a "drift" in the value of $r(t)$ that changes by an amount $\bar{v} \cdot T$ every T units of time. We rewrite the averaged Hamiltonian (9.4.15) as

$$\begin{aligned}
\overline{\mathcal{H}}(q,p) &= \frac{1}{2}p^T\overline{\mathcal{M}^{-1}}p - \overline{v^T\mathcal{A}\mathcal{M}^{-1}}p + \frac{1}{2}\overline{v^T\mathcal{A}\mathcal{M}^{-1}}\left(\overline{\mathcal{M}^{-1}}\right)^{-1}\overline{\mathcal{M}^{-1}\mathcal{A}^T v} \\
&\quad + \frac{1}{2}\overline{v^T\mathcal{A}\mathcal{M}^{-1}\mathcal{A}v} - \frac{1}{2}\overline{v^T\mathcal{A}\mathcal{M}^{-1}}\left(\overline{\mathcal{M}^{-1}}\right)^{-1}\overline{\mathcal{M}^{-1}\mathcal{A}^T v} + \overline{\mathcal{V}}_a \\
&= \underbrace{\frac{1}{2}(\overline{\mathcal{M}^{-1}}p - \overline{\mathcal{M}^{-1}\mathcal{A}^T v})^T\left(\overline{\mathcal{M}^{-1}}\right)^{-1}(\overline{\mathcal{M}^{-1}}p - \overline{\mathcal{M}^{-1}\mathcal{A}^T v})}_{\text{averaged kinetic energy}} \\
&\quad + \underbrace{\frac{1}{2}\overline{v^T\mathcal{A}\mathcal{M}^{-1}\mathcal{A}^T v} - \frac{1}{2}\overline{v^T\mathcal{A}\mathcal{M}^{-1}}\left(\overline{\mathcal{M}^{-1}}\right)^{-1}\overline{\mathcal{M}^{-1}\mathcal{A}^T v} + \overline{\mathcal{V}}_a}_{\text{averaged potential}}.
\end{aligned}$$

$$\tag{9.4.16}$$

The relationship between the dynamics prescribed by this averaged Hamiltonian and the dynamics of (9.4.16) has been studied in Weibel, Kaper and Baillieul [1997] and Weibel and Baillieul [1998b], which may be consulted for details. It is important to mention that for mechanical systems in which $\bar{v} \neq 0$ and $\mathcal{M}$ depends explicitly on r in (9.4.14), the averaging analysis of this chapter may not provide an adequate description of the dynamics. Indeed, in this case, $\|r(t)\|$ will not remain bounded as $t \to \infty$, and if $\mathcal{M}(r(t), q_2)$ also fails to remain bounded, the averaged potential will inherit a dependence on time that will make it difficult to apply the critical point analysis proposed below. Despite this cautionary remark, we shall indicate how our methods may be applied in many instances where $\bar{v} \neq 0$.

Our interest in the **averaged potential**

$$\mathcal{V}_A(q) = \frac{1}{2}\overline{v^T \mathcal{A}\mathcal{M}^{-1}\mathcal{A}^T v} - \frac{1}{2}\overline{v^T \mathcal{A}\mathcal{M}^{-1}}\left(\overline{\mathcal{M}^{-1}}\right)^{-1}\overline{\mathcal{M}^{-1}\mathcal{A}^T v} + \overline{\mathcal{V}_a} \quad (9.4.17)$$

is that for a broad class of systems, high-frequency oscillatory forcing of (9.4.13) produces motions that can be fairly completely characterized in terms of the critical point structure of $\mathcal{V}_A(\cdot)$. Indeed, we have the following **averaging principle**, which connects the dynamics (9.4.13) with the critical points of the averaged potential as follows:

1. If q^* is a strict local minimum of $\mathcal{V}_A(\cdot)$, then provided the frequency of the periodic forcing $u(\cdot)$ is sufficiently high, the system (9.4.13) will execute motions confined to a neighborhood of q^*.

2. If $(q, p) = (q^*, 0)$ is a hyperbolic fixed point of the averaged Hamiltonian system associated with (9.4.16), then there is a corresponding periodic orbit of the system defined by (9.4.14) such that the asymptotic stability properties of the fixed point $(Q^*, 0)$ associated with (9.4.16) are the same as those of the periodic orbit. ∎

Conditions under which this principle is valid are discussed in detail in Weibel and Baillieul [1998a]. This reference also describes a global averaging theory for Hamiltonian systems that provides conditions under which there is a close relationship between the phase space structures associated with (9.4.14) and (9.4.16).

An Area Rule for Averaged Acceleration-Controlled Lagrangian Systems. It was observed in Section 4.4 that if $(u_1(\cdot), u_2(\cdot)) : [0, \infty) \to \mathbb{R}^2$ is [a piecewise smooth] periodic function of period $T > 0$, then "second-order" averaging implies that the flow generated by the first-order affine differential equation,

$$\dot{x} = u_1(t)f_1(x) + u_2(t)f_2(x)$$

is approximated by

$$x(T) = x_0 + 2A \cdot \left[f_1(x_0), f_2(x_0)\right]T + o(T), \quad (9.4.18)$$

where A = the areas enclosed by the curve $\{(u_1(t), u_2(t)) : 0 \le t \le T\}$. Both the frequency $(1/T)$ and the area A are important parameters in this approximation. The frequency determines the sharpness (or accuracy) of the approximation, while it is in terms of the <u>area</u> of the input curve that we measure the effect of the *nonholonomy* in the differential equation (i.e., the noncommutativity of the vector fields f_1 and f_2). The formula (9.4.18) has appeared in several places, but was probably first written in the form in which it appears here in Brockett [1989]. In recognition of this, the formula is frequently referred to as *Brockett's area rule*.

For the second-order (controlled Lagrangian) systems treated in this chapter, there is a corresponding geometric interpretation. Consider the acceleration-controlled Lagrangian system with scalar input given by

$$\mathcal{L}(q, \dot{q}; v) = \frac{1}{2}\dot{q}^T \mathcal{M}(q)\dot{q} + \mathbf{a}(q)\dot{q}v - \mathcal{V}(q)$$

(cf. (9.4.12)). The equations of motion are written as

$$\mathcal{M}(q)\ddot{q} + \Gamma(q, \dot{q}) + \frac{\partial \mathcal{V}}{\partial q} + \mathbf{a}(q)\dot{v} + \left(\frac{\partial \mathbf{a}}{\partial q} - \frac{\partial \mathbf{a}^T}{\partial q} \right) v\dot{q} = 0.$$

According to the theory developed above, the dynamic response to a periodic input $(v(\cdot), \dot{v}(\cdot))$ may be understood in terms of the critical points of the *averaged potential*

$$\mathcal{V}_A(q) = \mathcal{V}(q) + \frac{1}{2}\mathbf{a}(q)^T \mathcal{M}(q)^{-1}\mathbf{a}(q)\sigma^2,$$

where

$$\sigma^2 = \frac{1}{T} \int_0^T v(t)^2 \, dt.$$

The simplest interpretation of the input-related term of the averaged potential is that it is simply the averaged rms value of the input wave form $v(\cdot)$. To make contact with the *area rule* (9.4.18), it is useful to recall how $\mathcal{L}(q, \dot{q}; v)$ arose from a reduction process described in obtaining equation (9.4.12). Thinking of the input v as the velocity of a (cyclic) exogenous variable r (i.e., $v = \dot{r}$), we may rewrite

$$\sigma^2 = \frac{1}{T} \int_0^T \dot{r}(t)^2 \, dt = \frac{1}{T} \oint_C \dot{r} \, dr, \qquad (9.4.19)$$

where the last quantity represents the line integral of $\dot{r}$ around the closed curve $C = \{(r(t), \dot{r}(t)) : 0 \le t \le T\}$. Appealing to Green's theorem in the plane, we find that this line integral is just the area enclosed by the curve in the $(r, \dot{r})$-phase plane.

Stability of Acceleration-Controlled Lagrangian Systems with Oscillatory Inputs. As described above, the Lagrangian (9.4.12) gives rise to the Lagrangian dynamics (9.4.13). To retain our general perspective, we continue to assume that the terms in these equations may depend on r.

The explicit form of this dependence will play no role in the present section, however, and hence we simplify our notation by omitting any further mention of the variable r. We seek to understand the stability of (9.4.13) in terms of the corresponding *averaged potential*

$$\mathcal{V}_A(q) = \frac{1}{2}\overline{v^T \mathcal{A}\mathcal{M}^{-1}\mathcal{A}^T v} - \frac{1}{2}\overline{v^T \mathcal{A}\mathcal{M}^{-1}}\left(\overline{\mathcal{M}^{-1}}\right)^{-1}\overline{\mathcal{M}^{-1}\mathcal{A}^T v} + \overline{\mathcal{V}}. \quad (9.4.20)$$

The *averaging principle* states that the effect of forcing (9.4.13) with an oscillatory input $v(\cdot)$ will be to produce stable motions confined to neighborhoods of relative minima of $\mathcal{V}_A(\cdot)$. While this principle appears to govern the dynamics encountered in a wide variety of systems, there is as yet no complete theory describing the observed behavior. Results reported in Baillieul [1995] show that for a certain class of systems (9.4.13) within a larger class of so-called *linear Lagrangian systems*, strict local minima of the averaged potential are Lyapunov stable rest points of (9.4.13) for all periodic forcing of a given amplitude and sufficiently high frequency. Here we shall review this result, and show that the extension of our analysis to arbitrary systems (9.4.13) is complicated by the fact that in general, a linearization of (9.4.13) fails to capture the stabilizing effects implied by an analysis of the averaged potential. Indeed, we shall show that the averaged potential depends on second-order jets of the coefficient functions $\mathcal{A}(q)$ and $\mathcal{M}(q)$. To simplify the presentation, we shall restrict our attention to the case of zero-mean oscillatory forcing in which $\overline{\mathcal{M}^{-1}\mathcal{A}^T v} = 0$.

Suppose q_0 is a strict local minimum of $\mathcal{V}_A$ in (9.4.20). Applying a high-frequency oscillatory input $v(\cdot)$, we shall look for stable motions of (9.4.13) in neighborhoods of $(q, \dot{q}) = (q_0, 0)$. It is important to note that even when there are such stable motions, $(q_0, 0)$ need not be a rest point of (9.4.13) for any choice of forcing function $v(\cdot)$. This will be illustrated below.

We may assume without loss of generality that $q_0 = 0$. (If this is not the case, we may always change coordinates to make it true.) Write

$$\mathcal{A}(q) = \mathcal{A}_0 + \mathcal{A}_1(q) + \mathcal{A}_2(q) + \text{h.o.t.},$$
$$\mathcal{M}(q) = \mathcal{M}_0 + \mathcal{M}_1(q) + \mathcal{M}_2(q) + \text{h.o.t.},$$

where the entries in the $n \times n$ matrix $\mathcal{M}_k(q)$ are homogeneous polynomials of degree k in the components of the vector q, and similarly for the $m \times n$ matrix $\mathcal{A}_k(q)$. It is easy to show that

$$\mathcal{M}^{-1}(q) = \mathcal{M}_0^{-1} - \mathcal{M}_0^{-1}\mathcal{M}_1(q)\mathcal{M}_0^{-1} + \mathcal{M}_0^{-1}\mathcal{M}_1(q)\mathcal{M}_0^{-1}\mathcal{M}_1(q)\mathcal{M}_0^{-1}$$
$$- \mathcal{M}_0^{-1}\mathcal{M}_2(q)\mathcal{M}_0^{-1} + \text{h.o.t.}$$

Using this, we write an expansion of $\mathcal{V}_A$ up through terms of order 2:

$$\mathcal{V}_A(q) = \overline{\mathcal{V}}_0 + \overline{\mathcal{V}}_1(q) + \overline{\mathcal{V}}_2(q) + \text{h.o.t.},$$

where as above, $\overline{\mathcal{V}}_k(q)$ denotes the sum of terms that are homogeneous polynomials of degree k in the components of the vector q, and "h.o.t." refers to a quantity that is of order $o(\|q\|^2)$. Explicitly, under our assumption that $\mathcal{M}^{-1}\mathcal{A}^T v = 0$,

$$\overline{\mathcal{V}}_0 = \frac{1}{2}\left(\overline{v^T \mathcal{A}_0 \mathcal{M}_0^{-1} \mathcal{A}_0^T v}\right) + \mathcal{V}_0,$$

$$\overline{\mathcal{V}}_1(q) = \frac{1}{2}\left(\overline{v^T(\mathcal{A}_1(q)\mathcal{M}_0^{-1}\mathcal{A}_0^T - \mathcal{A}_0\mathcal{M}_0^{-1}\mathcal{M}_1(q)\mathcal{M}_0^{-1}\mathcal{A}_0^T \mathcal{A}_0 \mathcal{M}_0^{-1}\mathcal{A}_1^T(q))v}\right)$$
$$+ \mathcal{V}_1 \cdot q,$$

and

$$\overline{\mathcal{V}}_2(q) = \frac{1}{2}\left(\overline{v^T \mathcal{A}_1(q)\mathcal{M}_0^{-1}\mathcal{A}_1^T(q)v}\right)$$
$$- \frac{1}{2}\left(\overline{v^T \mathcal{A}_1(q)\mathcal{M}_0^{-1}\mathcal{M}_1(q)\mathcal{M}_0^{-1}\mathcal{A}_0^T v}\right)$$
$$- \frac{1}{2}\left(\overline{v^T \mathcal{A}_0\mathcal{M}_0^{-1}\mathcal{M}_1(q)\mathcal{M}_0^{-1}\mathcal{A}_1^T(q)v}\right)$$
$$+ \frac{1}{2}\left(\overline{v^T \mathcal{A}_2(q)\mathcal{M}_0^{-1}\mathcal{A}_0^T v}\right)$$
$$+ \frac{1}{2}\left(\overline{v^T \mathcal{A}_0\mathcal{M}_0^{-1}\mathcal{M}_1(q)\mathcal{M}_0^{-1}\mathcal{M}_1(q)\mathcal{M}_0^{-1}\mathcal{A}_0^T v}\right)$$
$$- \frac{1}{2}\left(\overline{v^T \mathcal{A}_0\mathcal{M}_0^{-1}\mathcal{M}_2(q)\mathcal{M}_0^{-1}\mathcal{A}_0^T v}\right) + \frac{1}{2}\left(\overline{v^T \mathcal{A}_0\mathcal{M}_0^{-1}\mathcal{A}_2^T(q)v}\right)$$
$$+ q^T \mathcal{V}_2 q,$$

where $\mathcal{V}_0$, $\mathcal{V}_1$, and $\mathcal{V}_2$ define the jets of the potential $\mathcal{V}(q)$ of orders 0, 1, and 2, respectively. Writing $V_A(\cdot)$ in this way shows its dependence on jets of coefficient functions of (9.4.13) of order up to 2. This dependence implies that the observed stabilizing effects produced by high-frequency forcing cannot be understood in terms of a linearization of the dynamics (9.4.13). We shall examine this remark in a bit greater detail.

Having assumed that $q_0 = 0$ is a strict local minimum of $V_A(\cdot)$, it follows that

$$\frac{\partial V_A}{\partial q}(0) = \frac{\partial \overline{\mathcal{V}}_1}{\partial q} = 0.$$

There are two cases to consider here:

(i) $\partial \overline{\mathcal{V}}_1/\partial q = 0$ for a particular choice of oscillatory input $v(\cdot)$, and

(ii) $\partial \overline{\mathcal{V}}_1/\partial q \equiv 0$ independent of the choice of zero-mean oscillatory (periodic) forcing $v(\cdot)$.

Case (i) In this case, the location of the local minimum of $V_A(\cdot)$ depends on $v(\cdot)$, and it will not generally coincide with a rest point of (9.4.13). While the *averaging principle* suggests that there will be stable motions of (9.4.13) in neighborhoods of local minima of V_A, the analysis of this case has involved either the introduction of dissipation into the model (as was done in Baillieul [1993]) or the use of machinery from the theory of dynamical systems (as was done in the thesis of Weibel [1997]). The significant point to note here is that the dependence of $\mathcal{V}_A$ on $v(\cdot)$ implies that $\mathcal{A}_0 \neq 0$, which in turn implies both that $q_0 = 0$ will *not* be a rest point of (9.4.13) for any oscillatory input $v(\cdot)$, and also that the stabilizing effect of $v(\cdot)$ on motions of (9.4.13) will depend on jets of order up to 2 in the coefficient functions. For further details the reader is referred to Weibel and Baillieul [1998b].

Case (ii) If $\partial \overline{\mathcal{V}}_1 / \partial q \equiv 0$, by which we mean that the first partial derivatives of $\mathcal{V}_A$ evaluated at $q_0 = 0$ are zero independent of coefficients due to $v(\cdot)$, then we also find that $\mathcal{V}_1 = 0$. It then follows that

$$2\mathcal{A}_0 \mathcal{M}_0^{-1} \mathcal{A}_1^T(q) - \mathcal{A}_0 \mathcal{M}_0^{-1} \mathcal{M}_1(q) \mathcal{M}_0^{-1} \mathcal{A}_0^T \equiv 0,$$

and either of two subcases can occur:

(a) $\mathcal{A}_0 = 0$, or else

(b) there is a polynomial relationship among the coefficients in the zeroth- and first-order jets of $\mathcal{A}(q)$ and $\mathcal{M}(q)$.

In case (b), $q_0 = 0$ will correspond to a rest point of (9.4.13) in the absence of forcing, but it will not generally define a rest point when $v(t) \neq 0$. Stabilizing effects of the oscillatory input $v(\cdot)$ appear to again depend on jets of order up to two in the coefficients of the Lagrangian vector field (9.4.13). This case will not be treated further here. We shall consider case (a), $\mathcal{A}_0 = 0$. In this case, the averaged potential $\mathcal{V}_A$ depends on terms only up through first order in the coefficients of (9.4.13).

Slightly refining our notation, let $\mathcal{A}^\ell(q)$ denote the ℓth column of the $n \times m$ matrix $\mathcal{A}^T(q)$. Then we have

$$\mathcal{A}^\ell(q) = \mathcal{A}_1^\ell \cdot q + (\text{terms of order } \geq 2)$$

and

$$\mathcal{M}(q) = \mathcal{M}_0 + (\text{terms of order } \geq 1),$$

where we interpret $\mathcal{M}_0, \mathcal{A}_1^1, \ldots, \mathcal{A}_1^m$ as $n \times n$ coefficient matrices. The following result is now clear.

9.4.5 Proposition. *Suppose $v(\cdot)$ is an $\mathbb{R}^m$-valued piecewise continuous periodic function of period $T > 0$ such that*

$$\bar{v} = \frac{1}{T} \int_0^T v(s) \, ds = 0.$$

Suppose, moreover, that $\mathcal{A}_0 = 0$. Then the averaged potential of the La-grangian system (9.4.13) agrees up to terms of order 2 with the averaged potential associated with the linear Lagrangian system

$$\mathcal{M}_0 \ddot{q} + \sum_{\ell=1}^{m} \left(\dot{v}_\ell \mathcal{A}_1^\ell q + v_\ell (\mathcal{A}_1^\ell - \mathcal{A}_1^{\ell T}) \dot{q} \right) + \mathcal{V}_2 \cdot q = 0. \qquad (9.4.21)$$

Proof. The proof follows immediately from examining the above expansion of $\mathcal{V}_A$. ∎

A deeper connection with stability is now expressed in terms of the following theorem.

9.4.6 Theorem. *Suppose $w(\cdot)$ is an $\mathbb{R}^m$-valued piecewise continuous periodic function of period $T > 0$ such that $\bar{w} = \frac{1}{T} \int_0^T w(s) \, ds = 0$. Consider the linear Lagrangian system (9.4.21) with input $v(t) = w(\omega t)$, and suppose $\mathcal{A}_1^{\ell T} = \mathcal{A}_1^\ell$ for $\ell = 1, \ldots, m$. The averaged potential for this system is given by*

$$\mathcal{V}_A(q) = \tfrac{1}{2} q^T \left(\mathcal{V}_2 + \sum_{i,j=1}^{m} \sigma_{ij} \mathcal{A}_1^i \mathcal{M}_0^{-1} \mathcal{A}_1^{j T} \right) q, \qquad (9.4.22)$$

where $\sigma_{ij} = (1/T) \int_0^T w_i(s) w_j(s) \, ds$. If the matrix $\frac{\partial^2 \mathcal{V}_A}{\partial q^2}$ is positive definite, the origin $(q, \dot{q}) = (0, 0)$ of the phase space is stable in the sense of Lyapunov, provided that ω is sufficiently large.

Note the similarity here with the energy momentum method discussed in Section 3.12.

This theorem has been proved in Baillieul [1995]. In Baillieul [1993], it was shown that in the presence of dissipation, the positive definiteness of the Hessian matrix $\frac{\partial^2 \mathcal{V}_A}{\partial q^2}(0)$ is a sufficient condition for (9.4.13) to execute stable motions in a neighborhood of $(q_0, 0)$. Theorem 9.4.6 shows that in Case (ii), it is precisely the conditions of the averaging principle from which we may infer the Lyapunov stability of (9.4.21) based on the positive definiteness of the Hessian of the averaged potential. Clearly, this result is special and related to the property that the averaged potential depends only on first-order jets of the coefficients of (9.4.13) when $\mathcal{A}(q_0) = 0$. In Weibel and Baillieul [1998b] it has been shown that the condition $\mathcal{A}(q_0) = 0$ is also necessary and sufficient for the local minimum q_0 of the averaged potential to define a corresponding fixed point (rather than a periodic orbit) of the *forced* (nonautonomous) Hamiltonian system associated with (9.4.12).

Comparing Force Control and Acceleration Control of Mechanical Systems. Corresponding to the averaged Hamiltonian (9.4.16) there is an *averaged Lagrangian*

$$\bar{\mathcal{L}} = \frac{1}{2} \dot{q}^T \mathcal{M} \dot{q} - \mathcal{V}_A(q).$$

The Euler–Lagrange equations

$$\frac{d}{dt}\frac{\partial \bar{\mathcal{L}}}{\partial \dot{q}} - \frac{\partial \bar{\mathcal{L}}}{\partial q} = 0 \qquad (9.4.23)$$

provide a simple description of the averaged system dynamics. This means that the net averaged effect on (9.4.23) of a high-frequency oscillatory input $r(\cdot)$ is felt as a conservative (potential) force that is a component of $\frac{\partial \mathcal{V}_A}{\partial q}$. The net averaged effect on (9.1.3) of a high-frequency oscillatory input $u(\cdot)$ applied to (9.1.2), however, is

$$\frac{\partial V}{\partial q} \longmapsto M(q) \sum_{a,b=1}^{m} \langle Y_a : Y_b \rangle \overline{v_a v_b} \qquad (9.4.24)$$

which may be represented as the gradient of a potential function only when certain symmetry conditions are satisfied. We refer to Bullo [2001] for a discussion of such symmetry conditions.

The control design approach in both the force and acceleration control cases involves choosing stabilizing inputs from a parametrized family of controls. For acceleration-controlled systems, stabilizing control laws may be found through a bifurcation analysis of the averaged potential. In the case of force-controlled systems, the situation is more complex. The force field (9.4.24) may not arise as the gradient of a potential. Moreover, the influence on the q-dynamics (9.1.3) will be determined jointly by the r-dynamics (9.1.2) *and* the choice of input $u(\cdot)$. To illustrate the differences between the two cases, we return to the pendulum-on-a-cart example (9.1.4).

Suppose the cart is subjected to a zero-mean periodic forcing $u(\cdot)$. Then the averaged equation (9.4.10) specializes to

$$M(\theta)\begin{pmatrix}\ddot{s}\\ \ddot{\theta}\end{pmatrix} + \begin{pmatrix}-\beta\sin(\theta-\psi)\dot{\theta}^2\\ D\sin\theta\end{pmatrix} + M(\theta)G(\theta)\sigma^2 = \begin{pmatrix}0\\ 0\end{pmatrix}, \qquad (9.4.25)$$

where

$$M(\theta) = \begin{pmatrix}\gamma & \beta c(\theta)\\ \beta c(\theta) & \alpha\end{pmatrix},$$

$$G(\theta) = \begin{pmatrix}\dfrac{\alpha\beta^3 s(\theta)c^2(\theta)}{[\alpha\gamma - \beta^2 c^2(\theta)]^3}\\[2ex] \dfrac{-\alpha\gamma\beta^2 s(\theta)c(\theta)}{[\alpha\gamma - \beta^2 c^2(\theta)]^3}\end{pmatrix},$$

$c(\theta) = \cos(\theta - \psi)$, and $s(\theta) = \sin(\theta - \psi)$. Here σ^2 is the rms value of the periodic input:

$$\sigma^2 = \frac{1}{T}\int_0^T v(t)\,dt,$$

where $v(t) = \int^t u(\tau)\,d\tau$.

To facilitate comparison with the acceleration-controlled reduced Lagrangian, we multiply (9.4.25) through by $M^{-1}(\theta)$; this yields a second-order equation in θ that has no dependence on s:

$$\ddot{\theta} + \frac{\beta^2 \sin(\theta - \psi)\cos(\theta - \psi)\dot{\theta}^2 + \gamma D \sin\theta}{\alpha\gamma - \beta^2 \cos^2(\theta - \psi)}$$

$$-\frac{\alpha\gamma\beta^2 \sin(\theta - \psi)\cos(\theta - \psi)}{[\alpha\gamma - \beta^2 \cos^2(\theta - \psi)]^3}\sigma^2 = 0 \qquad (9.4.26)$$

Reintroducing the length and mass parameters, this equation may be rewritten

$$[M + m\sin^2(\theta - \psi)]\ddot{\theta} + m\sin(\theta - \psi)\cos(\theta - \psi)\dot{\theta}^2$$

$$+ \frac{(M + m)g}{\ell}\sin\theta - \frac{(M + m)\sin(\theta - \psi)\cos(\theta - \psi)}{\ell^2[M + m\sin^2(\theta - \psi)]^2}\sigma^2 = 0,$$

which is an equation of the form

$$\frac{d}{dt}\frac{\partial L}{\partial\dot{\theta}} - \frac{\partial L}{\partial\theta} = 0,$$

with

$$L(\theta, \dot{\theta}) = \frac{1}{2}\left[M + m\sin^2(\theta - \psi)\right]\dot{\theta}^2 - V_{A_1}(\theta),$$

where

$$V_{A_1}(\theta) = -\frac{(M + m)g}{\ell}\cos\theta + \frac{(M + m)\sigma^2}{\ell^2 m\,(2M + m(1 + \cos 2\theta))}.$$

The reduced Lagrangian (9.4.12) for the cart–pendulum model is

$$\mathcal{L}(\theta, \dot{\theta}; \dot{s}) = \frac{1}{2}\alpha\dot{\theta}^2 + \beta\cos(\theta - \psi)\dot{s}\dot{\theta} + D\cos\theta. \qquad (9.4.27)$$

This gives rise to the acceleration-controlled θ-dynamics

$$\alpha\ddot{\theta} + \beta\cos(\theta - \psi)\ddot{s} + D\sin\theta = 0, \qquad (9.4.28)$$

where the acceleration variable $\ddot{s}$ is regarded as the control input. Looking at the general form of (9.4.12), we note that in (9.4.27) $\dot{s}$ plays the role of $\dot{r} = v$, $\mathcal{A}(\theta) = \beta\cos(\theta - \psi)$, $\mathcal{M}(\theta) = \alpha$, and $\mathcal{V}_a(\theta) = -D\cos\theta$. (There are no time-dependent terms in $\mathcal{V}_a$ in this case.) The averaged potential (9.4.17) thus specializes to

$$V_{A_2}(\theta) = \frac{1}{2}m\cos^2(\theta - \psi)\sigma^2 - mg\ell\cos\theta,$$

where now

$$\sigma^2 = \frac{1}{T} \int_0^T \dot{s}(t)^2 \, dt.$$

Because the inputs in the force-controlled and acceleration-controlled systems ((9.1.4) and (9.4.28), respectively) are physically different, the rms parameter σ^2 does not have the same meaning in V_{A_1} and V_{A_2}. Nevertheless, in both cases, it plays a similar role as a bifurcation parameter. Suppose $\psi = \pi/2$, for instance. For the choice of physical parameters ($m = M = 1$, $\ell = 1$, $g = 10$) defining the potential functions in Figure 9.4, we find that for σ^2 sufficiently large, $\theta = \pi$ is a local minimum of both averaged potentials, V_{A_1} and V_{A_2}. (See Figure 9.4.) Accordingly, it can be shown that for all sufficiently high high-frequency inputs ($u(\cdot)$ and $\ddot{s}(\cdot)$, respectively) the pendulum systems (9.1.4) and (9.4.28) will undergo stable motions in a neighborhood of the inverted equilibrium $\theta = \pi$. It can be verified (and the graphs in Figure 9.4 suggest) that as the bifurcation parameter σ^2 is increased, the equilibrium $\theta = \pi$ becomes a local minimum of V_{A_1} as a result of a *supercritical* pitchfork bifurcation, while $\theta = \pi$ becomes a local minimum of V_{A_2} as a result of a subcritical pitchfork bifurcation. This type of stabilization goes back to work of Stephenson [1908] and Kapitsa [1951], and of course there has been much interesting work since.

9.5   Dynamic Nonholonomic Averaging

We can analyze the motion planning problem in the dynamic nonholonomic context using the following perturbation approach of Ostrowski [2000] and extending the analysis in the kinematic setting discussed in Chapter 4.

As in Section 5.7 let us write the controlled dynamic nonholonomic equations in the coordinate form

$$\xi^a = (g^{-1}\dot{g})^a = -A^a_\alpha(r)\dot{r}^\alpha + (I^{-1})^{ab}p_b,$$

$$\frac{d}{dt}p_b = C^c_{ab}I^{ad}p_c p_d + \mathcal{D}^c_{b\alpha}\dot{r}^\alpha p_c + \mathcal{D}_{\alpha\beta b}\dot{r}^\alpha \dot{r}^\beta,$$

$$\dot{r}^\alpha = u^\alpha,$$

assuming that the base dynamical variables r are fully actuated.

Consider inputs of the form

$$r(t) = r_0 + \epsilon u(t),$$

where generally $u(t)$ will be chosen to be periodic. Thus, ϵ measures the size or amplitude of the oscillatory inputs. Assume that the system is initially at rest: $p(0) = 0$. Expanding p in the series $p = p^0 + \epsilon p^1 + \epsilon^2 p^2 + \cdots$ and

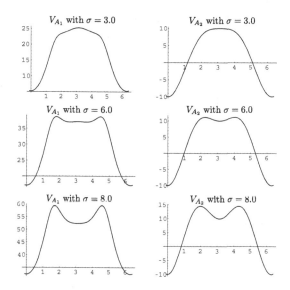

FIGURE 9.4.2. A comparison of parametric dependence (on σ^2) of the two *averaged* *potentials*, V_{A_1} (left column = force-controlled case) and V_{A_2} (right column = acceleration-controlled case). The figure illustrates the fact that in both cases for sufficiently large values of σ^2, the inverted pendulum equilibrium $\theta = \pi$ is a local minimum of the averaged potential. As described in the text, the qualitative features of the σ-dependent bifurcations differ, however.

Taylor expanding the $\mathcal{D}$ coefficients about r_0, we get (Ostrowski [2000])

$$p_a(t) = \epsilon^2 \mathcal{D}_{\alpha\beta a}(r_0) \int_0^t \dot{u}^\alpha \dot{u}^\beta d\tau$$

$$+ \epsilon^3 \left(\left(\frac{\partial \mathcal{D}_{\alpha\beta a}}{\partial r^k} - \mathcal{D}_{\alpha\gamma}^b \right) \bigg|_{r_0} \times \right. \tag{9.5.1}$$

$$\left. \int_0^t u^\gamma \dot{u}^\alpha \dot{u}^\beta d\tau + \mathcal{D}_{a\alpha}^b \mathcal{D}_{\beta\gamma b}(r_0) u^\alpha(t) \int_0^t \dot{u}^\beta \dot{u}^\kappa d\tau \right)$$

$$+ \cdots .$$

We note that to second order the momentum depends only on the term $\mathcal{D}_{\alpha\beta b}$. Note also that the cubic term of the form

$$\int_0^t u^\gamma \dot{u}^\alpha \dot{u}^\beta d\tau$$

will be equal to zero after one period for inputs of the form

$$u^\alpha = a^\alpha \sin 2\pi m t \quad \text{and} \quad u^\beta = a^\beta \sin 2\pi n t, \qquad m \neq n.$$

Application of the Exponential Expansion.   Now to obtain the position we use the Magnus expansion discussed in Chapter 4.

Let $g(t) = e^{z(t)}$. The solution to the equation $g^{-1}\dot{g} = \epsilon\xi(t)$ is given locally, for $\tilde{\xi}(t) = \int_0^t \xi(\tau)d\tau$, by

$$
z(t) = \epsilon \int_0^t \xi(\tau)d\tau + \frac{\epsilon^2}{2} \int_0^t [\tilde{\xi}(\tau), \xi(\tau)]d\tau
$$
$$
+ \frac{\epsilon^3}{4} \int_0^t \left[\int_0^\tau [\tilde{\xi}(\sigma), \xi(\sigma)]d\sigma \xi(\tau) \right] d\tau
$$
$$
+ \frac{\epsilon^3}{12} \int_0^t [\tilde{\xi}(\tau), [\tilde{\xi}(\tau),\ xi(\tau)]]d\tau + \cdots .
$$

Then again, using inputs $r(t) = r_0 + \epsilon u(t)$ for a body initially at rest, $(p(0) = 0)$, the curve in the Lie algebra that describes the body velocity is given by

$$
\xi^a(t) = -\epsilon A_\alpha^a(r_0)\dot{u}^\alpha + \epsilon^2 \left(-\frac{\partial A_\alpha^a}{\partial r^\beta}|_{r_0} \dot{u}^\alpha \dot{u}^\beta + (I^{-1}p^2)^\alpha(t) \right)
$$
$$
+ \epsilon^3 \left(-\frac{1}{2}\frac{\partial^2 A_\alpha^a}{\partial r^\beta r^\gamma}|_{r_0} u^\beta u^\gamma \dot{u}^\alpha + (I^{-1}p^3)^a + \frac{\partial(I^{-1}p^2)^a}{\partial r^\alpha}u^\beta \right) + \cdots .
$$

where $p(t) = p^0 + \epsilon p^1 + \epsilon^2 p^2 + \cdots$ is defined as above.

In exponential coordinates the group variables are given, for $z(0) = 0$, by

$$
z^a(t) = 0 + \epsilon \left(-A_\alpha^a u^\alpha|_0^t \right)
$$
$$
+ \epsilon^2 \left(-\frac{1}{2}DA_{\alpha\beta}^a \int_0^t u^\alpha \dot{u}^\beta d\tau + (I^{-1})^{ab}\mathcal{D}_{\alpha\beta b} \int_0^t \int_0^\tau \dot{u}^\alpha \dot{u}^\beta ds d\tau \right)
$$
$$
+ \epsilon^3 \left(-\frac{1}{3}\left(\frac{\partial DA_{\alpha\gamma}^a}{\partial r^\beta} - [A_\beta, DA_{\alpha\beta}] \right) \int_0^t u^\alpha u^\beta \dot{u}^\gamma d\tau \right.
$$
$$
\left. + \left(\frac{\partial(I^{-1})^{ab}}{\partial r^\beta} - [A_\beta, (I^{-1})^{ab}] \right) \int_0^t u^\beta p_b d\tau + (I^{-1})^{ab} \int_0^t p_b^3 d\tau \right)
$$
$$
- \epsilon^2 \left(\frac{1}{2}\frac{\partial A_\alpha^a}{\partial r^\beta}(u^\alpha u^\beta)|_0^t + [A_\alpha, A_\beta]^a u^\alpha(0)u^\beta|_0^t \right) + \epsilon^3 R + \cdots ,
$$

$$(9.5.2)$$

where the cubic term R integrates to 0 over one period when cyclic inputs are used. As Ostrowski observes, we note that the ϵ^2 term consists of an area term and a term driven by the momentum, thus decoupling into kinematic and dynamics pieces. We refer to Ostrowski [2000] for specific algorithms and examples generated using this formalism.

For related work on trajectory generation for nonholonomic systems see Morgansen [2001]. Also see recent work on locomotion for underactuated mechanical systems in Vela, Morgansen, and Burdick [2002] as well as Bullo [2001], Bullo and Lynch [2001], and references therein.

Using the expansions above one can choose controls giving local changes in position and momentum using small-scale cyclic inputs. These small changes can then be concatenated to drive the system toward a given goal. In Ostrowski [2000], to which we refer the reader for further details, steering of the snakeboard toward a given target is discussed.

References

Abraham, R. and J. E. Marsden [1978], *Foundations of Mechanics*, Addison-Wesley. Reprinted by Perseus Press, 1995.

Abraham, R., J. E. Marsden, and T. S. Ratiu [1988], *Manifolds, Tensor Analysis, and Applications*, Applied Mathematical Sciences. **75**, Springer-Verlag.

Aeyels, D. [1985], Stabilization by smooth feedback of the angular velocity of a rigid body, *Systems Control Lett.* **6**, 59–63.

Aeyels, D. [1989], Stabilizability and asymptotic stabilizability of the angular velocity of a rigid body, in *New trends in nonlinear control theory (Nantes, 1988)*, Springer, Berlin., 243–253.

Agrachev, A. A. and D. Liberzon [2001], Lie-algebraic stability criteria for switched systems, *SIAM J. Control Optim.* **40**, 253–269 (electronic).

Agrachev, A. A. and A. V. Sarychev [1996], Abnormal sub-Riemannian geodesics: Morse index and rigidity, *Ann. Inst. H. Poincaré Anal. Non Linéaire* **13**, 635–690.

Agrachev, A. A. and A. V. Sarychev [1998], On abnormal extremals for Lagrange variational problems, *J. Math. Systems Estim. Control* **8**, 87–118.

Anderson, I. and T. Duchamp [1980], On the existence of global variational principles, *Amer. J. Math.* **102**, 781–862.

Antman, S. S. [1998], The simple pendulum is not so simple, *SIAM Rev* **40**. No. 4. 927–930.

Antman, S. S. and J. E. Osborn [1979], The principle of virtual work and integral laws of motion, *Arch. Rat. Mech. An.* **69**, 231–262.

440   References

Appell, P. [1900], Sur l'intégration des équations du mouvement d'un corps pesant de révolution roulant par une arête circulaire sur un plan horizontal; cas parficulier du cerceau, *Rendiconti del circolo matematico di Palermo*, **14**, 1–6.

Appell, P. [1911], Sur les liaisons experimées par des relations non linéaires entre les vitesses, C. R. Acad. Sc. Paris **152**, 1197–1199.

Arms, J. M. [1981], The structure of the solution set for the Yang–Mills equations, *Math. Proc. Camb. Philos. Soc.* **90**, 361–372.

Arms, J. M., J. E. Marsden, and V. Moncrief [1981], Symmetry and bifurcations of momentum mappings, *Comm. Math. Phys.*, **78**, 455–478.

Arms, J. M., J. E. Marsden, and V. Moncrief [1982], The structure of the space solutions of Einstein's equations: II Several Killing fields and the Einstein–Yang–Mills equations, *Ann. of Phys.*, **144**, 81–106.

Arnold, V. I. [1966a], Sur la géométrie differentielle des groupes de Lie de dimension infinie et ses applications à l'hydrodynamique des fluides parfaits, *Ann. Inst. Fourier, Grenoble*, **16**, 319–361.

Arnold, V. I. [1966b], On an a priori estimate in the theory of hydrodynamical stability, *Izv. Vyssh. Uchebn. Zaved. Mat. Nauk* **54**, 3–5. English Translation: Amer. Math. Soc. Transl. **79** (1969), 267–269.

Arnold, V. I. [1983], *Geometrical Methods in the Theory of Ordinary Differential Equations*, Springer-Verlag.

Arnold, V. I. [1989], *Mathematical Methods of Classical Mechanics*, Second Edition, Springer-Verlag; First Edition 1978, Second Edition, 1989, Graduate Texts in Math, volume **60**.

Arnold, V. I., V. V. Kozlov, and A. I. Neishtadt [1988], *Dynamical Systems III*, Springer-Verlag, Encyclopedia of Math., **3**.

Astolfi, A. [1996], Discontinuous control of nonholonomic systems, *Systems Control Lett.* **27**, 37–45.

Atiyah, M. [1982], Convexity and commuting Hamiltonians, *Bull. London Math. Soc.* **14**, 1–5.

Auckly, D., L. Kapitanski, and W. White [2000], Control of nonlinear underactuated systems, *Comm. Pure Appl.Math.*, **53**, 354–369.

Auslander, L. and R. E. Mackenzie [1977], *Introduction to Differentiable Manifolds*, Dover Publications, New York.

Baillieul, J. [1975], *Some Optimization Problems in Geometric Control Theory*, Ph.D. thesis, Harvard University.

Baillieul, J. [1978], Geometric methods for nonlinear optimal control problems, *J. of Optimization Theory and Applications*, **25**, 519–548.

Baillieul, J. [1987], Equilibrium mechanics of rotating systems, in *Proceedings of the 26th IEEE Conf. Dec. and Control*, 1429–1434.

Baillieul, J. [1990], The behavior of super-articulated mechanisms subject to periodic forcing, in *Analysis of Controlled Dynamical Systems*, (Gauthier and Bride and Bonnard and Kupka, eds.), Birkhäuser.; Proceedings of a Conference held in Lyons, France 3–6 July, 1990.

Baillieul, J. [1993], Stable average motions of mechanical systems subject to periodic forcing, dynamics and control of mechanical systems: in *The Falling Cat and Related Problems* (Michael Enos, ed.) Fields Institute Communications, Am. Math. Soc. Providence. **1**, 1–23.

Baillieul, J. [1995], Energy methods for stability of bilinear systems with oscillatory inputs, in *International Journal of Robust and Nonlinear Control*, Special Issue on the Control of Mechanical Systems, (H. Nijmeijer and A. J. van der Schaft, guest eds.) **5**, 205–381.

Baillieul, J. [1998], The geometry of controlled mechanical systems, in *Mathematical Control Theory*, 322–354. Springer-Verlag, New York.

Baillieul, J. [2000], Kinematic asymmetries and the control of Lagrangian systems with oscillatory inputs, *Proceedings of the IFAC Workshop on Lagrangian and Hamiltonian Methods for Nonlinear Control*, 135–143, Pergamon.

Baillieul, J. and J. C. Willems [1999] (eds.) *Mathematical control theory* (Dedicated to Roger Ware Brockett on the occasion of his 60th birthday), Springer-Verlag, N.Y.

Barbashin, E. and N. N. Krasovskii [1952], Stability of motion in the large, *Doklady Mathematics (Translations of Proceedings of Russian Academy of Sciences)* **86**, 3, 453–456.

Barnett, S. [1978], *Introduction to Mathematical Systems Theory*, Oxford University Press, U.K.

Bates, L. [2002], Problems and progress in nonholonomic reduction, *Rep. Math. Phys.* **49**, 143–149. XXXIII Symposium on Mathematical Physics (Torún, 2001).

Bates, L. and R. Cushman [1999], What is a completely integrable nonholonomic dynamical system?, *Rep. Math. Phys.* **44**, 29–35.

Bates, L. and J. Sniatycki [1993], Nonholonomic Reduction, *Reports on Math. Phys.* **32**, 99–115.

Bates, L. [1998], Examples of singular nonholonomic reduction, *Reports in Mathematical Physics* **42**, 231–247.

Bedrosian, N. S. [1992], Approximate feedback linearization: the cart–pole example in *Proceedings 1992 IEEE International Conference on Robotics and Automation*, 1987–1992.

Berry, M. [1984], Quantal phase factors accompanying adiabatic changes, *Proc. Roy. Soc. London A* **392**, 45–57.

Berry, M. [1985], Classical adiabatic angles and quantal adiabatic phase, *J. Phys. A. Math. Gen.* **18**, 15–27.

Berry, M. [1990], Anticipations of the geometric phase, *Physics Today*, 34–40.

Blackall, C. J. [1941], On volume integral invariants of non-holonomic dynamical systems, *Am. J. of Math.* **63**, 155–168.

Blankenstein, G., R. Ortega, and A. J. van der Schaft [2002], The matching conditions of controlled Lagrangians and IDA-passivity based control, *Internat. J. Control*, **75**, 645–665.

Blankenstein, G. and A. J. van der Schaft [2001], Symmetry and reduction in implicit generalized Hamiltonian systems, *Rep. Math. Phys.* **47**, 57–100.

Bliss, G. [1930], The Problem of Lagrange in the Calculus of Variations. *Amer. J. Math.*, 673–744.

Bliss, G. A. [1946], *Lectures on Calculus of Variations*, Univ. of Chicago Press.

Bloch, A. M. [1990], Steepest descent, linear programming and Hamiltonian flows, *Contemp. Math. Amer. Math. Soc.* **114**, 77–88.

Bloch, A. M. [2000], Asymptotic Hamiltonian Dynamics: the Toda lattice, the three-wave interaction and the nonholonomic Chaplygin sleigh, *Physica D*, **141**, 297–315.

Bloch, A. M., R. Brockett, and P. Crouch [1997], Double bracket equations and geodesic flows on symmetric spaces, *Comm. Math. Phys.* **187**, 357–373.

Bloch, A. M., R. W. Brockett, and T. Ratiu [1990], A new formulation of the generalized Toda lattice equations and their fixed-point analysis via the moment map, *Bulletin of the AMS* **23**, 447–456.

Bloch, A. M., R. Brockett, and T. S. Ratiu [1992], Completely integrable gradient flows, *Comm. Math. Phys.* **147**, 57–74.

Bloch, A. M., D. Chang, N. Leonard, and J. E. Marsden [2001], Controlled Lagrangians and the stabilization of mechanical systems II: Potential shaping, *Trans IEEE on Auto. Control* **46**, 1556–1571.

Bloch, A. M., D. Chang, N. Leonard, J. E. Marsden, and C. Woolsey [2000], Asymptotic stabilization of Euler–Poincaré mechanical systems, *Proc. IFAC Workshop on Lagrangian and Hamiltonian Methods for Nonlinear Control*, 56–61.

Bloch, A. M. and P. Crouch [1992], On the dynamics and control of nonholonomic systems on Riemannian manifolds, *Proceedings of NOLCOS '92, Bordeaux*, 368–372.

Bloch, A. M. and P. E. Crouch [1993], Nonholonomic and vakonomic control systems on Riemannian manifolds, *Fields Institute Communications* **1**, 25–52.

Bloch, A. M. and P. E. Crouch [1994] Reduction of Euler–Lagrange problems for constrained variational problems and relation with optimal control problems, *The Proceedings of the 33rd IEEE Conference on Decision and Control*, 2584–2590, IEEE (1994).

Bloch, A. M., and P. E. Crouch [1995], Nonholonomic control systems on Riemannian manifolds, *SIAM J. on Control* **37**, 126–148.

Bloch, A. M. and P. Crouch [1996], Optimal control and geodesic flows, *Systems Control Lett.* **28**, no. 2, 65–72.

Bloch, A. M. and P. E. Crouch [1997] Optimal control and the full Toda lattice equations, *Proc 36th IEEE Conf. on Decision and Control*, 1736–1740.

Bloch, A. M. and P. E. Crouch [1998a], Newton's law and integrability of nonholonomic systems, *SIAM J. Control Optim.* **36**, 2020–2039.

Bloch, A. M. and P. E. Crouch [1998b] Optimal control, optimization and analytical mechanics, in *Mathematical Control Theory* (J. Baillieul and J. Willems, eds.), Springer, 268–321.

Bloch, A. M., P. Crouch, J. E. Marsden, and T. S. Ratiu [1998], Discrete rigid body dynamics and optimal control, *Proc. CDC* **37**, 2249–2254.

Bloch, A. M., P. E. Crouch, J. E. Marsden, and T. S. Ratiu [2000], An almost Poisson structure for the generalized rigid body equations, *Proceedings of the IFAC Workshop on Lagrangian and Hamiltonian Methods for Nonlinear Control*, 87–92, Pergamon.

Bloch, A. M., P. Crouch, J. E. Marsden, and T. S. Ratiu [2002], The symmetric representation of the rigid body equations and their discretization *Nonlinearity*. **15**, 1309–1341.

Bloch, A. M., P. E. Crouch, and T. S. Ratiu [1994] Sub-Riemannian optimal control problems *Fields Institute Communications, AMS*. **3**, 35–48.

Bloch, A. M. and S. V. Drakunov [1994] Stabilization of nonholonomic systems via sliding modes, in *Proceedings of the 33rd IEEE Conference on Decision and Control*, Orlando, Florida, 2961–2963.

Bloch, A. M. and S. V. Drakunov [1995] Tracking in nonholonomic dynamic systems via sliding modes, in *Proceedings of the 34th IEEE Conference on Decision and Control*, New Orleans, LA, 2103–2106.

Bloch, A. M. and S. V. Drakunov [1996], Stabilization and tracking in the nonholonomic integrator via sliding modes, *Systems and Control Letters* **29**, 91–99.

Bloch, A. M. and S. Drakunov [1998], Discontinuous Stabilization of Brockett's Canonical Driftless System, in *Essays in Mathematical Robotics, IMA*, 169–183.

Bloch, A. M., S. V. Drakunov, and M. Kinyon [1997], Stabilization of Brockett's generalized canonical driftless system, *Proceedings of the 36th IEEE Conference on Decision and Control*, 4260–4265.

Bloch, A. M., S. V. Drakunov, and M. Kinyon [2000], Stabilization of nonholonomic systems using isospectral flows, *SIAM Journal of Control and Optimization* **38**, 855–874,

Bloch, A. M., H. Flaschka, and T. S. Ratiu [1990], A convexity theorem for isospectral manifolds of Jacobi matrices in a compact Lie algebra, *Duke Math. J.* **61**, 41–65.

Bloch, A. M., P. S. Krishnaprasad, J. E. Marsden, and G. Sánchez De Alvarez [1992], Stabilization of rigid body dynamics by internal and external torques, *Automatica* **28**, 745–756.

Bloch, A. M., P. S. Krishnaprasad, J. E. Marsden, and R. Murray [1996], Nonholonomic mechanical systems with symmetry, *Arch. Rat. Mech. An.*, **136**, 21–99.

Bloch, A. M., P. S. Krishnaprasad, J. E. Marsden, and T. S. Ratiu [1994], Dissipation Induced Instabilities, *Ann. Inst. H. Poincaré, Analyse Nonlineare* **11**, 37–90.

Bloch, A. M., P. S. Krishnaprasad, J. E. Marsden, and T. S. Ratiu [1996], The Euler–Poincaré equations and double bracket dissipation, *Comm. Math. Phys.* **175**, 1–42.

Bloch, A. M., N. E. Leonard, and J. E. Marsden [1997], Stabilization of mechanical systems using controlled Lagrangians, *Proc CDC*, **36**, 2356–2361.

Bloch, A. M., N. E. Leonard, and J. E. Marsden [1998], Matching and stabilization by the method of controlled Lagrangians, *Proc CDC*, **37**, 1446–1451.

Bloch, A. M., N. E. Leonard, and J. E. Marsden [1999a], Stabilization of the pendulum on a rotor arm by the method of controlled Lagrangians, *Proc. of the International Conference on Robotics and Automation 1999*, IEEE 500–505.

Bloch, A. M., N. E. Leonard, and J. E. Marsden [1999b], Potential shaping and the method of controlled Lagrangians, *Proc CDC*, **38**, 1653–1657.

Bloch, A. M., N. Leonard, and J. E. Marsden [2000], Controlled Lagrangians and the stabilization of mechanical systems I: The first matching theorem, *IEEE Trans. Automat. Control*, **45**, 2253–2270.

Bloch, A. M., N. Leonard, and J. E. Marsden [2001], Controlled Lagrangians and the stabilization of Euler–Poincaré mechanical systems, *Int. J. of Robust and Nonlinear Control* **11**, 191–214.

Bloch, A. M. and J. E. Marsden [1989], Controlling homoclinic orbits, *Theoretical and Computational Fluid Dynamics* **1**, 179–190.

Bloch, A. M., J. E. Marsden, and G. Sánchez De Alvarez [1997], Stabilization of relative equilibria of mechanical systems with symmetry, in *Current and Future Directions in Applied Mathematics*, (M. Alber and B. Hu and J. Rosenthal, eds.), Birkhäuser., 43–64.

Bloch, A. M., J. E. Marsden and D. Zenkov [2005], Nonholonomic Dynamics, *Notices of the American Mathematical Society.* **52**, 324-333.

Bloch, A.M., J.E. Marsden D.V. Zenkov [2007] Quasicoordinates and symmetries in nonholonomic systems, to appear

Bloch, A. M., M. Reyhanoglu, and H. McClamroch [1992], Control and stabilization of nonholonomic systems, *IEEE Trans. Aut. Control* **37**, 1746–1757.

Bondi, H. [1986], The rigid body dynamics of unidirectional spin, *Proc. Roy. Soc. Lon.* **405**, 265–274.

Boothby, W. M. [1986], *An Introduction to Differentiable Manifolds and Riemannian Geometry, 2nd ed.*, Pure and applied mathematics series **120**, Academic Press, Orlando, FL.

Borisov, A.V. and I.S. Mamaev [2002], Rolling of a rigid body on plane and sphere *Regular and Chaotic Dynamics* **7**, 177-2000.

Borisov, A.V. and I.S. Mamaev [2003], Strange attractors in the dynamics of the rattleback, *Physics-Uspekhi*, **46**, 393–403.

Bou-Rabee, N. M., J. E. Marsden, and L. N. Romero [2004], Tippe top inversion as a dissipation induced instability, *SIAM J. on Appl. Dyn. Systems*, **3**, 352–377.

Bourbaki, N. [1971], *Groupes et Algèbres de Lie*, Diffusion C.C.L.S., Paris.

Brockett, R. W. [1970a], Systems theory on group manifolds and coset spaces, *SIAM J. Control* **10**, no. 2, 265–284.

Brockett, R. W. [1970b], *Finite Dimensional Linear Systems*, Wiley.

Brockett, R. W. [1973a], Lie theory and control systems defined on spheres, in *Geometric Methods in System Theory, SIAM J. Applied Math.* **10**, no. 2, 213–225.

Brockett, R. W. [1973b], Lie algebras and Lie groups in control theory, in *Geometric Methods in System Theory*, (D. Q. Mayne and R. W. Brockett, eds.), D. Reidel Publ. Co., Dordrecht, the Netherlands., 43–82.

Brockett, R. W. [1976a], Nonlinear systems and differential geometry, *Proc. of the IEEE.* **64**, 61–72.

Brockett, R. W. [1976b], Control theory and analytical mechanics, in *1976 Ames Research Center (NASA) Conference on Geometric Control Theory* (R. Hermann and C. Martin, eds.), Math. Sci. Press, Brookline, Massachussets, Lie Groups: History, Frontiers, and Applications. **VII**, 1–46.

Brockett, R. W. [1976c], Volterra series and geometric control theory, *Automatica—J. IFAC* **12**, 167–176.

Brockett, R. W. [1981], Control theory and singular Riemannian geometry, in *New Directions in Applied Mathematics* (P. J. Hilton and G. S. Young, eds.), Springer-Verlag, 11–27.

Brockett, R. W. [1983], Nonlinear control theory and differential geometry, in *Proc. International Congress of Mathematicians*, 1357–1368.

Brockett, R. W. [1989], On the rectification of vibratory motion, *Sensors and Actuators* **20**, 91–96.

Brockett, R. W. [1991], Dynamical systems that sort lists and solve linear programming problems, *Proc. 27th IEEE Conf. Dec. and Control*, 799–803; see also *Linear Algebra and Its Appl.* **146**, 79–91.

Brockett, R. W. [1994], The double bracket equation as a solution of a variational problem, *Fields Instit. Commun.* **3**, 69–76.

Brockett, R. W. [2000], Beijing lectures on nonlinear control systems, in *Lectures on systems, control, and information (Beijing, 1997)*, AMS/IP Stud. Adv. Math. **17**, 1–48.

Brockett, R. W. and L. Dai [1992], Nonholonomic kinematics and the role of elliptic functions in constructive controllability, in *Nonholonomic Motion Planning* (Z. Li and J. F. Canny, eds.), Kluwer, 1–22.

Brockett, R. W. and A. Rahimi [1972], Lie algebras and linear differential equations, in *Ordinary Differential Equations* (L. Weiss, ed.), Academic Press, 379–386.

Bryant, R. and P. Griffiths [1983], Reduction for constrained variational problems and $\int \kappa^2/2\,ds$, *Am. J. of Math.* **108**, 525–570.

Bryant, R. L. and L. Hsu [1993], Rigidity of integral curves of rank 2 distributions, *Inventiones Math.* **114**, 435–461.

Bullo, F. [2000], Stabilization of relative equilibria for underactuated systems on Riemannian manifolds, *Automatica J. IFAC* **36**, 1819–1834.

Bullo, F. [2001], Series expansions for the evolution of mechanical control systems, *SIAM J. Control Optim.* **40**, 166–190.

Bullo, F. [2002] Averaging and vibrational control of mechanical systems, *SIAM J. Control Optim.* **41**, 542-562.

Bullo, F., N. E. Leonard, and A. D. Lewis [2000], Controllability and motion algorithms for underactuated Lagrangian systems on Lie groups, *IEEE Trans. Automat. Control* **45**, 1437-1454. Mechanics and nonlinear control systems.

Bullo, F. and A. D. Lewis [2005], *Geometric control of mechanical systems*, volume 49 of *Texts in Applied Mathematics*. Springer-Verlag, New York. Modeling, analysis, and design for simple mechanical control systems.

Bullo, F., A. D. Lewis, and K. M. Lynch [2002] Controllable kinematic reductions for mechanical systems: concepts, computational tools, and examples *Proceedings of MTNS02*.

Bullo, F. and K. Lynch [2001], Kinematic controllability for decoupled trajectory planning in underactuated mechanical systems, *IEEE Trans. Robotics and Automation* **17**; no. 4, 401-412.

Bullo, F. and R. M. Murray [1999], Tracking for fully actuated mechanical systems: A geometric framework, *Automatica* **35**, 17-34.

Bullo, F. and M. Zefran [2002], On mechanical control systems with nonholonomic constraints and symmetries, *Systems and Control Letters* **45**, 133-142.

Burdick, J., B. Goodwine, and J. P. Ostrowski [1994], The Rattleback Revisited, *Preprint*.

Burke, W. L. [1985], *Applied Differential Geometry*, Cambridge University Press.

Byrnes, C. I. and A. Isidori [1988] Local stabilization of minimum-phase nonlinear systems *Systems and Control Letters* **11**, 9-17.

Camarinha, M., F. S. Leite, and P. Crouch [2001], On the geometry of Riemannian cubic polynomials, *Differential Geom. Appl.* **15**, 107-135.

Campion, G., B. D'andréa-Novel, and G. Bastin [1991], Controllability and state feedback stabilizability of nonholonomic mechanical systems. In *Advanced robot control (Grenoble, 1990)*, pages 106-124. Springer, Berlin.

Cantrijn, F. J. M. Cortés, M. de León, and M. de Diego [2002] On the geometry of generlized Chaplygin systems *Math. Proc. Cambridge Phil. Soc.* **132**, 323-351.

Cantrijn, F., M. de León, and M. de Diego [1999], On almost-Poisson structures in nonholonomic mechanics, *Nonlinearity* **12**, 721-737.

Cantrijn, F., M. de León, J. Marrero, and M. de Diego [2000], On almost-Poisson structures in nonholonomic mechanics. II. The time-dependent framework, *Nonlinearity* **13**, 1379-1409.

Cantrijn, F., M. de León, J. Marrero, and M. de Diego [1998], Reduction of nonholonomic mechanical systems with symmetries, *Rep. Math. Phys.* **42**, 25-45.

Canudas De Wit, C. and O. J. Sørdalen [1992], Exponential stabilization of mobile robots with nonholonomic constraints, *MIC—Modeling Identification Control* **13**, 3-14.

Carathéodory, C. [1933] Der Schlitten, *Z. Angew. Math. und Mech.* **13**, 71-76.

Carathéodory, C. [1967] *Calculus of Variations and Partial Differential Equations of the First Order*, Holden Day, San Francisco, CA.

Cardin, F. and M. Favretti [1996], On nonholonomic and vakonomic dynamics of mechanical systems with nonintegrable constraints, *J. Geom. and Phys.* **18**, 295–325.

Carr, J. [1981], *Applications of Centre Manifold Theory*, Springer-Verlag.

Cartan, E. [1923], Sur les variétés à connexion affines et la théorie de la relativité géneralisée, *Ann. Ecole Normale, Sup.* **40**, 325–412; See also, **41**, 1–25.

Cartan, E. [1952], Sur la représentation géométrique des systèmes matériels non holonomes, Gauthier-Villars, Paris, Collected Works, Oeuvres Complètes.

Cannas Da Silva, A. and A. Weinstein [1999], *Geometric Models for Noncommutative Alebras*, volume 10 of *Berkeley Mathematics Lecture Notes.* Amer. Math. Soc.

Cendra, H., D. D. Holm, J. E. Marsden, and T. S. Ratiu [1998], Lagrangian Reduction, the Euler–Poincaré Equations and Semidirect Products, *Amer. Math. Soc. Transl.*, **186**, 1–25.

Cendra, H., J. E. Marsden, and T. S. Ratiu [2001a], *Lagrangian reduction by stages*, Memoirs of the Amer. Math. Soc. **152**, Providence, R.I.

Cendra, H., J. E. Marsden, and T. S. Ratiu [2001b], Geometric mechanics, Lagrangian reduction and nonholonomic systems, in *Mathematics Unlimited-2001 and Beyond* (B. Enquist and W. Schmid, eds.), Springer-Verlag, New York., 221–273.

Chang, D., A. M. Bloch, N. Leonard, J. E. Marsden, and C. Woolsey [2002], The equivalence of controlled Lagrangian and controlled Hamiltonian systems, *Control and the Calculus of Variations (special issue dedicated to J.L. Lions)*, **8**, 393–422.

Chaplygin, S. A. [1897a], On the motion of a heavy body of revolution on a horizontal plane (Russian), *Physics Section of the Imperial Society of Friends of Physics, Anthropology and Ethnographics, Moscow*, **9**, 10–16; Reproduced in Chaplygin [1954], 413–425.

Chaplygin, S. A. [1897b], On some feasible generalization of the theorem of area, with an application to the problem of rolling spheres (Russian), *Mat. Sbornik* **XX**, 1–32; Reproduced in Chaplygin [1954], 434–454.

Chaplygin, S. A. [1903], On a Rolling Sphere on a Horizontal Plane, *Mat. Sbornik* **XXIV**, 139–168; (Russian) Reproduced in Chaplygin [1949], 72–99 and Chaplygin [1954], 455–471.

Chaplygin, S. A. [1911], On the Theory of the Motion of Nonholonomic Systems. Theorem on the Reducing Factor, *Mat. Sbornik* **XXVIII**, 303–314; (Russian). Reproduced in Chaplygin [1949], 28–38 and Chaplygin [1954], 426–433.

Chaplygin, S. A. [1949], *Analysis of the Dynamics of Nonholonomic Systems*, Moscow, Classical Natural Sciences; (Russian).

Chaplygin, S. A. [1954], *Selected Works on Mechanics and Mathematics*, State Publ. House, Technical-Theoretical Literature, Moscow; (Russian).

Chetaev, N. G. [1959], *The Stability of Motion*, Pergamon Press, New York.

Chow, S. N. and J. K. Hale [1982], *Methods of Bifurcation Theory*, Springer, New York.

Chern, S. J. and J. E. Marsden [1990], A Note on Symmetry and Stability for Fluid Flows, *Geo. Astro. Fluid. Dyn.* **51**, 1–4.

Chernoff, P. R. and J. E. Marsden [1974], *Properties of Infinite Dimensional Hamiltonian systems*, Springer-Verlag, New York, Lecture Notes in Math **124**.

Chorlton, F. [1983] *Textbook of Dynamics*, Ellis Horwood, London.

Chow, W. L. [1939] Über Systemen von linearen partiellen Differentialgleichungen erster Ordnung, *Math. Ann.* **117**, 98–105.

Coddington, E. A. and N. Levinson [1955], *Theory of Ordinary Differential Equations*, McGraw-Hill, New York.

Coron, J. M. [1990] A necessary condition for feedback stabilization, *Systems and Control Letters* **14**, 227–232.

Coron, J. M. [1992], Global asymptotic stabilization for controllable systems without drift, *Mathematics of Controls, Signals and Systems.* **5**, 295–312.

Cortés, J. M. [2002], *Geometric, control and numerical aspects of nonholonomic systems*, Ph.D. thesis, University Carlos III, Madrid (2001) and Springer Lecture Notes in Mathematics.

Cortés, J. M. and M. de León [1999], Reduction and reconstruction of the dynamics of nonholonomic systems, *J. Phys. A: Math. Gen.* **32**, 8615–8645.

Cortés, J. M., M. de León, M. de Diego, and S. Martinez [2001] Mechanical systems subject to generalized constraints, *R. Soc. London. Proc. Ser. A Math. Phys. Eng. Sci* **457**, 651–670.

Cortés, J., M. de León, D. Martín de Diego, and S. Martínez [2002], Geometric description of vakonomic and nonholonomic dynamics. Comparison of solutions, *SIAM J. Control Optim.*, **41**(5), 1389–1412.

Cortés. J. and S. Martinez [2000], Optimal control for nonholonomic systems with symmetry, *Proc. 39th IEEE. Conf. Dec. and Cont.*, IEEE, 5216–5218.

Cortés, J. and S. Martínez [2001], Non-holonomic integrators, *Nonlinearity* **14**, 1365–1392.

Cortés, J., S. Martínez, J. P. Ostrowski, and H. Zhang [2002], Simple mechanical control systems with constraints and symmetry, *SIAM J. Control Optim.*, **41**(3), 851–874.

Crabtree, H. [1909], *Spinning Tops and Gyroscopic Motion*, Chelsea.

Crampin, M. [1981], On the differential geometry of the Euler–Lagrange equations, and the inverse problem of Lagrangian dynamics, *J. Phys. A* **14**, 2567–2575.

Crouch, P E. [1977], *Dynamical Realizations of Finite Volterra Series*, Ph.D. thesis, Harvard University.

Crouch, P. E., [1981], Geometric structures in systems theory, *IEEE Proceedings* **128**, 242–252.

Crouch, P. E. [1981], Dynamical realizations of finite Volterra series, *SIAM J. Control Optim.* **19**, 177–202.

Crouch, P. E. [1984], Spacecraft attitude control and stabilization, *IEEE Trans. on Automatic Control*, **29**, 321–333.

Crouch, P. E. and F. Leite [1991], Geometry and the dynamic interpolation problem, *Proc. American Control Conference*, Boston, 1131–1136.

Crouch, P. E. and A. J. van der Schaft [1987], *Variational and Hamiltonian Control Systems*, Lecture Notes in Control and Information Sciences, Springer-Verlag, Berlin. **101**.

Crouch, P. E. and F. Silva-Leite [1991a], Geometry and the dynamic interpolation problem, *Proc. A.C.C.*, 1131–1136.

Cushman, R., D. Kemppainen, J. Śniatycki, and L. Bates [1995], Geometry of nonholonomic constraints, *Rep. Math. Phys.* **36**, 275–286.

Cushman, R., J. Hermans, and D. Kemppainen [1996], The rolling disc, in *Nonlinear Dynamical Systems and Chaos (Groningen, 1995)*, Birkhauser, Basel, Boston, MA, Progr. Nonlinear Differential Equations Appl. **19**, 21–60.

Cushman, R. and J. Śniatycki [2002], Nonholonomic reduction for free and proper actions, *Regul. Chaotic Dyn.* **7**, 61–72.

Das, T. and R. Mukherjee [2001], Dynamic analysis of rectilinear motion of a self-propelling disk with unbalanced masses, *Trans. of the ASME* **68**, 58–66.

Davis, M. W. [1987], Some aspherical manifolds, *Duke Math. J.* **5**, 105–139.

Decarlo R., S. Zak, and S. Drakunov [1996], Variable structure and sliding mode control, in *The Control Handbook*, CRC Press, Inc., The Electrical Engineering Handbook Series.

Deift, P., T. Nanda, and C. Tomei [1983], Differential equations for the symmetric eigenvalue problem, *SIAM J. on Numerical Analysis* **20**, 1–22.

De León, M. and M. de Diego [1996], On the geometry of nonholonomic Lagrangian systems, *J. Math. Phys.* **37**, 3389–3414

M. de León and P. Rodgrigues [1989] *Methods of Differential Geometry in Analytical Mechanics*, Mathematics Studies **158**, North Holland, Amsterdam.

Deprit, A. [1983], Elimination of the nodes in problems of N bodies, *Celestial Mech.* **30**, 181–195.

Drakunov, S. and V. I. Utkin [1992] Sliding mode control in dynamic systems, *Int. J. Control* **55**, 1029–1037.

Dubrovin, B. A., A. T. Fomenko and S. P. Novikov [1984], Modern Geometry—Methods and Applications. Part II.The Geometry and Topology of Manifolds, Springer-Verlag, Graduate Texts in Math. **104**.

Ebin, D. G. and J. E. Marsden [1970], Groups of diffeomorphisms and the motion of an incompressible fluid, *Ann. Math.*, **92**, 102–163.

Ehlers, K., J. Koiller, R. Montgomery, and P. M. Rios [2005], Nonholonomic systems via moving frames: Cartan equivalence and Chaplygin Hamiltonization, in *The breadth of symplectic and Poisson geometry*, volume 232 of *Progr. Math.*, 75–120, Birkhäuser Boston, Boston, MA.

Enos, M. J. [1993], *Dynamics and Control of Mechanical Systems*, American Mathematical Society, Providence, RI.; The falling cat and related problems, Papers from the Fields Institute Workshop; Waterloo, Ontario, March 1992.

Erlich, R. and J. Tuszynski [1995], Ball on a rotating turntable: comparison of theory and experiment, *Amer. J. Physics* **63**, 351–359.

Euler, L. [1744], Methodus inveniendi lineas curvas, *Lausannae, Genevae, apud Marcum–Michaelem Bosquet & Socias.*

Favretti, M. [1998], Equivalence of dynamics for nonholonomic systems with transverse constraints, *J. Dynam. Differential Equations* **10**, 511–536 and **29**, 1134–1142.

Faybusovich, L. E. [1988] Explicitly solvable optimal control problems. *Int. J. Control* **48**, no. 6, 235–250.

Fedorov, Y. N. [1985], Integration of a generalized problem on the rolling of a Chaplygin ball., *Geometry, differential equations and mechanics, Moscow*, 151–155.

Fedorov, Y. N. [1989], Two integrable nonholonomic systems in classical dynamics, *Vestnik Moskov. Univ. Ser. I Mat. Mekh.* **105**, 38–41.

Fedorov, Y. N. [1994], Generalized Poinsot interpretation of the motion of a multidimensional rigid body, *Trudy Mat. Inst. Steklov* **205**, 200–206.

Fedorov, Y. N. [1999], Systems with an invariant measure on Lie groups. In Simo, C., editor, *Hamiltonian Systems with Three or More Degrees of Freedom*, pages 350–357. Kluwer, NATO ASI Series C, Vol. 533.

Fedorov, Y. N. and B. Jovanović [2004], Nonholonomic LR systems as generalized Chaplygin systems with an invariant measure and flows on homogeneous spaces, *J. Nonlinear Sci.* **14**, 341–381.

Fedorov, Y. N. and V. V. Kozlov [1994], Integrable systems on a sphere with potentials of elastic interaction, *Math. Notes* **54**, 381–386.

Fedorov, Y. N. and V. V. Kozlov [1995], Various aspects of n-dimensional rigid body dynamics, *Amer. Math. Soc. Transl.* **168**, 141–171.

Fedorov, Y. N. and D. V. Zenkov [2005], Discrete nonholonomic LL systems on Lie groups, *Nonlinearity*, **18**(5), 2211–2241.

Ferrers [1871], N.M., *Quart. J. of Math.*, **XII**.

Feynman, R. P. [1989], *The Feynman Lectures on Physics*, Addison Wesley, New York.

Filippov A. F., [1988] *Differential equations with discontinuous right–hand sides*, Kluwer Academic Publishers, Boston.

Fischer, A. E., J. E. Marsden, and V. Moncrief [1980], The structure of the space of solutions of Einstein's equations, I: One Killing field, *Ann. Ins. H. Poincaré* **33**, 147–194.

Flaschka, H. [1974], The Toda Lattice, *Phys. Rev. B*, **9**, 1924–1925.

Fufaev, N.A. [1964] On the possibility of realizing a nonholonomic constraint by means of viscous friction forces, *Prikl. Math. Mech* **28**, 513–515.

Gabriel, S. and J. Kajiya [1988], Spline interpolation in curved space, *SIGGRAGH 85*, Course Notes.

Gantmacher, F. R. [1959], *Theory of Matrices*. 2 volumes. Chelsea.

Gauss, C. F. [1829] Über ein neues allgemeines Grundgesatz der Mechanik, *Journal für die Reine und Angewandte Mathematik* **4**, 232–235.

Gelfand, I. M. and S. V. Fomin [1963], *Calculus of Variations*. Prentice-Hall (reprinted by Dover, 2000).

Getz, N. H. and J. E. Marsden [1994], Symmetry and dynamics of the rolling disk, *CPAM Berkeley paper* **630**.

Getz, N. H. and J. E. Marsden [1995], Control for an autonomous bicycle. In *International Conference on Robotics and Automation, IEEE*, Nagoya, Japan.

Getz, N. and J. E. Marsden [1997], Dynamical methods for polar decomposition and inversion of matrices, *Linear Algebra and Its Appl.* **258**, 311–343.

Georgiou, T. T., L. Praly, E. D. Sontag, and A. Teel [1995], Input–output stability, in *The Control Handbook*, CRC Press, Baca Raton.

Giachetta, G. [1992], Jet methods in nonholonomic mechanics, *J. Math. Phys.* **33**, 1652–1665.

Gibbs, J. W. [1879] On the fundamental formulae of dynamics, *Am. J. of Math.* **II**, 49–64.

Godhavn, J.-M. and O. Egeland [1997], A Lyapunov approach to exponential stabilization of nonholonomic systems in power form, *IEEE Trans. Automat. Control* **42**, 1028–1032.

Goldman, W. M. and J. J. Millson [1990], Differential graded Lie algebras and singularities of level sets of momentum mappings, *Comm. Math. Phys.* **131**, 495–515.

Goldstein, H. [1980], *Classical Mechanics*, First Edition 1950, Second Edition 1980, Addison-Wesley.

Gozzi, E. and W. D. Thacker [1987], Classical adiabatic holonomy in a Grassmannian system, *Phys. Rev. D* **35**, 2388–2396.

Greenwood, D. T. [1977], *Classical Dynamics*, Prentice Hall.

Greenwood, D. [2003], *Advanced Dynamics* Cambridge University Press, 2003.

Griffiths, P. A. [1983], *Exterior Differential Systems*, Birkhäuser, Boston.

Grossman, R., P. S. Krishnaprasad, and J. E. Marsden [1988], The dynamics of two coupled rigid bodies, in *Dynamical Systems Approaches to Nonlinear Problems in Systems and Circuits* (Salam and Levi, eds.), SIAM., 373–378.

452   References

Guckenheimer J. and P. Holmes [1983], *Nonlinear Oscillations, Dynamical Systems and Vector Fields*, Springer-Verlag.

Guichardet, A. [1984], On rotation and vibration motions of molecules, *Ann. Inst. H. Poincaré* **40**, 329–342.

Guillemin, V. and A. Pollack [1974], *Differential Topology*, Prentice-Hall.

Guillemin, V. and S. Sternberg [1978], On the equations of motions of a classic particle in a Yang–Mills field and the principle of general covariance, *Hadronic J.* **1**, 1–32.

Guillemin, V. and S. Sternberg [1984], *Symplectic Techniques in Physics*, Cambridge University Press.

Gurvitz, L. [1992], Averaging approach to nonholonomic motion planning, in *Proc. IEEE Int. Conf. Robotics and Automation*, IEEE Press, Nice, France, 2541–2546.

Hamberg, J. [1999], General Matching Conditions in the Theory of Controlled Lagrangians, in *Proc. of the 38th IEEE Conference on Dec. and Control*, 2519–2523.

Hamel, G. [1904], Die Lagrange–Eulerschen Gleichungen der Mechanik, *Z. für Mathematik u. Physik* **50**, 1–57.

Hamel, G. [1949], *Theoretische Mechanik*, Springer-Verlag.

Hamilton, W. R. [1834], On a general method in dynamics, Part I, *Phil. Trans. Roy. Soc. Lond.*, 247–308.

Hamilton, W. R. [1835], On a general method in dynamics. Part II, *Phil. Trans. Roy. Soc. Lond.*, 95–144.

Hannay, J. [1985], Angle variable holonomy in adiabatic excursion of an integrable Hamiltonian, *J. Phys. A: Math. Gen.*, **18**, 221–230.

Hartman, P. [1982], *Ordinary Differential Equations*, Birkhäuser. (First Edition published by Wiley (1964)).

Helgason, S. [2001] *Differential geometry, Lie groups, and symmetric spaces*, American Mathematical Society, Providence, RI.

Helmke, U. and J. Moore [1994], *Optimization and Dynamical Systems*, Springer-Verlag, New York.

Hermann, R. [1962], Some differential geometric aspects of the Lagrange variational problem, *Indiana Math. J*, 634–673.

Hermann, R. [1963] On the accesibility problem in control theory, in *Nonlinear Differential Equations and Nonlinear Mechanics*, J. P. LaSalle and S. Lefshetz, eds., Academy Press, New York.

Hermann, R. [1977], *Differential Geometry and the Calculus of Variations*, Second Edition, Math. Sci. Press, Brookline, Mass., USA, Interdisciplinary Mathematics.

Hermann, R. and A. J. Krener [1977] Nonlinear controllability and observability *IEEE Trans. Aut. Control* **22**, 728–740.

Hermans, J. [1995], A symmetric sphere rolling on a surface, *Nonlinearity*, **8**, 1–23.

Hermes, H. [1980] On the Synthesis of a Stabilizing Feedback Control via Lie Algebraic Methods, *SIAM Journal on Control and Optimization* **18**, 352–361.

Hespanha, J. P. [1996] Stabilization of the non-holonomic Integrator via logic-based switchings, *Proc. 13th World Congress on Automatic Control, E: Nonlinear Systems*, 467–472.

Hertz, H.R. [1894] *Gessamelte Werke*, Band III. *Der Prinzipien der Mechanik in neuem Zusammenhange dargestellt*, Barth, Leipzig, 1894, English translation MacMillan, London, 1899, reprint Dover, NY, 1956.

Holm, D. D., J. E. Marsden, and T. S. Ratiu [1998], The Euler–Poincaré equations and semidirect products with applications to continuum theories, *Adv. in Math.* **137**, 1–81.

Holm, D. D., J. E. Marsden, T. S. Ratiu, and A. Weinstein [1985], Nonlinear stability of fluid and plasma equilibria, *Phys. Rep.* **123**, 1–196.

Holmes, P., R. J. Full, D. Koditschek and J. Guckenheimer [2005], The dynamics of legged locomotion: models, analyses and challenges, preprint.

Hubbard , M. and T. Smith [1999], Dynamics of golf ball-hole interactions: rolling around the rim, *Transactions of the ASME*, **121**, 88-95.

Husemoller, D. [1994], *Fibre Bundles,* Springer-Verlag, New York, Graduate Texts in Mathematics, vol. **20**.

Iserles, A. [2002], Expansions that grow on trees, *Notices of the Am. Math. Soc.* **49**, 430–444.

Iserles, A., S. P. Nørsett, and A. F. Rasmussen [2001], Time symmetry and high-order Magnus methods, *Appl. Numer. Math.* **39**, 379–401. Special issue: Themes in geometric integration.

Isidori, A. [1995], *Nonlinear Control Systems*, Springer-Verlag, Berlin, Communications and Control Engineering Series.

Iwai, T. [1987], A gauge theory for the quantum planar three-body system, *J. Math. Phys.* **28**, 1315–1326.

Jalnapurkar, S. M. [1994], Modeling of Constrained Systems, available from http://www.cds.caltech.edu/~smj. (A chapter in the Ph.D. thesis, UC Berkeley, May, 1999).

Jalnapurkar, S. M. and J. E. Marsden [1999], Stabilization of relative equilibria II, *Reg. and Chaotic Dyn.*, **3**, 161–179.

Jalnapurkar, S. M. and J. E. Marsden [2000], Stabilization of Relative Equilibria, *IEEE Trans. Automat. Control* **45**, 1483–1491.

Jiang, Z.-P. and H. Nijmeijer [1999], A recursive technique for tracking control of non-holonomic systems in chained form, *IEEE Trans. Automat. Control* **44**, 265–279.

Jovanović, B. [1998] Nonholonomic Geodesic Flows on Lie Groups and the Integrable Suslov Problem on $SO(4)$. *J. Phys. A: Math. Gen.* **31**, 1451–1422.

Jovanović, B. [2001], Geometry and integrability of Euler–Poincaré–Suslov equations, *Nonlinearity* **14**, 1555–1567.

Jovanović, B. [2003], Some multidimensional integrable cases of nonholonomic rigid body dynamics *Regular and Chaotic Dynamics* **8**, 125–132.

Jurdjevic, V. [1991], Optimal control problems on Lie groups, in *Analysis of Controlled Dynamical Systems*, B. Bonnard, B. Bride, J. P. Gauthier, and I. Kupka, eds., Birkhäuser, Boston.

Jurdjevic, V. [1993], The Geometry of the Plate–Ball Problem, *Arch. Rat. Mech. An.* **124**, 305–328.

Jurdjevic, V. [1997], *Geometric Control Theory*, Cambridge University Press, Studies in Advanced Mathematics. **52**.

Kalaba, R. E. and F. E. Udwadia [1994], Lagrangian mechanics, Gauss' principle, quadratic programming, and generalized inverses: new equations for non-holonomically constrained discrete mechanical systems, *Quart. Appl. Math.* **52**, 229–241.

Kane, T. R. and D. A. Levinson [1980], Formulation of the equations of motions for complex spacecraft, *J. Guidance and Control* **3**, 99–112.

Kane, T. and M. Scher [1969], A dynamical explanation of the falling cat phenomenon, *Int. J. Solids Structures* **5**, 663–670.

Kapitsa, P. L. [1951] Dynamic stability of a pendulum with a vibrating point of suspension, *Ehksp. Teor. Fiz.* **21**, 588–598.

Karapetyan, A. V. [1980], On the problem of steady motions of nonholonomic systems, *J. Appl. Math. Mech.* **44**, 418–426.

Karapetyan, A. V. [1981], On stability of steady state motions of a heavy solid body on an absolutely rough horizontal plane, *J. Appl. Math. Mech.* **45**, 604–608.

Karapetyan, A. V. [1983], Stability of steady motions of systems of a certain type, *Mechanics of Solids* **18**, 41–47.

Karapetyan, A. V. and V. V. Rumyantsev [1990], Stability of conservative and dissipative systems. In Mikhailov, G. K. and V. Z. Parton, editors, *Applied Mechanics: Soviet Reviews*, volume 1. Hemisphere, NY.

Kazhdan, D., B. Kostant, and S. Sternberg [1978], Hamiltonian group actions and dynamical systems of Calogero type, *Comm. Pure Appl. Math.* **31**, 481–508.

Kelly, S. D. and R. M. Murray [1995], Geometric Phases and Robotic Locomotion, *Journal of Robotic Systems* **12**, 417–431.

Khalil, H. K. [1992], *Nonlinear Systems*, Macmillan Publishing Company, New York.

Kharlamov, P. V. [1992] A critique of some mathematical models of mechanical systems with differential constraints, *J. Appl. Math. Mech.* **54** (4), 584–594.

Khennouf, H. and C. Canudas de Wit [1995] *On the construction of stabilizing discontinuous controllers for nonholonomic systems*, Proceedings of NOLCOS 95, 741–746.

Kobayashi, S. and K. Nomizu [1963], *Foundations of Differential Geometry*, Wiley.

Koiller, J. [1992], Reduction of Some Classical Nonholonomic Systems with Symmetry, *Arch. Rat. Mech. An.* **118**, 113–148.

Koiller, J., P. R. Rodrigues, and P. Pitanga [2001], Non-holonomic connections following Élie Cartan, *An. Acad. Brasil. Ciênc.* **73**, 165–190.

Kolmanovsky, I., M. Reyhanoglu, and N. H. McClamroch [1994], Discontinuous feedback stabilization of nonholonomic systems in extended power form, in *Proceedings of the Conf. Dec. and Control*, 3469–3474.

Komarov, A. È. [1990], On the control of nonholonomic systems. In *Problems in the mechanics of controllable motion (Russian)*, 59–64. Perm. Gos. Univ.

Koon, W. S. and J. E. Marsden [1997a], Optimal Control for Holonomic and Nonholonomic Mechanical Systems with Symmetry and Lagrangian Reduction, *SIAM J. Control and Optim.* **35**, 901–929.

Koon, W. S. and J. E. Marsden [1997b], The Hamiltonian and Lagrangian approaches to the dynamics of nonholonomic systems, *Reports on Math Phys.* **40**, 21–62.

Koon, W. S. and J. E. Marsden [1998], The Poisson reduction of nonholonomic mechanical systems, *Reports on Math. Phys.* **42**, 101–134.

Korteweg, D. [1899], Über eine ziemlich verbreitete unrichtige Behandlungsweise eines Problemes der rollenden Bewegung und insbesondere über kleine rollende Schwingungen um eine Gleichgewichtslage, *Nieuw Archiefvoor Wiskunde.* **4**, 130–155.

Kozlov, V. V. [1982a] The dynamics of systems with nonintegrable constraints. I. *Vestnik Moskov. Univ. Ser. I Mat. Mekh.* **3**, 92–100.

Kozlov, V. V. [1982b] The dynamics of systems with nonintegrable constraints. II. *Vestnik Moskov. Univ. Ser. I Mat. Mekh.* **4**, 70–76.

Kozlov, V. V. [1982c] The dynamics of systems with nonintegrable constraints. III. *Vestnik Moskov. Univ. Ser. I Mat. Mekh.* **3**, 102–111.

Kozlov, V. V. [1983], Realization of nonintegrable constraints in classical mechanics, *Sov. Phys. Dokl.* **28**, 735–737.

Kozlov, V. V. [1987] The dynamics of systems with nonintegrable constraints. IV. Integral principles. *Vestnik Moskov. Univ. Ser. I Mat. Mekh.* **5**, 76–83.

Kozlov, V. V. [1988a] The dynamics of systems with nonintegrable constraints. V. The freedom principle and a condition for ideal constraints. *Vestnik Moskov. Univ. Ser. I Mat. Mekh.* **6**, 51–54.

Kozlov, V. V. [1988b] Invariant measures of the Euler–Poincaré equations on Lie algebras. *Functional Anal. Appl.* **22**, 69–70.

Kozlov, V. V. [1992] The problem of realizing constraints in dynamics, *J. Appl. Math. Mech.* **56** (4), 594–600.

Kozlov, V. V. and N. N. Kolesnikov [1978], On Theorems of Dynamics, *J. Appl. Math. Mech.* **42**, 28–33.

Krasinskiy, A.Ya. [1988a] On the stability and stabilization of the points of equilibrium of nonholonomic system *J.Appl.Maths.Mechs* **52**, 194-202.

Krasinskiy, A.Ya. [1988b] On the stabilization of the steady motions of of nonholonomic mechanical systems *J.Appl.Maths.Mechs* **52**, 902-908.

456   References

Krasovskii, N. [1963], *Stability of Motion.* Stanford University Press (originally published 1959).

Krener, A. J. [1974], A generalization of Chow's theorem and the bang-bang theorem to nonlinear control systems *SIAM J. Control* **12**, 43–52.

Krener, A. J. [1977], The high order maximal principle and its application to singular extremals, *SIAM J. Control Optimization* **15**, 256–293.

Krener, A. J. [1999], Feedback linearization. In *Mathematical control theory*, pages 66–98. Springer, New York.

Krishnaprasad, P. S. [1985], Lie–Poisson structures, dual-spin spacecraft and asymptotic stability, *Nonl. Anal. Th. Meth. and Appl.* **9**, 1011–1035.

Krishnaprasad, P. S. [1989], Eulerian many-body problems, *Cont. Math. AMS* **97**, 187–208.

Krishnaprasad, P. S. and J. E. Marsden [1987], Hamiltonian structure and stability for rigid bodies with flexible attachments, *Arch. Rational Mech. Anal.* **98**, 137–158.

Krishnaprasad, P. and D. P. Tsakiris [2001], Oscillations, SE(2)-snakes and motion control: a study of the Roller Racer, *Dyn. Sys., An International J.* **16**, 347–397.

Krishnaprasad, P. S. and R. Yang [1991] Geometric Phases, Anholonomy and Optimal Movement *Proceedings of Conference on Robotics and Automation, 1991*, Sacramento, California, 2185–2189.

Krishnaprasad, P. S., R. Yang and W. Dayawansa [1991], Control problems on principal bundles and nonholonomic mechanics *Proc. 30th IEEE Conf. Decision and Control,* 1133–1138.

Kummer, M. [1981], On the construction of the reduced phase space of a Hamiltonian system with symmetry, *Indiana Univ. Math. J.* **30**, 281–291.

Lagrange, J. L. [1760], Différents Problémes de Dynamique, *Miscell. Taurin,* **II**; Oeuvres, I, Paris.

Lagrange, J. L. [1788], *Méchanique Analytique,* Chez la Veuve Desaint.

Lafferiere, G. and H. J. Sussman [1991], Motion Planning for Completely Nonholonomic Systems Without Drift, in *Proc. of the Conf. on Robotics and Automation 1991, Sacramento,* 1148–1153.

Laiou, M.-C. and A. Astolfi [1999], Discontinuous control of high-order generalized chained systems, *Systems Control Lett.* **37**, 309–322.

Lasalle, J. and S. Lefschetz [1961], *Stability by Liapunov's Direct Method, with Applications.* Academic Press, NY.

Lee, E. B. and L. Markus [1976] *Foundations of Optimal Control Theory,* J. Wiley and Sons, New York.

Leonard, N. E. and P. S. Krishnaprasad [1993], Averaging for attitude control and motion planning, in *Proceedings of the 32nd IEEE Conference on Decision and Control, December, 1993,* 3098–3104.

Leonard, N. E. and P. S. Krishnaprasad [1995], Motion control of drift-free, left-invariant systems on Lie groups *IEEE Transactions on Automatic Control* **40**, no. 9, 1539–1554.

Leonard, N. E. and J. E. Marsden [1997], Stability and drift of underwater vehicle dynamics: mechanical systems with rigid motion symmetry, *Physica D*, **105**, 130–162.

Levi, M. [1993], Geometric phases in the motion of rigid bodies, *Arch. Rat. Mech. Anal.*, **122**, 213-229.

Levi, M. and W. Weckesser [2002], Non-holonomic systems as singular limits for rapid oscillations, *Ergodic Theory and Dynamical Systems*, **22**, 1497-1506.

Lewis, A. [1995] Controllability of Lagrangian and Hamiltonian Systems, Ph.D. thesis, California Institute of Technology.

Lewis, A. [1996], The geometry of the Gibbs–Appell equations and Gauss' principle of least constraint, *Reports on Math. Phys.* **38**, 11–28.

Lewis, A. [1998], Affine connections and distributions with applications to nonholonomic systems, *Rep. Math. Phys* **42**, 135–164.

Lewis, A. [2000], Simple mechanical control systems with constraints, *IEEE Trans. Aut. Control* **45**, 1420–1436.

Lewis, D. [1995], Linearized dynamics of symmetric Lagrangian systems. In *Hamiltonian dynamical systems*, volume 63 of *IMA Math. Appl.*, pages 195–216. Springer, New York.

Lewis, A. and R. M. Murray [1995], Variational principles in constrained systems: theory and experiments, *Int. J. Nonlinear Mech.* **30**, 793–815.

Lewis, A. D. and R. M. Murray [1997], Controllability of simple mechanical control systems, *SIAM J. on Control and Optimization* **35**, 766–790.

Lewis, A. D. and R. M. Murray [1999], Configuration controllability of simple mechanical control systems, *SIAM Rev.* **41**, 555–574.

Lewis, A., J. P. Ostrowski, R. M. Murray, and J. Burdick [1994], Nonholonomic Mechanics and Locomotion: the Snakeboard Example, in *IEEE Intern. Conf. on Robotics and Automation.*

Lewis, D. and J. C. Simo [1990], Nonlinear stability of rotating pseudo-rigid bodies, *Proc. Roy. Soc. Lon. A* **427**, 281–319.

Li, Z. and J. Canny [1990], Motion of two rigid bodies with rolling constraint, *IEEE Transactions on Robotics and Automation* **6**, 62–71.

Li, Z. and J. Canny [1993], *Nonholonomic Motion Planning*, Kluwer, New York.

Li, Z. and R. Montgomery [1988], Dynamics and Optimal Control of a Legged Robot in Flight Phase, *Proceedings of Conference on Robotics and Automation*, Cincinnati, Ohio, 1816–1821.

Lie, S. [1890], *Theorie der Transformationsgruppen, Zweiter Abschnitt*, Teubner, Leipzig.

Libermann, P. and C. M. Marle [1987], *Symplectic Geometry and Analytical Mechanics.* Kluwer Academic Publishers.

Liu, W. [1992], *Averaging Theorems for Highly Oscillatory Differential Equations and the Approximation of General Paths by Admissable Trajectories for Nonholonomic Systems*, Ph.D. thesis, Rutgers University, NJ.

Liu, W. and H. J. Sussman [1995], Shortest paths for sub-Riemannian metrics on rank-two distributions, *Mem. Amer. Math. Soc.* **118**.

Lobry, C. [1972], *Quelques aspects qualitatifs de la théories de la commande*, Docteur es Sciences Mathématiques, L'Universite Scientifique et Médicale de Grenoble.

Lu, J.-H. and T. S. Ratiu [1991], On the nonlinear convexity theorem of Kostant, *Journ. Amer. Math. Soc.* **4**, 349–364.

Lum, K.-Y., and A. M. Bloch [1999], Generalized Serret–Andoyer transformation and applications for the controlled rigid body. *Dynam. Control* **9**, 39–66.

Lurie, A. I. [2002], *Analytical Mechanics* (translation of 1961 Russian edition), Springer Verlag.

Lyapunov, A. M. [1907], Problème général de la stabilité du mouvement, *Ann. Fac. Sci. Toulouse* **9**, 203–474.

Lyapunov [1992], The general problem of the stability of motion, *Int. J. Control* **55**, 531–773; translated into English by A. T. Fuller. Also published as a book by Taylor & Francis, London.

Maccullagh, J. [1840], On the rotation of a solid body, *Proc. Roy. Irish Acad. Dublin* **2**, 520–525, (see addenda, pages 542–545, and **3**, 370–371).

Magnus, W. [1954], On the exponential solution of differential equations for a linear operator, *Comm. Pure App. Math.* **7**, 649–673.

Malkin, I. G. [1938], On the Stability of Motion in a Sense of Lyapunov, *Mat. Sbornik*, **XXXV**, 47–101 (Russian).

Manakov, S. V. [1976], Note on the integration of Euler's equations of the dynamics of an n-dimensional rigid body, *Funct. Anal. and its Appl.* **10**, 328–329.

Markeev, A. P. [1983], On Dynamics of a Solid on an Absolutely Rough Plane, *J. Appl. Math. Mech.* **47**, 473–478.

Markeev, A. P. [1992], *The Dynamics of a body contiguous to a solid surface*, Nauka, Moscow (Russian).

Marle, C.-M. [1976], Symplectic manifolds, dynamical groups and Hamiltonian mechanics, in *Differential Geometry and Relativity* (M. Cahen and M. Flato, eds.), D. Reidel, Boston, 249–269.

Marle, C.-M. [1996], Kinematic and geometric constraints, servomechanism and control of mechanical systems, *Rend. Sem. Mat. Univ. Pol. Torino*, **54**, 353–364.

Marle, C.-M. [1998] Various approaches to conservative and nonconservative nonholonomic systems, *Reports on Mathematical Physics* **42**, 211–229.

Marsden, J. E. [1981], *Lectures on Geometric Methods in Mathematical Physics*, CBMS Series, SIAM, Philadelphia, PA.

Marsden, J. E. [1992], *Lectures on Mechanics*, Cambridge University Press, London Mathematical Society Lecture Note Series. **174**.

Marsden, J. E. and M. J. Hoffman [1993], *Elementary Classical Analysis*. W. H. Freeman and Co., NY, Second Edition.

Marsden, J. E. and T. J. R. Hughes [1994], *Foundations of Elasticity*, Prentice Hall, 1983, reprinted by Dover, 1994.

Marsden, J. E., P. S. Krishnaprasad, and J. C. Simo (editors) [1989], *Dynamics and Control of Multibody Systems*, Contemporary Mathematics, Am. Math. Soc. **97**.

Marsden, J. E., R. Montgomery, and T. S. Ratiu [1990], *Reduction, symmetry and phases in mechanics*, Memoirs of the Amer. Math. Soc., Providence, RI, vol **436**.

Marsden, J. E. and J. Ostrowski [1998], Symmetries in Motion: Geometric Foundations of Motion Control, *Nonlinear Sci. Today* (http://link.springer-ny.com).

Marsden, J. E., G. W. Patrick, and S. Shkoller [1998], Multisymplectic Geometry, Variational Integrators, and Nonlinear PDEs, *Commun. Math. Phys.* **1998**, 351–395.

Marsden, J. E., S. Pekarsky, and S. Shkoller [1999], Discrete Euler–Poincaré and Lie–Poisson equations, *Nonlinearity* **12**, 1647–1662.

Marsden, J. E. and T. S. Ratiu [1999], *Introduction to Mechanics and Symmetry*, Springer-Verlag, Texts in Applied Mathematics, **17**; First Edition 1994, Second Edition, 1999.

Marsden, J. E. and J. Scheurle [1993a], Lagrangian Reduction and the Double Spherical Pendulum, *ZAMP* **44**, 17–43.

Marsden, J. E. and J. Scheurle [1993b], The Reduced Euler–Lagrange Equations, *Fields Institute Comm.*, **1**, 139–164.

Marsden, J. E. and A. Weinstein [1974], Reduction of symplectic manifolds with symmetry, *Rep. Math. Phys.* **5**, 121–130.

Marsden, J. E. and M. West [2001], Discrete mechanics and variational integrators, *Acta Numerica* **10**, 357–514.

Martínez, S. and J. Cortés [2003], Motion control algorithms for simple mechanical systems with symmetry, *Acta Appl. Math.*, **76**(3), 221–264.

Martínez, S., J. Cortés, and M. De León [2000], The geometrical theory of constraints applied to the dynamics of vakonomic mechanical systems: the vakonomic bracket, *J. Math. Phys.* **41**, no. 4, 2090–2120.

Martínez, S., J. Cortés, and M. D. León [2001], Symmetries in vakonomic dynamics: applications to optimal control, *J. Geom. Phys.* **38**, 343–365.

Maupertuis [1740], Lois du Repos *Mem. de L'Acad. Royale des Sciences de Paris*.

M'Closkey, R. and R. Murray [1993] Convergence rates for nonholonomic systems in power form, in *Proceedings of the American Control Conference*, 2967–2972.

McMillan, W. D. [1936], *Dynamics of Rigid Bodies*, Duncan MacMillan Rowles, UK.

Merkin, D. R. [1997], *Introduction to the theory of stability*. Springer-Verlag, New York. Translated from the third (1987) Russian Edition, edited and with an introduction by Fred F. Afagh and Andrei L. Smirnov.

Meyer, K. R. [1973], Symmetries and integrals in mechanics, in *Dynamical Systems* (M. Peixoto, ed.), Academic Press, 259–273.

Mikhailov, G. K. and V. Z. Parton [1990], *Stability and Analytical Mechanics*, volume 1 of *Applied Mechanics, Soviet Reviews*, Hemisphere Publishing Co.

Milnor, J. [1963], *Morse Theory*, Princeton University Press.

Milnor, J. [1965], *Topology from the Differential Viewpoint*, University of Virginia Press.

Mishchenko, A. S. and A. T. Fomenko [1976], On the integration of the Euler equations on semisimple Lie algebras, *Sov. Math. Dokl.* **17**, 1591–1593.

Mishchenko, A. S. and A. T. Fomenko [1978], Generalized Liouville method of integration of Hamiltonian systems, *Funct. Anal. Appl.* **12**, 113–121.

Misner, C., K. Thorne, and J. Wheeler [1973], *Gravitation*, W. H. Freeman, San Fransisco.

Montaldi, J. A., R. M. Roberts, and I. N. Stewart [1988], Periodic solutions near equilibria of symmetric Hamiltonian systems, *Phil. Trans. R. Soc. Lond. A* **325**, 237–293.

Montgomery, R. [1984], Canonical formulations of a particle in a Yang–Mills field, *Lett. Math. Phys.* **8**, 59–67.

Montgomery, R. [1988], The connection whose holonomy is the classical adiabatic angles of Hannay and Berry and its generalization to the non-integrable case, *Comm. Math. Phys.* **120**, 269–294.

Montgomery, R. [1990], Isoholonomic problems and some applications, *Commun. Math. Phys*, **128**, 565–592.

Montgomery, R. [1991a], Optimal control of deformable bodies and its relation to gauge theory, in *The Geometry of Hamiltonian Systems* (T. Ratiu, ed.), Springer-Verlag, 403–438.

Montgomery, R. [1991b], How much does a rigid body rotate? A Berry's phase from the 18th century, *Amer. J. Phys.* **59**, 394–398.

Montgomery, R. [1993], Gauge theory of the falling cat, *Fields Inst. Commun.* **1**, 193–218.

Montgomery, R. [1994], Abnormal minimizers, *SIAM J. Control Optim.* **32**, 1605–1620.

Montgomery, R. [1995], A survey of singular curves in sub-Riemannian geometry, *J. Dynam. Control Systems* **1**, 49–90.

Montgomery, R. [2002], *A Tour of sub-Riemannian Geometries, their Geodesics and Applications*. Amer. Math. Soc. Mathematical Surveys and Monographs, volume 91.

Montgomery, R., J. E. Marsden, and T. S. Ratiu [1984], Gauged Lie–Poisson structures, in *Fluids and plasmas: geometry and dynamics (Boulder, Colo., 1983)*, Amer. Math. Soc., Providence, RI, 101–114.

Morandi, G., C. Ferrario, G. L. Vecchio, G. Marmo, and C. Rubano [1990], The inverse problem in the calculus of variations and the geometry of the tangent bundle, *Phys. Rep.* **188**, 147–284.

Morgansen, K. A. [2001], Controllability and trajectory tracking for classes of cascade form second-order nonholonomic systems, *Proc. of the 40th IEEE Conference on Dec. and Control* 3031–3036.

Morin, P., J.-B. Pomet, and C. Samson [1999], Design of homogeneous time-varying stabilizing control laws for driftless controllable systems via oscillatory approximation of Lie brackets in closed loop, *SIAM J. Control Optim.* **38**, 22–49.

Morin, P. and C. Samson [2000], Control of nonlinear chained systems: from the Routh–Hurwitz stability criterion to time-varying exponential stabilizers, *IEEE Trans. Automat. Control* **45**, 141–146.

Morse, A. S., editor, [1995], *Control using logic-based switching*, in *Trends in Control*, Papers from the workshop held on Block Island, RI, 1995, Springer-Verlag.

Moser, J. [1974], Finitely many mass points on the line under the influence of an exponential potential — an integrable system, *Springer Lecture Notes in Physics*, **38**, 467–497.

Moser, J. [1975], Three integrable Hamiltonian systems connected with isospectral deformations, *Adv. Math.* **16**, 197–220.

Moser, J. [1976], Periodic orbits near an equilibrium and a theorem by Alan Weinstein, *Comm. Pure Appl. Math.* **29**, 724–747; see also the addendum *Comm. Pure Appl. Math.* **31**, 529–530.

Moser, J. [1980] Various aspects of integrable Hamiltonian systems. In Proc. CIME Conference, Bressanone, Italy 1978, *Prog. Math.* **8**, Birkhäuser.

Moser, J. and A. Veselov [1991], Discrete versions of some classical integrable systems and factorization of matrix polynomials, *Commun. Math. Phys* **139**, 217–243.

Murray, R. [1995], Nonlinear control of mechanical systems: a Lagrangian perspective, in *Proc. IFAC Symposium on Nonlinear Control System Design (NOLCOS)*, 378–389.

Murray, R., Zexiang Li, S. Shankar Sastry [1994], *A mathematical introduction to robotic manipulation*, CRC Press, Boca Raton.

Murray, R. M. and S. Sastry [1993], Nonholonomic motion planning: steering using sinusoids, *IEEE Trans on Automatic Control* **38**, 700–716.

Neimark, J. I. and N. A. Fufaev [1966], On stability of stationary motions of holonomic and nonholonomic systems, *J. Appl. Math. Mech.* **30**, 293–300.

Neimark, J. I. and N. A. Fufaev [1972], *Dynamics of Nonholonomic Systems*, Translations of Mathematical Monographs, AMS, **33**.

Neumann, C. [1888], *Leipzig Berichte*, **XL**.

Newton, I. [1687], *Philosophi Naturalis Principia Mathematica*, Londini Societatis Regiae ac Typis, Josephi and Streater.

Nijmeijer, H. and A J. van der Schaft [1990], *Nonlinear Dynamical Control Systems*, Springer-Verlag, New York.

Noakes, L., G. Heinzinger, and B. Paden [1989], Cubic Splines on Curved Spaces, *IMA Journal of Mathematical Control and Information* **6**, 405–473.

Nomizu, K. [1954], Invariant Affine Connections on Homogeneous Spaces, *Amer. J. Math.* **76**, 33–65.

Oh, Y. G. [1987], A stability criterion for Hamiltonian systems with symmetry, *J. Geom. Phys.*, **4**, 163–182.

Oh, Y. G., N. Sreenath, P. S. Krishnaprasad, and J. E. Marsden [1989], The dynamics of coupled planar rigid bodies Part 2: bifurcations, periodic solutions and chaos, *Dynamics and Diff. Eqns* **1**, 269–298.

Olver, P. J. [1988], Darboux' theorem for Hamiltonian differential operators, *J. Diff. Eqn's* **71**, 10–33.

O'Reilly, O. M. [1996], The Dynamics of Rolling Disks and Sliding Disks, *Nonlinear Dynamics* **10**, 287–305.

Ortega, R., A. Loria, P. J. Nicklasson, and H. Sira-Ramirez [1998], *Passivity-based Control of Euler–Lagrange Systems*, Springer Series in Communications and Control Engineering, Springer-Verlag.

Ortega, R., A. J. van der Schaft, B. Maschke, and G. Escobar [1999], Energy-shaping of port-controlled Hamiltonian systems by interconnection, *Proc. 38th IEEE Conf. Decision Control*, IEEE, 1646–1651.

Ostrowski, J. P. [1995], *Geometric Perspectives on the Mechanics and Control of Undulatory Locomotion.* Ph.D. Thesis, California Institute of Technology.

Ostrowski, J. P. [1998], Reduced equations for nonholonomic mechanical systems with dissipative forces, *Rep. Math. Phys.* **42**, 185–209;

Ostrowski, J. P. [1999], Computing reduced equations for robotic systems with constraints and symmetries, *IEEE Trans. on Robotics and Automation* **15**, 111–113.

Ostrowski, J. P. [2000], Steering for a class of nonholonomic dynamic systems, *IEEE Trans. Aut. Control* **45**, 1492–1497.

Ostrowski, J. P. and J. W. Burdick [1996], Gait kinematics for a serpentine robot, in *IEEE Int. Conf. on Robotics and Automation*, 1294–1299.

Ostrowski, J. P. and J. W. Burdick [1997], Controllability for mechanical systems with symmetries and constraints, *Appl. Math. Comput. Sci.* **7**, 305–331.

Ostrowski, J. P., J. W. Burdick, A. D. Lewis, and R. M. Murray [1995], The mechanics of undulatory locomotion: The mixed kinematic and dynamic case, in *IEEE Intern. Conf. on Robotics and Automation*, 1945–1951.

Ostrowski, J., J. P. Desai, and V. Kumar, [1997], Optimal gait selection for nonholonomic locomotion systems, *Proc. IEEE International Conference on Robotics and Automation*, 786–791.

Papstavridis, J. G. [1998], A panoramic view overview of the principles and equations of advanced engineering dynamics *Applied Mechanics Reviews* **51**, 239–264.

Papastavridis, J. G. [1999], *Tensor calculus and analytical dynamics*, Library of Engineering Mathematics, CRC Press, Boca Raton, FL.

Papastavridis, J. G. [2002], *Analytical Mechanics - A Comprehensive Treatise on the Dynamics of Constrained Systems for Engineers, Physicists, and Mathematicians*, Oxford University Press.

Pars, L. A. [1965], *A Treatise on Analytical Dynamics*, Heinemann, London.

Pascal, M. [1983], Asymptptic solution of the equations of motion for a Celtic stone, *J. Appl. Math. Mech.* **47**, 269–276.

Pascal, M. [1986], The use of the method of averaging to study nonlinear oscillations of the Celtic stone, *J. Appl. Math. Mech.* **50**, 520–522.

Patrick, G. [1989], The dynamics of two coupled rigid bodies in three space, Amer. Math. Soc., Providence, RI, *Contemp. Math.* **97**, 315–336.

Patrick, G. [1992], Relative equilibria in Hamiltonian systems: The dynamic interpretation of nonlinear stability on a reduced phase space, *J. Geom. and Phys.* **9**, 111–119.

Patrick, G. [1995], Relative equilibria of Hamiltonian systems with symmetry: linearization, smoothness and drift, *J. Nonlinear Sci.* **5**, 373–418.

Polushin, I. G., A. L. Fradkov, and D. D. Khill [2000], Passivity and passification of nonlinear systems, *Avtomat. i Telemekh.*, 3–37.

Pontryagin, L. S. [1959], Optimal control processes, *Usp. Mat. Nauk.* **14**.

Pontryagin, L. S., V. G. Boltyanskii, R. V. Gamkrelidze, and E. F. Mischenko [1962], Averaging for attitude control and motion panning, in *The Mathematical Theory of Optimal Processes*, Wiley, New York.

Poincaré, H. [1885], Sur l'équilibre d'une masse fluide animée d'un mouvement de rotation, *Acta. Math.* **7**, 259.

Poincaré, H. [1892–1899]*Les Méthodes Nouvelles de la Mécanique Celeste.* 3 volumes. English translation, *New Methods of Celestial Mechanics*, History of Modern Physics and Astronomy **13**, Amer. Inst. Phys., 1993.

Poincaré, H. [1901], Sur une forme nouvelle des équations de la mécanique, *CR Acad. Sci.* **132**, 369–371.

Pomet, J.-B. [1992], Explicit design of time-varying stabilizing control laws for a class of controllable systems without drift, *Systems and Control Letters* **18**, 147–158.

Rabier, P. J. and W. C. Rheinboldt [2000], *Nonholonomic motion of rigid mechanical systems from a DAE viewpoint.* Society for Industrial and Applied Mathematics (SIAM), Philadelphia, PA.

Ramos, A. [2004], Poisson structures for reduced non-holonomic systems, *J. Phys. A* **37**, 4821–4842.

Rand, R. [2003], Review of *Analytical Mechanics*, by A. I. Lurie, *SIAM Review* **45**, 156-159.

Ratiu, T. [1980], The motion of the free *n*-dimensional rigid body, *Indiana U. Math. J.* **29**, 609–627.

Reyhanoglu, M., A. M. Bloch, and N. H. McClamroch [1995], Control of nonholonomic systems with extended base space dynamics, *Internat. J. Robust Nonlinear Control* **5**, 325–330.

Reyhanoglu, M., S. Cho, and N. H. McClamroch [2000], Discontinuous feedback control of a special class of underactuated mechanical systems, *Internat. J. Robust Nonlinear Control* **10**, 265–281.

Reyhanoglu, M., A. J. van der Schaft, N. H. McClamroch, and I. Kolmanovsky [1999], Dynamics and control of a class of underactuated mechanical systems, *IEEE Trans. Automat. Control* **44**, 1663–1671.

Riemann, B. [1860], Untersuchungen über die Bewegung eines flüssigen gleich-artigen Ellipsoides, *Abh. d. Königl. Gesell. der Wiss. zu Göttingen* **9**, 3–6.

Rosenberg, R. M. [1977], *Analytical Dynamics of Discrete Systems*, Plenum Press, NY.

Routh, E. J. [1860], *Treatise on the Dynamics of a System of Rigid Bodies*, Macmillan, London.

Routh, E. J. [1877], *Stability of a given state of motion*. Halsted Press, New York. Reprinted in *Stability of Motion* (1975), A. T. Fuller ed.

Ruina, A. [1998], Non-holonomic stability aspects of piecewise nonholonomic systems, *Reports in Mathematical Physics* **42**, 91–100.

Rumiantsev, V. [1966], On the stability of the steady motions, *J. Appl. Math. Mech.* **30**, 1090–1103.

Rumiantsev, V. [1978] On Hamilton's principle for nonholonomic systems *J. Appl. Math. Mech.* **42** No. 3, 387–399.

Rumiantsev, V. [1979] On the Lagrange and Jacobi principles for nonholonomic systems *J. Appl. Math. Mech.* **43** No. 4, 583–590.

Rumiantsev, V. [1982a] On integral principles for nonholonomic systems *J. Appl. Math. Mech.* **46** No. 1, 1–8.

Rumiantsev, V. [1982b], On stability problem of a top, *Rend. Sem. Mat. Univ. Padova* **68**, 119–128.

Rumiantsev, V. [1996], The general equations of analytical mechanics *J.Appl.Maths.Mechs*, **60**, 899-909.

Rund, H. [1966], *The Hamiltonian–Jacobi Theory in the Calculus of Variations*, Krieger, New York.

Ryan, E. P. [1990], Discontinuous feedback and universal adaptive stabilization. In *Control of uncertain systems (Bremen, 1989)*, 245–258. Birkhäuser Boston, MA.

Sánchez de Alvarez, G. [1986], *Geometric Methods of Classical Mechanics Applied to Control Theory*, Ph.D. thesis, Berkeley,

Sánchez de Alvarez, G. [1989], Controllability of Poisson control systems with symmetry, *Cont. Math. AMS* **97**, 399–412.

Sánchez De Alvarez, G. [1993], Poisson brackets and dynamics, in *Dynamical systems (Santiago, 1990)*, Longman Sci. Tech., Harlow, 230–249.

Santilli, R. M. [1978], *Foundations of Theoretical Mechanics, I*, Springer, New York.

Sanders, J. A. and F. Verhulst [1985], *Averaging Methods in Nonlinear Dynamical Systems*, Springer-Verlag, New York, *Applied Mathematical Sciences Series* **59**.

San Martin, L. and P. Crouch [1984], Controllability on principal fibre bundles with compact structure group, *Systems and Control Letters* **5** (1), 35–40.

Sattinger, D. H. and D. L. Weaver [1986], *Lie Groups and Lie Algebras in Physics, Geometry and Mechanics*, Springer-Verlag, New York.

Schneider, D. [2000] *Nonholonomic Euler–Poincaré equations and stability in Chaplygin's sphere*, Ph.D. thesis, University of Washington.

Schneider, D. [2002] Nonholonomic Euler–Poincaré equations and stability in Chaplygin's sphere, *Dynamical Systems*, **17**, no. 2, 87–130.

Schutz, B.F. [1980], *Geometrical Method of Mathematical Physics*, Cambridge University Press.

Seto, D. and J. Baillieul [1994], Control problems in super-articulated mechanical systems, *IEEE Transactions on Automatic Control* (H. Nijmeijer and A. J. van der Schaft, guest eds.) **39–12**, 2442–2453.

Shapere, A. and F. Wilczek [1987], Self-propulsion at low Reynolds number, *Phys. Rev. Lett.* **58**, 2051–2054.

Shapere, A. and F. Wilczek [1988], *Geometric Phases in Physics*, World Scientific, 1988.

Shen, J. [2002] *Nonlinear Control of Multibody Systems with Symmetries via Shape Change*, Ph.D. thesis, University of Michigan.

Shen, J., N. H. McClamroch, and A. M. Bloch [2004], Local equilibrium controllability of multibody systems controlled via shape change, *IEEE Trans. Automat. Control*, **49**(4), 506–520.

Shiriaev, A. S. and A. L. Fradkov [2000], Stabilization of invariant sets for nonlinear non-affine systems, *Automatica J. IFAC* **36**, 1709–1715.

Simo, J. C., D. R. Lewis, and J. E. Marsden [1991], Stability of relative equilibria I: The reduced energy–momentum method, *Arch. Rat. Mech. An.* **115**, 15–59.

Simo, J. C., T. A. Posbergh, and J. E. Marsden [1990], Stability of coupled rigid body and geometrically exact rods: block diagonalization and the energy-momentum method, *Physics Reports* **193**, 280–360.

Simo, J. C., T. A. Posbergh, and J. E. Marsden [1991], Stability of relative equilibria II: Three dimensional elasticity, *Arch. Rational Mech. Anal.* **115**, 61–100.

Simon, B. [1983], Holonomy, the quantum adiabatic theorem and Berry's phase, *Phys. Rev. Letters* **51**, 2167–2170.

Sira-Ramirez, H. [1995], On the sliding mode control of differentially flat systems, *Control-Theory and Advanced Technology* **10**, 1093–1113.

Sira-Ramirez, H. and P. Lischinsky-Arenas [1990], Dynamical discontinuous feedback control of nonlinear systems, *IEEE Trans. Automat. Control* **35**, 1373–1378.

Sjamaar, R. and E. Lerman [1991], Stratified symplectic spaces and reduction, *Ann. of Math.* **134**, 375–422.

Smale, S. [1970], Topology and Mechanics, *Inv. Math.* **10**, 305–331, and **11**, pp. 45–64.

Smart, E. H. [1951], *Advanced Dynamics, Volume II*, McMillan and Co., London.

Śniatycki, J. [1998], Non-holonomic Noether theorem and reduction of symmetries, *Rep. Math. Phys* **42**, 5–23.

Śniatycki, J. [2001], Almost Poisson spaces and nonholonomic singular reduction, *Rep. Math. Phys.* **48**, 235–248.

Śniatycki, J. [2002], The momentum equation and the second order differential equation condition, *Rep. Math. Phys.* **49**, 371–394.

Sommerfeld, A. [1952], *Mechanics, Lectures on Theoretical Physics Vol. 1*. First German Edition 1943, Academic Press.

Sontag, E. D. [1990], *Mathematical control theory. Deterministic finite-dimensional systems*, Springer-Verlag, New York, Texts in Applied Mathematics. **6**.

Sontag, E. [1998], Stability and stabilization: discontinuities and the effect of disturbances, *Proc. NATO Advanced Study Institute "Nonlinear Analysis, Differential Equations and Control,"* F. J. Clarke and R. J. Stern. eds. Kluwer, 551–598.

Sontag, E. D. [1999], Stability and stabilization: discontinuities and the effect of disturbances. In *Nonlinear analysis, differential equations and control (Montreal, QC, 1998)*, 551–598, Kluwer Acad. Publ., Dordrecht.

Sontag, E. D. and Y. Wang [1997], Output-to-state stability and detectability of nonlinear systems, *Systems Control Lett.* **29**, 279–290.

Sontag, E. D. and Y. Wang [1996], New characterizations of input-to-state stability, *IEEE Trans. Automat. Control* **41**, 1283–1294.

Sørdalen, O. J. and O. Egeland [1995], Exponential stabilization of nonholonomic chained systems, *IEEE Trans. Automat. Control* **40**, 35–49.

Spivak, M. [1979], *A Comprehensive Introduction to Differential Geometry*, Publish or Perish, Inc., Wilmington, Del.

Spong M. W., Underactuated mechanical systems, in *Control Problems in Robotics and Automation*, B. Siciliano and K. P. Valavanis (eds), *Lecture Notes in Control and Information Sciences* **230** Spinger-Verlag. [Presented at the International Workshop on Control Problems in Robotics and Automation: Future Directions Hyatt Regency, San Diego, California, Dec. 1997.]

Spong, M. W. [1998], On feedback linearization of robot manipulators and Riemannian curvature. In *Essays on Mathematical robotics*, IMA Volumes in Mathematics and Its Applications, Vol. **104**, J. Baillieul, S. S. Sastry, and H. J. Sussmann, eds., Springer-Verlag, NY.

Sreenath, N., Y. G. Oh, P. S. Krishnaprasad, and J. E. Marsden [1988], The dynamics of coupled planar rigid bodies. Part 1: Reduction, equilibria and stability, *Dyn. and Stab. of Systems* **3**, 25–49.

Steenrod, N. E. [1951], *The Topology of Fibre Bundles*, Princeton University Press.

Stephenson, A. [1908] On a new type of dynamical stability, *Mem. Proc. Manch. Lit. Phil. Soc.* **52**, 1–10.

Sternberg, S. [1977], Minimal coupling and the symplectic mechanics of a classical particle in the presence of a Yang–Mills field, *Proc. Nat. Acad. Sci.* **74**, 5253–5254.

Stoker, J. J. [1950], *Nonlinear Vibrations in Mechanical and Electrical Systems*, New York; Republished 1992 in Wiley Classics Library Edition.

Strichartz, R. [1983], Sub-Riemannian geometry, *J. Diff. Geom.* **24**, 221–263; see also *J. Diff. Geom.* **30**, (1989), 595–596.

Strichartz, R. S. [1987] The Campbell–Baker–Hausdorff Dynkin formula and solutions of differential equations, *Journal of Functional Analysis* **72**, 320–345.

Sumbatov, A. S. [1992], Developments of some of Lagrange's ideas in the works of Russian and Soviet mechanicians, *La Mécanique Analytique de Lagrange et son Héritage, Atti della Accademia delle Scienze di Torino, Suppl. 2* **126**, 169–200.

Sussmann, H. J. [1973], Orbits of families of vector fields and integrability of distributions, *Trans. Amer. Math. Soc.* **180**, 171–180.

Sussmann, H. J. [1979] Subanalytic sets and feedback control, *J. Differential Equations* **31**, 31–52.

Sussmann, H. J. [1986], A product expansion for the Chen series. In *Theory and applications of nonlinear control systems (Stockholm, 1985)*, pages 323–335. North-Holland, Amsterdam.

Sussmann, H. J. [1987] A general theorem on local controllability, *SIAM J. on Control and Optim.* **25**, 158–194.

Sussmann, H. J. [1996], A cornucopia of four-dimensional abnormal sub-Riemannian minimizers. In *Sub-Riemannian geometry*, Edited by A. Bellaïche and J.-J. Risler. Progress in Mathematics, Volume 144, 341–364. Birkhäuser, Basel.

Sussmann, H. J. [1998a], An introduction to the coordinate-free maximum principle, in *Geometry of Feedback and Optimal Control* (B. Jackubczyk and W. Respondek, eds.), Monographs Textbooks Pure. Appl. Math. **207** M. Dekker, 463–557.

Sussmann, H. J. [1998b], Geometry and optimal control. In *Mathematical Control Theory*, J. Baillieul & J. C. Willems, eds., Springer-Verlag, New York, 1998, 140–198.

Sussmann, H. J. and V. Jurdjevic [1972], Controllability of nonlinear systems, *J. Diff. Eqns.* **12**, 95–116; see also Control systems on Lie groups, *J. Differential Equations* **12**, 313–329.

Sussmann, H. J. and W. Liu [1991], Limits of highly oscillatory controls and the approximation of general paths by admissable trajectories, *Proc. 30th IEEE Conf. Decision and Control*, 1190–1195.

Sussmann, H. J. and J. C. Willems [1997], 300 years of optimal control: from the Brachystrochrone to the Maximum Principle, *IEEE Control System Magazine.* **17**, 32–44.

Symes, W. W. [1980] Hamiltonian group actions and integrable systems. *Physica D* **1**, 339–376.

Symes, W. W. [1982], The QR algorithm and scattering for the nonperiodic Toda lattice, *Physica D* **4**, 275–280.

Synge, J. L. and B. A. Griffiths [1959] *Principles of Mechanics*, Third Edition, McGraw Hill.

Tai, M. [2001], *Lagrange Reduction of Nonholonomic Systems on Semidirect Products*, M.A. thesis, Univ. of California, Berkeley.

Tavares, J. N. [2003], About Cartan geometrization of non-holonomic mechanics, *J. Geom. Phys.* **45**, 1–23.

Teel, A. R. [1996], On graphs, conic relations, and input–output stability of nonlinear feedback systems, *IEEE Trans. Automat. Control* **41**, 702–709.

Teel, A., R. Murray, and G. Walsh [1995], Nonholonomic control systems: from steering to stabilization with sinusoids, *International Journal of Control* **62** (4), 849–870.

Terra, G. and M. H. Kobayashi [2002], On classical mechanical systems with constraints, *Preprint*.

Thimm, A. [1981], Integrable Hamiltonian systems on homogeneous spaces, *Ergodic Theory and Dynamical Systems* **1**, 495–517.

Thomson, W (Lord Kelvin) and P. G. Tait [1879], *Treatise on Natural Philosophy*, Cambridge University Press.

Toda, M. [1975], Studies of a non-linear lattice, *Phys. Rep. Phys. Lett.* **8**, 1–125.

Tomei, C. [1984], The topology of isospectral manifolds of diagonal matrices, *Duke Math. J* **51**, 981–996.

Truesdell, C. [1977], *A First Course in Rational Continuum Mechanics*, Academic Press, New York.

Tsakiris, D. P. [1995] *Motion control and planning for nonholonomic kinematic chains*, Ph.D. thesis, Systems Research Institute, University of Maryland.

Tulczyjew, W. M. [1977], The Legendre transformation, *Ann. Inst. Poincaré* **27**, 101–114.

Utkin V. I. [1978], *Sliding Modes and Their Application in Variable Structure Systems*. Moscow: MIR, 1978.

Utkin, V. I. [1992] *Sliding modes in control and optimization*, Springer-Verlag, New York.

Vaisman, I. [1994], *Lectures on the Geometry of Poisson Manifolds*, volume 118 of *Progress in Mathematics*. Birkhäuser, Boston.

van der Schaft, A. J. [1982], Hamiltonian dynamics with external forces and observations, *Mathematical Systems Theory* **15**, 145–168.

van der Schaft, A. J. [1983], *System Theoretic Descriptions of Physical Systems*, doct. dissertation, University of Groningen; also *CWI Tract #3*, CWI, Amsterdam.

van der Schaft, A. J. [1986], Stabilization of Hamiltonian systems, *Nonlinear An., TMA* **10**, 1021–1035.

van der Schaft, A. J. L_2-Gain and Passivity Techniques in Nonlinear Control, Springer-Verlag, Communication & Control Engineering Series.

van der Schaft, A. J. and B. M. Maschke [1994], On the Hamiltonian formulation of nonholonomic mechanical systems, Rep. on Math. Phys. **34**, 225–233.

Vela, P. A., K. A. Morgansen, and J. Burdick [2002] Second order averaging methods for oscillatory control of underactuated mechanical systems, Proc. ACC, 4672–4677.

Vershik, A. M. [1984], Classical and nonclassical dynamics with constraints (Russian), Novoe Global. Anal., 23–48; see also Springer Lecture Notes in Mathematics, **1108**, 278–301.

Vershik, A. M. and L. D. Faddeev [1981], Lagrangian mechanics in invariant form, Sel. Math. Sov. **4**, 339–350.

Vershik, A. M. and V. Ya. Gershkovich [1988], Nonholonomic problems and the theory of distributions, Acta Applicandae Mathematicae **12**, 181–209.

Vershik, A. M. and V. Y. Gershkovich [1994], Nonholonomic dynamical systems, geometry of distributions and variational problems, in Dynamical Systems VII, 1–81. Springer-Verlag, New York,

Vierkandt, A. [1892], Über gleitende und rollende Bewegung, Monatshefte der Math. und Phys., **III**, 31–54.

Voss, A [1885], Ueber die Differentialgleichungen der Mechanik, Mathematische Annalen **25**, 258–286.

Vranceneau, G. [1936], Les Espaces Nonholonomes, Gauthier-Villars, Paris

Walsh, G. C. and L. G. Bushnell [1995], Stabilization of multiple input chained form control systems, Systems Control Lett. **25**, 227–234.

Wang, L. S. and P. S. Krishnaprasad [1992], Gyroscopic control and stabilization, J. Nonl. Sci. **2**, 367–415.

Walker, G. T. [1896], On a dynamical top, Quart. J. Pure Appl. Math. **28**, 175–184.

Warner, F. W. [1983], Foundations of Differentiable Manifolds and Lie Groups, Graduate Texts in Methematics. **94**, Springer-Verlag.

Weber, R. W. [1986], Hamiltonian systems with constraints and their meaning in mechanics, Arch. Rat. Mech. Anal., **91**, 309–335.

Weibel, S. [1997], Applications of Qualitative Methods in the Nonlinear Control of Superarticulated Mechanical Systems, PhD thesis, Boston University.

Weibel, S. and J. Baillieul [1998a], Averaging and energy methods for robust open-loop control of mechanical systems, in Essays on Mathematical Robotics, J. Baillieul, S.S. Sastry, and H.J. Sussmann, Eds. IMA Series of Springer-Verlag, 203–269.

Weibel, S. and J. Baillieul [1998b], Open-loop stabilization of an n–Pendulum, Int. J. of Control, **71**, 931–957.

Weibel, S., T. Kaper, and J. Baillieul [1997], Global dynamics of a rapidly forced cart and pendulum, Nonlinear Dynamics, **13**, 131–170.

Wei, J. and E. Norman [1964], On global representations of the solution of linear differential equations as a product of exponentials, *Proc. of the Amer. Math. Soc.*, **15**, 327–334.

Weckesser, W. [1997], A ball rolling on a freely spinning turntable *Am. J. of Phys.* **65**, 736–738.

Weinstein, A. [1971], Symplectic manifolds and their Lagrangian submanifolds, *Advances in Mathematics*, **6**, 329–346.

Weinstein, A. [1973], Normal modes for nonlinear Hamiltonian systems, *Inv. Math.* **20**, 47–57.

Weinstein, A. [1978a], Bifurcations and Hamilton's principle, *Math. Zeit.* **159**, 235–248.

Weinstein, A. [1978b], A universal phase space for particles in Yang–Mills fields, *Lett. Math. Phys.*, **2**, 417–420.

Weinstein, A. [1990], Connections of Berry and Hannay type for moving Lagrangian submanifolds, *Adv. in Math.*, **82**, 133–159.

Weltner, K. [1979], Stable circular orbits of freely moving balls on rotating discs *Am. J. Physics* **47**, 984–986.

Whittaker, E. T. [1988], *A Treatise on the Analytical Dynamics of Particles and Rigid Bodies*, Fourth Edition, Cambridge University Press; First Edition 1904, Fourth Edition, 1937, Reprinted by Dover 1944 and Cambridge University Press, 1988.

Willems, J.C. [1979] System theoretic models for the analysis of physical systems, *Ricerche di Automatica* **10**, 71–106.

Willems, J.C. [1986] From time series to linear systems, I and II *Automatica* **22**, 561–580, 675–694.

Wolf, J. A. [1972], *Spaces of Constant Curvature*, Publish or Perish, Inc., Boston, MA.

Yang, R. [1980], Fibre bundles and the physics of the magnetic monopole, in *Proc. Conference in honor of S.S. Chern*, Springer-Verlag, New York.

Yang, R. [1992], *Nonholonomic Geometry, Mechanics and Control*, Ph.D. thesis, University of Maryland.

Yang, R. and P. S. Krishnaprasad [1994], On the geometry and dynamics of floating four-bar linkages, *Dynam. Stability Systems* **9**, 19–45.

Yang, R., P. S. Krishnaprasad, and W. Dayawansa [1993], Chaplygin dynamics and Lagrangian reduction, *Proc. 2nd Int. Cong. on Nonlinear Mechanics*, (W. Z. Chien, Z. H. Guo, and Y. Z. Guo, eds.), Peking University Press, 745–749.

Yang, R., P. S. Krishnaprasad, and W. Dayawansa [1996], Optimal control of a rigid body with two oscillators. In *Mechanics day (Waterloo, ON, 1992)*, 233–260. Amer. Math. Soc., Providence, RI.

Yano, K. and S. Ishihara [1973], *Tangent and cotangent bundles: differential geometry.* Marcel Dekker Inc., New York. Pure and Applied Mathematics, No. 16.

Yoshimura, H. and J.E. Marsden [2005] Variational principles, Dirac structures, and implicit Lagrangian systems, preprint.

Zabczyk, J. [1989] Some comments on stabilizability, *Applied Math. Optimization* **19**, 1–9.

Zenkov, D. V. [1995], The geometry of the Routh problem, *J. Nonlinear Sci.* **5**, 503–519.

Zenkov, D. V. [1998], *Integrability and Stability of Nonholonomic Systems*, Ph.D. Thesis, The Ohio State University.

Zenkov, D. V. and A. M. Bloch [2000], Dynamics of the n-dimensional Suslov problem, *Journal of Geometry and Physics* **34**, 121–136

Zenkov, D. V. and A. M. Bloch [2003], Invariant measures of nonholonomic flows with internal degrees of freedom, *Nonlinearity* **16**, 1793–1807.

Zenkov, D., A. M. Bloch, N. E. Leonard, and J. E. Marsden [2000] Matching and stabilization for the unicycle with rider, *Proc. IFAC Workshop on Lagrangian and Hamiltonian Methods in Nonlinear Control*, 187–188.

Zenkov, D. V., A. M. Bloch, and J. E. Marsden [1998], The energy momentum method for the stability of nonholonomic systems, *Dyn. Stab. of Systems* **13**, 123–166.

Zenkov, D. V., A. M. Bloch, and J. E. Marsden [1999], Stabilization of the unicycle with rider, *Proc. of the 38th IEEE Conference on Dec. and Control*, 3470–3471.

Zenkov, D. V., A. M. Bloch, and J. E. Marsden [2002a], The Lyapunov–Malkin theorem and stabilization of the unicycle with rider, *Systems and Control Lett.* **45**, 293–300.

Zenkov, D. V., A. M. Bloch, and J. E. Marsden [2002b], Flat nonholonomic matching, *Proc ACC*, 2812–2817.

Zenkov, D. V. and V. V. Kozlov [1988], Poinsot's geometric representation in the dynamics of a multidimensional rigid body, *Trudy Sem. Vektor. Tenzor. Anal.*, **23**, 202–204.

Zhang, M. and R. M. Hirschorn [1997], Discontinuous feedback stabilization of nonholonomic wheeled mobile robots, *Dynam. Control* **7**, 155–169.

Index

Interdisciplinary Applied Mathematics